PARTIAL DIFFERENTIAL EQUATIONS OF PARABOLIC TYPE

PRENTICE-HALL INTERNATIONAL, INC., *London*
PRENTICE-HALL OF AUSTRALIA, PTY., LTD., *Sydney*
PRENTICE-HALL OF CANADA, LTD., *Toronto*
PRENTICE-HALL OF JAPAN, INC., *Tokyo*
PRENTICE-HALL DE MEXICO, S.A., *Mexico City*

PARTIAL DIFFERENTIAL EQUATIONS OF PARABOLIC TYPE

AVNER FRIEDMAN

Professor of Mathematics
Northwestern University

PRENTICE-HALL, INC.
Englewood Cliffs, N. J.

Library of Congress Catalog Card Number 64-7571
Printed in the United States of America
65013C

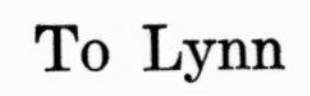

To Lynn

PREFACE

In recent years there have been so many developments in the theory of partial differential equations that one can hardly write a book for readers with no background in this field that will encompass the main achievements of the theory. It is possible, however, to divide the theory into several fairly independent domains, the main ones being: (i) elliptic and parabolic equations, (ii) hyperbolic equations, and (iii) linear differential equations of any type. The present book is concerned with the first topic. The general theory of elliptic and parabolic equations will be developed to a level where the reader may proceed without difficulty to reading research papers in the field.

There are many parallels between elliptic and parabolic equations. Although in certain problems, such as the construction of fundamental solutions, the elliptic case is more difficult to handle, in general it is the parabolic case that is more difficult and that necessitates additional considerations. In this book we shall usually present detailed proofs only for the parabolic case. At the end of almost every chapter a section deals with elliptic equations and states definitions and results in detail; the reader may find the proofs in papers indicated in the bibliographical remarks.

The main topics considered in the book are existence and uniqueness theorems for the Cauchy problem and for the first and second boundary value problems, fundamental solutions, and properties of solutions such as differentiability, asymptotic behavior, maximum principles, etc. The introduction to each chapter briefly describes the results and methods of that chapter. Here we shall simply comment on the interdependence of the various chapters.

Chapters 1–5 form a unit in which the main facts of the general theory (i.e., existence, uniqueness, and differentiability of solutions) of second-order linear parabolic equations are developed in detail; the analogous results for second-order linear elliptic equations are stated without proofs. The method of potential theory (i.e., the method revolving around evaluating "potentials") is the underlying method in most considerations, except those of Chap. 2.

In Chaps. 6–8 other aspects of second-order parabolic equations are considered. Chapters 6 and 7 depend on the material of Chaps. 1–5, but Chap. 8 is essentially independent of all the other chapters of the book.

In Chap. 9 the results of Chap. 1 are extended to systems of parabolic equations. In this sense Chaps. 1 and 9 form one unit. Chapter 9 is independent of the material of Chaps. 2–8.

In Chap. 10 the questions of existence and uniqueness for the first boundary value problem and the differentiability of solutions are considered for elliptic equations of any order and (then) for parabolic equations of any order. This chapter is essentially independent of all the previous chapters. The methods used are entirely different from those of the preceding chapters.

The bibliography and bibliographical remarks are reserved to the end of the book. The bibliography is by no means exhaustive of the field of elliptic and parabolic equations. We have included only works that are directly connected with the material of the book.

Additional material not occurring in the text is included in problems that appear at the end of each chapter. The problems are rather easy and some of them are used in the text.

The theory of real variables is the only requirement for Chaps. 1–9. In Chap. 10 the reader is assumed to be familiar also with the elements of the theory of Hilbert spaces.

I wish to thank Miss Vera Fisher for her excellent typing of the manuscript.

While working on the book I was partially supported by the National Science Foundation and by the Alfred P. Sloan Foundation.

Avner Friedman

GENERAL NOTATION

1. Unless the contrary is explicitly stated:

   (a) all the functions in Chaps. 1–8 and in the appendix are real-valued;
   (b) all the functions in Chaps. 9, 10 are complex-valued;
   (c) all the parabolic and elliptic equations occurring in Chaps. 1–8 are of the second order;
   (d) all the parabolic and elliptic equations occurring in Chaps. 9, 10 are of arbitrary order.

2. Hilbert spaces are always understood to be separable (they occur only in Chap. 10).

3. The union of two sets A and B is denoted by either $A \cup B$ or $A + B$. The complement of B in A is denoted by $A - B$ (A need not contain B).

4. We denote by $\bar{D}$ the closure of D.

5. We usually refer to a vector-function (and also to a matrix of functions) briefly as a function.

6. When in a chapter k a reference is made to a formula ($m.n$), this formula is to be found in Sec. m of the same chapter k; by ($h.m.n$) we refer to formula ($m.n$) of Chap. h ($h \neq k$).

7. When a reference is made in a chapter to Theorem n, Sec. m (or Lemma n, Sec. m), this theorem n (or this lemma n) is to be found in Sec. m of the same chapter.

CONTENTS

PARTIAL DIFFERENTIAL EQUATIONS OF PARABOLIC TYPE

CHAPTER 1

FUNDAMENTAL SOLUTIONS AND THE CAUCHY PROBLEM

Introduction. Consider the *heat equation*

$$\frac{\partial^2 u}{\partial x_1^2} + \cdots + \frac{\partial^2 u}{\partial x_n^2} - \frac{\partial u}{\partial t} = 0. \tag{0.1}$$

The function

$$\Gamma(x, t; \xi, \tau) = \frac{1}{(2\sqrt{\pi})^n} (t - \tau)^{-n/2} \exp\left[-\frac{\Sigma\,(x_i - \xi_i)^2}{4(t - \tau)}\right] \qquad (t > \tau) \tag{0.2}$$

satisfies (0.1) for every fixed (ξ, τ). Furthermore, for any continuous function $f(x)$ which is bounded by $0\{\exp[h\,\Sigma\, x_i^2]\}$ for some $h > 0$, the integral

$$u(x, t) \equiv \int_{-\infty}^{\infty} \cdots \int_{-\infty}^{\infty} \Gamma(x, t; \xi, 0)\, d\xi_1 \cdots d\xi_n \tag{0.3}$$

exists for $0 < t \leq T$ if $T < 1/4h$ and

$$u(x, t) \to f(x) \qquad \text{if } t \to 0. \tag{0.4}$$

$\Gamma(x, t; \xi, \tau)$ is called a *fundamental solution* of the heat equation. The function $u(x, t)$ is a solution of the *Cauchy problem*, i.e., the problem of finding a solution to (0.1) for $0 < t \leq T$ which satisfies the *initial condition* $u(x, 0) = f(x)$.

The *adjoint equation* of the heat equation is defined as

$$\frac{\partial^2 v}{\partial x_1^2} + \cdots + \frac{\partial^2 v}{\partial x_n^2} + \frac{\partial v}{\partial t} = 0. \tag{0.5}$$

One easily verifies that for each fixed (x, t) $\Gamma(x, t; \xi, \tau)$ satisfies (0.5) as a function of (ξ, τ). This fact may be used in order to prove the uniqueness of solutions of the Cauchy problem (for (0.1)) under the condition that

$$u(x, t) = 0\{\exp[k\,\Sigma\, x_i^2]\} \tag{0.6}$$

for some $k > 0$.

In this chapter we construct a fundamental solution $\Gamma(x, t; \xi, \tau)$ for second-order parabolic equations with Hölder continuous coefficients. It is then used to solve the Cauchy problem. Under some additional differentiability assumptions on the coefficients of the equation we construct a fundamental solution $\Gamma^*(x, t; \xi, \tau)$ for the adjoint equation such that $\Gamma(x, t; \xi, \tau) = \Gamma^*(\xi, \tau; x, t)$. We use this relation in order to prove the uniqueness of the solution of the Cauchy problem under the condition (0.6).

1. Definitions

We denote by R^n the real n-dimensional euclidean space. The distance of a point $x = (x_1, \ldots, x_n)$ of R^n to the origin (i.e., the norm of x) is defined by $|x| = (\sum_{i=1}^{n} x_i^2)^{1/2}$. A function $f(x)$ defined on a bounded closed set S of R^n is said to be *Hölder continuous of exponent* α $(0 < \alpha < 1)$ in S if there exists a constant A such that

$$(1.1) \qquad |f(x) - f(y)| \leq A|x - y|^\alpha$$

for all x, y in S. The smallest A for which (1.1) holds is called the *Hölder coefficient.* If S is an unbounded set whose intersection with every bounded closed set B is closed, then $f(x)$ is said to be Hölder continuous of exponent α in S if for every bounded closed set B (1.1) holds in $S \cap B$ with some A which may depend on B. If A can be taken independently of B, then we say that $f(x)$ is *uniformly* Hölder continuous (of exponent α).

If S is an open set then $f(x)$ is said to be *locally* Hölder continuous (exponent α) in S if (1.1) holds in every bounded closed set $B \subset S$, with A which may depend on B. If A is independent of B then $f(x)$ is *uniformly* Hölder continuous (exponent α).

If f depends on a parameter λ, i.e., $f = f(x, \lambda)$, and if the Hölder coefficient is independent of λ, then we say that $f(x, \lambda)$ is Hölder continuous in x, *uniformly* with respect to λ.

Sometimes it is convenient to define Hölder continuity with respect to different norms $|x|$; see, for instance, (1.4) below; also Chaps. 4, 5.

The reader may easily verify that if $f_1, \ldots, f_m$ are Hölder continuous (exponent α), then any rational function of $f_1, \ldots, f_m$ with nonvanishing denominator is also Hölder continuous (exponent α).

If in (1.1) $\alpha = 1$ then we say that $f(x)$ is *Lipschitz continuous.*

Consider the differential equation

$$(1.2) \quad Lu \equiv \sum_{i,j=1}^{n} a_{ij}(x, t) \frac{\partial^2 u}{\partial x_i \partial x_j} + \sum_{i=1}^{n} b_i(x, t) \frac{\partial u}{\partial x_i} + c(x, t)u - \frac{\partial u}{\partial t} = 0$$

where the coefficients a_{ij}, b_i, c are defined in a cylinder

$$\Omega \equiv \overline{D} \times [T_0, T_1] \equiv \{(x, t); x \in \overline{D}, T_0 \leq t \leq T_1\}.$$

$\overline{D}$ is the closure of a bounded domain $D \subset R^n$. We always take $(a_{ij}(x, t))$

to be a symmetric matrix, i.e., $a_{ij} = a_{ji}$. If the matrix $(a_{ij}(x, t))$ is positive definite, i.e., if for every real vector $\xi = (\xi_1, \ldots, \xi_n) \neq 0$, $\sum a_{ij}(x, t)\xi_i\xi_j > 0$ then we say that the operator L is *of parabolic type* (or that L is *parabolic*) at the point (x, t). If L is parabolic at all the points of Ω then we say that L is parabolic in Ω. If there exist positive constants $\bar{\lambda}_0$, $\bar{\lambda}_1$ such that, for any real vector ξ,

$$\bar{\lambda}_0|\xi|^2 \leq \sum_{i,j=1}^{n} a_{ij}(x, t)\xi_i\xi_j \leq \bar{\lambda}_1|\xi|^2 \tag{1.3}$$

for all $(x, t) \in \Omega$ then we say that L is *uniformly parabolic* in Ω.

Throughout this chapter we shall assume:

(A_1) L is parabolic in Ω;

(A_2) the coefficients of L are continuous functions in Ω and, in addition, for all $(x, t) \in \Omega$, $(x^0, t^0) \in \Omega$,

$$|a_{ij}(x, t) - a_{ij}(x^0, t^0)| \leq A(|x - x^0|^\alpha + |t - t^0|^{\alpha/2}), \tag{1.4}$$

$$|b_i(x, t) - b_i(x^0, t)| \leq A|x - x^0|^\alpha, \tag{1.5}$$

$$|c(x, t) - c(x^0, t)| \leq A|x - x^0|^\alpha. \tag{1.6}$$

Since the $a_{ij}(x, t)$ are continuous functions in the bounded set Ω, (A_1) implies that L is also uniformly parabolic in Ω, i.e., (1.3) holds for some positive constants $\bar{\lambda}_0$, $\bar{\lambda}_1$ independent of $(x, t) \in \Omega$.

Definition. We say that $u = u(x, t)$ is a *solution* of $Lu = 0$ in some region Δ if all the derivatives of u which occur in Lu (i.e., $\partial u/\partial x_i$, $\partial^2 u/\partial x_i\,\partial x_j$, $\partial u/\partial t$) are continuous functions in Δ and $Lu(x, t) = 0$ at each point (x, t) of Δ. A similar definition holds for any differential operator L.

Definition. A *fundamental solution* of $Lu = 0$ (in Ω) is a function $\Gamma(x, t; \xi, \tau)$ defined for all $(x, t) \in \Omega$, $(\xi, \tau) \in \Omega$, $t > \tau$, which satisfies the following conditions:

(i) for fixed $(\xi, \tau) \in \Omega$ it satisfies, as a function of (x, t) ($x \in D$, $\tau < t \leq T_1$) the equation $Lu = 0$;

(ii) for every continuous function $f(x)$ in $\bar{D}$, if $x \in D$ then

$$\lim_{t \searrow \tau} \int_D \Gamma(x, t; \xi, \tau) f(\xi)\, d\xi = f(x). \tag{1.7}$$

Here $d\xi = d\xi_1 \cdots d\xi_n$. The domain D is always assumed to be Lebesgue measurable; since this is an obvious assumption, we shall usually not mention it in what follows.

Sections 2–4 are devoted to the construction of a fundamental solution. In Sec. 5 we shall study some of its properties. The results of Secs. 2–5 are extended in Sec. 6 to the case of unbounded domains D.

2. The Parametrix Method

Let $(a^{ij}(x, t))$ be the inverse matrix to $(a_{ij}(x, t))$ and set

$$\vartheta^{y,\sigma}(x, \xi) = \sum_{i,j=1}^{n} a^{ij}(y, \sigma)(x_i - \xi_i)(x_j - \xi_j) \tag{2.1}$$

where $y = (y_1, \ldots, y_n)$. From (1.3), (1.4) it follows that

$$\lambda_0|x - \xi|^2 \leq \vartheta^{y,\sigma}(x, \xi) \leq \lambda_1|x - \xi|^2, \tag{2.2}$$

$$|a^{ij}(x, t) - a^{ij}(x^0, t^0)| \leq A'(|x - x^0|^\alpha + |t - t^0|^{\alpha/2}) \tag{2.3}$$

where λ_0, λ_1, A' are positive constants depending only on $\bar{\lambda}_0$, $\bar{\lambda}_1$, A.

We introduce the functions, for $t > \tau$,

$$w^{y,\sigma}(x, t; \xi, \tau) = (t - \tau)^{-n/2} \exp\left[-\frac{\vartheta^{y,\sigma}(x, \xi)}{4(t - \tau)}\right], \tag{2.4}$$

$$Z(x, t; \xi, \tau) = C(\xi, \tau)w^{\xi,\tau}(x, t; \xi, \tau), \tag{2.5}$$

where

$$C(x, t) = (2\sqrt{\pi})^{-n}[\det (a^{ij}(x, t))]^{1/2}. \tag{2.6}$$

For each fixed (ξ, τ) the function $Z(x, t; \xi, \tau)$ satisfies the equation with constant coefficients

$$L_0 u(x, t) \equiv \sum_{i,j=1}^{n} a_{ij}(\xi, \tau) \frac{\partial^2 u(x, t)}{\partial x_i \, \partial x_j} - \frac{\partial u(x, t)}{\partial t} = 0. \tag{2.7}$$

It also follows from Theorem 1 below that (1.7) is also satisfied for $\Gamma = Z$. Thus, $Z(x, t; \xi, \tau)$ is a fundamental solution of $L_0 u = 0$. In order to construct a fundamental solution for $Lu = 0$ we look upon L_0 as a "first approximation" to L and we view Z as a "principal part" of the fundamental solution Γ of $Lu = 0$. We then try to find Γ in the form

$$\Gamma(x, t; \xi, \tau) = Z(x, t; \xi, \tau) + \int_\tau^t \int_D Z(x, t; \eta, \sigma)\Phi(\eta, \sigma; \xi, \tau) \, d\eta \, d\sigma \tag{2.8}$$

where Φ is to be determined by the condition that Γ satisfies the equation $Lu = 0$.

This procedure is called the *parametrix method* (of E. E. Levi). Z is called the *parametrix*. In Chap. 9 we shall use this method also to construct fundamental solutions for systems of parabolic equations of any order.

Theorem 1. *Let $f(x, t)$ be a continuous function in Ω. Then*

$$J(x, t, \tau) \equiv \int_D Z(x, t; \xi, \tau)f(\xi, \tau) \, d\xi \tag{2.9}$$

is a continuous function in (x, t, τ), $x \in \bar{D}$, $T_0 \leq \tau < t \leq T_1$ and

$$\lim_{\tau \to t} J(x, t, \tau) = f(x, t) \tag{2.10}$$

uniformly with respect to (x, t), $x \in S$, $T_0 < t \leq T_1$, *where* S *is any closed subset of* D.

Proof. Consider first the case where f and the a_{ij} are all constants. The linear substitution $\zeta = P(x - \xi)$ reduces $\Sigma\, a^{ij}(x_i - \xi_i)(x_j - \xi_j)$ into $\Sigma\, \zeta_i^2$ provided $P^*P = (a^{ij})$, where P^* is the transpose of the matrix P. Denoting by D^* the image of D by this substitution and noting that

$$\frac{\partial(\xi_1, \ldots, \xi_n)}{\partial(\zeta_1, \ldots, \zeta_n)} = \frac{1}{\det P} = [\det (a^{ij})]^{-1/2},$$

we get

$$J(x, t, \tau) = \int_{D^*} \frac{f}{(2\sqrt{\pi})^n} (t - \tau)^{-n/2} \cdot \exp\left[-\frac{|\zeta|^2}{4(t - \tau)}\right] d\zeta \equiv J_R + J_R^0 \tag{2.11}$$

where J_R is the part of the integral taken over a ball $|\zeta| \leq R$ with radius R sufficiently small (so that the ball is contained in D^*), and J_R^0 is the complementary part.

Introducing polar coordinates, we get

$$\begin{aligned} J_R &= \frac{f}{(2\sqrt{\pi})^n} \omega_n (t - \tau)^{-n/2} \int_0^R \exp\left[-\frac{r^2}{4(t - \tau)}\right] r^{n-1}\, dr \\ &= \frac{f}{(2\sqrt{\pi})^n} 2^{n-1}\omega_n \int_0^{R^2/4(t-\tau)} \sigma^{(n/2)-1} e^{-\sigma}\, d\sigma \end{aligned} \tag{2.12}$$

where

$$\omega_n = \frac{2\pi^{n/2}}{\Gamma\left(\dfrac{n}{2}\right)}$$

is the surface area of the unit hypersphere. It follows that

$$\lim_{\tau \to t} J_R = f. \tag{2.13}$$

Since R is fixed, the integrand of J_R^0 tends to 0 as $\tau \to t$. Hence $J_R^0 \to 0$ as $\tau \to t$. Combining this remark with (2.13) it follows, by (2.11), that $J(x, t, \tau) \to f$ as $\tau \to t$.

Consider now the general case where f and the a_{ij} are not constants. Writing

$$\begin{aligned} J(x, t, \tau) &= f(x, \tau) \int_D C(x, t) w^{x,t}(x, t; \xi, \tau)\, d\xi \\ &\quad + f(x, \tau) \int_D [C(\xi, \tau) w^{\xi,\tau}(x, t; \xi, \tau) - C(x, t) w^{x,t}(x, t; \xi, \tau)]\, d\xi \\ &\quad + \int_D C(\xi, \tau) w^{\xi,\tau}(x, t; \xi, \tau)[f(\xi, \tau) - f(x, \tau)]\, d\xi \\ &\equiv J_1 + J_2 + J_3, \end{aligned} \tag{2.14}$$

we may treat the integral of J_1 as J in the previous case of constant coefficients. Hence,

$$\lim_{\tau \to t} J_1 = f(x, t). \tag{2.15}$$

As for J_2,

$$\begin{aligned} &|w^{\xi,\tau}(x, t; \xi, \tau) - w^{x,t}(x, t; \xi, \tau)| \\ &\quad \leq n^2|x - \xi|^2(t - \tau)^{-(n+2)/2} \max_{i,j} |a^{ij}(\xi, \tau) - a^{ij}(x, t)| \exp\left[-\frac{\lambda_0|x - \xi|^2}{4(t - \tau)}\right]. \end{aligned}$$

Writing the integrand of the integral of J_2 in the form

$$\begin{aligned} I \equiv [C(\xi, \tau) - C(x, t)]w^{\xi,\tau}(x, t; \xi, \tau) \\ + C(x, t)[w^{\xi,\tau}(x, t; \xi, \tau) - w^{x,t}(x, t; \xi, \tau)] \end{aligned}$$

and using the previous inequality we find that for any $\epsilon > 0$ there exist R and δ sufficiently small such that the expression I is bounded by

$$\epsilon(t - \tau)^{-n/2}[1 + |x - \xi|^2(t - \tau)^{-1}] \exp\left[-\frac{\lambda_0|x - \xi|^2}{4(t - \tau)}\right] \tag{2.16}$$

if $|\xi - x| \leq R$ and $t - \tau < \delta$. Dividing the integral of J_2 into two parts, say $I_1 + I_2$ where for I_1 the integration is taken over $|\xi - x| \leq R$, and estimating I_1 by using the bound (2.16) for its integrand and then using polar coordinates, as in (2.12), and substituting $\sigma = r^2/(t - \tau)$, we find that $|I_1| \leq C\epsilon$ if $t - \tau < \delta$, where C is a constant independent of ϵ. Now, if R is fixed and $\tau \to t$ then $I_2 \to 0$ since its integrand tends to zero. It follows that $|J_2| < C'\epsilon$ if τ is sufficiently close to t, where C' is a constant independent of ϵ, τ. Hence

$$\lim_{\tau \to t} J_2 = 0. \tag{2.17}$$

J_3 can be treated in a similar way. We break it into a sum $J_{31} + J_{32}$ where the integration in J_{31} is taken over a ball $|\xi - x| \leq R$. Since $f(x, t)$ is a continuous function, for any $\epsilon > 0$ the integrand of J_{31} is bounded by

$$\epsilon(t - \tau)^{-n/2} \exp\left[-\frac{\lambda_0|x - \xi|^2}{4(t - \tau)}\right],$$

provided R and $t - \tau$ are sufficiently small. Introducing polar coordinates, as in (2.12), and substituting $\sigma = r^2/(t - \tau)$ we find that $|J_{31}| \leq C\epsilon$, C being a constant independent of ϵ. For R fixed, $J_{32} \to 0$ as $\tau \to t$. We thus obtain: $J_3 \to 0$ if $\tau \to t$. Combining this with (2.17), (2.15) and recalling (2.14), we get (2.10).

The assertion concerning the uniform convergence follows from the previous proof.

3. Volume Potentials

Given a function $f(x, t)$ in Ω we consider the function

$$V(x, t) = \int_{T_0}^{t} \int_D Z(x, t; \xi, \tau) f(\xi, \tau) \, d\xi \, d\tau \tag{3.1}$$

and call it the *volume potential* of f (with respect to the parametrix Z). We shall study in this section some differentiability properties of V. These will be used in the construction of a fundamental solution in the following section.

Note that the volume potential is an improper integral, the integrand having a singularity at $\xi = x$, $\tau = t$. The singularity, however, is integrable. Indeed, writing $w^{y,\tau}$ in the form

$$\begin{aligned} &w^{y,\tau}(x, t; \xi, \tau) \\ &\quad = (t - \tau)^{-\mu} (\vartheta^{y,\tau})^{\mu - n/2} \left[\frac{\vartheta^{y,\tau}}{t - \tau}\right]^{(n/2) - \mu} \exp\left[-\frac{\vartheta^{y,\tau}}{4(t - \tau)}\right] \end{aligned} \tag{3.2}$$

where $\vartheta^{y,\tau} = \vartheta^{y,\tau}(x, \xi)$, we get

$$|w^{y,\tau}(x, t; \xi, \tau)| \le \frac{\text{const.}}{(t - \tau)^{\mu} |x - \xi|^{n - 2\mu}} \qquad (0 < \mu < 1) \tag{3.3}$$

and the right-hand side is integrable.

Before proceeding to the actual study of V, we give a useful elementary lemma.

Lemma 1. *Let $f(x, y)$ be a continuous function of (x, y) when x, y vary in a compact domain S of R^m and $x \neq y$, and let*

$$\int_{S(x,\epsilon)} |f(x, y)| \, dy \to 0 \qquad \text{as } \epsilon \to 0,$$

uniformly with respect to x in S, where $S(x, \epsilon)$ is the intersection of S with the ball with center x and radius ϵ. Then, for any bounded measurable function $g(y)$ in S, the (improper) integral

$$\varphi(x) = \int_S f(x, y) g(y) \, dy$$

is a continuous function in S.

Proof. Given $\epsilon > 0$ choose δ_1 such that

$$\int_{S(x,\delta)} |f(x, y) g(y)| \, dy < \epsilon$$

for all $x \in S$, $\delta \le \delta_1$. Since $f(x, y)$ is a uniformly continuous (and hence a bounded) function of (x, y) when $x \in S$, $y \in S$, $|x - y| \ge \delta_1/2$, there exists a $\delta_2 < \delta_1/4$ such that

$$\int_{S(z,\delta_2)} |f(x, y) g(y)| \, dy < \epsilon \tag{3.4}$$

for all $x \in S$, $z \in S$. Denoting the complement in S of $S(z, \delta_2)$ by $S_2(z)$ we obtain, from (3.4),

$$\left| \int_{S_2(z)} f(x, y)g(y)\, dy - \int_{S_2(z^0)} f(x, y)g(y)\, dy \right| < 2\epsilon. \tag{3.5}$$

Using the uniform continuity of $f(x, y)$ for $x \in S$, $y \in S$, $|x - y| \geq \delta_2/2$ we have

$$\left| \int_{S_2(x')} f(x', y)g(y)\, dy - \int_{S_2(x')} f(x'', y)g(y)\, dy \right| < \epsilon$$

provided $|x' - x''| < \delta$ for some δ sufficiently small ($\delta < \delta_2/2$). Hence, together with (3.5) (for $z = x'$, $z^0 = x''$, $x = x''$) we get

$$\left| \int_{S_2(x')} f(x', y)g(y)\, dy - \int_{S_2(x'')} f(x'', y)g(y)\, dy \right| < 3\epsilon.$$

Combining this with (3.4) (for $z = x = x'$ and $z = x = x''$) we derive the inequality $|\varphi(x') - \varphi(x'')| < 5\epsilon$, and the proof is completed.

If we define $Z(x, t; \xi, \tau) = 0$ for $t < \tau$ then we can apply Lemma 1 to the volume potential, and thus conclude:

Theorem 2. *If $f(x, t)$ is a bounded measurable function in Ω then the volume potential $V(x, t)$ is a continuous function in Ω.*

We next prove:

Theorem 3. *If $f(x, t)$ is a continuous function in Ω then $V(x, t)$ has first continuous derivatives with respect to x for $x \in D$, $T_0 < t \leq T_1$ and*

$$\frac{\partial V(x, t)}{\partial x_i} = \int_{T_0}^{t} \int_D \frac{\partial}{\partial x_i} Z(x, t; \xi, \tau) f(\xi, \tau)\, d\xi\, d\tau. \tag{3.6}$$

Proof. Writing $\partial w^{y,\tau}(x, t; \xi, \tau)/\partial x_i$ analogously to (3.2) we find that

$$\left| \frac{\partial}{\partial x_i} w^{y,\tau}(x, t; \xi, \tau) \right| \leq \frac{\text{const.}}{(t - \tau)^{\mu} |x - \xi|^{n+1-2\mu}} \quad \left(\frac{1}{2} < \mu < 1\right). \tag{3.7}$$

Thus, the singularity of the integrand in (3.6) is integrable. If we define $\partial Z(x, t; \xi, \tau)/\partial x_i = 0$ for $t < \tau$ then we can apply Lemma 1 and thus conclude that the integral of (3.6) is a continuous function. It remains to verify (3.6). Set

$$J(x, t, \tau) = \int_D Z(x, t; \xi, \tau) f(\xi, \tau)\, d\xi. \tag{3.8}$$

Then

$$V(x, t) = \int_{T_0}^{t} J(x, t, \tau)\, d\tau. \tag{3.9}$$

By Theorem 1, Sec. 2, $J(x, t, \tau)$ is a continuous function in (x, t, τ) when $x \in D$, $T_0 \leq \tau \leq t \leq T_1$, provided we define

$$J(x, t, t) = \lim_{\tau \to t} J(x, t, \tau). \tag{3.10}$$

Clearly, if $t > \tau$

$$(3.11)\qquad \frac{\partial J(x, t, \tau)}{\partial x_i} = \int_D \frac{\partial}{\partial x_i} Z(x, t; \xi, \tau) f(\xi, \tau)\, d\xi.$$

Using (3.7) it follows that

$$(3.12)\qquad \left|\frac{\partial J(x, t, \tau)}{\partial x_i}\right| \le \frac{\text{const.}}{(t - \tau)^\mu} \qquad \left(\frac{1}{2} < \mu < 1\right).$$

Consequently, the improper integral

$$(3.13)\qquad \tilde{V}_i(x, t) = \int_{T_0}^t \frac{\partial}{\partial x_i} J(x, t, \tau)\, d\tau$$

is absolutely uniformly convergent.

Let $x^h = (x_1, \ldots, x_{i-1}, x_i + h, x_{i+1}, \ldots, x_n)$ and consider

$$(3.14)\qquad I \equiv \frac{V(x^h, t) - V(x, t)}{h} - \tilde{V}_i(x, t)$$

$$= \int_{T_0}^t \left[\frac{J(x^h, t, \tau) - J(x, t, \tau)}{h} - \frac{\partial}{\partial x_i} J(x, t, \tau)\right] d\tau$$

$$= \int_{T_0}^t \left[\frac{\partial}{\partial x_i} J(x^*, t, \tau) - \frac{\partial}{\partial x_i} J(x, t, \tau)\right] d\tau$$

where x^* is some point in the interval connecting x^h to x. Using (3.12) it follows that, for any $\epsilon > 0$,

$$(3.15)\qquad \int_{t_\epsilon}^t \left|\frac{\partial}{\partial x_i} J(x, t, \tau)\right| d\tau < \epsilon \qquad \text{for all } x \in D$$

provided $t - t_\epsilon$ is sufficiently small (independently of x). Once t_ϵ is fixed ($t_\epsilon < t$) we can find $\delta = \delta(t_\epsilon, \epsilon)$ such that if $\tau < t_\epsilon$, $x \in D$,

$$(3.16)\qquad \left|\frac{\partial}{\partial x_i} J(x^*, t, \tau) - \frac{\partial}{\partial x_i} J(x, t, \tau)\right| < \frac{\epsilon}{T_1 - T_0}$$

provided $|h| < \delta$.

Writing I in the form

$$I = \int_{T_0}^{t_\epsilon} \left[\frac{\partial}{\partial x_i} J(x^*, t, \tau) - \frac{\partial}{\partial x_i} J(x, t, \tau)\right] d\tau$$

$$+ \int_{t_\epsilon}^t \frac{\partial}{\partial x_i} J(x^*, t, \tau)\, d\tau - \int_{t_\epsilon}^t \frac{\partial}{\partial x_i} J(x, t, \tau)\, d\tau$$

and using (3.15), (3.16), we get $|I| < 3\epsilon$ if $|h| < \delta$. Thus $\partial V/\partial x_i$ exists and is equal to $\tilde{V}_i$, i.e., (3.6) holds.

Theorem 4. *Let $f(x, t)$ be a continuous function in Ω and locally Hölder continuous (exponent β) in $x \in D$, uniformly with respect to t. Then $V(x, t)$ has second continuous derivatives with respect to x, for $x \in D$, $T_0 < t \le T_1$, and*

$$(3.17)\qquad \frac{\partial^2 V(x, t)}{\partial x_i\, \partial x_j} = \int_{T_0}^t d\tau \int_D \frac{\partial^2 Z(x, t; \xi, \tau)}{\partial x_i\, \partial x_j} f(\xi, \tau)\, d\xi.$$

Observe that whereas in (3.6) the order of integration is immaterial since the singularity of the integrand is absolutely integrable in each variable separately (by (3.7)), the integral in (3.17) is a repeated integral and thus only the τ-integral is taken as an improper integral. In contrast to (3.7) we now have the inequality (whose proof is similar to that of (3.7)),

$$\left| \frac{\partial^2}{\partial x_i \, \partial x_j} w^{y,\tau}(x, t; \xi, \tau) \right| \leq \frac{\text{const.}}{(t - \tau)^\mu |x - \xi|^{n+2-2\mu}} \tag{3.18}$$

so that we cannot assert that the singularity of the integrand of (3.17) is absolutely integrable in Ω.

Proof. By Theorem 3,

$$\frac{\partial V(x, t)}{\partial x_i} = \int_{T_0}^t \frac{\partial}{\partial x_i} J(x, t, \tau) \, d\tau \tag{3.19}$$

where J is defined by (3.8). Write $\partial J/\partial x_i$ in the form

$$\begin{aligned} \frac{\partial J(x, t, \tau)}{\partial x_i} &= f(y, \tau) C(y, \tau) \int_D \frac{\partial}{\partial x_i} w^{y,\tau}(x, t; \xi, \tau) \, d\xi \\ &\quad + f(y, \tau) \int_D \left[C(\xi, \tau) \frac{\partial}{\partial x_i} w^{\xi,\tau}(x, t; \xi, \tau) \right. \\ &\quad \left. - C(y, \tau) \frac{\partial}{\partial x_i} w^{y,\tau}(x, t; \xi, \tau) \right] d\xi \\ &\quad + \int_D \frac{\partial}{\partial x_i} Z(x, t; \xi, \tau) [f(\xi, \tau) - f(y, \tau)] \, d\xi \end{aligned} \tag{3.20}$$

where y is a fixed point. Let K be a fixed ball contained in D and denote by ∂K the boundary of K and by K^* the complement of K in D. Breaking the first integral on the right-hand side of (3.20) into two integrals, namely, $\int_D = \int_K + \int_{K^*}$ and applying the divergence theorem to the first integral, we obtain

$$\int_{\partial K} w^{y,\tau}(x, t; \eta, \tau) \cos(\nu, \eta_i) \, dS_\eta + \int_{K^*} \frac{\partial}{\partial x_i} w^{y,\tau}(x, t; \xi, \tau) \, d\xi \tag{3.21}$$

where ν is the outwardly directed normal to ∂K and dS_η is the surface element on ∂K. Substituting (3.21) into (3.20), then differentiating (3.20) once with respect to x_j and choosing $y = x$, we get

$$\begin{aligned} \frac{\partial^2 J(x, t, \tau)}{\partial x_i \, \partial x_j} &= f(x, \tau) C(x, \tau) \\ &\quad \cdot \left[\int_{\partial K} \frac{\partial}{\partial x_j} w^{y,\tau}(x, t; \eta, \tau) \cos(\nu, \eta_i) \, dS_\eta \right]_{y=x} \\ &\quad + f(x, \tau) C(x, \tau) \left[\int_{K^*} \frac{\partial^2}{\partial x_i \, \partial x_j} w^{y,\tau}(x, t; \xi, \tau) \, d\xi \right]_{y=x} \end{aligned} \tag{3.22}$$

$$+ f(x, \tau) \int_D \left[C(\xi, \tau) \frac{\partial^2}{\partial x_i \, \partial x_j} w^{\xi,\tau}(x, t; \xi, \tau) \right.$$

$$\left. - C(x, \tau) \frac{\partial^2}{\partial x_i \, \partial x_j} w^{y,\tau}(x, t; \xi, \tau) \right]_{y=x} d\xi$$

$$+ \int_D \left[\frac{\partial^2}{\partial x_i \, \partial x_j} Z(x, t; \xi, \tau) \right] [f(\xi, \tau) - f(x, \tau)] \, d\xi.$$

Let x be a fixed point lying in the interior of K. Then each of the first two integrals on the right-hand side of (3.22) is a bounded function of its variables which tends uniformly to zero if $\tau \to t$. To estimate the last integral I_4 on the right-hand side of (3.22), we use (3.18) and the Hölder continuity of f. We find that

$$|I_4| \le \frac{\text{const.}}{(t - \tau)^\mu} \int_D \frac{d\xi}{|x - \xi|^{n+2-2\mu-\beta}} \le \frac{\text{const.}}{(t - \tau)^\mu}$$

if $1 - (\beta/2) < \mu < 1$, where the constant is independent of (x, t, τ) provided x is restricted to a closed subset of D.

To evaluate the third integral I_3 on the right-hand side of (3.22) we need the explicit formula

$$\text{(3.23)} \quad \frac{\partial^2}{\partial x_i \, \partial x_j} w^{y,\tau}(x, t; \xi, \tau) = \frac{1}{4} (t - \tau)^{-2-n/2} \exp\left[-\frac{\vartheta^{y,\tau}(x, \xi)}{4(t - \tau)} \right]$$

$$\times \left[-2(t - \tau) a^{ij}(y, \tau) + \sum_{h=1}^{n} a^{ih}(y, \tau)(x_h - \xi_h) \sum_{k=1}^{n} a^{jk}(y, \tau)(x_k - \xi_k) \right].$$

Using (2.2), (2.3), and the mean value theorem we find that

$$\left| \exp\left[-\frac{\vartheta^{\xi,\tau}(x, \xi)}{4(t - \tau)} \right] - \exp\left[-\frac{\vartheta^{x,\tau}(x, \xi)}{4(t - \tau)} \right] \right|$$

$$\le \frac{\text{const.}}{t - \tau} \exp\left[-\frac{\lambda_0 |x - \xi|^2}{4(t - \tau)} \right] |x - \xi|^{2+\alpha}.$$

Using this inequality and employing once more (2.2), (2.3) we find from the formula (3.23) that

$$\text{(3.24)} \quad H \equiv \left| \frac{\partial^2}{\partial x_i \, \partial x_j} w^{\xi,\tau}(x, t; \xi, \tau) - \left[\frac{\partial^2}{\partial x_i \, \partial x_j} w^{y,\tau}(x, t; \xi, \tau) \right]_{y=x} \right|$$

$$\le \text{const.} \, (t - \tau)^{-(n/2)-1}$$

$$\cdot \exp\left[-\frac{\lambda_0 |x - \xi|^2}{4(t - \tau)} \right] \left[1 + \frac{|x - \xi|^2}{t - \tau} + \frac{|x - \xi|^4}{(t - \tau)^2} \right] |x - \xi|^\alpha$$

$$\le \text{const.} \, (t - \tau)^{-(n/2)-1} \exp\left[-\frac{\lambda_0^* |x - \xi|^2}{4(t - \tau)} \right] |x - \xi|^\alpha,$$

for any $\lambda_0^* < \lambda_0$, from which it follows (as in the proof of (3.3), (3.7)) that

$$(3.25)\qquad H \le \frac{\text{const.}}{(t-\tau)^\mu} \frac{1}{|x-\xi|^{n+2-2\mu-\alpha}} \qquad (1-(\alpha/2)<\mu<1).$$

Since $|C(x, \tau) - C(\xi, \tau)| \le \text{const.}\ |x - \xi|^\alpha$, using (3.18) we conclude that the integrand of I_3 is also bounded by the right-hand side of (3.25) (with a different constant). Hence $|I_3| \le \text{const.}\ (t-\tau)^{-\mu}$ if $1 - (\alpha/2) < \mu < 1$. Since the same bound was established for I_4, and since, as has already been noted, the first two terms on the right-hand side of (3.22) are bounded functions of (t, τ), we conclude that

$$(3.26)\qquad \left|\frac{\partial^2}{\partial x_i\, \partial x_j} J(x, t, \tau)\right| \le \frac{\text{const.}}{(t-\tau)^\mu} \qquad \left(1-\frac{\delta}{2}<\mu<1\right),$$

where $\delta = \min(\alpha, \beta)$.

We can now proceed as in the proof of Theorem 3; namely, we set

$$\tilde{V}_{ij}(x, t) = \int_{T_0}^t \frac{\partial^2}{\partial x_i\, \partial x_j} J(x, t, \tau)\, d\tau$$

and prove that the integral (3.19) satisfies

$$\frac{\partial}{\partial x_j}\left(\frac{\partial V}{\partial x_i}\right) = \tilde{V}_{ij},$$

i.e., (3.17) holds. The continuity of $\partial^2 V/\partial x_i\, \partial x_j$ follows by applying the method of proof of Lemma 1 to the right-hand side of (3.17).

Theorem 5. *Let $f(x, t)$ be as in Theorem 4. Then $\partial V(x, t)/\partial t$ exists and is continuous for $x \in D$, $T_0 < t \le T_1$, and*

$$(3.27)\qquad \frac{\partial V(x, t)}{\partial t} = f(x, t) + \int_{T_0}^t d\tau \int_D \sum_{i,j=1}^n a_{ij}(\xi, \tau) \frac{\partial^2 Z(x, t; \xi, \tau)}{\partial x_i\, \partial x_j} f(\xi, \tau)\, d\xi.$$

Proof. Since $Z(x, t; \xi, \tau)$ is a solution of (2.7), if $t > \tau$

$$(3.28)\qquad \frac{\partial}{\partial t} J(x, t, \tau) = \int_D \frac{\partial}{\partial t} Z(x, t; \xi, \tau) f(\xi, \tau)\, d\xi$$

$$= \sum_{i,j=1}^n \int_D a_{ij}(\xi, \tau) \frac{\partial^2}{\partial x_i\, \partial x_j} Z(x, t; \xi, \tau) f(\xi, \tau)\, d\xi.$$

Each term on the right-hand side can be treated similarly to $\partial^2 J/\partial x_i\, \partial x_j$ in the proof of Theorem 4. Hence we get (compare (3.26))

$$(3.29)\qquad \left|\frac{\partial}{\partial t} J(x, t, \tau)\right| \le \frac{\text{const.}}{(t-\tau)^\mu} \qquad \left(1-\frac{\delta}{2}<\mu<1\right).$$

We shall prove that $\partial V(x, t)/\partial t$ exists and

$$(3.30)\qquad \frac{\partial V(x, t)}{\partial t} = J(x, t, t) + \int_{T_0}^t \frac{\partial}{\partial t} J(x, t, \tau)\, d\tau.$$

Taking $h > 0$ we consider the finite difference

$$\text{(3.31)} \quad \frac{V(x, t + h) - V(x, t)}{h}$$
$$= \frac{1}{h} \int_t^{t+h} J(x, t + h, \tau)\, d\tau + \int_{T_0}^t \frac{\partial}{\partial t} J(x, t^*, \tau)\, d\tau$$

where $t < t^* < t + h$. As $h \to 0$ the first term on the right-hand side converges to $J(x, t, t)$. To evaluate the second term, consider

$$\text{(3.32)} \qquad H \equiv \int_{T_0}^t \frac{\partial}{\partial t} J(x, t^*, \tau)\, d\tau - \int_{T_0}^t \frac{\partial}{\partial t} J(x, t, \tau)\, d\tau.$$

Using (3.29) it follows that, for any $\epsilon > 0$,

$$\text{(3.33)} \qquad \int_{t_\epsilon}^t \left| \frac{\partial}{\partial \bar{t}} J(x, \bar{t}, \tau) \right| d\tau < \epsilon$$

for some $t_\epsilon < t$ and for all $t \leq \bar{t} \leq t + h$ provided h is sufficiently small, say $h < \delta_1$. Since $\partial J(x, \bar{t}, \tau)/\partial \bar{t}$ is a continuous function if $\bar{t} - \tau \geq t - t_\epsilon$, there exists a $\delta \leq \delta_1$ such that if $h < \delta$ then

$$\text{(3.34)} \qquad \left| \frac{\partial}{\partial t} J(x, t^*, \tau) - \frac{\partial}{\partial t} J(x, t, \tau) \right| < \frac{\epsilon}{T_1 - T_0},$$

for all $\tau < t_\epsilon$.

Breaking each of the integrals in (3.32) into two integrals by

$$\int_{T_0}^t = \int_{T_0}^{t_\epsilon} + \int_{t_\epsilon}^t$$

and using (3.33), (3.34), it follows that $|H| < 3\epsilon$ if $h < \delta$. Using this in (3.31), we obtain (3.30) for the right-hand t-derivative. The considerations for $h < 0$ are similar.

In view of (2.10) and the fact that $J(x, t, \tau)$ is a solution of (2.7), (3.30) reduces to (3.27). The continuity of $\partial V(x, t)/\partial t$ follows by applying the method of proof of Lemma 1.

Combining Theorems 4, 5 we obtain:

Theorem 6. *Let $f(x, t)$ be as in Theorem 4. Then $V(x, t)$ satisfies the equation* ($x \in D$, $T_0 < t \leq T_1$)

$$\text{(3.35)} \quad \sum_{i,j=1}^n a_{ij}(x, t) \frac{\partial^2 V(x, t)}{\partial x_i\, \partial x_j} - \frac{\partial V(x, t)}{\partial t}$$
$$= -f(x, t) + \int_{T_0}^t \int_D \sum_{i,j=1}^n [a_{ij}(x, t) - a_{ij}(\xi, \tau)] \frac{\partial^2 Z(x, t; \xi, \tau)}{\partial x_i\, \partial x_j}\, d\xi\, d\tau.$$

Note that the singularity of the integrand on the right-hand side of (3.35) is absolutely and separately integrable in ξ and τ.

4. Construction of Fundamental Solutions

We shall construct a fundamental solution $\Gamma(x, t; \xi, \tau)$ of $Lu = 0$ in the form (2.8). If Φ is such that Theorem 6 applies for $f(x, t) = \Phi(x, t; \xi, \tau)$ then the equation $L\Gamma = 0$ becomes

$$\Phi(x, t; \xi, \tau) = LZ(x, t; \xi, \tau) + \int_\tau^t \int_D LZ(x, t; y, \sigma) \cdot \Phi(y, \sigma; \xi, \tau)\, dy\, d\sigma. \tag{4.1}$$

Thus, for each fixed (ξ, τ) $\Phi(x, t; \xi, \tau)$ is a solution of a Volterra integral equation with a singular kernel $LZ(x, t; y, \sigma)$.

From

$$\begin{aligned} LZ(x, t; y, \sigma) = \sum_{i,j=1}^{n} [a_{ij}(x, t) - a_{ij}(y, \sigma)] \frac{\partial^2}{\partial x_i\, \partial x_j} Z(x, t; y, \sigma) \\ + \sum_{i=1}^{n} b_i(x, t) \frac{\partial}{\partial x_i} Z(x, t; y, \sigma) + c(x, t) Z(x, t; y, \sigma) \end{aligned} \tag{4.2}$$

we obtain the inequality

$$|LZ(x, t; y, \sigma)| \leq \frac{\text{const.}}{(t - \sigma)^\mu} \frac{1}{|x - y|^{n+2-2\mu-\alpha}} \qquad \left(1 - \frac{\alpha}{2} < \mu < 1\right). \tag{4.3}$$

Hence the singularity is integrable.

We shall prove that there exists a solution Φ of (4.1) of the form

$$\Phi(x, t; \xi, \tau) = \sum_{\nu=1}^{\infty} (LZ)_\nu(x, t; \xi, \tau), \tag{4.4}$$

where $(LZ)_1 = LZ$ and

$$(LZ)_{\nu+1}(x, t; \xi, \tau) = \int_\tau^t \int_D [LZ(x, t; y, \sigma)](LZ)_\nu(y, \sigma; \xi, \tau)\, dy\, d\sigma. \tag{4.5}$$

We shall prove the convergence of the series in (4.4). The following elementary lemma will be needed.

Lemma 2. *If G is a bounded domain in R^n and $0 < \alpha < n$, $0 < \beta < n$, then for any $x \in G$, $z \in G$, $x \neq z$,*

$$\int_G \frac{dy}{|x - y|^\alpha |y - z|^\beta} \leq \begin{cases} \text{const. } |x - z|^{n-\alpha-\beta} & \text{if } \alpha + \beta > n, \\ \text{const.} & \text{if } \alpha + \beta < n. \end{cases}$$

The proof is obtained by breaking G into three sets corresponding to $|y - z| < |x - z|/2$, $|y - x| < |x - z|/2$ and the complementary set, and estimating the corresponding integrals separately.

Using Lemma 2 and (4.3) we get, if $2\mu < 1$ and $2(n + 2 - 2\mu - \alpha) < n$,

$$|(LZ)_2(x, t; \xi, \tau)| \leq \frac{\text{const.}}{(t - \tau)^{\mu+(\mu-1)}} \frac{1}{|x - \xi|^{n+2-2\mu-\alpha+(2-2\mu-\alpha)}}.$$

Since $\mu < 1$, $2 < 2\mu + \alpha$, the singularity of $(LZ)_2$ is weaker than that of LZ. Proceeding similarly to evaluate $(LZ)_3$, $(LZ)_4$, etc., we arrive at some ν_0 for which

$$|(LZ)_{\nu_0}(x, t; \xi, \tau)| \leq \text{const.} \tag{4.6}$$

We proceed to prove by induction on m that

$$|(LZ)_{m+\nu_0}(x, t; \xi, \tau)| \leq K_0 \frac{[K(t-\tau)^{1-\mu}]^m}{\Gamma((1-\mu)m+1)}, \tag{4.7}$$

where K_0, K are some constants and $\Gamma(t)$ is the gamma function. For $m = 0$ this follows from (4.6). Assuming now that (4.7) holds for some integer $m \geq 0$ and using (4.3) we get

$$|(LZ)_{m+1+\nu_0}(x, t; \xi, \tau)| \leq \text{const. } K_0 \frac{K^m}{\Gamma((1-\mu)m+1)} \cdot \int_\tau^t (t-\sigma)^{-\mu}(\sigma-\tau)^{(1-\mu)m} \, d\sigma.$$

Substituting $\rho = (\sigma - \tau)/(t - \tau)$ and using the formula

$$\int_0^1 (1-\rho)^{a-1}\rho^{b-1} \, d\rho = \frac{\Gamma(a)\Gamma(b)}{\Gamma(a+b)},$$

(4.7) follows for $m + 1$, provided the constant K is appropriately chosen.

From (4.7) it follows that the series expansion of $\Phi(x, t; \xi, \tau)$ is convergent and that the integral in (4.1) is equal to

$$\sum_{\nu=1}^{\infty} \int_\tau^t \int_D LZ(x, t; y, \sigma) \cdot (LZ)_\nu(y, \sigma; \xi, \tau) \, dy \, d\sigma.$$

Hence Φ is a solution of (4.1). We also have

$$|\Phi(x, t; \xi, \tau)| \leq \frac{\text{const.}}{(t-\tau)^\mu} \frac{1}{|x-\xi|^{n+2-2\mu-\alpha}} \qquad \left(1 - \frac{\alpha}{2} < \mu < 1\right). \tag{4.8}$$

In order to study Φ in more detail we shall need the following lemma.

Lemma 3. *If* $-\infty < \alpha < \frac{n}{2} + 1$, $-\infty < \beta < \frac{n}{2} + 1$, *then*

$$\int_\sigma^t \int_{R^n} (t-\tau)^{-\alpha} \exp\left[-\frac{h|x-\xi|^2}{4(t-\tau)}\right] (\tau-\sigma)^{-\beta} \exp\left[-\frac{h|\xi-y|^2}{4(\tau-\sigma)}\right] d\xi \, d\tau$$
$$= \left(\frac{4\pi}{h}\right)^{n/2} B\left(\frac{n}{2} - \alpha + 1, \frac{n}{2} - \beta + 1\right) (t-\sigma)^{(n/2)+1-\alpha-\beta} \cdot \exp\left[-\frac{h|x-y|^2}{4(t-\sigma)}\right].$$

Proof. Substitute

$$z_i = \left(h\frac{t-\sigma}{t-\tau}\right)^{1/2} \frac{\xi_i - y_i}{2(\tau-\sigma)^{1/2}} + \left(h\frac{\tau-\sigma}{t-\tau}\right)^{1/2} \frac{y_i - x_i}{2(t-\sigma)^{1/2}}$$

and note that

$$\frac{h(x_i-\xi_i)^2}{4(t-\tau)} + \frac{h(\xi_i-y_i)^2}{4(\tau-\sigma)} = \frac{h(x_i-y_i)^2}{4(t-\sigma)} + z_i^2.$$

If we proceed as in (3.2) but then write

$$\exp\left[-\frac{\vartheta^{y,\tau}}{4(t-\tau)}\right] = \exp\left[-\epsilon\frac{\vartheta^{y,\tau}}{4(t-\tau)}\right]\exp\left[-(1-\epsilon)\frac{\vartheta^{y,\tau}}{4(t-\tau)}\right] \qquad (\epsilon > 0)$$

and use the inequality $\sigma^{n/2-\mu}e^{-\epsilon\sigma} \leq$ const. for $0 \leq \sigma < \infty$, then we obtain a bound from which the following inequality for Z follows:

$$\text{(4.9)} \qquad |Z(x, t; \xi, \tau)| \leq \frac{\text{const.}}{(t-\tau)^{\mu}|x-\xi|^{n-2\mu}} \exp\left[-\frac{\lambda_0^*|x-\xi|^2}{4(t-\tau)}\right]$$

for any $\lambda_0^* < \lambda_0$, $0 \leq \mu \leq n/2$. In a similar way it can be proved that

$$\text{(4.10)} \quad \sum_i \left|\frac{\partial Z(x, t; \xi, \tau)}{\partial x_i}\right| \leq \frac{\text{const.}}{(t-\tau)^{\mu}|x-\xi|^{n+1-2\mu}} \exp\left[-\frac{\lambda_0^*|x-\xi|^2}{4(t-\tau)}\right],$$

$$\text{(4.11)} \quad \left|\frac{\partial Z(x, t; \xi, \tau)}{\partial t}\right| + \sum_{i,j}\left|\frac{\partial^2 Z(x, t; \xi, \tau)}{\partial x_i\, \partial x_j}\right|$$

$$\leq \frac{\text{const.}}{(t-\tau)^{\mu}|x-\xi|^{n+2-2\mu}} \exp\left[-\frac{\lambda_0^*|x-\xi|^2}{4(t-\tau)}\right],$$

$$\text{(4.12)} \qquad \left|\frac{\partial^3 Z(x, t; \xi, \tau)}{\partial x_i\, \partial x_j\, \partial x_h}\right| \leq \frac{\text{const.}}{(t-\tau)^{(n+3)/2}} \exp\left[-\frac{\lambda_0^*|x-\xi|^2}{4(t-\tau)}\right],$$

$$\text{(4.13)} \qquad |LZ(x, t; \xi, \tau)| \leq \frac{\text{const.}}{(t-\tau)^{\mu}|x-\xi|^{n+2-2\mu-\alpha}} \exp\left[-\frac{\lambda_0^*|x-\xi|^2}{4(t-\tau)}\right]$$

for any $\lambda_0^* < \lambda_0$; in (4.10) $0 \leq \mu \leq (n+1)/2$, in (4.11) $0 \leq \mu \leq (n+2)/2$, and in (4.13) $0 \leq \mu \leq (n+2-\alpha)/2$.

Using (4.13) with $\mu = (n+2-\alpha)/2$ and applying Lemma 3, we find that

$$|(LZ)_2(x, t; \xi, \tau)| \leq \frac{\text{const.}}{(t-\tau)^{(n+2-2\alpha)/2}} \exp\left[-\frac{\lambda_0^*|x-\xi|^2}{4(t-\tau)}\right].$$

Proceeding by induction one easily establishes, with the aid of Lemma 3, the inequalities

$$\text{(4.14)} \quad |(LZ)_m(x, t; \xi, \tau)| \leq \frac{H_0 H^m}{\Gamma(m\alpha)} (t-\tau)^{m\alpha-(n/2)-1} \exp\left[-\frac{\lambda_0^*|x-\xi|^2}{4(t-\tau)}\right]$$

where H_0, H are some positive constants. From the definition of Φ we then conclude that

$$\text{(4.15)} \qquad |\Phi(x, t; \xi, \tau)| \leq \frac{\text{const.}}{(t-\tau)^{(n+2-\alpha)/2}} \exp\left[-\frac{\lambda_0^*|x-\xi|^2}{4(t-\tau)}\right].$$

From this inequality it also follows (compare the derivation of (4.9)) that

$$\text{(4.16)} \quad |\Phi(x, t; \xi, \tau)| \leq \frac{\text{const.}}{(t-\tau)^{\mu}|x-\xi|^{n+2-2\mu-\alpha}} \exp\left[-\frac{\lambda_0^*|x-\xi|^2}{4(t-\tau)}\right]$$

for $0 \leq \mu \leq (n+2-\alpha)/2$, where λ_0^* is any number less than the λ_0^* appearing in (4.13) and, consequently, it can be taken to be any number $<\lambda_0$. (4.16) is an improvement of (4.8).

By using (4.14) and employing the method of proof of Lemma 1, Sec. 3, one can show that the $(LZ)_\nu(x, t; \xi, \tau)$ are continuous functions of (x, t), uniformly with respect to (ξ, τ) if $t - \tau \geq$ const. > 0, and are continuous functions of (ξ, τ), uniformly with respect to (x, t) if $t - \tau \geq$ const. > 0. It follows that the $(LZ)_\nu(x, t; \xi, \tau)$ are continuous functions of $(x, t; \xi, \tau)$. From (4.4), (4.14) one then concludes that $\Phi(x, t; \xi, \tau)$ is also a continuous function of $(x, t; \xi, \tau)$.

Theorem 7. $\Phi(x, t; \xi, \tau)$ *is Hölder continuous in* x; *more precisely, for any* $0 < \beta < \alpha$,

$$(4.17)\quad |\Phi(x, t; \xi, \tau) - \Phi(y, t; \xi, \tau)| \leq \frac{\text{const. } |x - y|^\beta}{(t - \tau)^{(n+2-\gamma)/2}} \left\{ \exp\left[-\frac{\lambda^* |x - \xi|^2}{t - \tau} \right] + \exp\left[-\frac{\lambda^* |y - \xi|^2}{t - \tau} \right] \right\}$$

where $\gamma = \alpha - \beta$ *and where* λ^* *is a positive constant.*

Proof. We first prove the inequality

$$(4.18)\quad |LZ(x, t; \xi, \tau) - LZ(y, t; \xi, \tau)| \leq \frac{\text{const. } |x - y|^\beta}{(t - \tau)^{(n+2-\gamma)/2}} \left\{ \exp\left[-\frac{k|x - \xi|^2}{t - \tau} \right] + \exp\left[-\frac{k|y - \xi|^2}{t - \tau} \right] \right\}$$

where k is a positive constant. Consider first the case where

$$(4.19)\qquad |x - y|^2 < t - \tau,$$

and take the term

$$(4.20)\qquad F(x, t; \xi, \tau) = [a_{ij}(x, t) - a_{ij}(\xi, \tau)] \frac{\partial^2}{\partial x_i\, \partial x_j} Z(x, t; \xi, \tau)$$

of $LZ(x, t; \xi, \tau)$ (see (4.2)). We have

(4.21)

$$\begin{aligned} F(x, t; \xi, \tau) - F(y, t; \xi, \tau) &= [a_{ij}(x, t) - a_{ij}(y, t)] \frac{\partial^2}{\partial x_i\, \partial x_j} Z(x, t; \xi, \tau) \\ &\quad + \left[\frac{\partial^2}{\partial x_i\, \partial x_j} Z(x, t; \xi, \tau) - \frac{\partial^2}{\partial y_i\, \partial y_j} Z(y, t; \xi, \tau) \right] \\ &\quad \cdot [a_{ij}(y, t) - a_{ij}(\xi, \tau)] \\ &\equiv F_1 + F_2. \end{aligned}$$

Using (4.11) with $\mu = (n + 2)/2$ we get

$$(4.22)\qquad |F_1| \leq \frac{\text{const. } |x - y|^\alpha}{(t - \tau)^{(n+2)/2}} \exp\left[-\frac{\lambda_0^* |x - \xi|^2}{4(t - \tau)} \right].$$

Using the mean value theorem and (4.12), and noting that, because of (4.19), for any point ζ in the interval (x, y)

$$\exp\left[-c \frac{|\zeta - \xi|^2}{t - \tau} \right] \leq \text{const. } \exp\left[-c' \frac{|y - \xi|^2}{t - \tau} \right] \qquad (\text{for any } 0 < c' < c),$$

we get

$$(4.23) \quad \left| \frac{\partial^2}{\partial x_i \, \partial x_j} Z(x, t; \xi, \tau) - \frac{\partial^2}{\partial y_i \, \partial y_j} Z(y, t; \xi, \tau) \right| \leq \frac{\text{const.} \, |x - y|}{(t - \tau)^{(n+3)/2}} \exp\left[-k_1 \frac{|y - \xi|^2}{t - \tau} \right]$$

where k_i are used to denote appropriate positive constants. Hence,

$$(4.24) \qquad |F_2| \leq \frac{\text{const.} \, |x - y|}{(t - \tau)^{(n+3-\alpha)/2}} \exp\left[-k_2 \frac{|y - \xi|^2}{t - \tau} \right].$$

Combining (4.24), (4.22) and using (4.19), we get

$$|F(x, t; \xi, \tau) - F(y, t; \xi, \tau)| \leq \frac{\text{const.} \, |x - y|^\beta}{(t - \tau)^{(n+2-\gamma)/2}} \exp\left[-k_3 \frac{|y - \xi|^2}{t - \tau} \right].$$

The estimation of the Hölder exponent and coefficient for the lower-order terms in LZ is similar to the estimation for F. Combining these estimates, the inequality (4.18) follows.

If (4.19) is not satisfied then, by (4.13) (with $\mu = (n + 2 - \alpha)/2$),

$$|LZ(x, t; \xi, \tau)| \leq \frac{\text{const.} \, (t - \tau)^{\beta/2}}{(t - \tau)^{(n+2-\gamma)/2}} \exp\left[-\frac{\lambda_0^* |x - \xi|^2}{4(t - \tau)} \right].$$

A similar inequality holds for $LZ(y, t; \xi, \tau)$. Since $(t - \tau)^{\beta/2} \leq |x - y|^\beta$, (4.18) follows.

Denoting the integral on the right-hand side of (4.1) by $\Psi(x, t; \xi, \tau)$, it remains to prove that (4.18) holds with LZ replaced by Ψ (with a possibly different k). Writing

$$\Psi(x, t; \xi, \tau) - \Psi(y, t; \xi, \tau) = \int_\tau^t \int_D [LZ(x, t; \zeta, \sigma) - LZ(y, t; \zeta, \sigma)] \cdot \Phi(\zeta, \sigma; \xi, \tau) \, d\zeta \, d\sigma$$

and using (4.18), (4.15), we get

$$(4.25) \quad |\Psi(x, t; \xi, \tau) - \Psi(y, t; \xi, \tau)| \leq \text{const.} \, |x - y|^\beta [I(x, t; \xi, \tau) + I(y, t; \xi, \tau)]$$

where

$$I(x, t; \xi, \tau) = \int_\tau^t \int_D \frac{1}{(t - \sigma)^{(n+2-\gamma)/2}} \exp\left[-\frac{k|x - \zeta|^2}{t - \sigma} \right] \frac{1}{(\sigma - \tau)^{(n+2-\alpha)/2}} \cdot \exp\left[-\frac{\lambda_0^* |\zeta - \xi|^2}{4(\sigma - \tau)} \right] d\zeta \, d\sigma.$$

By Lemma 3,

$$I(x, t; \xi, \tau) \leq \frac{\text{const.}}{(t - \tau)^{(n+2-\gamma-\alpha)/2}} \exp\left[-k_4 \frac{|x - \xi|^2}{t - \tau} \right].$$

A similar inequality holds for $I(y, t; \xi, \tau)$. Substituting these inequalities into (4.25) we find that (4.18) is satisfied if LZ is replaced by Ψ and if k is replaced by k_4. This completes the proof of Theorem 7.

By reading carefully the proof of Theorem 7 one deduces:

Corollary. *The inequality* (4.17) *holds for any* $\lambda^* < \lambda_0/4$.

Theorem 8. *The function* $\Gamma(x, t; \xi, \tau)$, *defined by* (2.8), *is a fundamental solution of* $Lu = 0$ *in* Ω.

Proof. We first prove that for each fixed (ξ, τ) $L\Gamma = 0$. Write Γ in the form

$$(4.26) \quad \Gamma(x, t; \xi, \tau) = Z(x, t; \xi, \tau) + \int_\tau^{t_0} \int_D Z(x, t; \eta, \sigma)\Phi(\eta, \sigma; \xi, \tau)\, d\eta\, d\sigma + \int_{t_0}^{t} \int_D Z(x, t; \eta, \sigma)\Phi(\eta, \sigma; \xi, \tau)\, d\eta\, d\sigma,$$

where t_0 is a fixed number satisfying $\tau < t_0 < t$. For the first integral, the first two x-derivatives of $Z(x, t; \eta, \sigma)$ are continuous functions in $(x, t; \eta, \sigma)$ whereas $\Phi(\eta, \sigma; \xi, \tau)$ is absolutely integrable in (η, σ) (by (4.8) or (4.16)). Hence, by a standard theorem of calculus, the first two x-derivatives of the integral exist, and the order of any x-differentiation (up to the second order) and of the integration may be changed.

As for the second integral on the right-hand side of (4.26), $\Phi(\eta, \sigma; \xi, \tau)$ (for (ξ, τ) fixed) is uniformly Hölder continuous in η with any exponent $<\alpha$, as follows by Theorem 7. By Theorems 3, 4 of Sec. 3 it follows that the first two x-derivatives of the integral exist, and the order of any x-differentiation (up to the second order) and of the integration may be changed. We conclude that the first two x-derivatives of Γ exist, and

$$(4.27) \quad \frac{\partial^2 \Gamma(x, t; \xi, \tau)}{\partial x_i\, \partial x_j} = \frac{\partial^2 Z(x, t; \xi, \tau)}{\partial x_i\, \partial x_j} + \int_\tau^{t_0} \int_D \frac{\partial^2 Z(x, t; \eta, \sigma)}{\partial x_i\, \partial x_j} \Phi(\eta, \sigma; \xi, \tau)\, d\eta\, d\sigma + \int_{t_0}^{t} d\sigma \int_D \frac{\partial^2 Z(x, t; \eta, \sigma)}{\partial x_i\, \partial x_j} \Phi(\eta, \sigma; \xi, \tau)\, d\eta;$$

a similar formula holds for $\partial\Gamma/\partial x_i$.

The existence of $\partial\Gamma(x, t; \xi, \tau)/\partial t$ can be established by the same considerations as for $\partial\Gamma/\partial x_i$, $\partial^2\Gamma/\partial x_i\, \partial x_j$, making use of Theorem 5, Sec. 3. The formula

$$(4.28) \quad \frac{\partial \Gamma(x, t; \xi, \tau)}{\partial t} = \frac{\partial Z(x, t; \xi, \tau)}{\partial t} + \int_\tau^{t_0} \int_D \frac{\partial Z(x, t; \eta, \sigma)}{\partial t} \Phi(\eta, \sigma; \xi, \tau)\, d\eta\, d\sigma - \Phi(x, t; \xi, \tau) + \int_{t_0}^{t} d\sigma \int_D \frac{\partial Z(x, t; \eta, \sigma)}{\partial t} \Phi(\eta, \sigma; \xi, \tau)\, d\eta$$

is valid.

Combining (4.28), (4.27) and the analogue of (4.27) for $\partial\Gamma/\partial x_i$, we get

$$L\Gamma(x, t; \xi, \tau) = LZ(x, t; \xi, \tau) - \Phi(x, t; \xi, \tau) + \int_\tau^t \int_D LZ(x, t; \eta, \sigma)\cdot\Phi(\eta, \sigma; \xi, \tau)\, d\eta\, d\sigma.$$

Since Φ satisfies the integral equation (4.1), $L\Gamma = 0$.

Using (2.8) and the method of proof of Lemma 1, Sec. 3, one can show that $\Gamma(x, t; \xi, \tau)$ is a continuous function of (x, t), uniformly with respect to (ξ, τ) if $t - \tau \geq$ const. > 0, and it is a continuous function of (ξ, τ) uniformly with respect to (x, t) if $t - \tau \geq$ const. Hence $\Gamma(x, t; \xi, \tau)$ is a continuous function of $(x, t; \xi, \tau)$. Here x and ξ vary in $\bar{D}$ and $T_0 \leq \tau < t \leq T_1$.

Using (4.28), (4.27) and the analogue of (4.27) for $\partial\Gamma/\partial x_i$, one can similarly show that

$$\frac{\partial\Gamma(x, t; \xi, \tau)}{\partial x_i}, \quad \frac{\partial^2\Gamma(x, t; \xi, \tau)}{\partial x_i\, \partial x_j}, \quad \frac{\partial\Gamma(x, t; \xi, \tau)}{\partial t}$$

are continuous functions of $(x, t; \xi, \tau)$, where x, ξ vary in D and $T_0 \leq \tau < t \leq T_1$. This fact, however, will not be needed in the present chapter.

It remains to prove that for any continuous function $f(x)$ in $\bar{D}$,

$$\lim_{t\to\tau} \int_D \Gamma(x, t; \xi, \tau) f(\xi)\, d\xi = f(x) \qquad \text{for } x \in D. \tag{4.29}$$

In view of Theorem 1, Sec. 2, it suffices to show that

$$I \equiv \int_D \int_\tau^t \int_D |Z(x, t; \zeta, \sigma)\Phi(\zeta, \sigma; \xi, \tau)|\, d\zeta\, d\sigma\, d\xi \to 0 \qquad \text{if } t \to \tau. \tag{4.30}$$

Using (4.9) (with $\mu = n/2$), (4.15), and Lemma 3, we get

$$I \leq \text{const.} \int_D (t - \tau)^{(\alpha-n)/2} \exp\left[-\frac{\lambda_0^*|x - \xi|^2}{4(t - \tau)}\right] d\xi.$$

Substituting $\rho = |x - \xi|(t - \tau)^{-1/2}$ we find that

$$I \leq \text{const.}\ (t - \tau)^{\alpha/2},$$

from which (4.30) follows.

5. Properties of Fundamental Solutions

In Sec. 3 we have introduced the concept of volume potentials with respect to the parametrix $Z(x, t; \xi, \tau)$ and established some differentiability properties (Theorems 2–6). The purpose of the present section is to make a similar study of volume potentials with respect to the fundamental solution $\Gamma(x, t; \xi, \tau)$. Thus, we shall consider functions of the form

$$W(x, t) = \int_{T_0}^t \int_D \Gamma(x, t; \xi, \tau) f(\xi, \tau)\, d\xi\, d\tau. \tag{5.1}$$

For simplicity, f is always assumed to be continuous in $\Omega \equiv \bar{D} \times [T_0, T_1]$.

If we substitute for Γ its expression from (2.8), then we find that

$$W(x, t) = V(x, t) + U(x, t) \tag{5.2}$$

where $V(x, t)$ is the potential (3.1) and $U(x, t)$ can be written (after changing the order of integration) in the form

$$U(x, t) = \int_{T_0}^{t} \int_D Z(x, t; \xi, \tau)\hat{f}(\xi, \tau)\, d\xi\, d\tau \tag{5.3}$$

where

$$\hat{f}(x, t) = \int_{T_0}^{t} \int_D \Phi(x, t; \xi, \tau) f(\xi, \tau)\, d\xi\, d\tau. \tag{5.4}$$

Defining $\Phi(x, t; \xi, \tau) = 0$ if $t < \tau$ and recalling (4.8), it follows that Lemma 1, Sec. 3, can be applied to the right-hand side of (5.4). Hence $\hat{f}(x, t)$ is a continuous function. $\hat{f}$ is also uniformly Hölder continuous in x with any exponent $\beta < \alpha$. Indeed, using (4.17) we get

$$|\hat{f}(x, t) - \hat{f}(y, t)| \leq \text{const. } |x - y|^{\beta}[A(x, t) + A(y, t)] \tag{5.5}$$

where

$$A(x, t) = \int_{T_0}^{t} \int_D (t - \tau)^{-(n+2-\gamma)/2} \exp\left[-\frac{\lambda^*|x - \xi|^2}{t - \tau} \right] d\xi\, d\tau.$$

Substituting (for τ fixed) $\rho = |x - \xi|(t - \tau)^{-1/2}$ we find that

$$A(x, t) \leq \text{const.} \int_{T_0}^{t} (t - \tau)^{(\gamma-2)/2}\, d\tau \leq \text{const.}$$

A similar inequality holds for $A(y, t)$. Substituting these inequalities into (5.5), the uniform Hölder continuity (exponent β) of $\hat{f}(x, t)$ in x follows.

Theorem 9. *If $f(x, t)$ is a continuous function in Ω then $W(x, t)$ is a continuous function in Ω and $\partial W/\partial x_i$ are continuous functions for $x \in D$, $T_0 < t \leq T_1$. If $f(x, t)$ is also locally Hölder continuous in $x \in D$, uniformly with respect to t, then $\partial^2 W/\partial x_i\, \partial x_j$ and $\partial W/\partial t$ are continuous functions for $x \in D$, $T_0 < t \leq T_1$, and*

$$LW(x, t) = -f(x, t). \tag{5.6}$$

Proof. All the assertions of the theorem, except for (5.6), follow immediately from the formulas (5.2)–(5.4) and from the Hölder continuity of $\hat{f}(x, t)$ in x, by applying Theorems 2–5 of Sec. 3.

(5.6) is a consequence of Theorem 6, Sec. 3 and (4.1). Indeed,

$$\begin{aligned} LW(x, t) &= -f(x, t) - \hat{f}(x, t) + \int_{T_0}^{t} \int_D LZ(x, t; \xi, \tau) f(\xi, \tau)\, d\xi\, d\tau \\ &\quad + \int_{T_0}^{t} \int_D LZ(x, t; \eta, \sigma)\hat{f}(\eta, \sigma)\, d\eta\, d\sigma \\ &= -f(x, t) + \int_{T_0}^{t} \int_D f(\xi, \tau) \Big\{ -\Phi(x, t; \xi, \tau) + LZ(x, t; \xi, \tau) \\ &\quad + \int_{\tau}^{t} \int_D LZ(x, t; \eta, \sigma) \cdot \Phi(\eta, \sigma; \xi, \tau)\, d\eta\, d\sigma \Big\}\, d\xi\, d\tau \\ &= -f(x, t). \end{aligned}$$

6. Fundamental Solutions in Unbounded Domains

In this section we shall extend the results of Secs. 1–5 to the case where D is an unbounded domain in R^n. The special case $D = R^n$ is of particular importance. Since most of the arguments are similar to those for the case where D is bounded, we shall only describe the necessary modifications.

If D is unbounded we always assume in this chapter that L satisfies the following assumptions (which coincide with (A_1), (A_2) of Sec. 1 if D is bounded):

$(A_1)'$ L is uniformly parabolic in $\Omega \equiv \bar{D} \times [T_0, T_1]$;

$(A_2)'$ the coefficients of L are bounded continuous functions in Ω and (1.4), (1.5), (1.6) hold throughout Ω.

The definition of a fundamental solution $\Gamma(x, t; \xi, \tau)$ is the same as in Sec. 1 except that in (ii) we require that $f(x)$ satisfies the inequality

$$|f(x)| \leq \text{const. } \exp\,[h|x|^2] \tag{6.1}$$

for some positive constant h. The integral in (1.7) exists only if $4h(t - \tau) < \lambda_0$ (where λ_0 is as in (2.2)). If

$$h < \frac{\lambda_0}{4(T_1 - T_0)} \tag{6.2}$$

then the integral in (1.7) exists for all $T_0 \leq \tau < t \leq T_1$.

The inequalities (4.9)–(4.13) obviously remain true also in case D is unbounded. The inequalities (4.14)–(4.16) and (4.17) also remain true without any change in the proofs. The proof of Theorem 8, Sec. 4, can also be extended with obvious modifications, provided Theorems 2–6 of Secs. 2, 3 are valid for the case where D is unbounded. In order to prove these theorems in this case, we break integrals $\int g(x, t; \xi, \tau)\, d\xi$ into two parts by dividing the domain of integration into two domains. The first domain D_0 is bounded and contains x, and the complementary domain D_1 is unbounded. The first integral can be treated as in Secs. 2, 3, whereas the second integral contains no singularity and can be treated by standard theorems of calculus.

Thus in the proof of Theorem 1, Sec. 2, we write $J = J' + J''$ where in J' the ξ-integration is taken over a bounded domain D_0 containing x such that $|\xi - x| \geq 1$ if $\xi \notin D_0$. J' can be treated exactly as J in the proof of Theorem 1. As for J'', assuming that

$$|f(x, t)| \leq \text{const. } \exp\,[h|x|^2] \tag{6.3}$$

where h satisfies (6.2) and noting that the integral

$$I \equiv \int_{|\xi - x| \geq 1} \exp\,[h|\xi|^2] \exp\left[-\frac{\lambda_0^*|x - \xi|^2}{4(t - \tau)}\right] d\xi$$

is convergent for all $T_0 < \tau < t \leq T_1$ if λ_0^* is sufficiently close to λ_0, and that, furthermore,

$$I \leq \text{const. } \exp\left[-\frac{\lambda_0^*}{8(t-\tau)}\right] \qquad \text{if } t - \tau < \frac{\lambda_0^*}{8h},$$

we conclude that J'' exists for all $T_0 \leq \tau < t \leq T_1$ and is continuous, and $J'' \to 0$ as $\tau \to t$. The assertion of Theorem 1 then follows.

We proceed to extend the results of Sec. 3. In view of (6.3), (6.2), (4.9), the integral (3.1) exists. The continuity of $V(x, t)$ follows by breaking the D-integral into two parts and treating each part separately. The continuity of the integral corresponding to the unbounded part D_1 of D (x is bounded away from D_1) follows by a standard theorem of calculus, whereas the continuity of the integral corresponding to the bounded part D_0 of D follows by employing Lemma 1, Sec. 3.

To extend Theorem 3, Sec. 3, we again break V into two integrals $V' + V''$. V' corresponds to D_0 and is treated as in Sec. 3. As for V'', the integrand has no singularity and therefore the proof that $\partial V''/\partial x_i$ exists and is equal to the integral of the x_i-derivative of the integrand follows by a standard theorem of calculus. The continuity of $\partial V/\partial x_i$ follows from the continuity of $\partial V'/\partial x_i$, $\partial V''/\partial x_i$.

By the same considerations we extend Theorems 4, 5 of Sec. 3.

Having extended the results of Secs. 2–4 to the case where D is unbounded, the results of Sec. 5 now also extend to this case without any change in the proofs. For the sake of later reference we sum up some of the results (in the case of an arbitrary domain D) in the following theorem.

Theorem 10. *Let D be any domain in R^n and assume that* $(A_1)'$, $(A_2)'$ *hold. Then there exists a fundamental solution $\Gamma(x, t; \xi, \tau)$ of $Lu = 0$ given by* (2.8), (4.1). *If $f(x, t)$ is any continuous function in Ω satisfying* (6.3), *where h satisfies* (6.2), *then the function $W(x, t)$ defined in* (5.1) *is a uniformly continuous function in Ω. If $f(x, t)$ is also locally Hölder continuous in $x \in D$, uniformly with respect to t, then $\partial W/\partial x_i$, $\partial^2 W/\partial x_i\, \partial x_j$, $\partial W/\partial t$ exist and are continuous functions for $x \in D$, $T_0 < t \leq T_1$, and $LW(x, t) = -f(x, t)$.*

Let $(A_1)'$, $(A_2)'$ hold with $D = R^n$, and consider the function

$$\tilde{W}(x, t) = \int_{R^n} \Gamma(x, t; \xi, T_0)\varphi(\xi)\, d\xi \tag{6.4}$$

where $\varphi(x)$ is a continuous function in R^n satisfying

$$|\varphi(x)| \leq \text{const. } \exp\,[h|x|^2] \qquad \left(h < \frac{\lambda_0}{4(T_1 - T_0)}\right). \tag{6.5}$$

We can decompose $\tilde{W}$ analogously to (5.1): $\tilde{W} = \tilde{V} + \tilde{U}$ where

$$\tilde{V}(x, t) = \int_{R^n} Z(x, t; \xi, T_0)\varphi(\xi)\, d\xi, \tag{6.6}$$

$$\tilde{U}(x, t) = \int_{T_0}^{t} \int_{R^n} Z(x, t; y, \sigma)\hat{\varphi}(y, \sigma)\, dy\, d\sigma \tag{6.7}$$

where

$$\hat{\varphi}(x, t) = \int_{R^n} \Phi(x, t; \xi, T_0)\varphi(\xi)\, d\xi. \tag{6.8}$$

Using (6.5) and (4.17) (which, by the corollary to Theorem 7, holds for any $\lambda^* < \lambda_0/4$), we find that $\hat{\varphi}(x, t)$ is locally Hölder continuous in x, uniformly with respect to t in intervals $T_0 + \epsilon \leq t \leq T_1$, $\epsilon > 0$. It follows that

$$\begin{aligned} L\tilde{W} = {} & \int_{R^n} LZ(x, t; \xi, T_0)\varphi(\xi)\, d\xi \\ & + \int_{T_0}^{t} d\sigma \int_{R^n} LZ(x, t; y, \sigma)\hat{\varphi}(y, \sigma)\, dy - \hat{\varphi}(x, t). \end{aligned} \tag{6.9}$$

Substituting $\hat{\varphi}$ from (6.8) and changing the order of integration, we get

$$\begin{aligned} L\tilde{W} &= \int_{R^n} \Big\{ LZ(x, t; \xi, T_0) + \int_{T_0}^{t} d\sigma \int_{R^n} LZ(x, t; y, \sigma)\cdot\Phi(y, \sigma; \xi, T_0)\, dy \\ & \qquad -\Phi(x, t; \xi, T_0) \Big\} \varphi(\xi)\, d\xi \\ &= 0. \end{aligned}$$

We thus have:

Theorem 11. *Let the assumptions* $(A_1)'$, $(A_2)'$ *be satisfied with* $D = R^n$, *and let* $\varphi(x)$ *be a continuous function in* R^n *satisfying* (6.5). *Then the function* $\tilde{W}$, *defined by* (6.4), *exists for* $T_0 < t \leq T_1$ *and satisfies:*

$$L\tilde{W}(x, t) = 0 \qquad \text{for } T_0 < t \leq T_1, \tag{6.10}$$

$$\tilde{W}(x, t) \to \varphi(x) \qquad \text{as } t \to T_0. \tag{6.11}$$

The following inequalities will be needed later on:

$$|\Gamma(x, t; \xi, \tau)| \leq \text{const.}\ (t - \tau)^{-n/2} \exp\left[-\frac{\lambda_0^*|x - \xi|^2}{4(t - \tau)} \right], \tag{6.12}$$

$$\left| \frac{\partial \Gamma(x, t; \xi, \tau)}{\partial x_i} \right| \leq \text{const.}\ (t - \tau)^{-(n+1)/2} \exp\left[-\frac{\lambda_0^*|x - \xi|^2}{4(t - \tau)} \right], \tag{6.13}$$

for any $\lambda_0^* < \lambda_0$.

To prove (6.12) we use (4.9) (with $\mu = n/2$) and (4.15), and thus obtain, with the aid of Lemma 3, Sec. 4, a bound on the integral on the right-hand side of (2.8). Combining this bound with (4.9) (for $\mu = n/2$), and recalling (2.8), the inequality (6.12) follows. (6.13) follows in a similar manner, making use also of (4.10) (with $\mu = (n + 1)/2$).

We conclude this section with a remark concerning the condition (1.4). In Chap. 9 we shall construct a fundamental solution for parabolic systems of any order. If we specialize the results to the parabolic equation (1.2) then we find that all the results of Secs. 2–6 are valid also if (1.4) is replaced by the weaker condition

$$|a_{ij}(x, t) - a_{ij}(x^0, t)| \leq A|x - x^0|^\alpha. \tag{6.14}$$

The reason that we are able to obtain the same results under this weaker condition lies in the fact that in Chap. 9 we take for the parametrix a fundamental solution Z of parabolic equations with coefficients depending on t. Z will not be given explicitly; this will necessitate some additional analysis not encountered in the present chapter.

7. The Cauchy Problem

Given a function $f(x, t)$ in $\Omega \equiv R^n \times [0, T]$ and a function $\varphi(x)$ in R^n, the problem of finding a function $u(x, t)$ satisfying the parabolic equation

$$Lu(x, t) = f(x, t) \qquad \text{in } \Omega_0 \equiv R^n \times (0, T], \tag{7.1}$$

where L is defined in (1.2), and the *initial condition*

$$u(x, 0) = \varphi(x) \text{ on } R^n \tag{7.2}$$

is called a *Cauchy problem* (in the strip $0 \leq t \leq T$). The solution is always required to be continuous in Ω.

$f(x, t)$ and $\varphi(x)$ will be assumed to satisfy the boundedness conditions

$$|f(x, t)| \leq \text{const. } \exp\,[h|x|^2], \tag{7.3}$$

$$|\varphi(x)| \leq \text{const. } \exp\,[h|x|^2] \tag{7.4}$$

where h is any positive constant satisfying

$$h < \frac{\lambda_0}{4T}\cdot \tag{7.5}$$

Theorem 12. *Suppose that L satisfies $(A_1)'$, $(A_2)'$ (with $D = R^n$, $T_0 = 0$, $T_1 = T$) and let $f(x, t)$, $\varphi(x)$ be continuous functions in Ω and R^n respectively, satisfying* (7.3), (7.4). *Assume also that $f(x, t)$ is locally Hölder continuous (exponent α) in $x \in R^n$, uniformly with respect to t. Then the function*

$$u(x, t) = \int_{R^n} \Gamma(x, t; \xi, 0)\varphi(\xi)\, d\xi - \int_0^t \int_{R^n} \Gamma(x, t; \xi, \tau) f(\xi, \tau)\, d\xi\, d\tau \tag{7.6}$$

is a solution of the Cauchy problem (7.1), (7.2) *and*

$$|u(x, t)| \leq \text{const. } [k|x|^2] \qquad \text{for } (x, t) \in \Omega, \tag{7.7}$$

where k is a constant depending only on h, λ_0, T.

Proof. In view of Theorems 10, 11 of Sec. 6, it only remains to establish (7.7). We shall need the following lemma.

Lemma 4. *For any $h > 0$, $\epsilon > 0$ there exists a $C = C(h, \epsilon)$ such that the inequality*

$$h|x - \xi|^2 - (h + \epsilon)|\xi|^2 \leq C|x|^2 \tag{7.8}$$

holds for all $x \in R^n$, $\xi \in R^n$.

Proof. If $\xi = 0$ then (7.8) holds for any $C \geq h$. If $\xi \neq 0$ set $y = x/|\xi|$, $e = \xi/|\xi|$. Then (7.8) takes the form

$$h|y - e|^2 - (h + \epsilon) \leq C|y|^2. \tag{7.9}$$

Now if $|y - e|^2 \leq (h + \epsilon)/h$ then the left-hand side is ≤ 0 and hence (7.9) holds. If on the other hand $|y - e|^2 > (h + \epsilon)/h$ then

$$|y| > [(h + \epsilon)/h]^{1/2} - 1 \equiv \vartheta > 0$$

and (7.9) clearly holds if C is sufficiently large, depending only on h, ϑ.

Returning to the proof of (7.7), we first obtain from (6.12) the inequality

$$|\Gamma(x, t; \xi, \tau)| \leq \frac{\text{const.}}{(t - \tau)^{n/2}} \exp\left[-\frac{\epsilon|x - \xi|^2}{t - \tau}\right] \exp\left[-(h + \epsilon)|x - \xi|^2\right]$$

for some $\epsilon > 0$ sufficiently small, making use of the assumption (7.5). Using this inequality and (7.3) to evaluate the second integral on the right-hand side of (7.6) we get, after substituting $x - \xi$ for ξ, the bound

$$\text{const.}\left\{\int_0^t \int_{R^n} \frac{\exp\left[-\epsilon|\xi|^2/(t - \tau)\right]}{(t - \tau)^{n/2}}\, d\xi\, d\tau\right\} \sup_{\xi}$$

$$\{\exp\left[h|x - \xi|^2 - (h + \epsilon)|\xi|^2\right]\} \leq \text{const.} \exp\left[C|x|^2\right],$$

where use has been made of (7.8). The first integral on the right-hand side of (7.6) is estimated in the same way.

8. The Adjoint Equation

The *adjoint equation* of (1.2) is, by definition, the equation

$$L^*v \equiv \sum_{i,j=1}^{n} a_{ij}(x, t) \frac{\partial^2 v}{\partial x_i\, \partial x_j} + \sum_{i=1}^{n} b_i^*(x, t) \frac{\partial v}{\partial x_i} + c^*(x, t)v + \frac{\partial v}{\partial t} = 0 \tag{8.1}$$

where

$$b_i^* = -b_i + 2 \sum_{j=1}^{n} \frac{\partial a_{ij}}{\partial x_j},$$

(8.2)

$$c^* = c - \sum_{i=1}^{n} \frac{\partial b_i}{\partial x_i} + \sum_{i,j=1}^{n} \frac{\partial^2 a_{ij}}{\partial x_i\, \partial x_j}.$$

Thus, we may also write

$$L^*v = \sum_{i,j=1}^{n} \frac{\partial^2}{\partial x_i\, \partial x_j}(a_{ij}v) - \sum_{i=1}^{n} \frac{\partial}{\partial x_i}(b_i v) + cv + \frac{\partial v}{\partial t}. \tag{8.3}$$

When we speak of L^* we always assume that $\partial a_{ij}/\partial x_h$, $\partial^2 a_{ij}/\partial x_h \partial x_k$, $\partial b_i/\partial x_h$ exist and are, say, continuous functions.

If u and v are smooth functions in Ω (in this context "smoothness" means that the first two x-derivatives and the first t-derivative exist and are continuous) then, as can easily be verified,

$$(8.4)\quad vLu - uL^*v = \sum_{i=1}^{n} \frac{\partial}{\partial x_i}\left[\sum_{j=1}^{n}\left(va_{ij}\frac{\partial u}{\partial x_j} - ua_{ij}\frac{\partial v}{\partial x_j} - uv\frac{\partial a_{ij}}{\partial x_j}\right) + b_i uv\right] - \frac{\partial}{\partial t}(uv).$$

This identity is called *Green's identity* for the operator L.

If u and v vanish in some neighborhood of the boundary of Ω then, by integrating both sides of (8.4), we get

$$(8.5)\qquad \int_\Omega (vLu - uL^*v)\, dx\, dt = 0.$$

Definition. *A linear partial differential operator (in the independent variables $x_1, \ldots, x_m$) is a finite sum*

$$\sum a_{k_1\cdots k_m}(x)\frac{\partial^{k_1+\cdots+k_m}}{\partial x_1^{k_1}\cdots\partial x_m^{k_m}}.$$

The $a_{k_1\cdots k_m}(x)$ are called the *coefficients* of the operator.

Theorem 13. *If L^* exists and if $\hat{L}$ is a linear partial differential operator in the variables (x, t) with, say, continuous coefficients statisfying*

$$(8.6)\qquad \int_\Omega (vLu - u\hat{L}v)\, dx\, dt = 0$$

for all smooth u, v which vanish in some neighborhood of the boundary of Ω, then $\hat{L} \equiv L^*$, i.e., the coefficients of $\hat{L}$ coincide with the coefficients of L^*.

Indeed, it follows from (8.5), (8.6) that

$$\int_\Omega u(L^* - \hat{L})v\, dx\, dt = 0$$

for all u, v as in (8.6). Hence $(L^* - \hat{L})v = 0$. We can now employ the following theorem, the proof of which is left to the reader (see Problem 2).

Theorem 14. *If M is a linear partial differential operator in the variables $x_1, \ldots, x_m$, with coefficients defined in a domain G and if $Mv = 0$ for all infinitely differentiable functions in G with a compact support, then $M \equiv 0$, i.e., the coefficients of M are identically zero.*

From Theorem 13 it follows that the adjoint L^* of L can be defined, equivalently, by the condition (8.5). Writing (8.5) in the form

$$\int_\Omega (uL^*v - vLu)\, dx\, dt = 0$$

we conclude that

$$(8.7)\qquad (L^*)^* = L.$$

Definition. A *fundamental solution* of $L^*v = 0$ is a function $\Gamma^*(x, t; \xi, \tau)$ defined for all $(x, t) \in \Omega$, $(\xi, \tau) \in \Omega$, $t < \tau$ which satisfies the following conditons:

(i) for each fixed (ξ, τ), it satisfies the equation $L^*v = 0$ (as a function of (x, t), $x \in D$, $T_0 \leq t < \tau$);

(ii) for every continuous function $f(x)$ in $\bar{D}$ (satisfying (6.1) if D is unbounded)

$$\lim_{t \nearrow \tau} \int_D \Gamma^*(x, t; \xi, \tau) f(\xi)\, d\xi = f(x). \tag{8.8}$$

We shall now specialize to the case $D = R^n$ and assume, in addition to $(A_1)'$ of Sec. 6, the following condition:

$(A_3)'$ The functions

$$a_{ij}, \quad \frac{\partial}{\partial x_h} a_{ij}, \quad \frac{\partial^2}{\partial x_h\, \partial x_k} a_{ij}; \quad b_i, \quad \frac{\partial}{\partial x_h} b_i; \quad c \tag{8.9}$$

are bounded continuous functions in $\Omega \equiv R^n \times [T_0, T_1]$; they satisfy a uniform Hölder condition (exponent α) in $x \in R^n$, uniformly with respect to t, and (1.4) holds throughout Ω.

We can then construct a fundamental solution Γ^* of $L^*v = 0$ by the method used to construct Γ. Thus, for a parametrix we take

$$Z^*(x, t; \xi, \tau) = C(\xi, \tau)(\tau - t)^{-n/2} \exp\left[-\frac{\vartheta^{\xi,\tau}(x, \xi)}{4(\tau - t)}\right] \qquad (t < \tau). \tag{8.10}$$

Z^* satisfies the equation $L_0^*v = 0$ where L_0 is defined in (2.7). Instead of (2.8) we now take

$$\Gamma^*(x, t; \xi, \tau) = Z^*(x, t; \xi, \tau) + \int_t^\tau \int_D Z^*(x, t; \eta, \sigma)\Phi^*(\eta, \sigma; \xi, \tau)\, d\eta\, d\sigma \tag{8.11}$$

where Φ^* is the solution of the integral equation

$$\Phi^*(x, t; \xi, \tau) = L^*Z^*(x, t; \xi, \tau) + \int_t^\tau \int_D L^*Z^*(x, t; \eta, \sigma)\cdot\Phi^*(\eta, \sigma; \xi, \tau)\, d\eta\, d\sigma. \tag{8.12}$$

An analogue of Theorem 10, Sec. 6, is valid and we also have, analogously to (6.12), (6.13),

$$|\Gamma^*(x, t; \xi, \tau)| \leq \text{const.}\ (\tau - t)^{-n/2} \exp\left[-\frac{\lambda_0^*|x - \xi|^2}{4(\tau - t)}\right], \tag{8.13}$$

$$\left|\frac{\partial}{\partial x_i}\Gamma^*(x, t; \xi, \tau)\right| \leq \text{const.}\ (\tau - t)^{-(n+1)/2} \exp\left[-\frac{\lambda_0^*|x - \xi|^2}{4(\tau - t)}\right]. \tag{8.14}$$

Theorem 15. *Under the assumptions* $(A_1)'$, $(A_3)'$ *a fundamental solution* Γ^* *of* $L^*v = 0$ *exists and*

$$\Gamma(x, t; \xi, \tau) = \Gamma^*(\xi, \tau; x, t) \qquad (t > \tau). \tag{8.15}$$

Proof. The existence of Γ^* has already been discussed. It remains to establish (8.15). Consider the functions

$$u(y, \sigma) = \Gamma(y, \sigma; \xi, \tau),$$
$$v(y, \sigma) = \Gamma^*(y, \sigma; x, t),$$

for $y \in R^n$, $\tau < \sigma < t$. Integrating Green's identity (8.4) over the domain $|y| < R$, $\tau + \epsilon < \sigma < t - \epsilon$ ($R > 0$, $\epsilon > 0$) and using $Lu = 0$, $L^*v = 0$ we obtain

$$(8.16) \quad \int_{|y|<R} u(y, t-\epsilon)v(y, t-\epsilon)\, dy - \int_{|y|<R} u(y, \tau+\epsilon)v(y, \tau+\epsilon)\, dy = I_{\epsilon,R}$$

where

$$I_{\epsilon,R} = \sum_{i=1}^{n} \int_{\tau+\epsilon}^{t-\epsilon} \int_{|y|=R} \left[\sum_{j=1}^{n} \left(va_{ij} \frac{\partial u}{\partial y_j} - ua_{ij} \frac{\partial v}{\partial y_j} - uv \frac{\partial a_{ij}}{\partial y_j} \right) + b_i uv \right] \cdot \cos(\nu, y_i)\, dS_y\, d\sigma,$$

ν is the outwardly directed normal to $|y| = R$, and dS_y is the surface element on $|y| = R$. Using (6.12), (6.13) and (8.13), (8.14) we find that $I_{\epsilon,R} \to 0$ as $R \to \infty$. Hence,

$$\int_{R^n} u(y, t-\epsilon)\Gamma^*(y, t-\epsilon; x, t)\, dy = \int_{R^n} v(y, \tau+\epsilon)\Gamma(y, \tau+\epsilon; \xi, \tau)\, dy.$$

Taking $\epsilon \to 0$ and using the second property in the definition of fundamental solutions we get $u(x, t) = v(\xi, \tau)$, i.e., (8.15) holds.

9. Uniqueness for the Cauchy Problem

In Sec. 7 we proved the existence of a solution to the Cauchy problem (7.1), (7.2) satisfying (7.7). We now prove a uniqueness theorem.

Theorem 16 *Let L satisfy the assumptions* $(A_1)'$, $(A_3)'$ *in* $\Omega \equiv R^n \times [0, T]$. *Then there exists at most one solution to the Cauchy problem* (7.1), (7.2) *satisfying the boundedness condition*

$$(9.1) \qquad \int_0^T \int_{R^n} |u(x, t)| \exp[-k|x|^2]\, dx\, dt < \infty$$

for some positive number k.

It follows that under the assumptions of Theorem 12, Sec. 7, and the additional assumption $(A_3)'$, there exists a unique solution to the Cauchy problem (7.1), (7.2) satisfying (7.7).

Proof. We have to prove that if $Lu(x, t) \equiv 0$, $u(x, 0) \equiv 0$ then $u(x, t) \equiv 0$. We shall first prove that $u(x, t) \equiv 0$ if $0 \le t < \delta$ for some δ sufficiently small. Let $(\bar{x}, \bar{t})$ be an arbitrary fixed point with $0 < \bar{t} < \delta$. We have to prove that $u(\bar{x}, \bar{t}) = 0$. Denote by B_R a ball in R^n with center $\bar{x}$ and radius R and let B'_R be the complement of B_R in B_{R+1}.

There exists a function $h(\xi)$ twice continuously differentiable in R^n and having the following properties (see Problem 4):

$$h(\xi) = \begin{cases} 1 & \text{if } \xi \in B_R, \\ 0 & \text{if } \xi \notin B_{R+1}, \end{cases} \tag{9.2}$$

$$0 \le h(\xi) \le 1, \qquad \sum_i \left| \frac{\partial h(\xi)}{\partial \xi_i} \right| + \sum_{i,j} \left| \frac{\partial^2 h(\xi)}{\partial \xi_i \, \partial \xi_j} \right| \le H \qquad (\xi \in R^n) \tag{9.3}$$

where H is a constant independent of R.

Integrating Green's identity (8.4) with $u = u(\xi, \tau)$, $v = h(\xi)\Gamma(\bar{x}, \bar{t}; \xi, \tau)$ over the region $\xi \in B_{R+1}$, $0 < \tau < \bar{t} - \epsilon$ where $\epsilon > 0$, and using $u(x, 0) \equiv 0$ we obtain, as $\epsilon \to 0$,

$$\lim_{\tau \nearrow \bar{t}} \int_{B_{R+1}} h(\xi)\Gamma(\bar{x}, \bar{t}; \xi, \tau)u(\xi, \tau)\, d\xi = \int_0^{\bar{t}} \int_{B_{R+1}} uL^*v \, d\xi \, d\tau; \tag{9.4}$$

the integrals over the boundary of B_{R+1} do not appear since $h(\xi) \equiv 0$ outside B_{R+1}. The left-hand side of (9.4) is equal to $h(\bar{x})u(\bar{x}, \bar{t}) = u(\bar{x}, \bar{t})$. Using (9.2) and $L^*\Gamma(\bar{x}, \bar{t}; \xi, \tau) = 0$ (which follows by (8.15)) we find that $L^*v = 0$ if $\xi \in B_R$. Hence

$$u(\bar{x}, \bar{t}) = \int_0^{\bar{t}} \int_{B'_R} u \left[2 \sum a_{ij} \frac{\partial h}{\partial \xi_i} \frac{\partial \Gamma}{\partial \xi_j} + \sum a_{ij} \frac{\partial^2 h}{\partial \xi_i \, \partial \xi_j} \Gamma + \sum b_i \frac{\partial h}{\partial \xi_i} \Gamma \right] d\xi \, d\tau. \tag{9.5}$$

Now, from (8.15) and (8.13), (8.14) it follows that

$$|\Gamma(\bar{x}, \bar{t}; \xi, \tau)| + \sum_i \left| \frac{\partial}{\partial \xi_i} \Gamma(\bar{x}, \bar{t}; \xi, \tau) \right|$$
$$\le \text{const.}\, (\bar{t} - \tau)^{-(n+1)/2} \exp\left[-\frac{\lambda_0^* |\bar{x} - \xi|^2}{4(\bar{t} - \tau)} \right].$$

Substituting this into (9.5) we get

$$|u(\bar{x}, \bar{t})| \le \text{const.} \exp\left[-\frac{\lambda_0^* R^2}{4\delta} \right] \int_0^{\bar{t}} \int_{B'_R} |u(\xi, \tau)| \, d\xi \, d\tau. \tag{9.6}$$

We shall now make use of the condition (9.1). It implies that

$$\int_0^{\bar{t}} \int_{R^n} |u(\xi, \tau)| \exp\left[-2k|\bar{x} - \xi|^2\right] d\xi \, d\tau < \infty.$$

Hence,

$$\int_0^{\bar{t}} \int_{B'_R} |u(\xi, \tau)| \exp\left[-2k|\bar{x} - \xi|^2\right] d\xi \, d\tau \to 0 \qquad \text{as } R \to \infty. \tag{9.7}$$

Comparing the integral of (9.7) with the right-hand side of (9.6) we see that if $\delta < \lambda_0^*/8k$ then the right-hand side of (9.6) also tends to zero as $R \to \infty$. Thus, $u(\bar{x}, \bar{t}) = 0$ if $\bar{x} \in R^n$, $0 \le \bar{t} < \delta$ and $\delta = \lambda_0^*/9k$.

We can now proceed in the same way to prove that $u(x, t) \equiv 0$ in the strips $\delta \le t < 2\delta$, $2\delta \le t < 3\delta$, etc., and the proof is thereby completed.

We conclude this section by giving an example of a solution $u(x, t) \not\equiv 0$ of the heat equation ($n = 1$)

$$u_{xx} - u_t = 0 \qquad (-\infty < x < \infty, 0 \le t \le 1)$$

satisfying the initial condition

$$u(x, 0) \equiv 0 \qquad (-\infty < x < \infty)$$

and the boundedness condition

$$\int_0^1 \int_{-\infty}^{\infty} |u(x, t)| \exp\,[-k|x|^{2+\epsilon}]\, dx\, dt < \infty \tag{9.8}$$

where ϵ is any given positive number.

It is known (see [80]) that for any $\delta > 0$ there exist infinitely differentiable functions $f(t) \not\equiv 0$ $(-\infty < t < \infty)$ satisfying the conditions: $f(t) = 0$ if $t < 0$ and if $t > 1$ and

$$|f^{(m)}(t)| \le C^m m^{(1+\delta)m} \qquad (m = 1, 2, \ldots)$$

for $0 \le t \le 1$, where C is a constant. Now take

$$u(x, t) = \sum_{m=0}^{\infty} \frac{f^{(m)}(t)}{(2m)!} x^{2m}. \tag{9.9}$$

The series and its first two derivatives are uniformly convergent in bounded sets of (x, t), provided $\delta < 1$. Since

$$|u(x, t)| \le C_1 + \sum_{m=1}^{\infty} \frac{C_2^m x^{2m}}{m^{2m-(1+\delta)m}} \le C_3 \exp\,[C_4|x|^\eta]$$

where the C_i are constants and $\eta = 2/(1 - \delta)$, we can take δ sufficiently small so that $\eta < 2 + \epsilon$ and, consequently, (9.8) is satisfied.

The above example shows that Theorem 16 is not true if in the condition (9.1) we replace $\exp\,[-k|x|^2]$ by $\exp\,[-k|x|^{2+\epsilon}]$, $\epsilon > 0$.

PROBLEMS

1. Let A, B be disjoint closed sets in R^n and let A be bounded. Prove that there exists an infinitely differentiable function $h(x)$ in R^n such that $h(x) = 1$ if $x \in A$, $h(x) = 0$ if $x \in B$, and $0 \le h(x) \le 1$ for all $x \in R^n$.
   [*Hint:* Let $\delta = \text{dist.}\,(A, B)$ and denote by A_0 the set of all points whose distance to A is $\le \delta/2$. Take
   $$h(x) = \int_{A_0} \rho_\epsilon(x - y)\, dy \qquad (\epsilon = \delta/3)$$
   where
   $$\rho_\epsilon(x) = \begin{cases} 0 & \text{if } |x| \ge \epsilon, \\ \dfrac{k}{\epsilon^n} \exp\left[-\dfrac{\epsilon^2}{\epsilon^2 - |x|^2}\right] & \text{if } |x| < \epsilon.] \end{cases}$$
2. Prove Theorem 14, Sec. 8.
   [*Hint:* For any $x^0 \in G$ take (by Problem 1) an infinitely differentiable function h with compact support in G, satisfying $h(x^0) \ne 0$. Use the relation $M(h \exp [\lambda_1 x_1 + \cdots + \lambda_n x_n]) = 0$.]

3. For any linear differential operator

$$Lu = \sum a_{k_1 \cdots k_n}(x) \frac{\partial^{k_1 + \cdots + k_n} u}{\partial x_1^{k_1} \cdots \partial x_n^{k_n}}$$

with sufficiently smooth coefficients in a domain G, we define its *adjoint* by

$$L^* v = \sum (-1)^{k_1 + \cdots + k_n} \frac{\partial^{k_1 + \cdots + k_n}}{\partial x_1^{k_1} \cdots \partial x_n^{k_n}} [a_{k_1 \cdots k_n}(x) v].$$

If $L^* = L$ then we say that L is *self-adjoint*. Prove that this definition of adjoint is equivalent to the following one: L^* is a linear differential operator for which

$$\int_G (vLu - uL^*v)\, dx = 0$$

for all infinitely differentiable functions u, v having a compact support in G.

4. Prove that there exists a function $h(\xi)$ satisfying (9.2), (9.3). [*Hint:* Use the h in Problem 1 with $A_0 = B_{R+(1/2)}$, $\epsilon = 1/4$.]

5. Let $\Gamma(x, t; \xi, \tau)$ be the fundamental solution of $Lu = 0$ in $R^n \times [T_0, T_1]$ (under the assumptions $(A_1)'$, $(A_2)'$). Prove:

$$\Gamma(x, t; \xi, \tau) = \int_{R^n} \Gamma(x, t; y, \sigma)\Gamma(y, \sigma; \xi, \tau)\, dy \qquad (\tau < \sigma < t).$$

6. Show that the assumption $h < \lambda_0/4T$ made in Theorem 12 cannot be relaxed. [*Hint:* Take $u_{xx} - u_t = 0$, $u(x, t) = (1 - 4ht)^{-1/2} \exp [hx^2/(1 - 4ht)]$.]

7. Verify that all the results of Chap. 1 remain true if the conditions (1.5), (1.6) are replaced by the following weaker conditions: For every compact subset K of D there exists a constant A', depending on K, such that

$$|b_i(x, t) - b_i(x^0, t)| \le A'|x - x^0|^\alpha,$$
$$|c(x, t) - c(x^0, t)| \le A'|x - x^0|^\alpha,$$

for all $x \in K$, $x^0 \in K$, $T_0 \le t \le T_1$. Verify also that the assumption that b_i, c are continuous functions throughout Ω may be replaced by the assumption that b_i, c are bounded continuous functions in $D \times (T_0, T_1]$.

CHAPTER 2

THE MAXIMUM PRINCIPLE AND SOME APPLICATIONS

Introduction. Let D be a 2-dimensional domain bounded by an open interval B on $t = 0$, an open interval B_T on $t = T$ and two continuous curves C_i: $x = \gamma_i(t)$ $(i = 1, 2)$ defined for $0 < t \leq T$. Let u be a continuous function in $\bar{D}$ (the closure of D) satisfying $\partial^2 u/\partial x^2 - \partial u/\partial t \geq 0$ in $D + B_T$. The weak maximum principle (for the heat operator) asserts that the maximum of u in $\bar{D}$ is attained on the part $\bar{B} + C_1 + C_2$ of the boundary of D. The strong maximum principle asserts that unless $u \equiv$ const. it cannot assume its maximum in $\bar{D}$ at any point of $D + B_T$.

Consider the *first initial-boundary value problem*

$$(0.1) \qquad \begin{aligned} u_{xx} - u_t &= f \quad && \text{in } D, \\ u &= h \quad && \text{on } B + C_1 + C_2 \end{aligned}$$

where f, h are any given functions. From the weak maximum principle it follows that there exists at most one solution to this problem. The *second initial-boundary value problem* is the problem of finding a solution to

$$(0.2) \qquad \begin{aligned} u_{xx} - u_t &= f \quad && \text{in } D, \\ u &= \varphi \quad && \text{on } B, \\ \frac{\partial u}{\partial \nu} + \beta u &= g \quad && \text{on } C_1 + C_2, \end{aligned}$$

where β is a fixed function, ν is the inward normal, and f, φ, g are any given functions. With the aid of the maximum principle one can establish the uniqueness of solutions under some assumptions on C_1, C_2, provided $\beta \leq 0$.

In this chapter we prove for general second-order parabolic equations results similar to those mentioned above for the heat operator. Additionally, we give some applications to the Cauchy problem; we prove uniqueness theorems and also some "positivity theorems," i.e., theorems asserting that $u(x, 0) \geq 0$ implies $u(x, t) \geq 0$. In the last section we extend some of the results to second-order elliptic equations.

In most of the theorems of this chapter the coefficients of the differential equation are assumed to be continuous functions. The reader may notice, however, that everything remains true if the coefficients are only assumed to be bounded functions and the operators are assumed to be locally uniformly parabolic (elliptic, in Sec. 7).

1. The Maximum Principle

Consider the operator

$$(1.1) \quad Lu \equiv \sum_{i,j=1}^{n} a_{ij}(x, t) \frac{\partial^2 u}{\partial x_i \, \partial x_j} + \sum_{i=1}^{n} b_i(x, t) \frac{\partial u}{\partial x_i} + c(x, t)u - \frac{\partial u}{\partial t}$$

in an $(n + 1)$-dimensional domain D. We list some assumptions that will be needed in the future:

(A) L is parabolic in D, i.e., for every $(x, t) \in D$ and for any real vector $\xi \neq 0$, $\Sigma\, a_{ij}(x, t)\xi_i\xi_j > 0$;
(B) the coefficients of L are continuous functions in D;
(C) $c(x, t) \leq 0$ in D.

The functions u in (1.1) are always assumed to have two continuous x-derivatives and one continuous t-derivative in D.

Notation. For any point $P^0 = (x^0, t^0)$ in D, we denote by $S(P^0)$ the set of all points Q in D which can be connected to P^0 by a simple continuous curve in D along which the t-coordinate is nondecreasing from Q to P^0. By $C(P^0)$ we denote the component (in $t = t^0$) of $D \cap \{t = t^0\}$ which contains P^0. Note that $S(P^0) \supset C(P^0)$.

The *strong maximum principle* asserts the following:

Theorem 1. *Let* (A), (B), (C) *hold. If* $Lu \geq 0$ $(Lu \leq 0)$ *in* D *and if* u *has in* D *a positive maximum (negative minimum) which is attained at a point* $P^0(x^0, t^0)$ *then* $u(P) = u(P^0)$ *for all* $P \in S(P^0)$.

Some extensions of Theorem 1 will be given in Sec. 2.

It suffices to prove the theorem in the case of positive maximum since the case of negative minimum then follows by applying the first case to $-u$. For the sake of clarity we shall first establish several lemmas (under the assumptions (A), (B), (C)).

Lemma 1. *Assume that either* $Lu > 0$ *throughout* D *or that* $Lu \geq 0$ *and* $c(x, t) < 0$ *throughout* D. *Then* u *cannot have a positive maximum in* D.

Proof. Assuming that u has a positive maximum in D we shall derive a contradiction. Let $P^0 = (x^0, t^0)$ be a point where the maximum is attained. We claim that

$$(1.2) \qquad \sum_{i,j=1}^{n} a_{ij}(x^0, t^0) \frac{\partial^2 u(x^0, t^0)}{\partial x_i \, \partial x_j} \leq 0.$$

Indeed, by a linear transformation $y = Cx$, D is transformed into a domain D^* and the inequality (1.2) becomes

$$\sum_{i,j=1}^{n} b_{ij} \frac{\partial^2 v(y^0, t^0)}{\partial y_i \, \partial y_j} \leq 0 \tag{1.3}$$

where $v(y, t) = u(x, t)$, $y_0 = Cx_0$, $(b_{ij}) = C(a_{ij})C^*$ (C^* = transpose of C). Taking C such that (b_{ij}) becomes the unit matrix and noting that $v(y^0, t^0)$ is the positive maximum of v in D^*, so that $\partial^2 v(y^0, t^0)/\partial y_i^2 \leq 0$ holds, we see that (1.3) is satisfied; hence also (1.2).

To complete the proof of the lemma note that $\partial u/\partial x_i = 0$, $\partial u/\partial t = 0$ at the point (x^0, t^0). Hence

$$Lu(x^0, t^0) \leq c(x^0, t^0)u(x^0, t^0).$$

Since also $u(x^0, t^0) > 0$, we get a contradiction in each of the cases $Lu > 0$, $c \leq 0$ and $Lu \geq 0$, $c < 0$.

Lemma 2. *Let $Lu \geq 0$ in D and let u have a positive maximum M in D. Suppose that D contains a closed solid ellipsoid E:*

$$\sum_{i=1}^{n} \lambda_i (x_i - x_i^*)^2 + \lambda_0 (t - t^*)^2 \leq R^2 \qquad (\lambda_j > 0, R > 0)$$

and that $u < M$ in the interior of E and $u(\bar{x}, \bar{t}) = M$ at some point $\bar{P} = (\bar{x}, \bar{t})$ on the boundary ∂E of E. Then $\bar{x} = x^*$, where $x^* = (x_1^*, \ldots, x_n^*)$.

Proof. We may assume that $\bar{P}$ is the only point on ∂E where $u = M$ since otherwise we can confine ourselves to a smaller closed ellipsoid lying in E and having $\bar{P}$ as the only common point with ∂E. Suppose that $\bar{x} \neq x^*$ and let C be an $(n + 1)$-dimensional closed ball contained in D, with center $\bar{P}$ and radius $< |\bar{x} - x^*|$. Then

$$|x - x^*| \geq \text{const.} > 0 \qquad \text{for all } (x, t) \in C. \tag{1.4}$$

The boundary of C is composed of a part ∂C_1 lying in E and a part ∂C_2 lying outside E. Clearly, for some $\delta > 0$,

$$u < M - \delta \qquad \text{on } \partial C_1. \tag{1.5}$$

Introduce the function

$$h(x, t) = \exp\left\{-\alpha\left[\sum_{i=1}^{n} \lambda_i (x_i - x_i^*)^2 + \lambda_0 (t - t^*)^2\right]\right\} - \exp\{-\alpha R^2\} \qquad (\alpha > 0). \tag{1.6}$$

$h > 0$ in the interior of E, $h = 0$ on ∂E and $h < 0$ outside E. Next,

$$\exp\left\{\alpha\left[\sum_{i=1}^{n} \lambda_i (x_i - x_i^*)^2 + \lambda_0 (t - t^*)^2\right]\right\} Lh(x, t)$$

$$= \left\{4\alpha^2 \sum_{i,j=1}^{n} a_{ij} \lambda_i \lambda_j (x_i - x_i^*)(x_j - x_j^*)\right.$$

$$-2\alpha\left[\sum_{i=1}^{n} a_{ii}\lambda_i + \sum_{i=1}^{n} b_i\lambda_i(x_i - x_i^*) - \lambda_0(t - t^*)\right] + c\Bigg\}$$

$$-c \exp\{-\alpha R^2\} \exp\left\{\alpha\left[\sum_{i=1}^{n} \lambda_i(x_i - x_i^*)^2 + \lambda_0(t - t^*)^2\right]\right\}.$$

Using (1.4) we see that the expression in the first braces on the right-hand side is positive in C if α is sufficiently large. The last term is ≥ 0 since $c \leq 0$. Thus,

$$Lh > 0 \quad \text{in } C. \tag{1.7}$$

Consider the function $v = u + \epsilon h$ $(\epsilon > 0)$ in C. If ϵ is sufficiently small then $v < M$ on ∂C_1, by (1.5). On ∂C_2, $u \leq M$ and $h < 0$; hence $v < M$ on ∂C_2. Thus, $v < M$ on the boundary of C and $v(\overline{P}) = u(\overline{P}) = M$. It follows that v assumes a positive maximum in the interior of C. Since (1.7) is satisfied, we get a contradiction to Lemma 1; hence $\bar{x} = x^*$.

Lemma 3. *If $Lu \geq 0$ in D and if u has a positive maximum in D which is attained at a point $P^0 = (x^0, t^0)$, then $u(P) = u(P^0)$ for all $P \in C(P^0)$.*

Proof. If the lemma is not true then there exists a point $P^1 = (x^1, t^0)$ in $C(P^0)$ such that $u(P^1) < u(P^0)$. Connect P^1 to P^0 by a simple continuous curve γ lying in $C(P^0)$. On γ there exists a point $P^* = (x^*, t^0)$ with $u(P^*) = u(P^0)$ such that $u(\overline{P}) < u(P^0)$ for all $\overline{P} = (\bar{x}, t^0)$ lying on γ between P^1 and P^*. Take a point $\overline{P}$ on γ, between P^1 and P^*, such that the distance from $\overline{P}$ to the boundary of D is $\geq 2\overline{\overline{P}P^*}$.

Since $u(\overline{P}) < u(P^*)$, there exists a sufficiently small interval σ given by $x = \bar{x}$, $t^0 - \epsilon \leq t \leq t^0 + \epsilon$ such that

$$u(P) < u(P^*) \qquad \text{for all } P \in \sigma. \tag{1.8}$$

Consider the family of ellipsoids E_λ:

$$|x - \bar{x}|^2 + \lambda(t - t^0)^2 \leq R^2 \qquad (\lambda > 0).$$

Take $R^2 = \lambda\epsilon^2$ so that the end points of σ lie on the boundary of E_λ. If $\lambda \to 0$, $E_\lambda \to \sigma$. As λ increases, $E_\lambda \cap \{t = t^0\}$ increases indefinitely and, because of (1.8), there exists a first value of λ, say $\lambda = \lambda_0$, such that $u < u(P^*)$ in the interior of E_{λ_0} and $u = u(P^*)$ at some boundary point $Q = (y, t^0)$ of E_{λ_0}. Since, by (1.8), Q does not lie on σ, i.e., $y \neq \bar{x}$, we have obtained a contradiction to Lemma 2.

Lemma 4. *Let R be a rectangle*

$$x_i^0 - a_i \leq x_i \leq x_i^0 + a_i \quad (i = 1, \ldots, n), \quad t^0 - a_0 \leq t \leq t^0$$

contained in D and let $Lu \geq 0$ in D. If u has a positive maximum in R which is attained at the point $P^0 = (x^0, t^0)$, where $x^0 = (x_1^0, \ldots, x_n^0)$, then $u(P) = u(P^0)$ for all $P \in R$.

Proof. If the lemma is not true then there exists a point Q in R such that $u(Q) < u(P^0)$. Since $u < u(P^0)$ also in a neighborhood of Q, we may assume that Q does not lie on $t = t^0$. On the straight segment γ connecting Q to P^0 there exists a point P^1 such that $u(P^1) = u(P^0)$ and $u(\overline{P}) < u(P^1)$ for all the points of γ between Q and P^1. We may assume that $P^1 = P^0$ and that Q lies on $t = t^0 - a_0$ since otherwise we can confine ourselves to a smaller rectangle.

Let R_0 denote the set R minus the top face $t = t_0$. Since, for every point Q' of R_0, $C(Q')$ contains some point of γ, and since $u < u(P^0)$ on γ, Lemma 3 implies that $u(Q') < u(P^0)$.

Introduce the function

$$h(x, t) = t^0 - t - K|x - x^0|^2 \qquad (K > 0). \tag{1.9}$$

$h = 0$ on the paraboloid M: $t^0 - t = K|x - x^0|^2$, $h < 0$ above M, and $h > 0$ below M. Also,

$$Lh = -2K \sum_{i=1}^{n} a_{ii} - 2K \sum_{i=1}^{n} b_i(x_i - x_i^0) + c[t^0 - t - K|x - x^0|^2] + 1 > 0 \qquad \text{in } R,$$

if K is such that $4K \sum a_{ii} \leq 1$ in R, and if the dimensions of R are sufficiently small (which we may assume).

M divides R into two regions. Denote by R' the lower region. The upper boundary B' of R' touches $t = t^0$ only at the point P^0. Hence, on the complementary part B'' of the boundary of R', $u \leq u(P^0) - \delta$ for some $\delta > 0$ (recall that $u < u(P^0)$ in R_0). It follows that for any ϵ positive and sufficiently small, $v \equiv u + \epsilon h < u(P^0)$ on B''. Next, $v = u < u(P^0)$ for all points of B' with the exception of P^0, and $v(P^0) = u(P^0)$. Since $Lv = Lu + \epsilon Lh > 0$ in R', it follows, by Lemma 1, that the positive maximum of v in R' is attained on the boundary of R'. By the previous remarks, then, the maximum is attained at the point P^0.

It follows that $\partial v(P^0)/\partial t \geq 0$. Hence, since $\partial h(P^0)/\partial t = -1 < 0$,

$$\frac{\partial u(P^0)}{\partial t} > 0. \tag{1.10}$$

On the other hand, from the assumption that the positive maximum of u is attained at P^0 it follows that (1.2) holds, $\partial u(P^0)/\partial x_i = 0$ and $c(P^0)u(P^0) \leq 0$. Hence, $0 \leq Lu \leq -\partial u/\partial t$ at P^0, i.e., $\partial u(P^0)/\partial t \leq 0$; this contradicts (1.10).

We can now easily complete the proof of Theorem 1.

Proof of Theorem 1. Suppose that $u \not\equiv u(P^0)$ in $S(P^0)$. Then there exists a point $Q \in S(P^0)$ such that $u(Q) < u(P^0)$. Connect Q to P^0 by a simple continuous curve γ lying in $S(P^0)$ such that t-coordinate is nondecreasing from Q to P^0. On γ there exists a point P^1 such that $u(P^1) =$

$u(P^0)$ and $u(\overline{P}) < u(P^1)$ for all points $\overline{P}$ on γ lying between Q and P^1. Denote by γ_0 the subarc of γ lying between Q and P^1. Construct a rectangle

$$x_i^1 - a \leq x_i \leq x_i^1 + a \quad (i = 1, \ldots, n), \quad t^1 - a < t \leq t^1$$

where $P^1 = (x_1^1, \ldots, x_n^1, t^1)$ and a is sufficiently small so that the rectangle lies in D. Applying Lemma 4 it follows that $u \equiv u(P^1)$ in this rectangle. Hence $u \equiv u(P^1)$ on the segment of γ_0 lying in the rectangle. This however contradicts the definition of P^1.

2. Extensions of the Maximum Principle

Consider the differential operator

$$(2.1) \quad Mu \equiv \sum_{i,j=1}^{n} a_{ij}(x, t) \frac{\partial^2 u}{\partial x_i \, \partial x_j} + \sum_{i,j=1}^{m} b_{ij}(x, t) \frac{\partial^2 u}{\partial t_i \, \partial t_j} + \sum_{i=1}^{n} a_i(x, t) \frac{\partial u}{\partial x_i} + \sum_{i=1}^{m} b_i(x, t) \frac{\partial u}{\partial t_i} + c(x, t)u$$

with continuous coefficients in an $(n + m)$-dimensional domain D, where $(x, t) = (x_1, \ldots, x_n, t_1, \ldots, t_m)$. Assume that $(a_{ij}(x, t))$ is a positive definite matrix and that $(b_{ij}(x, t))$ is a positive semidefinite matrix for all $(x, t) \in D$. Let $c(x, t) \leq 0$ in D.

Theorem 2. *Under the foregoing assumptions on M, if $Mu \geq 0$ ($Mu \leq 0$) in D and if u has a positive maximum (negative minimum) in D which is attained at a point $P^0 = (x^0, t^0)$, then $u(P) = u(P^0)$ for all $P \in C(P^0)$.*

Proof. We first observe that Lemma 1 and its proof extend to the operator M. Lemma 2 also extends to the operator M, the ellipsoid now being

$$\sum_{i=1}^{n} \lambda_i(x_i - x_i^*)^2 + \sum_{i=1}^{m} \mu_i(t_i - t_i^*) \leq R^2.$$

We can now proceed similarly to the proof of Lemma 3.

Theorem 3. *Let M be as in Theorem 2, but omit the assumption $c \leq 0$. If u is nonpositive (nonnegative) in D and vanishes at a point P^0 of D, and if $Mu \geq 0$ ($Mu \leq 0$) in D, then $u \equiv 0$ in $C(P^0)$.*

Proof. It suffices to consider the case $u \leq 0$. The function $v = u \exp[-\alpha x_1]$ satisfies the inequality

$$(2.2) \qquad \tilde{M}v \equiv (M - c)v + 2\alpha \sum_{i=1}^{n} a_{1i} \frac{\partial v}{\partial x_i} \geq -(a_{11}\alpha^2 + a_1\alpha + c)v.$$

Since $v \leq 0$, for any neighborhood N of P^0 ($N \subset D$) there exists an α sufficiently large such that $\tilde{M}v \geq 0$ in N. Since the coefficient of v in $\tilde{M}v$ is zero, we can apply Theorem 2 (to, say, $v + 1$) and conclude that $v \equiv 0$

in $N \cap C(P^0)$. Hence also $u \equiv 0$ in $N \cap C(P^0)$. Thus the set in $C(P^0)$ where $u = 0$ is an open set. Since it is also closed, and since $C(P^0)$ is a connected set, it follows that $u \equiv 0$ in $C(P^0)$.

The following theorem is a stronger version of Theorem 1, Sec. 1.

Theorem 4. *Let* (A), (B), (C) *hold. Suppose that at some point P^0 of D u has a positive maximum (negative minimum) in the set $S(P^0)$ and that it assumes it at P^0. Suppose further that $Lu \geq 0$ ($Lu \leq 0$) in $S(P^0)$. Then $u(P) = u(P^0)$ for all $P \in S(P^0)$.*

The proof is the same as for Theorem 1, since in proving Theorem 1 we have only made use of the fact that the $u(P^0)$ is the maximum of u in $S(P^0)$ (and not necessarily in D).

Let F be an open domain lying on a hyperplane $t = t_1$ and suppose that F is a part of the boundary of D, and that all the points of D which lie in some (x, t)-neighborhood of F belong to the half-space $t < t_1$. Assume that $u(x, t)$ is continuous in $D + F$ together with its first two x-derivatives and its first t-derivative. Then the proof of Theorem 4 remains true if $P^0 \in F$ provided $Lu \geq 0$ in $S(P^0)$ where $S(P^0)$ is now being defined with respect to $D + F$ (thus $S(P^0)$ contains F if $P^0 \in F$). We formulate this remark in the following theorem.

Theorem 4′. *Let* (A), (B), (C) *hold in $D + F$. Suppose that at some point P^0 of $D + F$ u has a positive maximum (negative minimum) in the set $S(P^0)$ and that it assumes it at P^0. Suppose further that $Lu \geq 0$ ($Lu \leq 0$) in $S(P^0)$. Then $u(P) = u(P^0)$ for all $P \in S(P^0)$.*

By combining Theorem 3 and the device used in its proof together with the proof of Theorem 1, Sec. 1 (from Lemma 3 on), we get the following result.

Theorem 5. *Let* (A), (B) *hold. If $u \leq 0$ ($u \geq 0$) in $S(P^0)$, $Lu \geq 0$ ($Lu \leq 0$) in $S(P^0)$ and $u(P^0) = 0$, then $u \equiv 0$ in $S(P^0)$.*

As in the case of Theorem 4, Theorem 5 also extends to $D + F$.

We shall finally state a useful consequence of Theorem 4 which is referred to as the *weak maximum principle:*

Theorem 6. *Let* (A), (B), (C) *hold and let D be bounded. Assume that u is a continuous function in $\bar{D}$ and let $Lu \geq 0$ ($Lu \leq 0$) in D. Then for each $P \in D$ for which u has a positive maximum (negative minimum) in $\overline{S(P)}$, that maximum (minimum) is obtained at some point lying in the complement of $S(P)$.*

Note that Theorem 6 does not assert that the maximum (minimum) cannot be obtained also at points of $S(P)$. For a direct proof of Theorem 6, see Problem 1.

Remark. If in Theorems 1, 2, 4, 4′, 6 $c(x, t) \equiv 0$ then, for any constant A, $L(u + A) = Lu$. Consequently, all the assertions remain true also if the maximum (minimum) of u is not assumed to be positive (negative).

3. The First Initial-boundary Value Problem

Let D be a bounded $(n + 1)$-dimensional domain in R^{n+1}, and let $(x, t) = (x_1, \ldots, x_n, t)$ be a variable point in R^{n+1}. Let the boundary ∂D of D consist of the closure of a domain B lying on $t = 0$, a domain B_T lying on $t = T$ $(T > 0)$ and a (not necessarily connected) manifold S lying in the strip $0 < t \leq T$. For any set G, we denote by $\overline{G}$ the closure of G. The part $S + \overline{B}$ of the boundary of D is called the *normal boundary* of D.

We set

$$D_\tau = D \cap \{0 < t < \tau\}, \quad B_\tau = D \cap \{t = \tau\}, \quad S_\tau = S \cap \{0 < t \leq \tau\}$$

and assume that, for every $0 < \tau < T$, B_τ is a domain. We also assume that there exists a simple continuous curve γ connecting B to B_T along which the t-coordinate is nondecreasing. It follows that for every $(x, \tau) \in D$, $0 < \tau < T$,

$$(3.1) \qquad S((x, \tau)) = D_\tau + B_\tau, \qquad \overline{S((x, \tau))} - S((x, \tau)) = \overline{B} + S_\tau.$$

The *first initial-boundary value problem* consists of solving the differential equation

$$(3.2) \qquad Lu(x, t) = f(x, t) \qquad \text{in } D + B_T$$

under the *initial condition*

$$(3.3) \qquad u(x, 0) = \varphi(x) \qquad \text{on } \overline{B}$$

and the *boundary condition*

$$(3.4) \qquad u(x, t) = g(x, t) \qquad \text{on } S,$$

where f, φ, g are given functions and L is the parabolic operator (1.1). We can combine (3.3), (3.4) into one condition

$$(3.5) \qquad u(x, t) = h(x, t) \qquad \text{on } \overline{B} + S.$$

Unless the contrary is explicitly stated, the function h is assumed to be continuous on $\overline{B} + S$ and by a solution u we mean a continuous function in $\overline{D}$, having two continuous x-derivatives and one continuous t-derivative in $D + B_T$ such that (3.2), (3.5) are satisfied. If, however, h is discontinuous at some points $\overline{P}$, then u cannot be required to be continuous at $\overline{P}$. Instead we require that

$$\liminf_{Q \in \overline{B}+S, Q \to \overline{P}} h(Q) = \liminf_{P \in D, P \to \overline{P}} u(P) \leq \limsup_{P \in D, P \to \overline{P}} u(P) = \limsup_{Q \in \overline{B}+S, Q \to \overline{P}} h(Q).$$

Remark. If for an i_0 and all $j \neq i_0$ the coefficients of $\partial^2 u/\partial x_{i_0}\,\partial x_j$ in Lu vanish identically, then the concept of a solution can be weakened by not requiring that the derivatives $\partial^2 u/\partial x_{i_0}\,\partial x_j$ exist. All the results of this chapter as well as all the results of Chaps. 1, 3–7 remain true if this weaker concept of a solution is adopted. (In the proof of Lemma 1, Sec. 1, the transformation C should be chosen so that all the variables x_{i_0} for which $a_{i_0 j} = 0$ for all $j \neq i_0$ are not changed except possibly by multiplication by a constant.)

Theorem 7. *Let L satisfy* (A), (B). *Then there exists at most one solution to the first initial-boundary value problem* (3.2), (3.5).

In Chap. 3 we shall prove the existence of a solution to the first initial-boundary value problem.

Proof. If $c \leq 0$ the assertion follows from the weak maximum principle (Theorem 6, Sec. 2) upon recalling (3.1). The general case can now be reduced to the case where $c \leq 0$ by using the transformation $v = e^{-\epsilon t}u$ $(\epsilon \geq c)$ which carries $Lu = 0$ into $(L - \epsilon)v = 0$.

Consider the nonlinear differential operator

$$Lu \equiv F\left(x, t, u, \frac{\partial u}{\partial x_i}, \frac{\partial^2 u}{\partial x_i\,\partial x_j}\right) - \frac{\partial u}{\partial t} \tag{3.6}$$

where F is a nonlinear function of its arguments. We say that L is *parabolic* at a point (x^0, t^0) if for any $p, p_1, \ldots, p_n, p_{11}, \ldots, p_{nn}$, the matrix

$$\left(\frac{\partial F(x^0, t^0, p, p_i, p_{ij})}{\partial p_{hk}}\right) \tag{3.7}$$

is positive definite.

If $Lu^1 = Lu^2$ in the domain D then, by the mean value theorem,

$$\begin{aligned}\frac{\partial(u^1 - u^2)}{\partial t} &= F\left(x, t, u^1, \frac{\partial u^1}{\partial x_i}, \frac{\partial^2 u^1}{\partial x_i\,\partial x_j}\right) - F\left(x, t, u^2, \frac{\partial u^2}{\partial x_i}, \frac{\partial^2 u^2}{\partial x_i\,\partial x_j}\right)\\ &= \Sigma\, a_{hk}\frac{\partial^2(u^1 - u^2)}{\partial x_h\,\partial x_k} + \Sigma\, b_h\frac{\partial(u^1 - u^2)}{\partial x_h} + c(u^1 - u^2)\end{aligned}$$

where a_{hk}, b_h, c are continuous functions provided $\partial F/\partial p$, $\partial F/\partial p_h$, $\partial F/\partial p_{hk}$ are continuous functions. Furthermore, (a_{hk}) is a positive definite matrix. Applying Theorem 7 we conclude:

Theorem 8. *Theorem* 7 *remains true for the nonlinear parabolic operator* (3.6).

In the remaining part of this section we derive bounds on solutions u of the equation $Lu = f$ in D, which are continuous in $\bar{D}$.

(a) Let (A), (B), (C) hold. If $Lu = 0$ in D then

$$\max_{\bar{D}} |u| \leq \max_{\bar{B}+S} |u|. \tag{3.8}$$

This follows by applying the weak maximum principle to u and to $-u$.

(b) Let (A), (B) hold and let $c(x, t) \leq \epsilon$. If $Lu = 0$ in D then

$$\max_{\overline{D}} |u| \leq e^{\epsilon T} \max_{\overline{B}+S} |u|. \tag{3.9}$$

Indeed, apply (a) to $v = ue^{-\epsilon t}$.

(c) Let (A), (B), (C) hold and assume that $a_{11}\lambda^2 + b_1\lambda \geq 1$ in D, for some positive constant λ. If $Lu = f$ in D then

$$\max_{\overline{D}} |u| \leq \max_{\overline{B}+S} |u| + (e^{\lambda d} - 1) \max_{\overline{D}} |f|, \tag{3.10}$$

where d is the breadth of $\overline{D}$ in the x_1-direction, i.e., d is the smallest positive number such that, for some x_1^0,

$$x_1^0 - d \leq x_1 \leq x_1^0$$

for all x_1 such that $(x_1, x_2, \ldots, x_n, t) \in \overline{D}$.

To prove (3.10), introduce the function

$$h(x) = 1 - \exp [\lambda(x_1 - x_1^0)]. \tag{3.11}$$

Then

$$Lh = -(a_{11}\lambda^2 + b_1\lambda) \exp [\lambda(x_1 - x_1^0)] + ch \leq -\exp [\lambda(x_1 - x_1^0)] \leq -e^{-\lambda d}.$$

Consider the function

$$w = e^{\lambda d}(\max_{\overline{D}} |f|)h + \max_{\overline{B}+S} |u| \pm u.$$

Clearly $Lw \leq 0$ and $w \geq 0$ on $\overline{B} + S$. Hence $w \geq 0$ in D, from which (3.10) follows.

(d) If in (c) the assumption (C) is replaced by $c(x, t) \leq \epsilon$ then

$$\max_{\overline{D}} |u| \leq e^{\epsilon T}[\max_{\overline{B}+S} |u| + (e^{\lambda d} - 1) \max_{\overline{D}} |f|]. \tag{3.12}$$

This follows by applying (c) to $v = ue^{-\epsilon t}$.

4. Positive Solutions of the Cauchy Problem

The interest in positive solutions is not only a mathematical one but also a physical one since in the physical experience the solutions of parabolic equations usually represent positive quantities such as temperature, density, probability distribution, etc.

Throughout this section we use the notation:

$$\Omega_0 = R^n \times (0, T], \qquad \Omega = R^n \times [0, T].$$

The functions u are always assumed to be continuous in Ω.

The following lemma will be frequently used.

Lemma 5. *Let L satisfy the assumptions* (A), (B) *in* Ω_0 *and let* $c(x, t)$ *be bounded from above. If* $Lu \leq 0$ *in* Ω_0, $u(x, 0) \geq 0$ *in* R^n *and*

$$\liminf_{|x|\to\infty} u(x, t) \geq 0$$

uniformly with respect to t $(0 \leq t \leq T)$ *then* $u(x, t) \geq 0$ *in* Ω.

Proof. We may assume that $c \leq 0$ since otherwise we first perform a transformation $v = ue^{-\gamma t}$ where $\gamma \geq c$. Now, for any $\epsilon > 0$, $u(x, t) + \epsilon > 0$ on $t = 0$ and on $|x| = R$, $0 \leq t \leq T$ provided R is sufficiently large. Since $L(u + \epsilon) \leq c\epsilon \leq 0$, $u(x, t) + \epsilon > 0$ if $|x| \leq R$, $0 \leq t \leq T$. Taking (x, t) to be fixed and $\epsilon \to 0$, it follows that $u(x, t) \geq 0$, i.e., $u \geq 0$ in Ω.

We shall need the following assumptions: for $i, j = 1, \ldots, n$; $(x, t) \in \Omega_0$,

$$|a_{ij}(x, t)| \leq M, \quad |b_i(x, t)| \leq M(|x| + 1), \quad c(x, t) \leq M(|x|^2 + 1). \tag{4.1}$$

Theorem 9. *Let L be a parabolic operator with continuous coefficients in* Ω_0, *and let* (4.1) *be satisfied. Assume that* $Lu \leq 0$ *in* Ω_0 *and that*

$$u(x, t) \geq -B \exp\,[\beta|x|^2] \qquad \text{in } \Omega \tag{4.2}$$

for some positive constants B, β. *If* $u(x, 0) \geq 0$ *in* R^n *then* $u(x, t) \geq 0$ *in* Ω.

Proof. The function

$$H(x, t) = \exp\left[\frac{k|x|^2}{1 - \mu t} + \nu t\right] \qquad \left(0 \leq t \leq \frac{1}{2\mu}\right) \tag{4.3}$$

satisfies:

$$\frac{LH}{H} = \frac{4k^2}{(1 - \mu t)^2} \Sigma\, a_{ij}x_ix_j + \frac{2k}{1 - \mu t} \Sigma\, a_{ii} + \frac{2k}{1 - \mu t} \Sigma\, b_ix_i + c - \frac{\mu k|x|^2}{(1 - \mu t)^2} - \nu.$$

Using (4.1) we obtain

$$\frac{LH}{H} \leq (16k^2n^2M^2 + 8knM + M - \mu k)|x|^2 + (8knM + M - \nu).$$

Thus, given any $k > 0$ we can choose sufficiently large positive numbers μ, ν such that

$$\frac{LH}{H} \leq 0. \tag{4.4}$$

Consider the function v defined by $u = Hv$ where H is defined by (4.3) with a fixed $k > \beta$ and with μ, ν such that (4.4) holds for $0 \leq t \leq 1/2\mu$. From (4.2) it follows that

$$\liminf_{|x|\to\infty} v(x, t) \geq 0,$$

uniformly with respect to t, $0 \leq t \leq 1/2\mu$.

v satisfies the equation

$$\bar{L}v \equiv \sum_{i,j=1}^{n} a_{ij} \frac{\partial^2 v}{\partial x_i \, \partial x_j} + \sum_{i=1}^{n} \bar{b}_i \frac{\partial v}{\partial x_i} + \bar{c}v - \frac{\partial v}{\partial t} = \bar{f},$$

where $\bar{f} = (Lu)/H \leq 0$ and

$$\bar{b}_i = b_i + 2 \sum_{j=1}^{n} a_{ij} \frac{\partial H/\partial x_j}{H}, \qquad \bar{c} = \frac{LH}{H}.$$

Since, by (4.4), $\bar{c} \leq 0$, we can apply Lemma 5 and thus conclude that $v(x, t) \geq 0$ in $R^n \times [0, 1/2\mu]$. The same is therefore true of $u(x, t)$. We can now proceed step by step to prove the positivity of u in Ω.

From Theorem 9 we obtain the following uniqueness theorem to the Cauchy problem

$$\begin{aligned} Lu &= f(x, t) && \text{in } R^n \times (0, T], \\ u(x, 0) &= \varphi(x) && \text{in } R^n. \end{aligned} \tag{4.5}$$

Theorem 10. *Let L be a parabolic operator with continuous coefficients in $R^n \times (0, T]$ and let (4.1) be satisfied. Then there exists at most one solution to the Cauchy problem (4.5) satisfying*

$$|u(x, t)| \leq B \exp [\beta |x|^2] \tag{4.6}$$

for some positive constants B, β.

Proof. We have to show that if $f \equiv 0$, $\varphi \equiv 0$ then $u \equiv 0$. This follows by applying Theorem 9 to u and to $-u$.

It is interesting to compare Theorem 10 with Theorem 16, Chap. 1, Sec. 9. The assumptions on the coefficients, in Theorem 10, are much weaker, and the boundedness condition (4.6) is slightly more restrictive than the corresponding condition (1.9.1). A more significant difference lies in the method of proof: in proving Theorem 10 we have used the maximum principle whereas in proving Theorem 16 of Chap. 1 we have employed fundamental solutions. Now the maximum principle is known only for second-order parabolic equations and, consequently, the proof of Theorem 10 cannot be extended to more general parabolic equations. On the other hand, fundamental solutions can be constructed for parabolic systems of any order and Theorem 16, Chap. 1, can also be extended to such systems (see Sec. 5, Chap. 9).

As another application of the maximum principle we shall prove a uniqueness theorem for solutions which only satisfy a boundedness condition from below (Theorem 13 below). We first establish a general property of the fundamental solution $\Gamma(x, t; \xi, \tau)$ which was constructed in Chap. 1 under the assumptions $(A_1)'$, $(A_2)'$ (see Chap. 1, Sec. 6).

Theorem 11. *Let L satisfy the assumptions $(A_1)'$, $(A_2)'$ of Chap. 1, Sec. 6. Then $\Gamma(x, t; \xi, \tau)$ is a positive function.*

Proof. For any nonnegative continuous function $f(x)$ on R^n, having a compact support, introduce the function

$$v(x, t) = \int_{R^n} \Gamma(x, t; \xi, \tau) f(\xi) \, d\xi.$$

Since $Lv = 0$ for $\tau < t \leq T$, $v(x, \tau) = f(\xi) \geq 0$ and $\liminf v(x, t) \geq 0$, it follows, by Lemma 5, that $v(x, t) \geq 0$.

Take a sequence of continuous nonnegative functions $\{f_m(\xi)\}$ such that

$$f_m(\xi) = 0 \quad \text{if } |\xi - \xi^0| > \frac{1}{m}, \quad \text{and} \quad \int_{R^n} f_m(\xi) \, d\xi = 1;$$

for instance, $f_m(\xi) = \rho_{1/m}(\xi - \xi^0)$ where $\rho_\epsilon(x)$ is the function defined in Chap. 1, Problem 1. Denote by $v_m(x, t)$ the function $v(x, t)$ corresponding to $f = f_m$, $\tau = \tau^0$. Then, for any fixed (x^0, t^0), (ξ^0, τ^0), $t^0 > \tau^0$,

$$\lim_{m \to \infty} v_m(x^0, t^0) = \Gamma(x^0, t^0; \xi^0, \tau^0).$$

Since $v_m(x^0, t^0) \geq 0$ for all m, it follows that $\Gamma(x^0, t^0; \xi^0, \tau^0) \geq 0$. Noting finally that, for every fixed (ξ, τ), $\Gamma(x, t; \xi, \tau) \not\equiv$ const. and using Theorem 5, Sec. 2, the strict positivity of Γ follows.

For some parabolic equations Γ is not only positive but is also bounded from below as follows:

$$\Gamma(x, t; \xi, \tau) \geq C \exp[-\gamma|x - \xi|^2] \qquad \text{if } |x - \xi| \geq 1/\delta, \ t - \tau > \delta, \tag{4.7}$$

where δ is an arbitrary positive number and C, γ are positive constants (depending on δ). Thus, for the fundamental solution $\Gamma_0(x, t; \xi, \tau)$ of the heat equation the inequality (4.7) follows from the explicit formula

$$\Gamma_0(x, t; \xi, \tau) = \frac{1}{(2\sqrt{\pi})^n} (t - \tau)^{-n/2} \exp\left\{-\frac{|x - \xi|^2}{4(t - \tau)}\right\}.$$

Observe that if Γ is a fundamental solution of (1.1), then for any smooth function $f(x, t)$ such that

$$f(x, t) - f(\xi, \tau) = o(|x - \xi|^2) \qquad (\text{as } |x - \xi| \to \infty) \tag{4.8}$$

the function

$$\Gamma'(x, t; \xi, \tau) \equiv \Gamma(x, t; \xi, \tau) \exp[f(\xi, \tau) - f(x, t)]$$

is a fundamental solution of the equation

$$\begin{aligned} L'v \equiv & \sum_{i,j=1}^n a_{ij} \frac{\partial^2 v}{\partial x_i \, \partial x_j} + \sum_{i=1}^n \left(b_i + 2 \sum_{j=1}^n a_{ij} f_j\right) \frac{\partial v}{\partial x_i} \\ & + \left(c + \sum_{i=1}^n b_i f_i + \sum_{i,j=1}^n a_{ij} f_{ij} + \sum_{i,j=1}^n a_{ij} f_i f_j - f_t\right) v - \frac{\partial v}{\partial t} \\ = & \ 0 \end{aligned}$$

where $f_i = \partial f/\partial x_i$, $f_t = \partial f/\partial t$, $f_{ij} = \partial^2 f/\partial x_i\, \partial x_j$. If the property (4.7) holds for Γ, then it holds also for Γ'.

Taking L to be the heat operator and $f(x, t) = \Sigma\, b_i x_i/2$, b_i constants, we conclude that the fundamental solution of

$$\sum_{i=1}^{n} \frac{\partial^2 u}{\partial x_i^2} + \sum_{i=1}^{n} b_i \frac{\partial u}{\partial x_i} - \frac{\partial u}{\partial t} = 0$$

satisfies (4.7). By a linear transformation on the x's it also follows that (4.7) holds for the fundamental solution of

$$(4.9) \qquad L_0 u \equiv \sum_{i,j=1}^{n} a_{ij} \frac{\partial^2 u}{\partial x_i\, \partial x_j} + \sum_{i=1}^{n} b_i \frac{\partial u}{\partial x_i} - \frac{\partial u}{\partial t} = 0 \ (a_{ij} \text{ constants}).$$

Again we make a general observation: if (4.7) holds for the fundamental solution of $Lu = 0$ then it holds also for the fundamental solution Γ_e of $(L + e)u = 0$ for any positive function e. Indeed, consider the function

$$v(x, t) = \int_{R^n} [\Gamma(x, t; \xi, \tau) - \Gamma_e(x, t; \xi, \tau)] f(\xi)\, d\xi$$

where $f(\xi)$ is as in the proof of Theorem 11. Clearly, $\lim v(x, t) = 0$ as $|x| \to \infty$, and $v(x, 0) = 0$. Since

$$L(\Gamma - \Gamma_e) = L\Gamma - (L + e)\Gamma_e + e\Gamma_e = e\Gamma_e \geq 0,$$

it follows that also $Lv \geq 0$. By Lemma 5, $v(x, t) \leq 0$ for $\tau < t \leq T$. Taking a sequence $\{f_m\}$ as in the proof of Theorem 11 we conclude that $\Gamma - \Gamma_e \leq 0$ and, therefore, (4.7) holds also for Γ_e.

Since the transformation $u = ve^{-\lambda t}$ takes $Lu = 0$ into $(L + \lambda)v = 0$ and a fundamental solution $\Gamma(x, t; \xi, \tau)$ of $Lu = 0$ into a fundamental solution $e^{\lambda(t-\tau)}\Gamma(x, t; \xi, \tau)$ of $(L + \lambda)v = 0$ it follows that if in the previous observation e is not assumed to be positive, then the assertion is still valid. Thus, *if* (4.7) *holds for a fundamental solution of* $Lu = 0$ *then it also holds for a fundamental solution of* $(L + e)u = 0$, *for any bounded continuous function* $e(x, t)$ (which is uniformly Hölder continuous in x).

Applying this statement to L_0 defined in (4.9) we arrive at the following result.

Theorem 12. *The inequality* (4.7) *holds for the parabolic equation*

$$(4.10) \qquad \sum_{i,j=1}^{n} a_{ij} \frac{\partial^2 u}{\partial x_i\, \partial x_j} + \sum_{i=1}^{n} b_i \frac{\partial u}{\partial x_i} + c(x, t)u - \frac{\partial u}{\partial t} = 0,$$

where a_{ij}, b_i *are constants and* $c(x, t)$ *is a continuous bounded function, uniformly Hölder continuous in* x.

We shall now prove a uniqueness theorem for solutions of the Cauchy problem which are bounded from below.

Theorem 13. *Let* L *satisfy the assumptions* $(A_1)'$, $(A_3)'$ *of Chap.* 1, *Secs.* 6, 8 *in* $\Omega = R^n \times [0, T]$ *and let* (4.7) *be satisfied. If* $Lu \equiv 0$ *in* $\Omega_0 =$

$R^n \times (0, T]$ and $u(x, 0) \equiv 0$ on R^n, and if $u(x, t)$ satisfies (4.2) for some positive constants B, β then $u(x, t) \equiv 0$ in Ω.

Recalling Theorem 12 we obtain the following:

Corollary 1. *If u is a solution of (4.10) in Ω_0 and $u(x, 0) \equiv 0$ in R^n, and if u satisfies (4.2), then $u(x, t) \equiv 0$ in Ω.*

Proof of Theorem 13. From Theorem 9 it follows that $u(x, t) \geq 0$ in Ω. For any $R > 0$, introduce the function

$$U_R(x, t) = \int_{|\xi|<R} \Gamma(x, t; \xi, \tau) u(\xi, \tau)\, d\xi.$$

From the second property of fundamental solutions (see (1.1.7)) it follows that $U_R(x, t) \to u(x, \tau)$ as $t \to \tau$, provided $|x| < R$ and $U_R(x, t) \to 0$ as $t \to \tau$, provided $|x| > R$. Since, further, $U_{R+1}(x, t) \to u(x, t)$ if $|x| = R$ and $U_R \leq U_{R+1}$ (since $\Gamma > 0$, $u \geq 0$), it follows that, for any $y \in R^n$,

$$\liminf U_R(x, t) \leq u(y, t) \quad \text{as } x \to y,\ t \to \tau.$$

Thus the function $v(x, t) = u(x, t) - U_R(x, t)$ satisfies:

$$\liminf_{x \to y, t \to \tau} v(x, t) \geq 0.$$

Since $U_R(x, t) \to 0$ as $|x| \to \infty$ (uniformly with respect to t, $\tau < t \leq T$) and since $u(x, t) \geq 0$ in Ω, it also follows that

$$\liminf_{|x| \to \infty} v(x, t) \geq 0$$

uniformly with respect to t, $\tau < t \leq T$. By the proof of Lemma 5 we conclude that $v(x, t) \geq 0$, i.e.,

$$\int_{|\xi|<R} \Gamma(x, t; \xi, \tau) u(\xi, \tau)\, d\xi \leq u(x, t).$$

Since this inequality holds for any $R > 0$, and since the integrand is nonnegative, we conclude that

$$\int_{R^n} \Gamma(x, t; \xi, \tau) u(\xi, \tau)\, d\xi \leq u(x, t). \tag{4.11}$$

Integrating with respect to τ and taking $x = 0$ we get

$$\int_0^{t-\delta} \int_{R^n} \Gamma(0, t; \xi, \tau) u(\xi, \tau)\, d\xi\, d\tau \leq (t - \delta) u(0, t). \tag{4.12}$$

We take δ to be any positive number $< t$. Using (4.7) we get

$$\int_0^{t-\delta} \int_{R^n} \exp\,[-\gamma|\xi|^2] u(\xi, \tau)\, d\xi\, d\tau < \infty.$$

Employing Theorem 16, Chap. 1, Sec. 9, we thus find that $u(\xi, \tau) \equiv 0$ in the strip $0 < \tau < t - \delta$. Since $t - \delta$ can be made arbitrarily close to T, $u \equiv 0$ in Ω.

Corollary 2. *If $Lu = 0$ in Ω_0 and $u(x, t) \geq 0$ in Ω, and if L satisfies all the assumptions made in Theorem 13 and, in addition, its coefficients are locally Hölder continuous in (x, t), then u has the representation*

$$u(x, t) = \int_{R^n} \Gamma(x, t; \xi, 0)u(\xi, 0)\, d\xi \qquad \text{in } \Omega_0. \tag{4.13}$$

Indeed, denoting the integral on the right-hand side by $v(x, t)$ we have, by (4.11), $u(x, t) \geq v(x, t)$. $v(x, t)$ is continuous in $R^n \times [0, \epsilon]$ for any $\epsilon > 0$ sufficiently small as follows from (1.6.12) and the inequality

$$\int \exp\,[-\gamma|\xi|^2]u(\xi, 0)\, d\xi < \infty \qquad \text{(using (4.7) with } x = 0,\ \tau = 0).$$

Since $u(x, 0) = v(x, 0)$, the assertion (4.13) would follow from Theorem 13 (applied to $u - v$) provided we prove that v is a solution of $Lv = 0$ in Ω_0. To prove this we recall that for each (x, t), $U_R(x, t) \to v(x, t)$ (here $\tau = 0$ in the definition of U_R). Furthermore, $U_R \nearrow$ if $R \nearrow$ and the U_R are bounded uniformly with respect to R for (x, t) in bounded sets (since $U_R(x, t) \leq u(x, t)$). We can now apply Theorem 15, Sec. 6, Chap. 3 (the assumption that the coefficients of L are Hölder continuous in (x, t) is being used here). It follows that $v(x, t)$ is a solution of $Lv = 0$.

The proof of Theorem 13 remains true if instead of the assumption (4.7) we only assume that

$$\Gamma(0, t; \xi, \tau) \geq A \exp\,[-\beta|\xi|^2] \qquad (\xi \in R^n,\ t - \tau > \delta) \tag{4.14}$$

where δ is an arbitrary positive number and A, β are positive constants (depending on δ). As will be shown in the Appendix (see Corollary 1 to Theorem 5′, Appendix), if the assumptions $(A_1)'$, $(A_3)'$ are satisfied then (4.14) holds for all sufficiently small $\delta > 0$. Hence, *Theorem* 13 *and Corollary* 2 *remain true even if the assumption* (4.7) *is omitted.* If $c \equiv 0$ then the assumption $(A_3)'$ in Theorem 13 can be weakened; see Corollary 2 to Theorem 5′, Appendix.

5. The Second Initial-boundary Value Problem

Definition. Let $P^0 = (x^0, t^0)$ be a point on the boundary ∂D of a domain D. If there exists a closed ball B with center $(\bar{x}, \bar{t})$ such that $B \subset \bar{D}$, $B \cap \partial D = \{P^0\}$, and if $\bar{x} \neq x^0$, then we say that P^0 has the *inside strong sphere property.*

Let Mu be the operator defined in (2.1) and let all the assumptions made on Mu in Theorem 2, Sec. 2, be satisfied not only in D but also in $\bar{D}$, and assume, further, that D is bounded. Suppose that u is a continuous function in $\bar{D}$ and that $Mu \geq 0$ in D. If u has a positive maximum M in $\bar{D}$ then, by Theorem 2, $u(P^0) = M$ for some point $P^0 = (x^0, t^0)$ on the boundary ∂D of D.

Theorem 14. *Let the foregoing assumptions be satisfied and let P^0 have the inside strong sphere property. Assume further that, for some neighborhood V of P^0, $u < M$ in $D \cap V$. Then, for any nontangential inward direction τ,*

$$\frac{\partial u}{\partial \tau} \equiv \liminf_{\Delta\tau \to 0} \frac{\Delta u}{\Delta \tau} < 0 \qquad \text{at } P^0. \tag{5.1}$$

By a nontangential inward direction we mean a direction pointing from P^0 into the interior of the ball B whose boundary touches ∂D at P^0.

Proof. We may assume that the interior of B lies in $D \cap V$. Denote the boundary of B by S. Let π be a hyperplane which divides the (x, t)-space into two half-spaces π^- and π^+ such that $(\bar{x}, \bar{t}) \in \pi^-$ and $(x^0, t^0) \in \pi^+$. Since $\bar{x} \neq x^0$ we can choose π such that $B^+ \equiv \pi^+ \cap B$ is not empty and such that $|\bar{x} - x| \geq \text{const.} > 0$ for all $(x, t) \in B^+$. The boundary of B^+ consists of one part C_1 lying on S and another part C_2 lying on π.

Introduce the function

$$h(x, t) = \exp\{-\alpha[|x - \bar{x}|^2 + |t - \bar{t}|^2]\} - \exp\{-\alpha R^2\}, \tag{5.2}$$

where R is the radius of S. $h = 0$ on C_1, $h \geq 0$ on $\overline{B^+}$, and $Mh > 0$ in B^+ if α is sufficiently large.

For any $\epsilon > 0$ sufficiently small the function $v = u + \epsilon h$ is smaller than M on C_2. For $P \in C_1$, $v(P) = u(P) < M$ if $P \neq P^0$, and $v(P^0) = u(P^0) = M$. Since $Mv = Mu + \epsilon Mh > 0$ in B^+, v cannot assume its positive maximum M in $\overline{B^+}$ in its interior (this follows from Theorem 2, but a direct proof can be given similar to the proof of Lemma 1). Hence $v < M$ in the interior of B^+. It follows that

$$\frac{\partial v}{\partial \tau} \equiv \liminf_{\Delta\tau \to 0} \frac{\Delta v}{\Delta \tau} \leq 0 \qquad \text{at } P^0. \tag{5.3}$$

Since $\partial h/\partial \nu > 0$ (ν is the inward normal to S at P^0) whereas $\partial h/\partial \sigma = 0$ (σ is any tangential direction to S at P^0), we conclude that $\partial h/\partial \tau > 0$. Combining this with (5.3) we find that $\partial u/\partial \tau = \partial v/\partial \tau - \epsilon\, \partial h/\partial \tau < 0$ at P^0.

Remark 1. The assumption that $u < M$ in $D \cap V$ is of course essential since otherwise u may be a constant in $D \cap V$ and hence $\partial u/\partial \tau = 0$.

Remark 2. The assumption $\bar{x} \neq x^0$ cannot be omitted. This is shown by the following counterexample:

$$P^0 = (0, 0), \quad Mu = \frac{\partial^2 u}{\partial x^2} - \frac{\partial u}{\partial t}, \quad u(x, t) = 1 - t^2$$

and D is the half-space $t > 0$.

Remark 3. If P^0 is a vertex point of ∂D then Theorem 14 is not valid. As a counterexample take D to be defined by

$$x^2 + t^2 < R^2, \quad t < \gamma_1 x, \quad t < \gamma_2 x \quad (\gamma_1 > 0 > \gamma_2)$$

and let

$$P^0 = (0, 0), \quad Mu = \frac{\partial^2 u}{\partial x^2} - \frac{\partial u}{\partial t}, \quad u(x, t) = (t - \gamma_1 x)(\gamma_2 x - t) + 1.$$

Then $u < 1$ in D, $u = 1$ at P^0, and

$$Mu = -2\gamma_1\gamma_2 + 0(|x| + |t|) > 0$$

if R is sufficiently small. Yet $\partial u/\partial \tau = 0$ at P^0 for any direction τ.

We shall now give an application to Theorem 14.

The notation of Sec. 3 will be adopted. Let β be a continuous function defined on the lateral boundary S of D and let τ be a direction defined at each point of S in a continuous manner. Given any functions f on D, φ on $\overline{B}$ and ψ on S, the problem of finding a solution to the parabolic equation

$$Lu(x, t) = f(x, t) \qquad \text{in } D + B_T \tag{5.4}$$

satisfying the initial condition

$$u(x, 0) = \varphi(x) \qquad \text{on } \overline{B} \tag{5.5}$$

and the boundary condition

$$\frac{\partial u(x, t)}{\partial \tau} + \beta(x, t)u(x, t) = \psi(x, t) \qquad \text{on } S \tag{5.6}$$

is called the *third initial-boundary value problem.* If τ is never tangent to S then the problem is said to be *regular.*

In the special case where D is a cylinder with base B and lateral boundary S, if $\nu = (\nu_1, \ldots, \nu_n, 0)$ is the inward normal to S at some point (x, t) of S, then the inward *conormal* to S at (x, t) is, by definition, the vector $\mu = (\mu_1, \ldots, \mu_n, 0)$ with $\mu_i = \sum_{j=1}^{n} a_{ij}(x, t)\nu_j$. Thus, if $a_{ij} = \delta_{ij}$ then the normal and the conormal coincide. If D is a cylinder and if the directions τ in (5.6) are the inward conormals then we call the problem (5.4)–(5.6) the *second initial-boundary value problem.*

We say that a direction τ (at $P^0 \in S$) *points into the interior of* D if there exists a finite ray with P^0 as its origin all of whose interior points lie in D. If further the inside strong sphere property holds at P^0 then it is always understood that τ, in fact, points into the interior of a touching hypersphere. This, of course, does not impose any further restrictions on τ if S has a unique tangent hyperplane at P^0.

The definition of $\partial u/\partial \tau$ will now be made clear. If τ points into the interior of D then we can define $\partial u/\partial \tau$ as a directional derivative, i.e.,

$$\frac{\partial u(x, t)}{\partial \tau} = \lim_{\Delta\tau \to 0} \frac{u(x, t) - u(\overline{x}, \overline{t})}{-\Delta\tau}, \tag{5.7}$$

where $(\bar{x}, \bar{t})$ lies on the ray issuing from (x, t) in the direction τ, the distance from $(\bar{x}, \bar{t})$ to (x, t) being $\Delta\tau$.

Another way to define $\partial u/\partial\tau$, for any direction τ, is by

$$\frac{\partial u(x, t)}{\partial\tau} = \lim_{(y,\sigma)\to(x,t)} \frac{\partial u(y, \sigma)}{\partial\tau}, \tag{5.8}$$

where $(y, \sigma) \in D$ and $(y, \sigma) \to (x, t)$ along any arc whose tangent at (x, t) points into the interior of D (when suitably oriented), the limit being independent of the arc.

If τ points into the interior of D then (5.8) implies (5.7).

From now on we always assume that the directions τ in (5.6) point into the interior of D. We also recall that (3.1) is assumed to hold. Finally, we shall take $\partial u/\partial\tau$ to be defined in the sense of (5.7).

Theorem 15. *Let L be a parabolic operator with continuous coefficients in $\bar{D}$ and assume that $c(x, t) \leq 0$, that the inside strong sphere property holds for each $P \in S$, and that $\beta \leq 0$. Then there exists at most one solution to the third initial-boundary value problem* (5.4)–(5.6). *If τ is independent of t then the assumption $c(x, t) \leq 0$ may be omitted.*

Proof. The last assertion follows by performing a transformation $v = e^{-\gamma t}u$ where $\gamma \geq c(x, t)$. We have to prove that if $f \equiv 0$, $\varphi \equiv 0$, $\psi \equiv 0$ then $u \equiv 0$. Now, if $u \not\equiv 0$ then we may assume that u has a positive maximum M in $\bar{D}$. If $u(P^0) = M$ then P^0 cannot belong to any B_t $(0 < t \leq T)$ since the strong maximum principle (Theorem 4′, Sec. 2) would imply that $u \equiv M$ in $S(P^0)$. But then, by (3.1), $u(x, 0) \equiv M$ on B which contradicts our assumption that $u(x, 0) \equiv 0$. It follows that the maximum of u can be attained only on S.

Thus, for some $P^0 \in S$, $u(P^0) = M$, and $u < M$ in $V \cap D$ where V is some neighborhood of P^0. Since the inside strong sphere property is satisfied, we can apply Theorem 14 and thus conclude that $\partial u/\partial\tau < 0$ at P^0. Since, however,

$$\frac{\partial u}{\partial\tau} = -\beta u \geq 0 \qquad \text{at } P^0,$$

we get a contradiction.

Remark. The assumption that the inside strong sphere property is satisfied at the points of $S \cap \{t = T\}$ is very restrictive. It is therefore worth noting that if this assumption is omitted and if $\beta < 0$ on $S \cap \{t = T\}$ then the assertion of Theorem 15 remains true. Similarly if the assumption concerning the inside strong sphere property (at all the points of S) is omitted, and if $\beta < 0$ on S, then the assertion of Theorem 15 remains true. The proof is obtained by minor modifications of the previous proof.

6. Comparison Theorems

The notation D, S, B, D_τ, B_τ, S_τ of Sec. 3 will be used in the present section. B_τ is a domain in R^n, for every $0 \leq \tau \leq T$, and D is a bounded domain in R^{n+1}.

Theorem 16. *Let $v(x, t)$, $w(x, t)$ be continuous functions in $\overline{D}$, and let the first two x-derivatives and the first t-derivative of v, w be continuous in D. Let $F(x, t, p, p_i, p_{ij})$ $(i, j = 1, \ldots, n)$ be a continuous function together with its first derivatives with respect to the p_{hk} in a domain E containing the closure of the set of points (x, t, p, p_i, p_{ij}) where*

$$(x, t) \in D, \quad p \in (v(x, t), w(x, t)), \quad p_i \in \left(\frac{\partial v(x, t)}{\partial x_i}, \frac{\partial w(x, t)}{\partial x_i}\right),$$

$$p_{ij} \in \left(\frac{\partial^2 v(x, t)}{\partial x_i \, \partial x_j}, \frac{\partial^2 w(x, t)}{\partial x_i \, \partial x_j}\right);$$

here (a, b) denotes the interval connecting a to b. Assume also that $(\partial F/\partial p_{hk})$ is a positive semidefinite matrix.

If

$$\frac{\partial v}{\partial t} > F\left(x, t, v, \frac{\partial v}{\partial x_i}, \frac{\partial^2 v}{\partial x_i \, \partial x_j}\right) \quad \text{in } D, \tag{6.1}$$

$$\frac{\partial w}{\partial t} \leq F\left(x, t, w, \frac{\partial w}{\partial x_i}, \frac{\partial^2 w}{\partial x_i \, \partial x_j}\right) \quad \text{in } D, \tag{6.2}$$

and if

$$v > w \qquad \text{on } \overline{B} + S, \tag{6.3}$$

then also $v > w$ in D.

Proof. Consider the set M of points σ in the interval $(0, T)$ such that $v(x, t) > w(x, t)$ for all $x \in \overline{B}_t$, $0 \leq t < \sigma$. If we prove that l.u.b. $\sigma = T$ then the proof of the theorem is completed. Let $t_0 =$ l.u.b. σ. By (6.3), $t_0 > 0$. If $t_0 < T$ then the function $z = v - w$ is positive in D_{t_0}, nonnegative on B_{t_0} and is equal to zero at some point $P^0 = (x^0, t^0)$ of $\overline{B}_{t_0}$. P^0 cannot belong to the boundary of B_{t_0} since, by (6.3), $v > u$ on S. Thus, P^0 must lie in B_{t_0} and it is a minimum point of z in the domain B_{t_0}. It follows that

$$\frac{\partial z(P^0)}{\partial x_i} = 0.$$

By an argument similar to that used in proving (1.2), we further have

$$\sum_{i,j=1}^{n} \frac{\partial F}{\partial p_{ij}} \frac{\partial^2 z(P^0)}{\partial x_i \, \partial x_j} \geq 0,$$

where the argument of F is any point of E. Since, finally, $v(P^0) = u(P^0)$, by the mean value theorem it follows that, at P^0, the right-hand side of

(6.1) is larger than or equal to the right-hand side of (6.2). Hence $\partial w/\partial t < \partial v/\partial t$ at P^0. Since, however, $z(P^0) < z(P)$ for all $P \in D_{t_0}$, $\partial z/\partial t \leq 0$ at P^0; a contradiction.

Theorem 17. *The assumptions of Theorem* 16 *remain the same except for the assumption* (6.3) *which is replaced by*

$$v > w \qquad \text{on } \overline{B}, \tag{6.4}$$

$$\frac{\partial v}{\partial \tau} + \beta(x, t, v) < \frac{\partial w}{\partial \tau} + \beta(x, t, w) \qquad \text{on } S, \tag{6.5}$$

where β is any function, $\tau = \tau(x, t)$ is pointing into $D_t + B_t$ and $\partial v/\partial \tau$, $\partial w/\partial \tau$ (defined as in (5.7)) *are assumed to exist. Then $v > w$ in D.*

The proof is similar to the previous proof, the only difference being that $P^0 = (x^0, t^0)$ cannot lie on the boundary of B_{t_0} because of (6.5) (instead of (6.3)).

7. Elliptic Equations

Consider the linear differential operator

$$Lu \equiv \sum_{i,j=1}^{n} a_{ij}(x) \frac{\partial^2 u}{\partial x_i \, \partial x_j} + \sum_{i=1}^{n} b_i(x) \frac{\partial u}{\partial x_i} + c(x)u \tag{7.1}$$

with coefficients defined in an n-dimensional domain D. L is said to be of *elliptic type* (or *elliptic*) at a point x^0 if the matrix $(a_{ij}(x^0))$ is positive definite, i.e., if for any real vector $\xi \neq 0$, $\Sigma\, a_{ij}(x^0)\xi_i\xi_j > 0$.

The *strong maximum principle* for elliptic operators is the following theorem.

Theorem 18. *Let L be an elliptic operator with continuous coefficients in a domain D and assume that $c(x) \leq 0$ and that $Lu \geq 0$ ($Lu \leq 0$) in D. If $u \not\equiv$ const. then u cannot have a positive maximum (negative minimum) in D.*

Proof. If u has a positive maximum M, then $u(x^0) = M$ for some point $x^0 \in D$. Since $u \not\equiv$ const. there is a point $\overline{x} \in D$ such that $u(\overline{x}) < M$. Proceeding similarly to the proof of Lemma 3 we find that there exists a closed ball B with a center x^* and boundary ∂B such that $u(x) < M$ for all $x \in B$ with the exception of one point $\overline{x}$ on ∂B where $u(\overline{x}) = M$. Consider a closed ball E with radius $< |\overline{x} - x^*|$ and center $\overline{x}$, lying in D. Its boundary is composed of two parts: E_1 lying in B and E_2 lying outside B. On E_1 $u \leq M - \delta$ for some $\delta > 0$.

The function

$$h(x) = \exp\,[-\alpha|x - x^*|^2] - \exp\,[-\alpha R^2] \qquad (R = |\overline{x} - x^*|)$$

satisfies: $h \geq 0$ in B, $h < 0$ outside B, $h = 0$ on ∂B, and $Lh > 0$ in E (if α is sufficiently large). Introducing the function $v = u + \epsilon h$, $\epsilon > 0$, we then see that if ϵ is sufficiently small then $v < M$ on E_1, $v < M$ on E_2, and $Lv > 0$ in E. By the argument of Lemma 1, Sec. 1, it follows that v cannot have a positive maximum in the interior of E. Since, however, $v(\bar{x}) = M$, we get a contradiction.

Remark. The reader may note that Theorem 18 can also be proved, indirectly, by showing that it is a corollary of the strong maximum principle for parabolic equations. Indeed, considering $u(x)$ to be a solution of $Lu - \partial u/\partial t = 0$ in a cylinder $D \times (0, 1)$, Theorem 18 is then a consequence of Theorem 1, Sec. 1. A similar remark applies to Theorem 21 below.

The following consequence of Theorem 18 is known as the *weak maximum principle:*

Theorem 19. *Let L be as in Theorem 18, let D be a bounded domain, and let u be continuous in* $\bar{D}$ *with* $Lu \geq 0$ $(Lu \leq 0)$ *in D. If u has a positive maximum (negative minimum) in* $\bar{D}$, *then*

$$\text{l.u.b.}_{x \in D}\, u(x) \leq \max_{x \in \partial D} u(x) \qquad \left(\text{g.l.b.}_{x \in D}\, u(x) \geq \min_{x \in \partial D} u(x)\right), \tag{7.2}$$

where ∂D *is the boundary of D.*

The problem of finding a solution u to the elliptic equation

$$Lu(x) = f(x) \qquad \text{in } D, \tag{7.3}$$

satisfying the *boundary condition*

$$u(x) = \varphi(x) \qquad \text{on } \partial D, \tag{7.4}$$

is known as the *first boundary value problem*, or the *Dirichlet problem.* Unless the contrary is explicitly stated, φ is always assumed to be a continuous function on ∂D. When φ is a continuous function the solution is always understood to be continuous in $\bar{D}$.

From the weak maximum principle we obtain the following uniqueness theorem.

Theorem 20. *Let L be elliptic operator with continuous coefficients in a bounded domain D and assume that* $c(x) \leq 0$. *Then there exists at most one solution to the Dirichlet problem* (7.3), (7.4).

Introducing the function (3.11) and assuming that

$$\lambda > 0, \qquad a_{11}\lambda^2 + b_1\lambda \geq 1 \qquad \text{in } D,$$

we can prove, similarly to (3.10), that the solution of (7.3), (7.4) satisfies the inequality

$$\max_{\bar{D}} |u| \leq \max_{\partial D} |\varphi| + (e^{\lambda d} - 1) \max_{\bar{D}} |f|. \tag{7.5}$$

Here d is the breadth of D in the x_1-direction, D is a bounded domain, and $c \leq 0$.

If the assumption $c \leq 0$ is replaced by

$$(e^{\lambda d} - 1)c(x, t) \leq \vartheta < 1, \tag{7.6}$$

then, introducing $c^- = \min(c, 0)$ and writing $Lu = f$ in the form $(L - c + c^-)u = \bar{f}$ where $\bar{f} = (c^- - c)u + f$ and applying (7.5) we easily get the inequality

$$\max_{\bar{D}} |u| \leq \frac{1}{1 - \vartheta} [\max_{\partial D} |\varphi| + (e^{\lambda d} - 1) \max_{\bar{D}} |f|]. \tag{7.7}$$

(7.7) implies a uniqueness theorem for the Dirichlet problem.

Definition. Let x^0 be any point on the boundary ∂D of a domain D. If there exists a closed ball B such that $B \subset \bar{D}$ and such that $B \cap \partial D = \{x^0\}$, then we say that x^0 has the *inside sphere property*.

We shall now state an analogue of Theorem 14, Sec. 5.

Theorem 21. *Let D be a bounded domain and assume that each of the points of ∂D has the inside sphere property. Let Lu be an elliptic operator with continuous coefficients in $\bar{D}$ and let $c(x) \leq 0$. If u is a continuous function in $\bar{D}$, if $Lu \geq 0$, and if u has a positive maximum in $\bar{D}$ and $u \not\equiv$* const., *then there exist points x^0 on the boundary ∂D such that*

$$u(x^0) = \max_{x \in \bar{D}} u(x),$$

and at each of these points

$$\frac{\partial u(x^0)}{\partial \tau} < 0$$

for any nontangential inward direction τ.

The definition of a nontangential inward direction is as in Theorem 14.

The proof is similar to the proof of Theorem 14 and is left to the reader.

Note that the theorem is not true if ∂D contains vertex points. As an example take

$$D: x_1^2 + x_2^2 < R^2, \quad x_2 < \gamma_1 x, \quad x_2 > \gamma_2 x \qquad (\gamma_1 > 0 > \gamma_2),$$

$$x^0 = (0, 0), \quad Lu = \frac{\partial^2 u}{\partial x_1^2} + A \frac{\partial^2 u}{\partial x_2^2} \quad \text{where } A > |\gamma_1 \gamma_2|.$$

Then $u = (x_2 - \gamma_1 x_1)(x_2 - \gamma_2 x_1) + 1$ satisfies: $u < 1$ in D, $u(x^0) = 1$, $Lu = 2\gamma_1\gamma_2 + 2A > 0$, but $\partial u / \partial \tau = 0$ at x^0 for any direction τ.

Given a continuous function β on ∂D and a direction τ which varies continuously on ∂D, consider the problem of finding a solution to (7.3) satisfying the boundary condition

$$\frac{\partial u(x)}{\partial \tau} + \beta(x)u(x) = \psi(x) \qquad \text{on } \partial D, \tag{7.8}$$

where f, ψ are given functions. It is called the *third boundary value problem.* If τ is never tangent to ∂D then we say that the problem is *regular.* If $\tau = \tau(x)$ is the inward *conormal*, i.e., if its components are $\Sigma\, a_{ij}(x)\nu_j$ where ν is the inward normal to ∂D at x, then we call (7.3), (7.8) the *second boundary value problem*, or the *Neumann problem.*

Theorem 21 yields the uniqueness of a solution of (7.3), (7.8) under the assumptions that τ points into the interior of D and

$$\beta \leq 0, \quad \beta \not\equiv 0 \qquad \text{on } \partial D. \tag{7.9}$$

If

$$\beta < 0 \qquad \text{on } \partial D, \tag{7.10}$$

then the uniqueness of solutions for (7.3), (7.8) follows without using Theorem 21. Indeed, if $f \equiv 0$, $\psi \equiv 0$ and $u \not\equiv 0$, then, by the maximum principle, u takes a positive maximum (or a negative minimum) at a point y on ∂D. This however contradicts (7.8) with $\psi \equiv 0$ since $\partial u/\partial \tau \geq 0$ (≤ 0) and $\beta u > 0$ (< 0) at y.

Note that in the last proof no assumption was made on ∂D.

PROBLEMS

1. Give a direct proof of Theorem 6, Sec. 2.
[*Hint:* If $u(P)$ is larger than $u(Q)$ for all Q in $\overline{S(P)} - S(P)$, then apply Lemma 1 to $u(x, t) + \epsilon \exp[-\alpha|x - x^0|^2]$ where $|x - x^0| \geq$ const. > 0 for all $(x, t) \in D$.]

2. L is a parabolic operator in $\Omega_0 = R^n \times (0, T]$ with continuous coefficients, and
$$|a_{ij}(x, t)| \leq M(|x|^2 + 1), \quad |b_i(x, t)| \leq M(|x| + 1), \quad c(x, t) \leq M$$
for some constant M. Prove: if $Lu \leq 0$ in Ω_0, $u(x, t) \geq -N(|x|^q + 1)$ in $\Omega = R^n \times [0, T]$ for some positive constants N, q then $u(x, 0) \geq 0$ (in R^n) implies $u(x, t) \geq 0$ in Ω.
[*Hint:* The function $w = \epsilon(|x|^2 + Kt)^p e^{\alpha t}$ $(\epsilon > 0, K > 0, \alpha > 0, 2p > q)$ satisfies $Lw < 0$ for appropriate K, α. Consider $w + u$.]

3. Let L be as in Problem 2. If $|u(x, 0)| \leq M_1$, $|Lu(x, t)| \leq M_2$, $c(x, t) \leq M_3$, and if u is bounded, then $|u(x, t)| \leq (M_1 + M_2 t) \exp[M_3 t]$.
[*Hint:* Consider $(M_1 + M_2 t) \exp[M_3 t] \pm u$ and use Problem 2.]

4. Give another proof of Theorem 9, Sec. 4, by employing the function
$$v(x, t) \equiv \exp[2\beta(|x|^2 + 1)e^{\alpha t}].$$
[*Hint:* $Lv < 0$ if $0 \leq t \leq 1/\alpha$ and α is sufficiently large. Consider first $u + \epsilon v$ for $0 \leq t \leq 1/\alpha$, and then proceed step by step.]

5. Let L be a parabolic operator in $\Omega_0 = R^n \times (0, T]$ with continuous coefficients satisfying (4.1) and let $c(x, t) \geq 0$. Suppose that $Lu \leq 0$ in Ω_0 and that $u(x, t) \geq -B \exp [\beta|x|^2]$ in $\Omega = R^n \times [0, T]$ (for some positive constants B, β). Then $u(x, 0) \geq N > 0$ (in R^n) implies that $u(x, t) \geq N$ in Ω (N a positive constant). [*Hint:* Apply Theorem 9, Sec. 4.]

6. Let L and u be as in Problem 5, and assume further that $c(x, t) \geq \alpha|x|^2 + \beta$ where α, β are constants and $\alpha > 0$. Deduce the stronger inequality $u(x, t) \geq N \exp [\lambda|x|^2 t + \nu t]$ where λ, ν are constants and $\lambda > 0$. [*Hint:* Define v by $u = v \exp [\lambda|x|^2 t + \nu t]$ and apply Problem 5 to v.]

7. Prove Corollary 2 of Theorem 13 without making the additional assumption of local Hölder continuity in (x, t) of the coefficients of L. [*Hint:* Prove (4.13) with $u(\xi, 0)$ replaced by $u(\xi, t_0)$ and $t_0 < t \leq t_0 + \epsilon$. Next proceed step by step, employing Problem 5, Chap. 1.]

8. Prove: if $\partial v/\partial t > \partial^2 v/\partial^2 x + Av^2$ for $0 < x < 1, 0 < t < 4N$ and $v(x, 0) > \mu/N$, $v(0, t) > \mu/N$, $v(1, t) > \mu/N$ for $0 \leq x \leq 1$, $0 \leq t \leq 4N$ (A, N, μ are positive constants) and if $A\mu > 8N + 1/4$ then $v(1/2, t) \to \infty$ as $t \to 4N$. [*Hint:* Use Theorem 16, Sec. 6, with $w(x, t) = \mu[N - tx(1 - x)]$.]

9. Show, by means of examples, that Theorem 1, Sec. 1, and Theorem 18, Sec. 7, are not true if the assumption $c \leq 0$ is omitted.

10. Let L be an elliptic operator with continuous coefficients in a domain D and assume that $Lu \geq 0$ ($Lu \leq 0$) in D, and that $u \leq 0$ ($u \geq 0$) in D. Prove: if for some point $x^0 \in D$ $u(x^0) = 0$ then $u \equiv 0$ in D.

CHAPTER 3

THE FIRST INITIAL-BOUNDARY VALUE PROBLEM

Introduction. In this chapter we prove the existence of a unique solution for the first initial-boundary value problem. The proof is given by the *method of continuity* which is roughly the following one: Denoting by L the parabolic operator under investigation, and by L_0 the heat operator, we connect L to L_0 by a family of parabolic operators $L_\lambda = \lambda L + (1 - \lambda)L_0$ $(0 \leq \lambda \leq 1)$. We then show that if the problem can be solved for all $\lambda < \sigma$ $(0 < \sigma \leq 1)$ then it can also be solved for $\lambda = \sigma$, and that if the problem can be solved for all $\lambda \leq \sigma$ $(0 \leq \sigma < 1)$ then it can also be solved for all $\lambda \leq \sigma + \epsilon$ for some $\epsilon > 0$. Since, finally, the problem can be shown to be solvable for L_0, the same is true for L.

The decisive tool in carrying out the actual proof is some a priori estimates on the derivatives of the solution (Sec. 2). The introduction of Banach spaces is convenient but not essential.

The use of barriers (Sec. 4) makes it possible to extend the existence theorems to domains whose boundary is not necessarily smooth.

Other applications of the a priori estimates are also given. Thus, in Sec. 5 it is proved that, roughly speaking, the solutions of a parabolic equation are as smooth as the coefficients.

The existence and differentiability theorems are used in Sec. 7 in studying Green's function.

Finally, in Sec. 8 we briefly describe the extension of the results of the previous sections to elliptic equations.

1. Banach Spaces and Metric Spaces

Given a linear space X over the real or the complex field, a rule which corresponds to each element $x \in X$ a nonnegative number $||x||$ is called a *norm* if it satisfies the following properties:

(1) $||x|| = 0$ if and only if $x = 0$,
(2) $||\lambda x|| = |\lambda|\,||x||$ for any scalar λ, and
(3) $||x + y|| \leq ||x|| + ||y||$.

A linear space X with a norm defined on it is called a *normed space.*

In a normed space X the *distance* between any two points x, y is defined by $||x - y||$. An *open* (*closed*) *ball* with center z and radius R ($R > 0$) is the set of all points whose distance to z is less than (less or equal to) R. A set X_0 is said to be *open* if together with each of its points it contains also an open ball with that point as its center. A set X_1 is said to be *closed* if its complement is open. It can easily be verified that the open (closed) ball is an open (a closed) set.

A sequence $\{x_n\}$ is said to be *convergent* if there exists a point x such that $||x_n - x|| \to 0$ as $n \to \infty$; x is called the *limit* of the sequence and the sequence is said to *converge* to x. A point x is called a *limit point* of a set X_0 if there exists a sequence $\{x_n\}$ whose elements belong to X_0 and which converges to x. The *closure* of a set is the union of the set with the set of its limit points. It is easy to show that a set X_0 is closed if and only if it coincides with its closure.

A sequence $\{x_n\}$ is called a *Cauchy sequence* if $||x_m - x_n|| \to 0$ as m, $n \to \infty$. Every convergent sequence is obviously also a Cauchy sequence. If, conversely, every Cauchy sequence is a convergent sequence, then we say that X is a *complete* space. A complete normed space is also called a *Banach space*.

Let T be a transformation defined on a set X_0 of a Banach space X and suppose that T maps X_0 into itself, i.e., $Tx \in X_0$ if $x \in X_0$. T is called a *contraction* (in X_0) if there exists a positive number $\vartheta < 1$ such that

$$||Tx - Ty|| \leq \vartheta ||x - y||$$

for all x, y in X_0.

Theorem 1. *Let T be a transformation which maps a closed set X_0 of a Banach space X into itself, and assume that T is a contraction in X_0. Then there exists a unique point $y \in X_0$ such that $Ty = y$.*

The point y is called a *fixed point* of the transformation T. Theorem 1 is referred to as a fixed point theorem.

Proof. Take any $x_0 \in X_0$ and define successively $x_{n+1} = Tx_n$ for $n = 0, 1, \ldots$. All the x_n belong to X_0. We further have

$$||x_{n+1} - x_n|| = ||Tx_n - Tx_{n-1}|| \leq \vartheta ||x_n - x_{n-1}|| \leq \cdots \leq \vartheta^n ||x_1 - x_0||.$$

Hence,

$$||x_{n+m} - x_n|| = ||\sum_{i=1}^{m} (x_{n+i} - x_{n+i-1})|| \leq \sum_{i=1}^{m} ||x_{n+i} - x_{n+i-1}||$$

$$\leq \left(\sum_{i=1}^{m} \vartheta^{n+i-1}\right) ||x_1 - x_0|| \leq \frac{\vartheta^n}{1 - \vartheta} ||x_1 - x_0||.$$

It follows that $\{x_n\}$ is a Cauchy sequence; hence, it converges to some point $y \in X$. Since X_0 is a closed set, $y \in X_0$. Finally,

$$||Ty - y|| = ||Ty - Tx_n + x_{n+1} - y|| \leq \vartheta||y - x_n|| + ||x_{n+1} - y|| \to 0$$

as $n \to \infty$, i.e., $Ty = y$.

To prove uniqueness, suppose z is another fixed point. Then

$$||z - y|| = ||Tz - Ty|| \leq \vartheta||z - y|| \leq \cdots \leq \vartheta^n||z - y|| \to 0$$

as $n \to \infty$, i.e., $z = y$.

Let X be any space of points. We say that X is a *metric space* with a *metric* d if to each pair of points x, y in X there corresponds a nonnegative number $d(x, y)$ satisfying the following properties:

(1) $d(x, y) = 0$ if and only if $x = y$,
(2) $d(x, y) = d(y, x)$, and
(3) $d(x, z) \leq d(x, y) + d(y, z)$ (the triangle inequality).

We define the *distance* between each two points x, y by $d(x, y)$. The definitions of ball, open set, closed set, convergence, completeness, etc., can now be given similarly to the case of a normed space. Thus, a sequence $\{x_n\}$ is a Cauchy sequence if $d(x_m, x_n) \to 0$ as $m, n \to \infty$, and it is convergent to x if $d(x_n, x) \to 0$ as $n \to \infty$. A transformation T mapping a subset X_0 of X into itself is called a *contraction* (in X_0) if

$$d(Tx, Ty) \leq \vartheta\, d(x, y)$$

for some $\vartheta < 1$ and for all x, y in X_0. The reader may easily extend Theorem 1 to the case of complete metric spaces.

We conclude this section with a general, well-known theorem concerning the extension of continuous functions.

Theorem 2. *Suppose that a function f is uniformly continuous on a set S in a euclidean space R. Then there exists an extension F of f which is continuous in the whole space R. F is uniquely determined on the closure $\bar{S}$ of S. If $\varphi(t)$ is a modulus of continuity of f which is concave and approaches zero with t, then the extension F may be constructed so as to have the same modulus of continuity. If f is bounded (and has a modulus of continuity as above) then the extension F can be constructed so that it is also bounded with the same bounds as f (and has the same modulus of continuity). Finally, if S is a closed set, then all the previous assertions remain true if f is only assumed to be continuous on S.*

For a detailed proof the reader is referred to [81]; see also [50; 117–118]. *Except for the last statement, the theorem is also valid for sets S in general metric spaces* (see [81]).

A special important case occurs when the modulus of continuity is $\varphi(t) = Ht^\alpha$, i.e., when f is uniformly Hölder continuous with exponent α.

2. A Priori Estimates of the Schauder Type

Throughout this chapter we adopt the following notation: D is a bounded $(n + 1)$-dimensional domain in R^{n+1}, $(x, t) = (x_1, \ldots, x_n, t)$ is a variable point in R^{n+1}. D is bounded by a domain B on $t = 0$, a domain B_T on $t = T$, and a manifold (not necessarily connected) lying in the strip $0 < t \leq T$. We set

$$B_\tau = D \cap \{t = \tau\}, \quad S_\tau = S \cap \{t \leq \tau\}, \quad D_\tau = D \cap \{t < \tau\}.$$

It is assumed that there exists a simple continuous curve γ connecting some point on B_T to some point on B so that t is nonincreasing along γ. It is also assumed that, for any $0 < \tau < T$, B_τ is a domain. It follows that (2.3.1) holds and that Theorem 7, Sec. 3, Chap. 2, which asserts uniqueness for the first initial-boundary value problem, is valid.

Some of the results of this chapter, in particular Theorems 5, 10, 11, 13–15, remain true without any change in the proof even for arbitrary bounded domains D.

It will be useful to introduce the following definition of distance:

$$d(P, Q) = [|x - \bar{x}|^2 + |t - \bar{t}|]^{1/2} \tag{2.1}$$

where $P = (x, t)$, $Q = (\bar{x}, \bar{t})$, and $|x|$ is the euclidean norm $(\Sigma\, x_i^2)^{1/2}$. $d(P, Q)$ is easily seen to satisfy the metric properties of Sec. 1. The concept of Hölder continuity will always be defined in this (and in the following) chapter with respect to the metric (2.1).

We introduce the following notation:

$$|u|_0^D = \operatorname*{l.u.b.}_{D} |u|,$$

$$\overline{H}_\alpha^D(u) = \operatorname*{l.u.b.}_{P,Q \in D} \frac{|u(P) - u(Q)|}{d(P, Q)^\alpha},$$

$$\overline{|u|}_\alpha^D = |u|_0^D + \overline{H}_\alpha^D(u). \tag{2.2}$$

$\overline{H}_\alpha^D(u) < \infty$ if and only if u is uniformly Hölder continuous (exponent α) in D, and then $\overline{H}_\alpha^D(u)$ is the Hölder coefficient of u. Denote by $\overline{C}_\alpha(D)$ the set of all functions u for which $\overline{|u|}_\alpha^D < \infty$. $\overline{C}_\alpha(D)$ is easily seen to be a normed space with the norm $|u|_\alpha^D$.

We denote by D_x^m any partial derivative of order m with respect to the variables $x_1, \ldots, x_n$, and let $D_t = \partial/\partial t$. If $D_x u$, $D_x^2 u$, $D_t u$ exist in D, then we define

$$\overline{|u|}_{2+\alpha}^D = \overline{|u|}_\alpha^D + \Sigma\, \overline{|D_x u|}_\alpha^D + \Sigma\, \overline{|D_x^2 u|}_\alpha^D + \overline{|D_t u|}_\alpha^D, \tag{2.3}$$

where the sums are taken over all the partial derivatives of the indicated order. The set of all functions u for which $\overline{|u|}_{2+\alpha}^D < \infty$ is denoted by

$\overline{C}_{2+\alpha}(D)$. $\overline{C}_{2+\alpha}(D)$ is easily seen to be a normed space with the norm given by (2.3).

Theorem 3. *$\overline{C}_{2+\alpha}(D)$ and $\overline{C}_{\alpha}(D)$ are Banach spaces.*

Proof. It is enough to give the proof for $\overline{C}_{2+\alpha}(D)$. We have only to prove that this space is complete. Let then $\{u_m\}$ be a Cauchy sequence in $\overline{C}_{2+\alpha}(D)$, i.e.,

$$\overline{|u_m - u_k|}{}^D_{2+\alpha} \to 0 \qquad \text{if } m,\, k \to \infty. \tag{2.4}$$

It follows that the sequence is bounded (in the norm), i.e.,

$$\overline{|u_m|}{}^D_{2+\alpha} \leq K \qquad \text{for all } m. \tag{2.5}$$

In particular,

$$\underset{D}{\text{l.u.b.}}\, |D_x^2 u_m| \leq K, \qquad \overline{H}{}^D_\alpha(D_x^2 u_m) \leq K.$$

Thus $\{D_x^2 u_m\}$ is a uniformly bounded and equicontinuous family in D. By the theorem of Ascoli-Arzela it follows that there exists a subsequence $\{u_{m'}\}$ such that $\{D_x^2 u_{m'}\}$ is uniformly convergent in D. By the same argument we can also proceed to find a subsequence of $\{u_{m'}\}$ which is uniformly convergent in D together with its first two x-derivatives and its first t-derivative. Denoting this subsequence by $\{u_{m''}\}$ and its limit by u, we then have:

$$u_{m''} \to u, \quad D_x u_{m''} \to D_x u, \quad D_x^2 u_{m''} \to D_x^2 u, \quad D_t u_{m''} \to D_t u. \tag{2.6}$$

We shall prove that $u \in \overline{C}_{2+\alpha}(D)$. From (2.5) it follows that for any two points P, Q in D,

$$\frac{|D_x^2 u_m(P) - D_x^2 u_m(Q)|}{d(P, Q)^\alpha} \leq K.$$

Taking $m = m'' \to \infty$ and using (2.6) we find that

$$\frac{|D_x^2 u(P) - D_x^2 u(Q)|}{d(P, Q)^\alpha} \leq K,$$

i.e., $\overline{H}{}^D_\alpha(D_x^2 u) < \infty$. Similarly one proves that each of the terms on the right-hand side of (2.3) is finite, i.e., $\overline{|u|}{}^D_{2+\alpha} < \infty$.

We finally have to prove $u_m \to u$ in the norm sense. Because of (2.4) it suffices to show that

$$\overline{|u_{m''} - u|}{}^D_{2+\alpha} \to 0 \qquad \text{as } m'' \to \infty. \tag{2.7}$$

Now from (2.4) it follows that for any $\epsilon > 0$ there exists an $N > 0$ such that

$$\overline{H}{}^D_\alpha(D_x^2 u_{m''} - D_x^2 u_{k''}) \leq \epsilon \qquad \text{if } m'' \geq N,\, k'' \geq N. \tag{2.8}$$

(2.8) implies that for any two points P, Q in D,

$$d(P, Q)^{-\alpha} |D_x^2 u_{m''}(P) - D_x^2 u_{k''}(P) - D_x^2 u_{m''}(Q) + D_x^2 u_{k''}(Q)| \leq \epsilon.$$

Taking $k'' \to \infty$ and using (2.6) we obtain

$$d(P, Q)^{-\alpha}|D_x^2 u_{m''}(P) - D_x^2 u(P) - D_x^2 u_{m''}(Q) + D_x^2 u(Q)| \leq \epsilon.$$

Since this holds for all P, Q in D, by taking the l.u.b. we get

$$\overline{H}_\alpha^D(D_x^2 u_{m''} - D_x^2 u) \leq \epsilon. \tag{2.9}$$

Proceeding in a similar way to evaluate the other terms in the norm of $\overline{|u - u_{m''}|}{}_{2+\alpha}^D$, the proof of (2.7) follows.

For any point $Q = (\xi, \tau)$ in D we denote by d_Q the distance from Q to $B + S_\tau$, i.e.,

$$d_Q = \operatorname*{g.l.b.}_{P \in B + S_\tau} d(P, Q)$$

where $d(P, Q)$ is defined by (2.1). For any two points P, Q in D we set

$$d_{PQ} = \min(d_P, d_Q).$$

Similarly to (2.2) we define

$$|u|_\alpha^D = |u|_0^D + H_\alpha^D(u), \tag{2.10}$$

where

$$H_\alpha^D(u) = \operatorname*{l.u.b.}_{P,Q \in D} d_{PQ}^\alpha \frac{|u(P) - u(Q)|}{d(P, Q)^\alpha}.$$

Thus, the difference between $\overline{H}_\alpha^D$ and H_α^D is that the latter allows u to grow to infinity (like $d_{PQ}^{-\alpha}$) near the normal boundary $\overline{B} + S$ of D. The space of all functions u with finite norm $|u|_\alpha^D$ is denoted by $C_\alpha(D)$. This is clearly a normed space.

Analogously to the norm (2.3) we define

$$|u|_{2+\alpha}^D = |u|_\alpha^D + \Sigma\, |dD_x u|_\alpha^D + \Sigma\, |d^2 D_x^2 u|_\alpha^D + |d^2 D_t u|_\alpha^D, \tag{2.11}$$

where

$$|d^m v|_0^D = \operatorname*{l.u.b.}_{P \in D} d_P^m |v(P)|,$$

$$H_\alpha^D(d^m v) = \operatorname*{l.u.b.}_{P,Q \in D} d_{PQ}^{m+\alpha} \frac{|v(P) - v(Q)|}{d(P, Q)^\alpha},$$

$$|d^m v|_\alpha^D = |d^m v|_0^D + H_\alpha^D(d^m v).$$

The space of all functions u with finite norm $|u|_{2+\alpha}^D$ is denoted by $C_{2+\alpha}(D)$. $C_{2+\alpha}(D)$ is obviously a normed space.

Theorem 4. *$C_\alpha(D)$ and $C_{2+\alpha}(D)$ are Banach spaces.*

The proof is similar to that of Theorem 3 and it is therefore left to the reader.

For the sake of simplicity, when there is no confusion about the domain D with respect to which the previous norms are defined, we shall usually omit the superscript D.

An *a priori estimate* is an estimate which can be established without a preliminary knowledge that the solutions, for which the estimate holds, do in fact exist. It turns out that some a priori estimates can be used to prove the existence of solutions. We shall state here some such a priori estimates. They are analogous to estimates derived by Schauder for elliptic equations. They will be used in the next two sections to solve the first initial-boundary value problem.

There are two kinds of these estimates. The first kind is in terms of the norms $|u|_\alpha$, $|u|_{2+\alpha}$ and the estimates are called *interior estimates*. The second kind is in terms of the norms $\overline{|u|}_\alpha$, $\overline{|u|}_{2+\alpha}$ and the estimates are called *boundary estimates*.

Consider the equation

$$(2.12)\quad Lu \equiv \sum_{i,j=1}^{n} a_{ij}(x, t) \frac{\partial^2 u}{\partial x_i\, \partial x_j} + \sum_{i=1}^{n} b_i(x, t) \frac{\partial u}{\partial x_i} + c(x, t)u - \frac{\partial u}{\partial t} = f(x, t) \qquad \text{in } D + B_T.$$

The following assumptions will be needed:

(A) The coefficients of L are locally Hölder continuous (exponent α) in D, and

$$(2.13)\qquad |a_{ij}|_\alpha \leq K_1, \quad |db_i|_\alpha \leq K_1, \quad |d^2c|_\alpha \leq K_1.$$

(B) For any $(x, t) \in D$ and for any real vector ξ,

$$(2.14)\qquad \sum_{i,j=1}^{n} a_{ij}(x, t)\xi_i\xi_j \geq K_2|\xi|^2 \qquad (K_2 > 0).$$

(C) f is locally Hölder continuous (exponent α) in D and

$$(2.15)\qquad |d^2 f|_\alpha < \infty.$$

We can now state the interior estimates.

Theorem 5. *Let* (A), (B), (C) *hold. There exists a constant K depending only on K_1, K_2 and on n, α such that for any solution u of* (2.12) *in D for which* $\text{l.u.b.}_D\, |u| < \infty$ *and u, $D_x u$, $D_x^2 u$, $D_t u$ are locally Hölder continuous (exponent α) in D, u must belong to $C_{2+\alpha}$ and*

$$(2.16)\qquad |u|_{2+\alpha} \leq K(|u|_0 + |d^2 f|_\alpha).$$

Definition. We say that D has the property ($\bar{\text{E}}$) if for every point Q of $\bar{S}$ there exists an $(n + 1)$-dimensional neighborhood V such that $V \cap \bar{S}$ can be represented, for some i $(1 \leq i \leq n)$ in the form

$$(2.17)\qquad x_i = h(x_1, \ldots, x_{i-1}, x_{i+1}, \ldots, x_n, t)$$

and h, $D_x h$, $D_x^2 h$, $D_t h$ are Hölder continuous (exponent α).

In particular it follows that the tangent hyperplanes to the points of $\bar{S}$ can never be of the form $t = \text{const}$.

Definition. A function ψ, defined on $\overline{B} + S$ is said to belong to $\overline{C}_{2+\alpha}(D)$ (or simply to $\overline{C}_{2+\alpha}$) if there exist functions Ψ in $\overline{C}_{2+\alpha}(D)$ such that $\Psi = \psi$ on $\overline{B} + S$. We then define

$$\overline{|\psi|}_{2+\alpha} = \underset{\Psi}{\text{g.l.b.}}\ \overline{|\Psi|}_{2+\alpha}$$

where the g.l.b. is taken with respect to all the Ψ's in $\overline{C}_{2+\alpha}(D)$ which coincide with ψ on $\overline{B} + S$.

If $(\bar{\text{E}})$ holds and if $\psi \in \overline{C}_{2+\alpha}$ then, for any extension Ψ of ψ, $\partial\Psi/\partial t$ is uniquely defined (by continuity) on the boundary ∂B of B, and the definition is independent of Ψ. We denote this function (on ∂B) by $\partial\psi/\partial t$. The other terms of $L\psi$ are also uniquely defined (by continuity) on ∂B.

The following assumptions will be needed:

$(\bar{\text{A}})$ The coefficients of L are uniformly Hölder continuous (exponent α) in D and

$$\overline{|a_{ij}|}_\alpha \le \overline{K}_1, \quad \overline{|b_i|}_\alpha \le \overline{K}_1, \quad \overline{|c|}_\alpha \le \overline{K}_1. \tag{2.18}$$

$(\bar{\text{C}})$ f is uniformly Hölder continuous (exponent α) in D, i.e.,

$$\overline{|f|}_\alpha < \infty. \tag{2.19}$$

We can now state the boundary estimates for the solutions of (2.12) which satisfy the initial-boundary condition

$$u = \psi \qquad \text{on } \overline{B} + S. \tag{2.20}$$

Theorem 6. *Let* $(\bar{\text{A}})$, (B), $(\bar{\text{C}})$ *hold and assume that D has the property* $(\bar{\text{E}})$ *and that* $\psi \in \overline{C}_{2+\alpha}$. *There exists a constant* $\overline{K}$ *depending only on* $\overline{K}_1$, $K_{2,\alpha}$ *and D such that if u is a solution of* (2.12), (2.20) *and if* $u \in \overline{C}_{2+\alpha}$ *then*

$$\overline{|u|}_{2+\alpha} \le \overline{K}(\overline{|\psi|}_{2+\alpha} + \overline{|f|}_\alpha). \tag{2.21}$$

The proofs of Theorems 5, 6 are technically difficult and lengthy. They will be given, in detail, in the following chapter.

3. Solution of the First Initial-boundary Value Problem

Definition. If D has the property $(\bar{\text{E}})$ and if, in addition, the functions D_xD_th, D_t^2h of the local representations of $\overline{S}$ exist and are continuous functions, then we say that D has the property $(\bar{\bar{\text{E}}})$.

In this section we shall prove the following existence theorem.

Theorem 7. *Let* $(\bar{\text{A}})$, (B), $(\bar{\text{C}})$ *hold and assume that D has the property* $(\bar{\bar{\text{E}}})$, *that* $\psi \in \overline{C}_{2+\alpha}$ *and that* $L\psi = f$ *on* ∂B. *Then there exists a unique solution u of the first initial-boundary value problem* (2.12), (2.20) *and, furthermore,* $u \in \overline{C}_{2+\alpha}$.

Proof. The uniqueness was proved in Chap. 2, Sec. 3.

In the sequel we shall need to use the rather obvious rules:

$$\overline{H}_\alpha(fg) \leq \overline{H}_\alpha(f)|g|_0 + |f|_0\overline{H}_\alpha(g), \tag{3.1}$$

$$\overline{|fg|}_\alpha \leq \overline{|f|}_\alpha \overline{|g|}_\alpha. \tag{3.2}$$

If $\Psi \in \overline{C}_{2+\alpha}$ then using $(\bar{\text{A}})$ and (3.2) one deduces that $\overline{|L\Psi|}_\alpha \leq K'|\Psi|_{2+\alpha}$ where K' depends only on n, $\overline{K}_1$. From this remark it follows that without loss of generality we may assume, in the proof of existence, that $\psi \equiv 0$, since otherwise we can consider $v = u - \Psi$ for some $\Psi \in \overline{C}_{2+\alpha}$ which coincides with ψ on $\overline{B} + S$, noting that

$$\overline{|Lv|}_\alpha \leq \overline{|f|}_\alpha + \overline{|L\Psi|}_\alpha \leq \overline{|f|}_\alpha + K'\overline{|\Psi|}_{2+\alpha}$$

and that $\overline{|\Psi|}_{2+\alpha} \leq 2\overline{|\psi|}_{2+\alpha}$ for some Ψ.

The outline of the proof is the following. Consider the one parameter family of parabolic operators

$$L_\lambda = \lambda L + (1 - \lambda)L_0 \qquad (0 \leq \lambda \leq 1), \tag{3.3}$$

where

$$L_0 = \sum_{i=1}^{n} \frac{\partial^2}{\partial x_i^2} - \frac{\partial}{\partial t}$$

is the heat operator. Denote by Σ the set of all values λ for which the problem

$$\begin{aligned} L_\lambda u &= f \qquad \text{in } D + B_T, \\ u &= 0 \qquad \text{on } \overline{B} + S \end{aligned} \tag{3.4}$$

has a (unique) solution u in $\overline{C}_{2+\alpha}$ for any f in $\overline{C}_\alpha$, $f = 0$ on ∂B. We want to show:

(a) Σ contains $\lambda = 0$;

(b) Σ is an open set in the interval $0 \leq \lambda \leq 1$, i.e., if $\lambda_0 \in \Sigma$ then there exists an $\epsilon > 0$ such that $\lambda \in \Sigma$ for all $|\lambda - \lambda_0| < \epsilon$, $0 \leq \lambda \leq 1$;

(c) Σ is a closed set.

It would then follow that Σ coincides with the interval $0 \leq \lambda \leq 1$. Since, in particular, $1 \in \Sigma$, the assertion of the theorem follows.

The method of proof just described is called the *method of continuity*.

The proof of (a) is based on lengthy calculations, most of which appear in the proofs of Theorems 5, 6 of Sec. 2. We therefore postpone it until the next chapter (see Chap. 4, Sec. 8).

Proof of (*b*). Write $L_\lambda u = f$ in the equivalent form

$$L_{\lambda_0} u = (L_{\lambda_0} u - L_\lambda u) + f \equiv F(u)$$

and consider the linear transformation $v = Au$ defined as follows: Given $u \in \overline{C}_{2+\alpha}$ which vanishes on $\overline{B} + S$, we define Au to be the (unique) solution of

$$\text{(3.5)} \qquad \begin{aligned} L_{\lambda_0}v &= F(u) \qquad && \text{in } D + B_T, \\ v &= 0 && \text{on } \overline{B} + S. \end{aligned}$$

Since $\lambda_0 \in \Sigma$ and $F(u) \in \overline{C}_\alpha$, $F = 0$ on ∂B, the transformation $v = Au$ is well defined for all $u \in \overline{C}_{2+\alpha}$ which vanishes on $\overline{B} + S$. If we prove that, for some such u, $Au = u$, then u is the (unique) solution of (3.4) and, consequently, $\lambda \in \Sigma$.

An easy calculation based on (3.2) shows that

$$\text{(3.6)} \qquad \overline{|F(u)|}_\alpha \leq K_3|\lambda - \lambda_0| \, \overline{|u|}_{2+\alpha} + \overline{|f|}_\alpha,$$

where K_3 is a constant independent of λ, u, f. Applying Theorem 6, Sec. 2, to the system (3.5), and using (3.6), we get

$$\text{(3.7)} \qquad \overline{|v|}_{2+\alpha} \leq \overline{K}K_3|\lambda - \lambda_0| \, \overline{|u|}_{2+\alpha} + \overline{K}\,\overline{|f|}_\alpha,$$

where $\overline{K}$ is independent of λ, u, f. Hence, if

$$\text{(3.8)} \qquad \overline{|u|}_{2+\alpha} \leq 2\overline{K}\,\overline{|f|}_\alpha$$

and if λ is such that $2\overline{K}K_3|\lambda - \lambda_0| \leq 1$, then also $\overline{|v|}_{2+\alpha} \leq 2\overline{K}\,\overline{|f|}_\alpha$. Thus, denoting by X_0 the set of u's vanishing on $\overline{B} + S$ and satisfying (3.8), it follows that A maps X_0 into itself. X_0 is a closed set in $\overline{C}_{2+\alpha}$.

Let $v_1 = Au_1$, $v_2 = Au_2$ where u_1, u_2 belong to the set X_0. Then

$$\begin{aligned} L_{\lambda_0}(v_1 - v_2) &= L_{\lambda_0}(u_1 - u_2) - L_\lambda(u_1 - u_2) \qquad && \text{in } D + B_T, \\ v_1 - v_2 &= 0 && \text{on } \overline{B} + S. \end{aligned}$$

Proceeding as before we find that

$$\overline{|v_1 - v_2|}_{2+\alpha} \leq \overline{K}K_3|\lambda - \lambda_0| \, \overline{|u_1 - u_2|}_{2+\alpha} \leq \frac{1}{2}\,\overline{|u_1 - u_2|}_{2+\alpha}$$

provided $2\overline{K}K_3|\lambda - \lambda_0| \leq 1$. Thus, if λ satisfies the last inequality then Av is a contraction in the closed set X_0. Since, by Theorem 3, Sec. 2, $\overline{C}_{2+\alpha}$ is a Banach space, Theorem 1, Sec. 1, can be used to conclude that there exists a function u in X_0 for which $Au = u$. This completes the proof of (b).

Proof of (c). Let $\lambda_m \in \Sigma$, $\lambda_m \to \sigma$. We have to prove that $\sigma \in \Sigma$, i.e., we have to prove that for any $f \in \overline{C}_\alpha$, $f = 0$ on ∂B, there exists a function $u \in \overline{C}_{2+\alpha}$ such that

$$\text{(3.9)} \qquad \begin{aligned} L_\sigma u &= f \qquad && \text{in } D + B_T, \\ u &= 0 && \text{on } \overline{B} + S. \end{aligned}$$

By assumption, for any m there exists a function u_m in $\overline{C}_{2+\alpha}$ such that

$$\text{(3.10)} \qquad \begin{aligned} L_{\lambda_m}u_m &= f \qquad && \text{in } D + B_T, \\ u_m &= 0 && \text{on } \overline{B} + S. \end{aligned}$$

Noting that the coefficients of L_λ satisfy the conditions $(\bar{\text{A}})$, (B) with constants $\overline{K}_1$, K_2 independent of λ and applying Theorem 6, we get

(3.11) $$\overline{|u_m|}_{2+\alpha} \leq \overline{K}\overline{|f|}_\alpha,$$

where $\overline{K}$ is independent of m.

We can now proceed by arguments used in the proof of Theorem 3 (using the theorem of Ascoli-Arzela) in order to conclude that there exists a subsequence of $\{u_m\}$, denote it again by $\{u_m\}$, such that

(3.12) $$\{u_m\}, \quad \{D_x u_m\}, \quad \{D_x^2 u_m\}, \quad \{D_t u_m\}$$

are uniformly convergent in D. Furthermore, if $u_m \to u$ in D, then the sequences of the derivatives of the u_m converge to the corresponding derivatives of u, and $u \in \overline{C}_{2+\alpha}$.

Taking $m \to \infty$ in (3.10) we find that u satisfies (3.9). This completes the proof of (c).

In the proof of Theorem 7 we have made use only of the a priori boundary estimates (i.e., of Theorem 6, Sec. 2). In the next section we shall instead use only the a priori interior estimates (i.e., Theorem 5, Sec. 2) and thus derive the existence of a solution which belongs only to $C_{2+\alpha}$, but under weaker conditions on f, L, D and ψ than in Theorem 7.

4. Solution of the First Initial-boundary Value Problem (continued)

Let L be a uniformly parabolic operator with uniformly bounded coefficients in D.

Definition. A function w_Q ($Q \in \overline{B} + S$) is called a *barrier* at the point Q (with respect to L) if it satisfies the following conditions:

(1) w_Q is continuous in $\overline{D}$,
(2) $w_Q(P) > 0$ if $P \in \overline{D}$, $P \neq Q$,
(3) $w_Q(Q) = 0$, and
(4) $Lw_Q \leq -1$ in $D + B_T$ (the derivatives $D_x w_Q$, $D_x^2 w_Q$, $D_t w_Q$ are assumed to exist and to be continuous in $D + B_T$).

Definition. S is said to have *local barriers* (with respect to L) if there exists an $\epsilon > 0$ such that for every domain

$$D_{\tau,\epsilon} = D \cap \{\tau < t < \min(\tau + \epsilon, T)\},$$

where $0 \leq \tau < T$, there exists a barrier (with respect to L) at each point of

$$S_{\tau,\epsilon} = S \cap \{\tau < t \leq \min(\tau + \epsilon, T)\}.$$

We shall give some examples.

(a) If $Q = (x^0, 0) \in \overline{B}$ then

(4.1) $$w_Q(x, t) = (|x - x^0|^2 + At)e^{\gamma t}$$

is a barrier, provided $\gamma \geq c(x, t)$ and A is sufficiently large.

(b) Let $Q = (x^0, t^0) \in S$ and assume that there exists a closed ball K with center $(\bar{x}, \bar{t})$ such that $K \cap D = \phi$, $K \cap \bar{D} = \{Q\}$. If

$$\bar{x} \neq x \qquad \text{for all } (x, t) \in \bar{D} \tag{4.2}$$

then there exists a barrier w_Q at Q, namely,

$$w_Q(x, t) = ke^{\gamma t}\left(\frac{1}{R_0^p} - \frac{1}{R^p}\right), \tag{4.3}$$

where $\gamma \geq c(x, t)$, $R_0 = [|x^0 - \bar{x}|^2 + (t^0 - \bar{t})^2]^{1/2}$, $R = [|x - \bar{x}|^2 + (t - \bar{t}^2)]^{1/2}$, and k, p are appropriate positive numbers.

Indeed, as is easily verified,

$$Lw_Q(x, t) = \frac{kp}{R^{p+4}} e^{\gamma t} \cdot$$

$$[-(p + 2)\Sigma a_{ij}(x_i - \bar{x}_i)(x_j - \bar{x}_j) + R^2\Sigma a_{ii} + R^2\Sigma b_i(x_i - \bar{x}_i) - (t - \bar{t})R^2] + (c - \gamma)w_Q.$$

Since, by (4.2), $|x - \bar{x}| \geq \text{const.} > 0$, the expression in brackets is negative if p is sufficiently large. Hence $Lw_Q < 0$ in $\bar{D}$. Taking next k to be sufficiently large we get $Lw_Q \leq -1$ in D. The other properties of a barrier clearly hold for w_Q.

(c) We shall say that S has the *outside strong sphere property* if for every $Q = (x^0, t^0) \in S$ there exists a ball K with center $(\bar{x}, \bar{t})$ such that $K \cap \bar{D} = \{Q\}$ and

$$|\bar{x} - x| \geq \mu(Q) > 0 \text{ for all } (x, t) \in \bar{D}, \qquad |t - t^0| < \epsilon \tag{4.4}$$

where ϵ is independent of Q. By constructing the barrier of (4.3) for each Q and using (4.4) we obtain the following result.

Theorem 8. *If S has the outside strong sphere property then S has local barriers.*

If $\bar{S}$ has local representations of the form (2.17) where h is twice continuously differentiable, then the osculating spheres at the boundary points of S, which are tangent to S from the outside of D, have radii bounded from below by a positive constant. Since (4.4) is easily verified, the outside strong sphere property is satisfied. For the sake of future reference we state the following consequence.

Theorem 8′. *If D has the property* $(\bar{\bar{\text{E}}})$ *then S has local barriers.*

We shall now prove an analogue of Theorem 7, using the a priori interior estimates and the tool of barriers.

Theorem 9. *Let* (A), (B), (C) *hold and assume that S has local barriers with respect to both L and L_0. Then for any continuous function ψ on $\bar{B} + S$ there exists a unique solution u of the first initial-boundary value problem* (2.12), (2.20), *and $u \in C_{2+\alpha}$.*

Proof. The proof will first be given in the special case where $\psi \equiv 0$, and the method of continuity will be used. We define L_λ by (3.3) and denote by Σ the set of all values λ for which the problem (3.4) has a (unique) solution u in $C_{2+\alpha}$ for any $f \in C_\alpha$. We want to show:

(a) Σ contains $\lambda = 0$;
(b) Σ is an open set in the interval $0 \leq \lambda \leq 1$;
(c) Σ is a closed set.

The proof of (a) will be given in Chap. 4, Sec. 8. The proof of (b) is similar to the proof of (b) in Sec. 3 if we use the interior estimates instead of the boundary estimates. It thus remains to prove (c).

Taking a sequence $\{\lambda_m\}$, $\lambda_m \in \Sigma$, which converges to a point σ, we wish to show that $\sigma \in \Sigma$. The solutions u_m of (3.10) satisfy

$$|u_m|_{2+\alpha} \leq K|f|_\alpha \tag{4.5}$$

where K is independent of m. Hence we can extract a subsequence which, for simplicity, we again denote by $\{u_m\}$, such that the sequences (3.12) converge uniformly in closed subsets of $D + B_T$. Defining $u = \lim u_m$ we thus see that $L_\sigma u = 0$ in $D + B_T$. From (4.5) it also follows that $|u|_{2+\alpha} < \infty$ (by using arguments employed in the proof of Theorem 3, Sec. 2).

It remains to show that u is continuous up to $\overline{B} + S$ and that $u = 0$ on $\overline{B} + S$. The tool of barriers will now be used. Assume first that $Q \in \overline{B} + S$ and that there exists a barrier w_Q at Q (with respect to both L and L_0). Then, for any $0 \leq \lambda \leq 1$, the function $v = (\operatorname{l.u.b.}_D |f|)w_Q$ satisfies $L_\lambda v \leq -|f|$ in $\overline{D}$, $v \geq 0$ on $\overline{B} + S$. Applying the maximum principle to $v \pm u_m$ we get $v \pm u_m \geq 0$ in $\overline{D}$, i.e.,

$$|u_m(P)| \leq (\operatorname{l.u.b.}_D |f|)w_Q(P) \qquad \text{if } P \in \overline{D}. \tag{4.6}$$

Taking $m \to \infty$ we obtain, for each $P \in D + B_T$,

$$|u(P)| \leq (\operatorname{l.u.b.}_D |f|)w_Q(P),$$

from which it follows that $u(P) \to 0$ as $P \to Q$.

Now if $Q \in \overline{B}$ then the w_Q defined by (4.1) is a barrier. On the other hand, if $Q \in S_{2\epsilon}$ and ϵ is sufficiently small, then there exists in $D_{2\epsilon}$ a barrier at Q. Thus, we can assert that Theorem 9 (for $\psi \equiv 0$) is valid in $D_{2\epsilon}$ if ϵ is sufficiently small.

Next we consider the equation (2.12) with the boundary values zero and the initial values given by $u(x, \epsilon)$ on $t = \epsilon$. As before, we can solve this problem in $D \cap \{\epsilon \leq t \leq 3\epsilon\}$. Because of uniqueness, the solution must coincide in $D \cap \{\epsilon \leq t \leq 2\epsilon\}$ with the previous solution $u(x, t)$. We have thus proved the assertion of Theorem 9 in $D_{3\epsilon}$. Proceeding in this manner step by step, the proof of Theorem 9 readily follows, in case $\psi \equiv 0$.

Suppose next that $\psi \not\equiv 0$, but assume that $\psi \in \overline{C}_{2+\alpha}$. Setting $v = u - \psi$ we then immediately obtain, by the previous result, the existence of a solution v to $Lv = f - L\Psi$ (Ψ is an extension of ψ which belongs to $\overline{C}_{2+\alpha}$) which vanishes on $\overline{B} + S$ and which belongs to $C_{2+\alpha}$. The assertions for u then follow.

We finally consider the general case where ψ is only assumed to be continuous. By Theorem 2, Sec. 1, there exists a continuous function Ψ in R^{n+1} which coincides with ψ on $\overline{B} + S$. Let N be a rectangle $|x_i| \leq \nu$ $(i = 1, \ldots, n)$, $|t| \leq \nu$ which contains $\overline{D}$. By the Weierstrass approximation theorem there exists a sequence of polynomials Ψ_m which approximate Ψ uniformly in N. By what we have already proved, there exist solutions u_m to the problem

$$\begin{aligned} Lu_m &= f && \text{in } D + B_T, \\ u_m &= \Psi_m && \text{on } \overline{B} + S, \end{aligned} \tag{4.7}$$

and they belong to $C_{2+\alpha}$. By Theorem 5, Sec. 2,

$$|u_m - u_k|_{2+\alpha} \leq K|u_m - u_k|_0,$$

and by the maximum principle (see (2.3.9))

$$|u_m - u_k|_0 \leq e^{\gamma T} \max_{B+S} |\Psi_m - \Psi_k| \tag{4.8}$$

if $c(x, t) \leq \gamma$.

It follows that $\{u_m\}$ is a Cauchy sequence in $C_{2+\alpha}$. Hence, by Theorem 4, Sec. 2, there exists a function u in $C_{2+\alpha}$ which is the limit of $\{u_m\}$ in the norm of $C_{2+\alpha}$. It follows that $Lu = f$ in $D + B_T$. Furthermore, (4.8) shows that $\{u_m\}$ is uniformly convergent in $\overline{D}$ to u. Hence u is continuous in $\overline{D}$ and $u = \psi$ on $\overline{B} + S$. This completes the proof of Theorem 9.

In view of Theorem 8, we can state the following corollary to Theorem 9.

Corollary 1. *Let* (A), (B), (C) *hold and assume that S has the outside strong sphere property. Then for any continuous function ψ on $\overline{B} + S$ there exists a unique solution u of* (2.12), (2.20), *and* $u \in C_{2+\alpha}$.

For the sake of future reference we also state the following corollary to Theorem 9.

Corollary 2. *Let* (A), (B), (C) *hold, let D be a cylinder, and assume that for every point $x^0 \in \partial B$ there exists a closed ball C in R^n such that $C \cap \overline{B} = \{x^0\}$. Then for any continuous function ψ on $\overline{B} + S$ there exists a unique solution u of* (2.12), (2.20), *and* $u \in C_{2+\alpha}$.

5. Differentiability of Solutions

We recall that by a solution of a parabolic equation $Lu = f$ in a domain D we understand a function which satisfies the equation $Lu = f$ in D, and all the derivatives of which appearing in Lu (i.e., $D_x u$, $D_x^2 u$, $D_t u$) are continuous functions in D. With the aid of Theorem 5, Sec. 2, we shall now prove that, roughly speaking, the solution is as smooth as the coefficients of the equation.

Theorem 10. *Let L be a parabolic operator in a domain D and assume that*

$$D_x^m a_{ij}, \quad D_x^m b_i, \quad D_x^m c, \quad D_x^m f \qquad (0 \leq m \leq p) \tag{5.1}$$

are Hölder continuous (exponent α) in D. If u is a solution of $Lu = f$ in D, then

$$D_x^m u, \quad D_t D_x^k u \qquad (0 \leq m \leq p + 2, 0 \leq k \leq p) \tag{5.2}$$

exist and are Hölder continuous (exponent α) in D.

Proof. We first prove the theorem for $p = 0$. Let D_0 be a cylinder whose base B_0 is a ball, and let $\overline{D}_0$ be contained in D. Denoting the lateral boundary of D_0 by S_0, consider the problem

$$\begin{aligned} Lv &= f && \text{in } D_0, \\ v &= u && \text{on } B_0 + S_0. \end{aligned}$$

By Corollary 2 to Theorem 9, Sec. 4, the solution v exists and belongs to $C_{2+\alpha}(D_0)$. Since, by uniqueness, $v \equiv u$ in D_0, it follows that the functions (5.2), for $p = 0$, are Hölder continuous (exponent α) in the interior of D_0. Since D_0 is an arbitrary cylinder, the proof is completed.

We next prove the theorem for $p = 1$. Let B, C be two open cylinders with bases lying on hyperplanes $t = \text{const.}$, such that

$$\overline{B} \subset C \subset \overline{C} \subset D.$$

For any point $P = (x_1, \ldots, x_n, t)$ denote by P_h the point

$$(x_1 + h, x_2, \ldots, x_n, t).$$

If $|h|$ is sufficiently small and P belongs to B then P_h belongs to C. Set

$$g^h(P) = \frac{g(P_h) - g(P)}{h}.$$

It is readily verified that

$$\begin{aligned} \sum_{i,j=1}^{n} a_{ij}(P) \frac{\partial^2 u^h(P)}{\partial x_i \partial x_j} + \sum_{i=1}^{n} b_i(P) \frac{\partial u^h(P)}{\partial x_i} + c(P)u^h(P) - \frac{\partial u^h(P)}{\partial t} \\ = f^h(P) - \sum_{i,j=1}^{n} a_{ij}^h(P) \frac{\partial^2 u(P_h)}{\partial x_i \partial x_j} - \sum_{i=1}^{n} b_i^h(P) \frac{\partial u(P_h)}{\partial x_i} - c^h(P)u(P_h) \\ \equiv F_h(P). \end{aligned} \tag{5.3}$$

Writing

$$f^h(P) = \frac{1}{h} \int_{x_1}^{x_1+h} D_\zeta f(\zeta, x_2, \ldots, x_n, t)\, d\zeta,$$

one can easily verify, with the aid of the mean value theorem, that

$$|d^2 f^h|_\alpha^B \leq M_1 |f|_{1+\alpha}^C < \infty, \tag{5.4}$$

where M_i are used to denote constants independent of h, and $|h|$ is assumed to be sufficiently small.

In a similar way we obtain

$$\begin{gathered} |a_{ij}^h|_\alpha^B \leq M_2 |a_{ij}|_{1+\alpha}^C < \infty, \qquad |db_i^h|_\alpha^B \leq M_3 |b_i|_{1+\alpha}^C < \infty, \\ |d^2 c^h|_\alpha^B \leq M_4 |c|_{1+\alpha}^C < \infty. \end{gathered} \tag{5.5}$$

Observing next that for any function g,

$$|d^2 g(P_h)|_\alpha^B \leq M_5 |d^2 g|_\alpha^C,$$

we get

$$|d^2 D_x^2 u(P_h)|_\alpha^B \leq M_5 |d^2 D_x^2 u|_\alpha^C < \infty, \tag{5.6}$$

since, by the result for $p = 0$, $D_x^2 u$ is Hölder continuous (exponent α) in D. Similarly we have

$$\begin{gathered} |d^2 D_t u(P_h)|_\alpha^B \leq M_5 |d^2 D_t u|_\alpha^C < \infty, \\ |dD_x u(P_h)|_\alpha^B \leq M_6 |dD_x u|_\alpha^C < \infty, \\ |u(P_h)|_\alpha^B \leq M_7 |u|_\alpha^C < \infty. \end{gathered} \tag{5.7}$$

We can now estimate F_h. Using the easily verified inequality

$$|d^2(gh)|_\alpha \leq |d^2 g|_\alpha |h|_\alpha \tag{5.8}$$

and (5.5), (5.6), we get

$$\left| d^2 \left[\sum_{i,j=1}^{n} a_{ij}^h(P) \frac{\partial^2 u(P_h)}{\partial x_i \, \partial x_j} \right] \right|_\alpha^B \leq M_8 < \infty.$$

The other terms in F_h are estimated in a similar manner, making use of (5.5), (5.7), and (5.4). We get

$$|d^2 F_h(P)|_\alpha^B \leq M_9 < \infty. \tag{5.9}$$

We now apply Theorem 5, Sec. 2, to the solution u^h of (5.3) in the domain B, noting that u^h belongs to $C_{2+\alpha}(B)$. Since

$$|u^h|_0^B \leq M_{10} |u|_1^C < \infty,$$

we obtain the inequality

$$|u^h|_{2+\alpha}^B \leq M_{11} < \infty. \tag{5.10}$$

The theorem of Ascoli-Arzela can now be used to show that for any sequence $\{h_m\}$ converging to zero there exists a subsequence $\{h_{m'}\}$ such that

$$u^{h_{m'}} \to v, \quad D_x u^{h_{m'}} \to D_x v, \quad D_x^2 u^{h_{m'}} \to D_x^2 v, \quad D_t u^{h_{m'}} \to D_t v$$

in B, as $m' \to \infty$, and the convergence is uniform in closed subsets of B. Since $u^h \to \partial u/\partial x_1$ (pointwise) as $h \to 0$,

$$\frac{\partial u}{\partial x_1}, \quad D\frac{\partial u}{\partial x_1}, \quad D_x^2\frac{\partial u}{\partial x_1}, \quad D_t\frac{\partial u}{\partial x_1}$$

exist (in B) and are equal to the corresponding derivatives of v, and

$$\left|\frac{\partial u}{\partial x_1}\right|_{2+\alpha}^{B} \leq M_{11}.$$

A similar result can be proved also for $\partial u/\partial x_j$, where $j = 2, \ldots, n$. This completes the proof of the theorem for $p = 1$.

To prove the theorem for $p = 2$ we differentiate the equation $Lu = f$ once with respect to x_j and then apply the result for $p = 1$ to the differentiated equation. The proof for any p is obtained in a similar manner, namely, we differentiate the equation once with respect to x_j and apply the inductive assumption to the differentiated equation.

Writing the parabolic equation in the form

$$\frac{\partial u}{\partial t} = \sum a_{ij}\frac{\partial^2 u}{\partial x_i\,\partial x_j} + \sum b_i\frac{\partial u}{\partial x_i} + cu - f \tag{5.11}$$

and assuming that, in addition to the assumptions made in Theorem 10, $\partial a_{ij}/\partial t$, $\partial b_i/\partial t$, $\partial c/\partial t$ and $\partial f/\partial t$ are Hölder continuous (exponent α) together with their first $p - 2$ x-derivatives, it follows that $D_x^m D_t^2 u$ are Hölder continuous, for $0 \leq m \leq p - 2$. Repeating this argument q times ($2q \leq p$) we arrive at the following theorem.

Theorem 11. *Let L be a parabolic operator in a domain D and assume that*

$$D_x^m D_t^k a_{ij}, \quad D_x^m D_t^k b_i, \quad D_x^m D_t^k c, \quad D_x^m D_t^k f \qquad (0 \leq m + 2k \leq p, k \leq q) \tag{5.12}$$

are Hölder continuous (exponent α) in D. If u is a solution of $Lu = f$ in D then

$$D_x^m D_t^k u \qquad (0 \leq m + 2k \leq p + 2, k \leq q + 1) \tag{5.13}$$

exist and are Hölder continuous (exponent α) in D.

If one reads carefully the proofs of Theorems 10, 11 one sees that the theorems remain true if the domain D is replaced everywhere by $D + B_T$ (The Hölder continuity on $D + B_T$ is taken in the local sense.). We state this as a corollary.

Corollary 1. *If L is parabolic in $D + B_T$ and if the functions* (5.12) *are Hölder continuous (exponent α) in $D + B_T$ then for any solution u of $Lu = f$ in $D + B_T$ the functions* (5.13) *are Hölder continuous (exponent α) in $D + B_T$.*

Corollary 2. *If f and all the coefficients of a parabolic operator L in a domain D (region $D + B_T$) are infinitely differentiable functions, then all the solutions of $Lu = f$ in D ($D + B_T$) are also infinitely differentiable functions.*

Theorem 11 can be extended to differentiability up to the boundary $B + S$ (for the differentiability on B_T see Corollary 1 above). The result is stated in the following theorem.

Theorem 12. *Let L be a parabolic operator in $\overline{D}$, D satisfying the property $(\overline{\text{E}})$, and assume that, for some $p \geq 0$, the functions*

$$D_x^m D_t^k a_{ij}, \quad D_x^m D_t^k b_i, \quad D_x^m D_t^k c, \quad D_x^m D_t^k f \qquad (0 \leq m + k \leq p) \tag{5.14}$$

are uniformly Hölder continuous (exponent α) in every domain whose closure is contained in $D + B + S + B_T$. Assume further that the functions h which appear in the local representations of S (see (2.17)) are such that

$$D_x^{m+2} D_t^k h, \quad D_x^m D_t^{k+1} h \qquad (m \geq -2, k \geq -1, m + k \leq p) \tag{5.15}$$

are Hölder continuous (exponent α). Assume finally that $\psi \in \overline{C}_{2+\alpha}$, that $L\psi = f$ on ∂B, and that, as a function of the local parameters of S, ψ is a function (of $(x_1, \ldots, x_{i-1}, x_{i+1}, \ldots, x_n, t)$) satisfying the condition that

$$D_x^{m+2} D_t^k \psi, \quad D_x^m D_t^{k+1} \psi \qquad (m \geq -2, k \geq -1, m + k \leq p) \tag{5.16}$$

are Hölder continuous (exponent α), whereas on B

$$D_x^{m+2} \psi \qquad (-2 \leq m \leq p) \tag{5.17}$$

are Hölder continuous (exponent α). If u is the solution of (2.12), (2.20), then in every domain whose closure is contained in $D + B + S + B_T$, the functions

$$D_x^{m+2} D_t^k u, \quad D_x^m D_t^{k+1} u \qquad (m \geq -2, k \geq -1, m + 2k \leq p) \tag{5.18}$$

are uniformly Hölder continuous (exponent α).

From the proof it will be seen that the assertion remains true if instead of the original assumption on the functions in (5.14) we only assume that for every domain D_0, with $\overline{D}_0 \subset D + B + S + B_T$, the functions (5.14) are uniformly Hölder continuous (exponent α) in the intersection of D_0 with some neighborhood of S, and the functions (5.12) (with $2q > p$) are uniformly Hölder continuous (exponent α) in the complementary part of D_0.

Proof. The proof for $p = 0$ follows from Theorem 7, Sec. 3, and the uniqueness of solutions. We proceed to prove the theorem for $p = 1$. We shall show that:

(a) for every $Q \in S$ there exists a neighborhood V such that the functions (5.18) (for $p = 1$) are uniformly Hölder continuous (exponent α) in $D \cap V$, and
(b) for every $Q \in B$ there exists a neighborhood W (in R^{n+1}) such that the functions (5.18) (for $p = 1$) are uniformly Hölder continuous (exponent α) in $D \cap W$.

If we then use Corollary 1 of Theorem 11, which asserts the uniform Hölder continuity of the functions (5.18) in every domain whose closure is contained in $D + B_T$, then the proof of Theorem 12 follows.

It suffices to prove (a) in case Q lies on a portion of S which is a domain on $x_1 = 0$. Indeed, this follows by making a transformation

$$(5.19) \qquad \begin{aligned} y_1 &= x_i - h(x_1, \ldots, x_{i-1}, x_{i+1}, \ldots, x_n, t), \\ y_i &= x_1 \qquad \text{if } i \neq 1, \\ y_j &= x_j \qquad \text{if } j \neq 1, j \neq i. \end{aligned}$$

Since all the considerations which will follow make use of the differentiability assumptions of the theorem only in $D \cap V_0$, where V_0 is some neighborhood of Q, all our assumptions remain unchanged in the new coordinates.

Next we may assume that $\psi = 0$ in $V_1 \cap S$ (V_1 being some neighborhood of Q) since otherwise we extend the definition of ψ to some neighborhood of Q by

$$\psi(x_1, x_2, \ldots, x_n, t) = \psi(x_2, \ldots, x_n, t)$$

and consider the function $u - \psi$.

Let S_* be a cube $x_i^1 < x_i < x_i^2$ $(i = 2, \ldots, n)$, $t_1 < t < t_2$ on S with Q in its interior and let D_0 be a cylinder with lower base B_0 and upper base B_1, such that $D_0 = B_0 \times (t_1, t_2)$ and such that S_* is a part of the lateral boundary S_0 of D_0. We take the boundary of B_0 to be three times continuously differentiable so that the property ($\bar{\text{E}}$) holds for D_0 (and Theorem 6 may therefore be applied). We further take D_0 to be such that its closure is contained in $D + \bar{S}_*$.

For the sake of clarity consider first the special case where

$$(5.20) \qquad u = 0 \qquad \text{in } W_0 \cap D_0,$$

W_0 being some neighborhood (in R^{n+1}) of $\bar{B}_0 + \bar{B}_1 + (S_0 - S_*)$.

We can then proceed similarly to the proof of Theorem 10. We define for any point $P = (x_1, \ldots, x_n, t)$ a translation

$$P_h = (x_1, \ldots, x_{i-1}, x_i + h, x_{i+1}, \ldots, x_n, t)$$

where $i > 1$, and for B, C we take subdomains of D_0 whose boundary consists of two parts: one part lies in the half-space $x_1 < 0$ (if D_0 lies in $x_1 < 0$) and is contained in W_0 and the other part lies in S_*. The equation (5.3) remains the same. We next apply to u^h the boundary estimates of Theorem 6, Sec. 2, noting that the initial and boundary values of u^h, in D_0, are zero. By calculation similar to that given in the proof of Theorem 10 we can also estimate $\overline{|F_h|}_\alpha$. The conclusion that

$$D_x^2 \frac{\partial u}{\partial x_i}, \qquad D_t \frac{\partial u}{\partial x_i}$$

are uniformly Hölder continuous (exponent α) in D_0 also follows by the same argument as in the proof of Theorem 10.

The previous finite-difference procedure can also be accomplished with respect to the variable t, i.e., we define

$$P_h = (x_1, \ldots, x_n, t + h)$$

and proceed as before. We then conclude that

$$D_x^2 D_t u, \qquad D_t^2 u \tag{5.21}$$

are Hölder continuous (exponent α).

In order to complete the proof of (a) we need to establish the existence and the Hölder continuity of $\partial^3 u/\partial x_1^3$. This is done using the equation $Lu = f$. Indeed, writing the equation in the form

$$a_{11} \frac{\partial^2 u}{\partial x_1^2} = -\sum_{(i,j) \neq (1,1)} a_{ij} \frac{\partial^2 u}{\partial x_i \, \partial x_j} - \sum b_i \frac{\partial u}{\partial x_i} - cu + \frac{\partial u}{\partial t} + f$$

we see that the right-hand side has an x_1-derivative which is Hölder continuous; since $a_{11} > 0$ in $\bar{D}$, the same is true of $\partial^2 u/\partial x_1^2$. The proof of (a), under the assumption (5.20), is thereby completed.

The previous considerations can be extended to the general case where the assumption (5.20) is not being made, provided instead of using Theorem 6, Sec. 2, we use the stronger result on boundary estimates given in Theorem 4, Sec. 7, Chap. 4 (with D being D_0, and R, R_0 being some neighborhoods of Q in S_*). Instead of establishing the uniform Hölder continuity in D_0 of the functions in (5.18) (with $p = 1$), we now establish their uniform Hölder continuity in the intersection of D_0 with some $(n + 1)$-dimensional neighborhood of Q.

The proof of (b) is similar. The finite differences are now taken with respect to $x_1, \ldots, x_n$. The existence and the Hölder continuity of the functions $D_x D_t u$ is finally deduced from the differential equation $Lu = f$.

Having proved the theorem for $p = 1$, we proceed to prove it for $p = 2$ as follows: first we note that it suffices to establish only the analogues of (a), (b) for $p = 2$. To prove the analogue of (a), for $p = 2$, we differentiate the equation $Lu = f$ once with respect to any one of the variables $x_2, \ldots, x_n, t$ and apply the result for $p = 1$. Then we use the differential equation $Lu = f$ to derive the Hölder continuity of the remaining derivative. In the case of (b) we differentiate the equation with respect to any of the variables $x_1, \ldots, x_n$, then apply the result for $p = 1$, and finally use the equation $Lu = f$ to derive the Hölder continuity of the remaining derivative.

We leave it to the reader to verify that in this way we obtain the Hölder continuity of all the functions in (5.18), for $p = 2$.

The proof of the theorem can be completed by induction from p to

$p + 1$, using the same arguments as in the passage from $p = 1$ to $p = 2$. The details are left to the reader.

From the proof of (a) (in particular, the Hölder continuity of the functions in (5.21)) and from the argument given for $p > 1$, we obtain the following corollary.

Corollary 1. *Let S_0, S_1 be two open submanifolds of S, $\overline{S}_0 \subset S_1$, and let V_0, V_1 be open neighborhoods of S_0, S_1 respectively, such that $\overline{V}_0 \subset V_1$. Set $D_0 = D \cap V_0$, $D_1 = D \cap V_1$ and assume that $\overline{D}_0 \subset D_1 + \overline{S}_0$. Assume that for some $p \geq 1$ the functions* (5.14) *are Hölder continuous (exponent α) in $\overline{D}_1$, and that the functions* (5.15), (5.16) *are Hölder continuous (exponent α)* on *S_1. If $u \in \overline{C}_{2+\alpha}(D_1)$ then the functions* (5.18), *for all $m \geq -2$, $k \geq -1$, $m + k \leq p$, are uniformly Hölder continuous (exponent α) in D_0.*

If we only assume that $u \in \overline{C}_{2+\epsilon}(D_1)$ for some $\epsilon > 0$ ($\epsilon < \alpha$) then from the Hölder continuity (exponent ϵ) of the functions (5.18) it follows that $u \in \overline{C}_{2+\alpha}(D_2)$ for any D_2 with $\overline{D}_2 \subset D_1 + S_1$. Hence, the assertion remains true if we only assume that $u \in \overline{C}_{2+\epsilon}(D_1)$ for some $\epsilon > 0$.

The assertion of the corollary is valid also if u is not assumed to belong to $\overline{C}_{2+\epsilon}(D_1)$; see Theorem 8, Sec. 8, Chap. 4.

Corollary 1 and the remarks following it extend also to differentiability near B. The precise statements are omitted.

Corollary 2. *Let L be a parabolic operator in $\overline{D}$, D having the property $(\overline{\overline{\text{E}}})$, and assume that f and the coefficients of L are infinitely differentiable functions in $\overline{D}$, that the functions h in the local representations of S are infinitely differentiable functions, and that ψ is infinitely differentiable on S and on B. Assume finally that $\psi \in \overline{C}_{2+\alpha}$ and that $L\psi = f$ on ∂B. Then the solution of* (2.12), (2.20) *is an infinitely differentiable function in $D + B + S + B_T$.*

We shall extend Theorems 10–12 to nonlinear parabolic equations

$$F\left(x, t, u, \frac{\partial u}{\partial x_i}, \frac{\partial^2 u}{\partial x_i \, \partial x_j}, \frac{\partial u}{\partial t}\right) = 0. \tag{5.22}$$

We say that (5.22) is a *parabolic equation in D with respect to a solution* $u(x, t)$ if $b(x, t) \equiv \partial F/\partial(\partial u/\partial t)$ is strictly negative in D, the argument of F being

$$\left(x, t, u(x, t), \frac{\partial u(x, t)}{\partial x_i}, \frac{\partial^2 u(x, t)}{\partial x_i \, \partial x_j}, \frac{\partial u(x, t)}{\partial t}\right), \tag{5.23}$$

and if the matrix (b_{hk}), where

$$b_{hk} = \left(\frac{\partial^2 F}{\partial(\partial^2 u/\partial x_h \, \partial x_k)}\right),$$

is positive definite in D, the argument of F being given by (5.23).

Substituting $u = u(x, t) + \epsilon v$ in (5.22) and then dividing by ϵ and letting $\epsilon \to 0$, we find the "first variation" (with respect to $u(x, t)$)

$$(5.24) \qquad Lv \equiv \sum a_{ij} \frac{\partial^2 v}{\partial x_i \, \partial x_j} + \sum b_i \frac{\partial v}{\partial x_i} + cv - \frac{\partial v}{\partial t} = 0$$

where $a_{ij} = -b_{ij}/b$, $b_i = -[\partial F/\partial(\partial u/\partial x_i)]/b$, $c = -(\partial F/\partial u)/b$. If $b < 0$, L is a parabolic operator if and only if (5.22) is parabolic (with respect to $u(x, t)$).

If (5.22) is parabolic in D with respect to every solution $u(x, t)$, then we say that it is *parabolic in* D.

In what follows, when we speak of Hölder continuity for functions $F(x, t, u_1, \ldots, u_N)$, we understand the distance between two points $(x, t, u_1, \ldots, u_N)$, $(\bar{x}, \bar{t}, \bar{u}_1, \ldots, \bar{u}_N)$ to be

$$[|x - \bar{x}|^2 + |t - \bar{t}| + (u_1 - \bar{u}_1)^2 + \cdots + (u_N - \bar{u}_N)^2]^{1/2}.$$

Theorem 13. *Let $u(x, t)$ be a solution of (5.22) in D and assume that $\partial F/\partial x$, $\partial F/\partial u_j$ $(j = 1, \ldots, N)$ are Hölder continuous (exponent α) in $(x, t, u_1, \ldots, u_N)$ together with all their derivatives up to order $p \geq 0$ with respect to $x, u_1, \ldots, u_N$ for $(x, t) \in D$, $-\infty < u_i < \infty$ $(i = 1, \ldots, N)$. If (5.22) is parabolic in D with respect to $u(x, t)$, and if $D_x u$, $D_x^2 u$, $D_t u$ are Hölder continuous in D with some exponent $\epsilon > 0$, then*

$$(5.25) \qquad D_x^m u, \quad D_t D_x^k u \qquad (0 \leq m \leq p + 3, 0 \leq k \leq p + 1)$$

exist and are Hölder continuous (exponent α) in D.

Proof. Beginning as in the proof of Theorem 10 and using the mean value theorem we get for u^h a linear parabolic equation $L_h u^h = f_h$ with coefficients expressible in terms of the first derivatives of F with respect to $x, u_1, \ldots, u_N$. Observing that a Hölder continuous (exponent α) function of another Hölder continuous (exponent β) function is also Hölder continuous (exponent $\alpha\beta$), we conclude that f_h and the coefficients of L_h are Hölder continuous functions of (x, t) with some exponent ϵ'. By the proof of Theorem 10 we get (see (5.10))

$$(5.26) \qquad |u^h|_{2+\epsilon'}^B \leq N_1,$$

where N_i are constants independent of h. We then conclude that

$$(5.27) \qquad \left| \frac{\partial u}{\partial x_j} \right|_{2+\epsilon'}^B \leq N_1$$

for $j = 1$. The proof for $j = 2, \ldots, n$ is similar.

From the equation $L_h u^h = f_h$ we also get

$$(5.28) \qquad |D_t u^h|_{\epsilon'}^B \leq N_2,$$

from which it follows that

$$(5.29) \qquad |D_t D_x u|_{\epsilon'}^B \leq N_3.$$

From (5.27), (5.28) we see that

$$u \in C_{2+1}, \tag{5.30}$$

where C_{2+1} is defined exactly as $C_{2+\alpha}$ but with $\alpha = 1$.

Using (5.30) we can re-examine f_h and the coefficients of L_h and find that they are Hölder continuous with exponent α. We can therefore use the previous proof with ϵ' replaced by α. We thus obtain (5.27), (5.29) with $\epsilon' = \alpha$ and the proof, for $p = 0$, is completed.

If $p \geq 1$ we differentiate (5.22) once with respect to any of the x_j and then apply Theorem 10 to the resulting linear parabolic equation.

We shall not stop to formulate the analogue of Theorem 11, but only mention the following special case.

Theorem 14. *If* (5.22) *is parabolic in* D *with respect to a solution* u *which belongs to* $\overline{C}_{2+\epsilon}$ $(\epsilon > 0)$ *in closed subsets of* D *and if* F *is infinitely differentiable for* $(x, t) \in D$, $-\infty < u_i < \infty$ $(i = 1, \ldots, N)$, *then* $u(x, t)$ *is infinitely differentiable in* D.

Theorem 12 and its proof also extend to nonlinear equations.

6. Families of Solutions

With the aid of Theorem 5, Sec. 2, one easily establishes the following theorem.

Theorem 15. *Let* $\{L_m\}$ *be a sequence of parabolic operators satisfying* (A), (B) *with constants* K_1, K_2 *independent of* m, *and let* $\{f_m\}$ *be a sequence of functions satisfying* $|f_m|_\alpha^D \leq K_3$ *where* K_3 *is independent of* m. *Suppose that* $\{u_m\}$ *is a sequence of functions satisfying*

$$L_m u_m = f_m \qquad \text{in } D + B_T.$$

If $|u_m|_0^D \leq K_4$ *where* K_4 *is independent of* m, *then for any subsequence* $\{u_{m'}\}$ *of* $\{u_m\}$ *there exists a subsequence of it, say* $\{u_{m''}\}$, *such that*

$$u_{m''}, \quad D_x u_{m''}, \quad D_x^2 u_{m''}, \quad D_t u_{m''} \tag{6.1}$$

are uniformly convergent in any subdomain of D, *whose closure lies in* $D + B_T$, *to some function* u *and its corresponding derivatives. Furthermore,* $u \in C_{2+\alpha}(D)$.

If, in particular, the coefficients of L_m *converge to the corresponding coefficients of* L *and* $\{f_m\}$ *converges to* f, *pointwise in* $D + B_T$, *then* $Lu = f$ *in* $D + B_T$.

Corollary. *If* (A), (B), (C) *hold and*

$$Lu_m = f \quad \text{in } D + B_T, \qquad u_m = \psi_m \quad \text{on } \overline{B} + S \tag{6.2}$$

and if $\psi_m \to \psi$ *uniformly on* $\overline{B} + S$, *then the sequence* $\{u_m\}$ *is uniformly con-*

vergent in $\bar{D}$ *to a function* u, *the derivatives* $D_x u_m$, $D_x^2 u_m$, $D_t u_m$ *converge uniformly in closed subsets of* $D + B_T$ *to the corresponding derivatives of* u, *and*

$$Lu = f \quad \text{in } D + B_T, \qquad u = \psi \quad \text{on } \bar{B} + S. \tag{6.3}$$

Indeed, by the maximum principle,

$$|u_m - u_k|_0^D \to 0 \qquad \text{as } m, k \to \infty.$$

Thus, $\lim u_m = u$ exists. Clearly $u = \psi$ on $\bar{B} + S$. By Theorem 15 $Lu = f$ in $D + B_T$. Since u is the unique limit of $\{u_m\}$, it follows that all the subsequences (6.1) converge to the same limits. This completes the proof of the corollary.

Results analogous to Theorem 15 and its corollary, where the norm $|\cdot|_\alpha$ is replaced by $\lceil\cdot\rceil_\alpha$, are also valid.

The corollary is analogous to Harnack's first theorem for harmonic functions.

7. Green's Function

We begin with a general useful lemma.

Lemma 1. *If* L *is a parabolic operator with Hölder continuous (exponent* α*) coefficients in* $\bar{D}$, *then the coefficients of* L *can be extended into uniformly Hölder continuous (exponent* α*) functions in the whole strip* $0 \le t \le T$ *such that the corresponding extension of* L *is uniformly parabolic in the strip* $0 \le t \le T$.

Proof. By the italicized statement following Theorem 2, Sec. 1, we can extend the coefficients of L into the strip $0 \le t \le T$ so that the extended functions are uniformly Hölder continuous in $0 \le t \le T$. Denote by L_1 the corresponding extension of L. L_1 is parabolic in a closed neighborhood D_0 of $\bar{D}$ in $0 \le t \le T$. Let ζ_1 be a continuously differentiable function in $0 \le t \le T$ such that $\zeta_1 \equiv 1$ on D and $\zeta_1 \equiv 0$ outside D_0 (see Chap. 1, Problem 1). Similarly let ζ_2 be a continuously differentiable function in the strip $0 \le t \le T$ such that $\zeta_2 \equiv 0$ in $\bar{D}$ and $\zeta_1 + \zeta_2 \ge$ const. > 0 in $0 \le t \le T$. If L_0 denotes the heat operator, then the desired extension of L is given by

$$\frac{\zeta_1}{\zeta_1 + \zeta_2} L_1 + \frac{\zeta_2}{\zeta_1 + \zeta_2} L_0.$$

Let L be a uniformly parabolic operator in a domain D. In Chap. 1 we have defined a fundamental solution of $Lu = 0$ in a domain D, provided D is a cylinder. If D is not a cylinder, we extend L (by Lemma 1) into a cylindrical domain D_0 in $0 < t < T$ and then we can construct a fundamental solution $\Gamma(x, t; \xi, \tau)$. It clearly satisfies $Lu = 0$ as a function of $(x, t) \in D + B_T$, and also

(7.1) $$\int_{B_\tau} \Gamma(x, t; \xi, \tau) f(\xi)\, d\xi \to f(\xi) \qquad \text{as } t \to \tau \qquad (\xi \in B_\tau)$$

for any continuous function f on $\overline{B}_\tau$.

From the construction of $\Gamma(x, t; \xi, \tau)$ in Chap. 1 it is seen that Γ, $D_x\Gamma$, $D_x^2\Gamma$, $D_t\Gamma$ are continuous functions of $(x, t; \xi, \tau)$ (see the paragraph preceding (1.4.29)). Hence the left-hand side of (7.1) is a solution of $Lu = 0$.

Definition. A function $G(x, t; \xi, \tau)$ defined and continuous for $(x, t; \xi, \tau) \in \overline{D} \times (D \cup B)$, $t > \tau$, is called a *Green's function* of $Lu = 0$ in D if for any $0 \leq \tau < T$ and for any continuous function f on B_τ having a compact support, the function

(7.2) $$u(x, t) = \int_{B_\tau} G(x, t; \xi, \tau) f(\xi)\, d\xi$$

is a solution of $Lu = 0$ in $D \cap \{\tau < t \leq T\}$ and it satisfies the initial and boundary conditions

(7.3) $$\lim_{t \searrow \tau} u(x, t) = f(x) \qquad \text{for } x \in \overline{B}_\tau,$$

(7.4) $$u(x, t) = 0 \qquad \text{on } S \cap \{\tau < t \leq T\}.$$

Theorem 16. *Let L satisfy* ($\bar{\text{A}}$), (B) *and let D have the property* ($\bar{\bar{\text{E}}}$). *Then there exists a unique Green's function $G(x, t; \xi, \tau)$. Furthermore, G, D_xG, D_x^2G, D_tG are continuous functions of $(x, t; \xi, \tau)$ in $(D + B_T) \times (D + B)$, $t > \tau$.*

Proof. To prove the existence of G we first extend the coefficients of L into $0 \leq t \leq T$, by using Lemma 1. By the results of Chap. 1, there exists a fundamental solution $\Gamma(x, t; \xi, \tau)$ of the extended L and, furthermore, Γ, $D_x\Gamma$, $D_x^2\Gamma$, $D_t\Gamma$ are continuous functions of $(x, t; \xi, \tau)$, $t > \tau$.

Let $V(x, t; \xi, \tau)$ denote the solution (if existing) of the problem

(7.5) $$\begin{aligned} LV &= 0 && \text{in } D \cap \{\tau < t \leq T\}, \\ V &= 0 && \text{on } \overline{B}_\tau, \\ V(x, t; \xi, \tau) &= -\Gamma(x, t; \xi, \tau) && \text{on } S \cap \{\tau < t \leq T\}. \end{aligned}$$

Since $\Gamma(x, t; \xi, \tau) \to 0$ as $t \to \tau$, provided $x \neq \xi$, the initial and boundary values of V form a continuous function. Theorems 8′, 9, Sec. 4, can therefore be used to conclude that a unique solution V in fact exists.

We shall next prove that the functions V, D_xV, D_x^2V, D_tV are continuous functions of $(x, t; \xi, \tau)$ in $(D + B_T) \times (D + B)$, $t > \tau$.

Let (ξ^*, τ^*) be a point in D whose distance to (ξ, τ) is less than δ. Assume, for definiteness, that $\tau^* \geq \tau$. Since the initial values of $V(x, t; \xi, \tau)$ vanish, and since the boundary values on $S \cap \{\tau \leq t \leq \tau^*\}$ can be made arbitrarily small if δ is sufficiently small, it follows (by the maximum principle) that, on $t = \tau^*$, $V(x, t; \xi, \tau) = V(x, t; \xi, \tau) - V(x, t; \xi^*, \tau^*)$ can be made arbitrarily small in absolute value, if δ is sufficiently small. The same is obviously true for $V(x, t; \xi, \tau) - V(x, t; \xi^*, \tau^*)$ on $S \cap \{\tau^* < t \leq T\}$. By the maximum principle it follows that

$$(7.6) \qquad |V(x, t; \xi, \tau) - V(x, t; \xi^*, \tau^*)| < \epsilon \qquad \text{in } D \cap \{\tau^* < t \leq T\}$$

if δ is sufficiently small. Note that δ is independent of (ξ, τ) provided (ξ, τ) is restricted to a closed subset of $D + B$.

The continuity of $D_x V$, $D_x^2 V$, $D_t V$ in (ξ, τ) now follows by applying the interior estimates (2.16) to $V(x, t; \xi, \tau) - V(x, t; \xi^*, \tau^*)$. Note that the continuity is uniform with respect to (x, t) in closed subsets of $D + B_T$ and with respect to (ξ, τ) in closed subsets of $D + B$, provided $t - \tau \geq \mu$ for some $\mu > 0$. The continuity of V is uniform also when (x, t) varies in closed subsets of $D + B_T + S$. Since finally, V, $D_x V$, $D_x^2 V$, $D_t V$ are continuous functions of (x, t) in $D + B_T$, $t > \tau$, for each fixed $(\xi, \tau) \in D$, they are also continuous functions of $(x, t; \xi, \tau)$ in $(D + B_T) \times (D + B)$ $(t > \tau)$ as claimed above.

It is now easily seen that the function

$$G(x, t; \xi, \tau) = \Gamma(x, t; \xi, \tau) + V(x, t; \xi, \tau)$$

is a Green's function. Furthermore, $D_x G$, $D_x^2 G$, $D_t G$ are continuous functions of $(x, t; \xi, \tau) \in (D + B_T) \times (D + B)$, $t > \tau$.

To prove uniqueness, suppose G_1 and G_2 are two Green's functions and consider the function

$$v(x, t) = \int_{B_\tau} [G_1(x, t; \xi, \tau) - G_2(x, t; \xi, \tau)] f(\xi) \, d\xi$$

where $f(x)$ is any continuous function with compact support in B_τ. Clearly, $Lv = 0$ and $v = 0$ on $S \cap \{\tau < t \leq T\}$ and on $\overline{B}_\tau$. Hence, by the maximum principle, $v \equiv 0$ in $D \cap \{\tau < t \leq T\}$.

Taking a sequence $\{f_m(\xi)\}$ as in the proof of Theorem 11, Sec. 4, Chap. 2, we then conclude that $G_1(x, t; \xi, \tau) = G_2(x, t; \xi, \tau)$.

Corollary 1. *As a function of $(x, t) \in D$, $LG = 0$. Furthermore, for each $(\xi, \tau) \in D + B$, $G(x, t; \xi, \tau) = 0$ on $S \cap \{\tau < t \leq T\}$ and $G(x, t; \xi, \tau) > 0$ in $D \cap \{\tau < t \leq T\}$.*

The positivity of G follows similarly to the proof of Theorem 11, Sec. 4, Chap. 2. The other assertions follow from the previous construction of G.

Using the italicized statement following Corollary 1 to Theorem 12, Sec. 5, we deduce from Corollary 1 the following:

Corollary 2. *For any fixed $(\xi, \tau) \in D + B$, $\epsilon > 0$, the functions $D_x G$, $D_x^2 G$, $D_t G$ are uniformly continuous functions of (x, t) for $(x, t) \in D$, $t - \tau > \epsilon$.*

If the adjoint L^* of L exists (see Chap. 1, Sec. 8) then *Green's function* $G^*(x, t; \xi, \tau)$ $(t < \tau)$ of $L^* v = 0$ in D is defined as a continuous function of

$(x, t; \xi, \tau) \in \bar{D} \times (D + B_T)$, $t < \tau$, such that for any f as in (7.2), the function

$$v(x, t) = \int_{B_\tau} G^*(x, t; \xi, \tau) f(\xi)\, d\xi$$

satisfies $L^*v = 0$ in $D \cap \{0 \leq t < \tau\}$ and

$$\lim_{t \nearrow \tau} v(x, t) = f(x) \qquad \text{for } x \in \bar{B}_\tau,\ v = 0 \qquad \text{on } S \cap \{0 \leq t < \tau\}.$$

If the coefficients of L^* satisfy the conditions $(\bar{\text{A}})$, (B) and if D satisfies the property $(\bar{\bar{\text{E}}})$, then the uniqueness, existence, and differentiability properties of G^* can be proved as for G, by first substituting $t \to -t$ in $L^*v = 0$. We also have, then,

$$L^*G^*(x, t; \xi, \tau) = 0 \qquad \text{for } (x, t) \in D \cap \{0 \leq t < \tau\},$$
$$G^*(x, t; \xi, \tau) = 0 \qquad \text{for } (x, t) \in \bar{S} \cap \{0 \leq t < \tau\},$$
$$G^*(x, t; \xi, \tau) > 0 \qquad \text{for } (x, t) \in D \cap \{0 \leq t < \tau\}.$$

Theorem 17. *If L is a parabolic operator in D, if a_{ij}, $D_x a_{ij}$, $D_x^2 a_{ij}$, b_i, $D_x b_i$, c are uniformly Hölder continuous (exponent α) in D, and if D satisfies $(\bar{\bar{\text{E}}})$, then $G^*(x, t; \xi, \tau)$ exists and for any two points (x, t), (ξ, τ) in D with $t > \tau$,*

$$G(x, t; \xi, \tau) = G^*(\xi, \tau; x, t). \tag{7.7}$$

Proof. We have only to prove (7.7). Setting

$$u(y, \sigma) = G(y, \sigma; \xi, \tau),$$
$$v(y, \sigma) = G^*(y, \sigma; x, t),$$

and integrating Green's identity (1.8.4) over the domain

$$D \cap \{\tau + \epsilon < \sigma < t - \epsilon\},$$

we get

$$\int_{B_{t-\epsilon}} u(y, t - \epsilon) G^*(y, t - \epsilon; x, t)\, dy = \int_{B_{\tau+\epsilon}} v(y, \tau + \epsilon) G(y, \tau + \epsilon, \xi, \tau)\, dy.$$

(7.7) now follows by taking $\epsilon \to 0$.

For the heat equation

$$\sum_{i=1}^{n} \frac{\partial^2 u}{\partial x_i^2} - \frac{\partial u}{\partial t} = 0$$

in a rectangle $-a < x_i < a$ $(i = 1, \ldots, n)$, $0 < t < T$, the Green function can be given explicitly. If $n = 1$,

$$G(x, t; \xi, \tau) = \sum_{m=-\infty}^{\infty} \Gamma(x + 4ma, t; \xi, \tau) - \sum_{m=-\infty}^{\infty} \Gamma(2a - x + 4ma, t; \xi, \tau), \tag{7.8}$$

where $\Gamma(x, t; \xi, \tau)$ is the fundamental solution (1.0.2).

If $n = 2$,

$$
\begin{aligned}
(7.9)\quad & G(x, t; \xi, \tau) \\
&= \sum_{j=-\infty}^{\infty} \left[\sum_{m=-\infty}^{\infty} \Gamma(x_1^j, x_2^m, t; \xi_1, \xi_2, \tau) - \sum_{m=-\infty}^{\infty} \Gamma(x_1^j, \tilde{x}_2^m, t; \xi_1, \xi_2, \tau) \right] \\
&\quad - \sum_{j=-\infty}^{\infty} \left[\sum_{m=-\infty}^{\infty} \Gamma(\tilde{x}_1^j, x_2^m, t; \xi_1, \xi_2, \tau) - \sum_{m=-\infty}^{\infty} \Gamma(\tilde{x}_1^j, \tilde{x}_2^m, t; \xi_1, \xi_2, \tau) \right],
\end{aligned}
$$

where $x_\lambda^m = x_\lambda + 4ma$, $\tilde{x}_\lambda^m = 2a - x_\lambda + 4ma$ $(\lambda = 1, 2)$. The formula for $n > 2$ follows the same pattern.

8. Elliptic Equations

In this section we shall briefly extend the results of Secs. 2–6 to elliptic operators

$$
(8.1) \qquad Lu = \sum_{i,j=1}^{n} a_{ij}(x) \frac{\partial^2 u}{\partial x_i \, \partial x_j} + \sum_{i=1}^{n} b_i(x) \frac{\partial u}{\partial x_i} + c(x)u
$$

in a bounded domain D of R^n. We begin with some definitions.

A function $u(x)$ is said to belong to the class $C^m(D)$ (or simply C^m) if all its first m partial derivatives exist and are continuous functions in D. If all the mth order derivatives are locally Hölder continuous (exponent α) in D, then we say that u belongs to $C^{m+\alpha}$. Denoting $\underset{D}{\text{l.u.b.}}\, |v|$ by $|v|_0$, we introduce the norms

$$
(8.2) \qquad \overline{|u|}_m = \sum_{j=0}^{m} \sum |D^j u|_0,
$$

$$
(8.3) \qquad \overline{|u|}_{m+\alpha} = \overline{|u|}_m + \sum \overline{H}_\alpha(D^m u)
$$

in C^m and $C^{m+\alpha}$ respectively; the sums with the unspecified limits are taken over all the partial derivatives of the indicated orders, and $\overline{H}_\alpha(v)$ is the Hölder coefficient of v in D. The normed space consisting of all the functions u in C^m $(C^{m+\alpha})$ for which $\overline{|u|}_m < \infty$ $(\overline{|u|}_{m+\alpha} < \infty)$ is denoted by $\overline{C}_m$ $(\overline{C}_{m+\alpha})$.

Denoting by d_x the distance of a point x to the boundary ∂D of D and setting $d_{xy} = \min (d_x, d_y)$, we define

$$
H_\alpha(d^k u) = \underset{x,y \in D}{\text{l.u.b.}}\, d_{xy}^{k+\alpha} \frac{|u(x) - u(y)|}{|x - y|^\alpha},
$$

$$
|d^k v|_0 = \underset{D}{\text{l.u.b.}}\, |d_x^k v(x)|.
$$

Next we introduce the norms

$$
(8.4) \qquad |u|_m = \sum_{j=0}^{m} \sum |d^j D^j u|_0,
$$

(8.5) $$|u|_{m+\alpha} = |u|_m + \Sigma\, H_\alpha(d^m D^m u).$$

The normed spaces consisting of all the functions u in C^m ($C^{m+\alpha}$) for which $|u|_m < \infty$ ($|u|_{m+\alpha} < \infty$) is denoted by C_m ($C_{m+\alpha}$).

The spaces $\overline{C}_m$, $\overline{C}_{m+\alpha}$, C_m, $C_{m+\alpha}$ are Banach spaces.

The proof is similar to the proof of Theorem 3, Sec. 2.

We proceed to state the interior Schauder estimates.

Interior Estimates. *If (with the notation of the present section)* (2.13), (2.14) *(with* $a_{ij}(x, t) \equiv a_{ij}(x)$*),* (2.15) *(with* $f = f(x)$*) hold, and if* $Lu = f$ *in* D *and* $u \in C^{2+\alpha}(D)$, $|u|_0 < \infty$, *then* $u \in C_{2+\alpha}$ *and* (2.16) *holds.*

We say that the boundary ∂D of D belongs to the class C^m, or $C^{m+\alpha}$, if there exist local representations of ∂D, in neighborhoods of each of its points, having the form

(8.6) $$x_i = h(x_1, \ldots, x_{i-1}, x_{i+1}, \ldots, x_n)$$

where the functions h belong (locally) to C^m, or $C^{m+\alpha}$, respectively.

A function ψ on ∂D is said to belong to C^m, or $C^{m+\alpha}$, if the same is true of ∂D and if in the local parameters of ∂D (given by $x_1, \ldots, x_{i-1}, x_{i+1}, \ldots, xn$, if (8.6) is a local representation of ∂D) the function ψ belongs to C^m, or $C^{m+\alpha}$, respectively.

A function ψ on ∂D is said to belong to $\overline{C}_{2+\alpha}$ if there exists a function Ψ in $\overline{C}_{2+\alpha}(D)$ which coincides with ψ on ∂D. We then define

$$\overline{|\psi|}_{2+\alpha} = \underset{\Psi}{\text{g.l.b.}}\ \overline{|\Psi|}_{2+\alpha}$$

where the g.l.b. is extended over all the functions Ψ in $\overline{C}_{2+\alpha}(D)$ which coincide with ψ on ∂D.

We can now state the boundary Schauder estimates.

Boundary Estimates. *If* (2.18), (2.14) *(with* $a_{ij}(x, t) = a_{ij}(x)$*),* (2.19) *(with* $f = f(x)$*) hold and if* ∂D *belongs to* $C^{2+\alpha}$ *and* ψ *belongs to* $\overline{C}_{2+\alpha}$, *then for any solution* u *of* $Lu = f$ *in* D, $u = \psi$ *on* ∂D, *which belongs to* $\overline{C}_{2+\alpha}(D)$, *the inequality*

(8.7) $$\overline{|u|}_{2+\alpha} \le \overline{K}(\overline{|\psi|}_{2+\alpha} + |u|_0 + \overline{|f|}_\alpha)$$

holds.

Note that if $c \le 0$ then, by (2.7.5),

$$|u|_0 \le \underset{\partial D}{\text{l.u.b.}}\ |\psi| + \text{const.}\ \underset{D}{\text{l.u.b.}}\ |f|,$$

and (8.7) reduces to (2.21).

We next state an existence theorem for the Dirichlet problem

(8.8) $$Lu = f \qquad \text{in } D,$$

(8.9) $$u = \psi \qquad \text{on } \partial D.$$

Theorem 18. *Let* L *be an elliptic operator in* $\overline{D}$ *with Hölder continuous (exponent* α*) coefficients in* $\overline{D}$ *and with* $c \le 0$, *and let* $\partial D \in C^{2+\alpha}$. *If*

$f \in \overline{C}_\alpha(D)$, $\psi \in \overline{C}_{2+\alpha}$, then there exists a unique solution u of the Dirichlet problem (8.8), (8.9) *and, furthermore, $u \in \overline{C}_{2+\alpha}(D)$.*

The proof of Theorem 18 is similar to the proof of Theorem 7, Sec. 3, and the details are left to the reader.

In Chap. 4, Sec. 9, we shall comment on the proof of the Schauder estimates as well as on the proof of the analogue of (a) (in the proof of Theorem 7) for the Laplace operator $\Sigma\, \partial^2 u/\partial x_i^2$.

A *barrier* w_y at a point $y \in \partial D$ is a continuous nonnegative function in $\overline{D}$, which vanishes only at the point y and for which $Lw_y \leq -1$.

Theorem 19. *Let L be an elliptic operator in $\overline{D}$ with $c \leq 0$, and let* (2.13), (2.15) *hold. If at each point of ∂D there exists a barrier, then for any continuous function ψ on ∂D there exists a unique solution u of the Dirichlet problem* (8.8), (8.9). *Furthermore, $u \in C_{2+\alpha}$.*

The proof is similar to that of Theorem 9, Sec. 4.

If there exists a closed ball K such that $K \cap D = \phi$, $K \cap \partial D = \{y\}$ then

$$w_y(x) = k(|x^0 - y|^{-p} - |x - y|^{-p}) \qquad (x^0 = \text{center of } K) \tag{8.10}$$

is a barrier at y provided k and p are sufficiently large.

Concerning the differentiability of solutions of elliptic equations we have the following result.

Theorem 20. *Let L be an elliptic operator in $\overline{D}$ and assume that f and the coefficients of L belong to $C^{p+\alpha}(D)$ $(p \geq 0)$. Then any solution of $Lu = f$ belongs to $C^{p+2+\alpha}(D)$.*

If f and the coefficients of L belong to $\overline{C}_{p+\alpha}(D)$, if ∂D belongs to $C^{p+2+\alpha}$, $\psi \in C^{p+2+\alpha}$ and $\psi \in \overline{C}_{2+\alpha}$, then any solution of (8.8), (8.9) *belongs to $\overline{C}_{p+2+\alpha}(D)$.*

The proof for $c \leq 0$ is based on the same arguments that were given, in Sec. 5, in the proofs of Theorems 10–12. For arbitrary c, the proof for $p \geq 1$ is the same as for $c \leq 0$ provided we already know the assertion for $p = 0$. To prove the assertion for $p = 0$ we need the existence theorems 18 and 19 to hold in the case of arbitrary c. This is not generally the case; however, the first part of Theorem 20 (for $p = 0$) is concerned with a local property. Therefore, if we take a sufficiently small subdomain D_0 of D so that (2.7.6) and, consequently, (2.7.7) hold, then the solution in D_0 is uniquely determined by its boundary values and the assertion of Theorem 18 holds. Hence, by the argument used in the proof of Theorem 10, Sec. 5 (for $p = 0$) it follows that the first part of Theorem 20 is true.

For the proof of the second part (in the case $p = 0$) we need an existence theorem in which ψ is a continuous function on the boundary, and it belongs to $C^{2+\alpha}$ on a portion S of the boundary, and the solution is in $C_{2+\alpha}(D)$ as well as in $\overline{C}_{2+\alpha}(D_1)$ in any subdomain D_1 of D whose closure is a closed subset of $D + S_0$, $\overline{S}_0$ lying in the interior of S. Such a theorem can be established by the method of continuity, using a more general form of Schauder's boundary estimates (see Sec. 9, Chap. 4). Since the existence theorem will hold also for arbitrary c if D is sufficiently small, we can complete the proof of the second part of the theorem (for $p = 0$) by breaking D into small subdomains.

We shall now show that *if $p \geq 1$ then the assumption that $\psi \in \overline{C}_{2+\alpha}$ is superfluous.*

At each point x^0 of ∂D draw an inwardly directed normal $\nu(x^0)$ and measure on it a length δ. Denote the endpoint by $\nu(x^0, \delta)$. For any $\partial D \in C^1$, if δ is sufficiently small then the function $x = \nu(x^0, \delta)$ is a one-to-one function of (x^0, δ) into R^n. Indeed, this follows by the theorem on implicit functions and from the formula

$$x_i = x_i^0(s) + \delta g_i(s)/g(s) \tag{8.11}$$

where $s = (s_1, \ldots, s_{n-1})$ is a local parameter on ∂D,

$$x_i = x_i^0(s) \qquad (i = 1, \ldots, n)$$

are the equations of ∂D, $g_i(s)$ is $(-1)^{i-1}$ times the determinant of the matrix obtained from the matrix $(\partial x_i^0/\partial s_j)$ (i indicates rows, j indicates columns) by cancelling out the ith row, and

$$g(s) = \left(\sum_{i=1}^{n} (g_i(s))^2\right)^{1/2}.$$

Note, incidentally, that the normals $\nu(x^0)$ are normal to each of the manifolds ∂D^δ generated by the points $\nu(x^0, \delta)$ when x^0 varies on ∂D.

Let $\zeta(t)$ be a three times continuously differentiable function such that $\zeta(0) = 1$, $\zeta(t) = 0$ if $|t| \geq \delta_0$. Then, if δ_0 is sufficiently small,

$$\Psi(x) = \begin{cases} \zeta(\delta)\psi(x^0) & \text{if } 0 \leq \delta < \delta_0 \quad (x = \nu(x^0, \delta)) \\ 0 & \text{otherwise} \end{cases}$$

extends ψ to D and belongs to $\overline{C}_{2+\alpha}$ if $\partial D \in C^{3+\alpha}$. This completes the proof that $\psi \in \overline{C}_{2+\alpha}$ if $p \geq 1$.

Theorem 20 and its proof can be extended to nonlinear elliptic equations. The analogue of Corollary 2 of Theorem 12, Sec. 5, is also valid.

PROBLEMS

1. Let $g(u)$ be a continuously differentiable function for $-\infty < u < \infty$ and let L, f, D be as in Theorem 7. Prove that there exists a unique solution to the problem

$$\begin{aligned} Lu &= f(x, t) + \epsilon g(u) && \text{in } D + B_T, \\ u &= 0 && \text{on } \overline{B} + S, \end{aligned}$$

provided ϵ is sufficiently small.
[*Hint:* Use arguments similar to those in part (b) of the proof of Theorem 7.]

2. Prove that if the first initial-boundary value problem $Lu = f$, $u = \psi$ has a solution for any continuous ψ and Hölder continuous f, then barriers exist at all the points of S.

3. With the notation of (4.3), show that if (4.2) holds then
$$ke^{\gamma t}\{\exp[-\lambda R_0^2] - \exp[-\lambda R^2]\}$$
is also a barrier if k and λ are sufficiently large.

4. Consider the family $\mathfrak{M}$ of all positive solutions of the heat equation $u_{xx} - u_t = 0$ ($n = 1$) in a rectangle $-b < x < b$, $-c < t < T$ ($c > 0$). Prove that for any $0 < a < b$ there exists a positive constant K such that
$$(\bigstar) \qquad u(\bar{\xi}, \bar{\tau}) \leq Ku(\xi, \tau)$$
holds for all $u \in \mathfrak{M}$, $-a < \bar{\xi}, \xi < a$, $0 < \bar{\tau} \leq \tau < T$. The inequality ($\bigstar$) is analogous to Harnack's inequality for harmonic functions.
[*Hint:* Derive the representation
$$u(\xi, \tau) = \int_{-a}^{a} u(x, 0)G^*(x, 0; \xi, \tau)\, dx + \Sigma \pm \int_0^\tau \left[u \frac{\partial G^*}{\partial x}\right]_{x=\mp a} dt$$
(G given by (7.8)). Use $[\partial G^*/\partial x] \geq 0$ (≤ 0) on $x = -a$ ($x = a$) and
$$\frac{G^*(x, 0; \xi, \tau)}{G^*(x, 0; \bar{\xi}, \bar{\tau})} \geq C > 0, \qquad \frac{\partial G^*(\pm a, t; \xi, \tau)/\partial x}{\partial G^*(\pm a, t; \bar{\xi}, \bar{\tau})/\partial x} \geq C > 0.]$$

5. Extend the result of Problem 4 to domains D bounded by $t = 0$, $t = T$, and two nonintersecting curves in the strip $0 \leq t \leq T$, i.e., show that for every closed subdomain G of D, ($\bigstar$) holds for all the positive solutions in D, $(\xi, \tau) \in G$, $(\bar{\xi}, \bar{\tau}) \in G$, $\tau \geq \bar{\tau}$, where K depends only on G, D.

6. Prove the following theorem (which is analogous to Harnack's second theorem for harmonic functions): If $\{u_m\}$ is a monotone sequence of solutions of $u_{xx} - u_t = 0$ in D, and if for some $Q \in D$ the sequence $\{u_m(Q)\}$ is convergent, then the sequence is uniformly convergent in closed subsets of $S(Q)$ (for definition of $S(Q)$ see Sec. 1, Chap. 2) to a solution u of $u_{xx} - u_t = 0$.

7. Let L be an elliptic operator with continuous coefficients and with $c \leq 0$ in $\bar{D}$, D a bounded domain, and let $\partial D \in C^1$, $Lu = f$ in D, $u = 0$ on ∂D and $u \in \bar{C}_1(D)$. Prove that if for every $y \in \partial D$ there exists a closed ball K with a radius R independent of y such that $K \cap \bar{D} = \{y\}$, then, for any $i = 1, \ldots, n$,
$$\operatorname*{l.u.b.}_{\partial D} \left|\frac{\partial u}{\partial x_i}\right| \leq H \operatorname*{l.u.b.}_{D} |f|, \qquad H \text{ depending only on } L, D.$$
[*Hint:* Take w_y as in (8.10) and show that $|u(x)| \leq w_y(x)|f|_0$ ($x \in D$) and, consequently, $|\partial u/\partial \nu| \leq |\partial w_y/\partial \nu|\, |f|_0$ at y.]

CHAPTER 4

DERIVATION OF A PRIORI ESTIMATES

Introduction. The main concern of this chapter is proving the a priori estimates stated in Theorems 5, 6 of Chap. 3. The statements (a) made in the proofs of Theorems 7, 9 of Chap. 3 will also be proved. The notation of Chap. 3, Sec. 2, will be freely used.

The general procedure for deriving the a priori estimates is very simple. First we establish them (or, actually, we establish a somewhat more useful version of them) for the special case of the heat operator, or for the somewhat more general parabolic operators of the form

$$L_0 u = \sum a_{ij} \frac{\partial^2 u}{\partial x_i \, \partial x_j} - \frac{\partial u}{\partial t} \qquad (a_{ij} \text{ constants}). \tag{0.1}$$

Next we consider the general parabolic equation of the form (1.1) below, and write it, near a point (x^0, t^0), in the form $L_0 u = \tilde{f}$ where $\tilde{f} = f + (L_0 u - Lu)$ and L_0 is given by (0.1) with $a_{ij} = a_{ij}(x^0, t^0)$. Using the estimates previously obtained for L_0, we then try to obtain the desired estimates for L.

This method of perturbation, based upon first deriving estimates in the case of equations with constant coefficients, has been used also in other problems of differential equations.

The analysis of this chapter is quite heavy. The greatest technical efforts are made in establishing the necessary estimates for the special case of the heat operator. The perturbation procedure by itself is not too involved.

The results in Secs. 3, 5, 6 concern only the case of the heat operator; the reader may perhaps find it more suitable not to go, in a first reading, through the detailed proofs of these sections.

1. Notation

In the definition (3.2.11) of $|u|_{2+\alpha}$ there appears the term $|d^2 D_t u|_\alpha$. Since the a priori estimates are concerned with solutions of the parabolic equation

$$\text{(1.1)}\quad Lu \equiv \sum_{i,j=1}^{n} a_{ij}(x, t) \frac{\partial^2 u}{\partial x_i \, \partial x_j} + \sum_{i=1}^{n} b_i(x, t) \frac{\partial u}{\partial x_i} + c(x, t)u - \frac{\partial u}{\partial t} = f(x, t),$$

if we get an estimate of the form

$$\text{(1.2)}\qquad |u|_\alpha + \Sigma\, |dD_x u|_\alpha + \Sigma\, |d^2 D_x^2 u|_\alpha \leq K(|u|_0 + |d^2 f|_\alpha),$$

under the assumptions (A), (B), (C) of Chap. 3, Sec. 2, then a similar estimate follows immediately also for $|d^2 D_t u|_\alpha$, by using the obvious rules (3.5.8) and

$$\text{(1.3)}\qquad |d^2(gh)| \leq |dg|_\alpha |dh|_\alpha.$$

Thus, in order to prove (3.2.16) it suffices to prove (1.2).

In Theorem 5, Chap. 3, D was bounded by $t = 0$, $t = T$, and a manifold S. It was also assumed that, for every $0 < \tau < T$, $D \cap \{t = \tau\}$ is a domain, and that there exists a continuous curve γ connecting the domain B_T to the domain B along which the t-coordinate is nonincreasing. In this chapter, however, we shall prove Theorem 5 for any bounded open set D. d_P, where $P = (\xi, \tau)$, is now defined by

$$d_P = \underset{Q \in \Delta_\tau}{\text{g.l.b.}}\; d(P, Q)$$

where Δ_τ is the intersection of the boundary of D with the half-space $t \leq \tau$.

Definition. N is a *semicube* with *top* $P = (x^0, t^0)$ and *edge* δ if N is defined by

$$x_i^0 - \delta \leq x_i \leq x_i^0 + \delta \qquad (i = 1, \ldots, n),$$
$$t^0 - \delta^2 \leq t \leq t^0,$$

where $x^0 = (x_1^0, \ldots, x_n^0)$.

If $Q \in N$ then clearly $d(P, Q) \leq (n + 1)^{1/2}\delta$.

It is convenient to introduce the following new norms, for any nonnegative integers p, m and for $0 < \alpha < 1$:

$$\text{(1.4)}\qquad |g|_{p,m} = \sum_{j=0}^{m} M_{p,j}[g],$$

$$\text{(1.5)}\qquad |g|_{p,m+\alpha} = |g|_{p,m} + \sum_{j=0}^{m} M_{p,j+\alpha}[g],$$

where

$$\text{(1.6)}\qquad M_{p,j}[g] = \Sigma\, |d^{p+j} D_x^j g|_0 = \Sigma \underset{P \in D}{\text{l.u.b.}}\; d_P^{p+j} |D_x^j g(P)|,$$

$$\text{(1.7)}\qquad M_{p,j+\alpha}[g] = \Sigma\, H_\alpha[d^{p+j} D_x^j g] = \Sigma \underset{P,Q \in D}{\text{l.u.b.}}\; d_{PQ}^{p+j+\alpha} \frac{|D_x^j g(P) - D_x^j g(Q)|}{d(P, Q)^\alpha}.$$

Since, as previously remarked, the inequality (3.2.16) is equivalent to (1.2), Theorem 5 of Chap. 3 is a consequence of the following theorem.

Theorem 1. *Assume that*

$$|a_{ij}|_{0,\alpha} \leq K_1, \quad |b_i|_{1,\alpha} \leq K_1, \quad |c|_{2,\alpha} \leq K_1, \tag{1.8}$$

that for any real vector ξ *and* $(x, t) \in D$,

$$\sum_{i,j=1}^{n} a_{ij}(x, t)\xi_i\xi_j \geq K_2|\xi|^2 \qquad (K_2 > 0), \tag{1.9}$$

and that $|f|_{2,\alpha} < \infty$. *There exists a constant* K *depending only on* K_1, K_2, n, α, *such that if* u *is any solution of* (1.1) *in* D *such that* $D_x u$, $D_x^2 u$, $D_t u$ *are Hölder continuous (exponent* α*) in* D *and such that* $|u|_0 < \infty$, *then* $u \in C_{2+\alpha}$ *and*

$$|u|_{0,2+\alpha} \leq K(|u|_0 + |f|_{2,\alpha}). \tag{1.10}$$

In Sec. 2 we shall prove some elementary lemmas concerning the norms (1.4), (1.5). In Sec. 3 a fundamental theorem will be proved concerning a priori estimates for the heat operator. This theorem will be used in Sec. 4 to complete the proof of Theorem 1 by a method of perturbation.

2. Preliminary Lemmas

Lemma 1. *Let* D_ϵ *be the subset of* D *consisting of all the points whose distance to the boundary of* D *is* $> \epsilon$, *and denote by* $M_{p,m}^{\epsilon}[u]$ *the expression* $M_{p,m}[u]$ *taken with respect to* D_ϵ. *If* $M_{p,m}^{\epsilon}[u] \leq M$, *where* M *is a constant independent of* ϵ, *then* $M_{p,m}[u]$ *is finite and* $\leq M$. *The same statement holds for* $M_{p,m+\alpha}[u]$, $|u|_{p,m}$, $|u|_{p,m+\alpha}$.

Proof. Denote by $d_{P,\epsilon}$ the function d_P taken with respect to D_ϵ. Then, for any fixed $P \in D_\epsilon$,

$$d_{P,\epsilon}^{p}|D_x^{p+m}u(P)| \leq M.$$

Letting $\epsilon \to 0$ we get

$$d_P^p|D_x^{p+m}u(P)| \leq M.$$

Taking the l.u.b. for $P \in D$, the assertion of the lemma, for $M_{p,m}[u]$, follows. The proofs for $M_{p,m+\alpha}[u]$, $|u|_{p,m}$, $|u|_{p,m+\alpha}$ are similar.

Lemma 2. *Let* $p \geq 0$, $m \geq 1$ *be integers. For any* $0 < \epsilon < 1$ *there exists a constant* C *depending only on* ϵ, p, m *such that the inequality*

$$|u|_{p,m-1} \leq \epsilon M_{p,m}[u] + CM_{p,0}[u] \tag{2.1}$$

holds for all u *for which* $M_{p,m}[u] < \infty$, $M_{p,0}[u] < \infty$.

Proof. For simplicity we set $M_{p,j} = M_{p,j}[u]$. Without loss of generality we may assume that $|u|_{p,m-1}$ is finite since, otherwise, we first derive (2.1) for D_δ (see Lemma 1) and thus get

$$|u|^{\delta}_{p,m-1} \le \epsilon M^{\delta}_{p,m} + CM^{\delta}_{p,0} \le \epsilon M_{p,m} + CM_{p,0};$$

then take $\delta \to 0$ and apply Lemma 1.

Let P be any fixed point in D and let N be a semicube with top P and edge μd_P. Take $\mu < 1/(n+1)^{1/2}$ so that $N \subset D$. By the mean value theorem, if P_1, P_2 are any two points on the upper base N^+ of N, then

$$D^j_x u(Q) = \frac{D^{j-1}_x u(P_1) - D^{j-1}_x u(P_2)}{\overline{P_1P_2}}$$

for some $Q \in N^+$. Here $\overline{P_1P_2}$ denotes the euclidean distance from P_1 to P_2. Taking P_1, P_2 to be at opposite vertices of N^+, we get

$$|D^j_x u(Q)| \le \frac{2 \operatorname*{l.u.b.}_{N^+} |D^{j-1}_x u|}{2n^{1/2}\mu d_P}, \tag{2.2}$$

for some $Q \in N^+$.

From the relation

$$D^j_x u(P) - D^j_x u(Q) = \int_Q^P D_\zeta D^j_x u \, d\zeta$$

where ζ varies in the interval connecting Q to P, we find, upon using the inequality (2.2),

$$|D^j_x u(P)| \le \frac{\operatorname*{l.u.b.}_{N^+} |D^{j-1}_x u|}{\mu d_P} + \mu n^{1/2} d_P \Sigma \operatorname*{l.u.b.}_{N^+} |D^{j+1}_x u|,$$

where the sum is taken over all the partial derivatives of order $j+1$. From the definitions of $M_{p,j-1}$, $M_{p,j+1}$ it then follows that

$$|D^j_x u(P)| \le (\mu d_P)^{-1} \frac{M_{p,j-1}}{[(1-\mu n^{1/2})d_P]^{p+j-1}} + \mu n^{1/2} d_P \frac{M_{p,j+1}}{[(1-\mu n^{1/2})d_P]^{p+j+1}}.$$

Multiplying both sides by d_P^{p+j}, and recalling that P is an arbitrary point in D, we conclude that, for any $\epsilon > 0$,

$$M_{p,j} \le \epsilon M_{p,j+1} + \frac{C_0}{\epsilon} M_{p,j-1} \tag{2.3}$$

where C_0 depends only on p, j.

Using (2.3) we shall next prove, by induction on j, that for all $0 \le i < j$,

$$M_{p,i} \le \epsilon M_{p,j} + \frac{C}{\epsilon^{i/(j-i)}} M_{p,0}, \tag{2.4}$$

where C depends only on p, j. For $j = 1$ this is obvious. Assuming that the assertion holds for all $j \le k$, we shall prove it for $j = k+1$. By (2.3),

$$M_{p,k} \le \frac{\epsilon}{2} M_{p,k+1} + \frac{C_1}{\epsilon} M_{p,k-1} \tag{2.5}$$

where C_i are used to denote constants depending only on p, k. By the inductive assumption

$$M_{p,k-1} \leq \frac{\epsilon}{2C_1} M_{p,k} + \frac{C_2}{\epsilon^{k-1}} M_{p,0}.$$

Substituting this into (2.5) we obtain

$$M_{p,k} \leq \epsilon M_{p,k+1} + \frac{C_3}{\epsilon^k} M_{p,0}, \tag{2.6}$$

i.e., (2.4) holds for $j = k + 1$, $i = k$.

If $i < k$ then, by the inductive assumption (with $\epsilon = \delta$) and (2.6) (with $\epsilon = \lambda$),

$$M_{p,i} \leq \delta M_{p,k} + \frac{C_4}{\delta^{i/(k-i)}} M_{p,0}$$

$$\leq \delta\lambda M_{p,k+1} + \left[\frac{C_3\delta}{\lambda^k} + \frac{C_4}{\delta^{i/(k-i)}}\right] M_{p,0}.$$

Taking $\delta = \epsilon^{(k-i)/(k+1-i)}$, $\lambda = \epsilon^{1/(k+1-i)}$, the proof of (2.4), for $j = k + 1$, is completed.

Recalling the definition (1.4), we see that (2.1) is a consequence of (2.4).

Lemma 3. *For any $a \geq 0$ and any nonnegative integers p, q, j with $q \geq j$,*

$$|D_x^j u|_{q,a} \leq C_1 |u|_{q-j,a+j}, \tag{2.7}$$

$$|uv|_{p+q,a} \leq C_2 |u|_{p,a} |v|_{q,a}, \tag{2.8}$$

$$|u|_{p,a+1} \leq C_3 |u|_{p,a-[a]} + C_4 \sum |D_x u|_{p+1,a}, \tag{2.9}$$

where $[a]$ is the largest integer $\leq a$ and C_k are constants depending only on a, p, q, j.

The proof is straightforward and is left to the reader.

In order to further familiarize the reader with the notation of Sec. 1, we give two more elementary lemmas even though they will not be used in the future.

Lemma 4. *Let N be a semicube in D with top P and edge*

$$r = d_P/2(n + 1)^{1/2}.$$

Then, for any $a \geq 0$ and integer $p \geq 0$,

$$r|u|_{p,a}^N \leq |u|_{p+1,a}^D. \tag{2.10}$$

Proof. Take one term of $r|u|_{p,a}^N$, say

$$r \operatorname*{l.u.b.}_{Q \in N} d_{Q,N}^{p+i} |D_x^i u(Q)| \tag{2.11}$$

where $d_{Q,N}$ is d_Q defined with respect to N. Since, for $Q \in N$,

$$r \leq (n+1)^{1/2}r = d_P - (n+1)^{1/2}r \leq d_Q,$$

and since $d_{Q,N} \leq d_Q$, the expression in (2.11) is less than or equal to

$$\underset{Q \in N}{\text{l.u.b.}}\, d_Q^{p+i+1}|D_x^i u(Q)|.$$

Treating the other terms in $r|u|_{p,a}^N$ in the same way, the inequality (2.10) follows.

The next lemma is a converse of the previous one.

Lemma 5. *Suppose that for some $a \geq 0$ and an integer $p \geq 0$, $r|u|_{p,a}^N \leq H$ where N is a semicube with top P and edge $r = d_P/2(n+1)^{1/2}$ and P varies in D, and H is a constant. Then $|u|_{p+1,a}^D$ is finite and*

$$|u|_{p+1,a}^D \leq CH \tag{2.12}$$

where C depends only on p, a.

Proof. For any $P \in D$, $0 \leq j \leq [a]$,

$$r d_{P,N}^{p+j}|D_x^j u(P)| \leq H$$

where $d_{P,N}$ is d_P defined with respect to N. Since

$$d_{P,N} = r = d_P/2(n+1)^{1/2},$$

we get

$$d_P^{p+j+1}|D_x^j u(P)| \leq H[2(n+1)^{1/2}]^{p+j+1}.$$

Taking the l.u.b. with respect to $P \in D$, the inequality

$$M_{p+1,j}[u] \leq H[2(n+1)^{1/2}]^{p+j+1}$$

follows. Treating the other terms in $|u|_{p+1,a}$ in the same manner, the proof is completed.

3. An Auxiliary Theorem

In this section N is a fixed semicube with top P and edge d. Various positive constants which depend only on n, α will be denoted by K. The notation

$$H_{Q,N}[g] = \underset{R \in N}{\text{l.u.b.}} \frac{|g(Q) - g(R)|}{d(Q, R)^\alpha}$$

will be used.

The decisive step in proving Theorem 1, Sec. 1, consists of establishing the following auxiliary theorem.

Theorem 2. *Let f be Hölder continuous (exponent α) in N, let u, $D_x u$, $D_x^2 u$ be Hölder continuous (exponent α) in N and, finally, let u satisfy the equation*

$$L_0 u \equiv \Delta u - \frac{\partial u}{\partial t} = f \qquad \text{in } N \tag{3.1}$$

where Δ is the Laplace operator $\sum_{i=1}^{n} \partial^2/\partial x_i^2$. There exists a constant K (depending only on n, α) such that for $i = 0, 1, 2$,

$$(3.2)\quad |D_x^i u(P)| \le d^{-i}K \operatorname*{l.u.b.}_N |u| + d^{2-i}K \operatorname*{l.u.b.}_N |f| + d^{2-i+\alpha}KH_{P,N}[f] \equiv KI_i,$$

$$(3.3)\quad d^\alpha \frac{|D_x^i u(P) - D_x^i u(Q)|}{d(P, Q)^\alpha} \le KI_i + Kd^{2-i+\alpha}H_{Q,N}[f] \qquad \text{if } d(P, Q) \le \frac{d}{4}\cdot$$

Proof. Denote by N_η the semicube with top P and edge ηd and let $\varphi(Q)$ be a three times continuously differentiable function in N satisfying

$$(3.4)\qquad \varphi(Q) = \begin{cases} 1 & \text{if } Q \in N_{1/2} \\ 0 & \text{if } Q \in (N - N_{3/4}), \end{cases}$$

and

$$(3.5)\qquad |D_x^k D_t^h \varphi(x, t)| \le Ad^{-k-2h} \qquad (0 \le k + h \le 2).$$

The existence of such a function follows by the construction in Problem 1, Chap. 1, with $A_0 = N_{2/3}$, $\epsilon = 1/12$.

Let $P = (x^0, t^0 + d^2)$ and denote by B the lower base of N. If we use Green's identity (1.8.4) with $u(x, t)$ and $\varphi(x, t)G(x - \xi, t - \tau)$ where

$$(3.6)\quad G(x - \xi, t - \tau) \equiv G(x, t; \xi, \tau) = \frac{(\tau - t)^{-n/2}}{(2\sqrt{\pi})^n} \exp\left[-\frac{|x - \xi|^2}{4(\tau - t)}\right]$$

is the fundamental solution of the adjoint equation $L_0^* u = 0$ of $L_0 u = 0$, we get, upon integration,

$$(3.7)\qquad \begin{aligned} u(\xi, \tau) &= -\int_{t^0}^{\tau}\int_B \varphi(x, t)f(x, t)G(x - \xi, t - \tau)\,dx\,dt \\ &\quad + \int_{t^0}^{\tau}\int_B u(x, t)L_0^*[\varphi(x, t)G(x - \xi, t - \tau)]\,dx\,dt \\ &\equiv -H_0 + J_0 \end{aligned}$$

for any $(\xi, \tau) \in N_{1/2}$.

We shall first prove (3.2), (3.3) for $i = 2$. Setting

$$(3.8)\qquad H(Q) = D_\xi^2 H_0, \quad J(Q) = D_\xi^2 J_0 \qquad \text{where } Q = (\xi, \tau),$$

we have

$$(3.9)\qquad D_\xi^2 u(Q) = -H(Q) + J(Q).$$

We begin with estimating $J = J(Q)$ for any point Q whose distance to P is less than $d/4$.

Denoting by Ω the domain $(N_{3/4} - N_{1/2}) \cap \{t^0 \le t \le \tau\}$, we can write

$$(3.10) \quad J = \int_\Omega u(x, t)\{(\Delta_x + D_t)[\varphi(x, t)D_\xi^2 G(x - \xi, t - \tau)]\}\, dx\, dt.$$

In what follows we shall need the inequalities

$$(3.11) \quad |D_\tau^k D_\xi^j G(x - \xi, t - \tau)| \le K(\tau - t)^{-(n+2k+j)/2} \exp\left[-\frac{|x - \xi|^2}{5(\tau - t)}\right] \qquad (0 \le k + j \le 4)$$

which can easily be verified by directly differentiating G and using the inequality $t^m e^{-\epsilon t} \le C$ for $0 < t < \infty$ (m, ϵ are any positive numbers and C depends only on m, ϵ).

Using (3.5), (3.11) and noting that $(\Delta_x + D_t)D_\xi^2 G = 0$, we obtain

$$(3.12) \quad |J| \le K(\operatorname*{l.u.b.}_N |u|) \sum_{i=0}^{1} d^{-2+i} \int_\Omega (\tau - t)^{-(n+2+i)/2} \cdot \exp\left[-\frac{|x - \xi|^2}{5(\tau - t)}\right] dx\, dt.$$

Decompose the last integral into $\int_{\Omega_1} + \int_{\Omega_2}$ where Ω_1 consists of all the points (x, t) in Ω for which

$$|x - x^0| > \frac{d}{2},$$

and $\Omega_2 = \Omega - \Omega_1$.

Using the inequality $|x - \xi| \ge d/4$ for $(x, t) \in \Omega_1$, and then substituting $z = d^2/(\tau - t)$, we obtain

$$(3.13) \quad \int_{\Omega_1} \le Kd^n \int_0^\infty \frac{z^{(n+2+i)/2}}{d^{n+2+i}} \exp[-Kz] \frac{d^2}{z^2}\, dz \le \frac{K}{d^i}.$$

Since $\tau - t > (d/4)^2$ if $(x, t) \in \Omega_2$ (recall that $d(P, Q) < d/4$),

$$(3.14) \quad \int_{\Omega_2} \le \frac{K}{d^{n+2+i}} \text{ vol. } \Omega_2 \le \frac{K}{d^i}.$$

Combining (3.13), (3.14) we conclude from (3.12) that

$$(3.15) \quad |J| \le Kd^{-2} \operatorname*{l.u.b.}_N |u|.$$

In estimating $H = H(Q)$ (for points Q with $d(P, Q) < d/4$), we shall use the following equality:

$$(3.16) \quad H = \int_{t^0}^{\tau} \int_B [D_\xi^2 G(x - \xi, t - \tau)][\varphi(x, t)f(x, t) - \varphi(\xi, \tau)f(\xi, \tau)]\, dx\, dt$$
$$+ [\varphi(\xi, \tau)f(\xi, \tau)]D_\xi \left[\int_{t^0}^{\tau} \int_B D_\xi G(x - \xi, t - \tau)\, dx\, dt\right]$$
$$\equiv H_1 + H_2.$$

Note that the inner integral of H_2 can also be written as a boundary integral, by using the divergence theorem. The verification of (3.16) follows from the results of Chap. 1, Sec. 3. Indeed, by Theorem 4, Sec. 3, Chap. 1,

$$H = \int_{t^0}^{\tau} dt \int_B [D_\xi^2 G(x - \xi, t - \tau)]\varphi(x, t)f(x, t)\, dx.$$

Writing $\varphi(x, t)f(x, t) = [\varphi(x, t)f(x, t) - \varphi(\xi, \tau)f(\xi, \tau)] + \varphi(\xi, \tau)f(\xi, \tau)$ we find that $H = H_1 + H_2'$ where

$$H_2' = \varphi(\xi, \tau)f(\xi, \tau) \int_{t^0}^{\tau} dt \int_B D_\xi^2 G(x - \xi, t - \tau)\, dx.$$

Using Theorems 3, 4 of Sec. 3, Chap. 1, with $f \equiv 1$, it follows that $H_2' = H_2$, and the proof of (3.16) is completed.

Using (3.11) and the inequality

$$(3.17)\quad |\varphi(x, t)f(x, t) - \varphi(\xi, \tau)f(\xi, \tau)| \\ \leq K(|x - \xi|^\alpha + |t - \tau|^{\alpha/2})H_{Q,N}[f] + K\left(\frac{|x - \xi|}{d} + \frac{|t - \tau|}{d^2}\right) \underset{N}{\text{l.u.b.}}\, |f|,$$

we can estimate H_1:

$$(3.18)\quad |H_1| \leq KH_{Q,N}[f] \int_{t^0}^{\tau} \int_B (\tau - t)^{-(n+2)/2} \exp\left[-\frac{|x - \xi|^2}{5(\tau - t)}\right] \\ \cdot [|x - \xi|^\alpha + |t - \tau|^{\alpha/2}]\, dx\, dt \\ + K(\underset{N}{\text{l.u.b.}}\, |f|) \int_{t^0}^{\tau} \int_B (\tau - t)^{-(n+2)/2} \exp\left[-\frac{|x - \xi|^2}{5(\tau - t)}\right] \\ \cdot \left[\frac{|x - \xi|}{d} + \frac{|t - \tau|}{d^2}\right] dx\, dt.$$

Substituting $|x - \xi| = \rho(\tau - t)^{1/2}$ (for each fixed t) in the first integral and recalling that $\tau - t^0 \leq d^2$, we find that the coefficient of $H_{Q,N}[f]$ is bounded by

$$K \int_{t^0}^{\tau} \int_0^{\infty} \frac{(\tau - t)^{\alpha/2}\rho^\alpha + (\tau - t)^{\alpha/2}}{(\tau - t)^{(n+2)/2}} \exp\,[-K\rho](\tau - t)^{n/2}\rho^{n-1}\, d\rho\, dt \\ \leq K \int_{t^0}^{\tau} \frac{dt}{(\tau - t)^{1-\alpha/2}} \leq K(\tau - t^0)^{\alpha/2} \leq Kd^\alpha.$$

Similarly, the coefficient of $\underset{N}{\text{l.u.b.}}\, |f|$ is found to be bounded by K. Hence,

$$(3.19)\qquad |H_1| \leq Kd^\alpha H_{Q,N}[f] + K\, \underset{N}{\text{l.u.b.}}\, |f|.$$

As for H_2, denoting by dS_x the surface element of the boundary ∂B of B, we have

$$(3.20)\qquad |H_2| \leq K(\underset{N}{\text{l.u.b.}}\, |f|) \int_{t^0}^{\tau} \int_{\partial B} |D_\xi G(x - \xi, t - \tau)|\, dS_x\, dt.$$

Using (3.11) and the inequality $|x - \xi| > 3d/4$ for $x \in \partial B$ (recall that $d(P, Q) < d/4$), and then substituting $z = d^2/(\tau - t)$, we get

$$(3.21)\quad \int_{t^0}^{\tau}\int_{\partial B} |D_\xi G(x-\xi, t-\tau)|\, dS_x\, dt \le Kd^{n-1}\int_\beta^\infty \frac{z^{(n+1)/2}}{d^{n+1}}\frac{d^2}{z^2}\exp[-Kz]\,dz \le K,$$

since $\beta = d^2/(\tau - t^0) \ge 1$. Hence,

$$|H_2| \le K \operatorname*{l.u.b.}_N |f|.$$

Combining this with (3.19), we obtain from (3.16)

$$(3.22)\quad |H| \le K\, d^\alpha H_{Q,N}[f] + K \operatorname*{l.u.b.}_N |f|.$$

From (3.22), (3.15), and (3.9) follows the inequality (3.2) for $i = 2$.

Proof of (3.3) *for* $i = 2$. Write

$$(3.23)\quad d^\alpha \frac{|D_\xi^2 u(P) - D_\xi^2 u(Q)|}{d(P,Q)^\alpha} \le d^\alpha \frac{|H(P) - H(Q)|}{d(P,Q)^\alpha} + d^\alpha \frac{|J(P) - J(Q)|}{d(P,Q)^\alpha} \equiv H' + J',$$

where $J(Q)$, $H(Q)$ are given by (3.10), (3.16) with $Q = (\xi, \tau)$. We proceed to estimate J'.

It will be convenient to set $P = (\xi, \tau)$, $Q = (\xi', \tau')$. Thus $\xi = x^0$, $\tau = t^0 + d^2$, $\tau \ge \tau'$. Using (3.11) and noting that $(\Delta_x + D_t)D_\xi^2 G = 0$, we obtain

$$(3.24)\quad |J'| \le \frac{d^\alpha}{d(P,Q)^\alpha} K(\operatorname*{l.u.b.}_N |u|)\Big\{\sum_{i=0}^{1}(\operatorname*{l.u.b.}_N |D_x^{2-i}\varphi|) \cdot \int_{\Omega_0} |D_x^{2+i}G(x-\xi, t-\tau) - D_x^{2+i}G(x-\xi', t-\tau')|\,dx\,dt$$
$$+ (\operatorname*{l.u.b.}_N |D_t\varphi|)\int_{\Omega_0} |D_x^2 G(x-\xi, t-\tau) - D_x^2 G(x-\xi', t-\tau')|\,dx\,dt$$
$$+ \sum_{i=0}^{1}(\operatorname*{l.u.b.}_N |D_x^{2-i}\varphi|)\int_{\Omega_*} |D_x^{2+i}G(x-\xi, t-\tau)|\,dx\,dt$$
$$+ (\operatorname*{l.u.b.}_N |D_t\varphi|)\int_{\Omega_*} |D_x^2 G(x-\xi, t-\tau)|\,dx\,dt\Big\}.$$

Here Ω_0 is the region $(N_{3/4} - N_{1/2}) \cap \{t^0 \le t \le \tau'\}$ and Ω_* is the region $(N_{3/4} - N_{1/2}) \cap \{\tau' < t \le \tau\}$.

In order to estimate the integral

$$(3.25)\quad M \equiv \int_{\Omega_0} |D_x^{2+i}G(x-\xi, t-\tau) - D_x^{2+i}G(x-\xi', t-\tau')|\,dx\,dt$$

we apply the mean value theorem and get

$$M \le \int_{\Omega_0} [|D_x^{2+i+1}G(x-\bar\xi, t-\bar\tau)|\,|\xi - \xi'| + |D_x^{2+i}D_t G(x-\bar\xi, t-\bar\tau)|(\tau - \tau')]\,dx\,dt,$$

where $\bar{\xi}$ lies in the interval (ξ, ξ') and $\bar{\tau}$ lies in the interval (τ, τ'). Using (3.11) and then dividing Ω_0 into two regions as in the calculations following (3.12), we get (compare (3.13))

$$M \leq \frac{K|\xi - \xi'|}{d^{i+1}} + \frac{K(\tau - \tau')}{d^{i+2}}.$$

Since

$$|\xi - \xi'| \leq d(P, Q) \leq d/4, \qquad \tau - \tau' \leq d(P, Q)^2 \leq d^2/16,$$

it follows that

$$M \leq \frac{d(P, Q)^\alpha}{d^\alpha} \frac{K}{d^i}. \tag{3.26}$$

Next we have to estimate the integral

$$M_1 \equiv \int_{\Omega_*} |D_x^{2+i} G(x - \xi, t - \tau)| \, dx \, dt. \tag{3.27}$$

Noting that $|x - \xi| > d/4$ if $(x, t) \in \Omega_*$ and using (3.11), we get

$$M_1 \leq K d^n \int_{\tau'}^{\tau} (\tau - t)^{-(n+2+i)/2} \exp\left[-\frac{Kd^2}{\tau - t}\right] dt.$$

Substituting $z = d^2/(\tau - t)$ we get

$$M_1 \leq K d^n \int_A^\infty \frac{z^{(n+2+i)/2}}{d^{n+2+i}} \exp\,[-Kz] \frac{d^2}{z^2} \, dz \leq \frac{K}{d^i} \frac{1}{A^\mu}$$

for any $\mu \geq 0$ (K depends on μ), where $A = d^2/(\tau - \tau')$. Taking $\mu = \alpha/2$ and using

$$\frac{1}{A^{\alpha/2}} = \frac{(\tau - \tau')^{\alpha/2}}{d^\alpha} \leq \frac{d(P, Q)^\alpha}{d^\alpha},$$

we get

$$M_1 \leq \frac{d(P, Q)^\alpha}{d^\alpha} \frac{K}{d^i}. \tag{3.28}$$

Substituting the inequalities (3.26), (3.28) into (3.24), and using (3.5) we find that

$$J' \leq K d^{-2} \operatorname*{l.u.b.}_N |u|. \tag{3.29}$$

Estimation of H', defined by (3.23), (3.16). We can write

$$H' \leq W_1 + W_2 + W_3 + W_4 \tag{3.30}$$

where

$$W_1 = \frac{d^\alpha}{d(P, Q)^\alpha} \left| \int_{t^0}^{\tau'} \int_B \{[D_\xi^2 G(x - \xi, t - \tau)][\varphi(x, t) f(x, t) - \varphi(\xi, \tau) f(\xi, \tau)] \right.$$
$$\left. - [D_{\xi'}^2 G(x - \xi', t - \tau')][\varphi(x, t) f(x, t) - \varphi(\xi', \tau') f(\xi', \tau')]\} \, dx \, dt \right|,$$

$$W_2 = \frac{d^\alpha}{d(P, Q)^\alpha} \left| \varphi(\xi, \tau) f(\xi, \tau) D_\xi \int_{t^0}^{\tau'} \int_B D_\xi G(x - \xi, t - \tau) \, dx \, dt \right.$$
$$\left. - \varphi(\xi', \tau') f(\xi', \tau') D_{\xi'} \int_{t^0}^{\tau'} \int_B D_{\xi'} G(x - \xi', t - \tau') \, dx \, dt \right|,$$

$$W_3 = \frac{d^\alpha}{d(P, Q)^\alpha} \bigg| \int_{\tau'}^{\tau} \int_B [D_\xi^2 G(x - \xi, t - \tau)][\varphi(x, t)f(x, t)$$

$$- \varphi(\xi, \tau)f(\xi, \tau)] \, dx \, dt \bigg|,$$

$$W_4 = \frac{d^\alpha}{d(P, Q)^\alpha} \left|\varphi(\xi, \tau)f(\xi, \tau)D_\xi \int_{\tau'}^{\tau} \int_B D_\xi G(x - \xi, t - \tau) \, dx \, dt\right|.$$

To estimate W_3 we use (3.11), (3.17) and then substitute $|x - \xi| = (\tau - t)^{1/2}\rho$ and note that $\tau - \tau' < d(P, Q)^2$. We obtain (compare the derivation of (3.19))

$$W_3 \leq Kd^\alpha H_{P,N}[f] + K \operatorname*{l.u.b.}_N |f|. \tag{3.31}$$

To estimate W_4 we use the divergence theorem to reduce the integration over B into integration over ∂B. We next use the inequality $|x - \xi| \geq d$ (for $x \in \partial B$) and substitute $z = d^2/(\tau - t)$. We get

$$W_4 \leq K \left(\operatorname*{l.u.b.}_N |f|\right) \frac{d^\alpha}{d(P, Q)^\alpha} d^{n-1} \int_A^\infty \frac{z^{(n+1)/2}}{d^{n+1}} \exp\,[-Kz] \frac{d^2}{z^2} \, dz \tag{3.32}$$

$$\leq K \left(\operatorname*{l.u.b.}_N |f|\right) \frac{d^\alpha}{d(P, Q)^\alpha} \frac{1}{A^\mu},$$

for any $\mu \geq 0$, where $A = d^2/(\tau - \tau')$. Taking $\mu = \alpha/2$ and using the inequalities $\tau - \tau' \leq d(P, Q)^2 \leq d^2/16$, we find that

$$W_4 \leq K \operatorname*{l.u.b.}_N |f|. \tag{3.33}$$

We proceed to estimate W_2. Since $\varphi(\xi, \tau) = \varphi(\xi', \tau') = 1$,

$$W_2 \leq W_{21} + W_{22} \tag{3.34}$$

where

$$W_{21} = \frac{d^\alpha}{d(P, Q)^\alpha} \left(\operatorname*{l.u.b.}_N |f|\right) \bigg| D_\xi \int_{t^0}^{\tau'} \int_B D_\xi G(x - \xi, t - \tau) \, dx \, dt$$

$$- D_{\xi'} \int_{t^0}^{\tau'} \int_B D_{\xi'} G(x - \xi', t - \tau') \, dx \, dt \bigg|,$$

$$W_{22} = \frac{d^\alpha}{d(P, Q)^\alpha} H_{P,N}[f] d(P, Q)^\alpha \left| D_\xi \int_{t^0}^{\tau'} \int_B D_\xi G(x - \xi, t - \tau) \, dx \, dt \right|.$$

The last factor in W_{22} is similar to the last factor in the definition of H_2 in (3.16) and can be estimated by using (3.21). The inequality

$$W_{22} \leq Kd^\alpha H_{P,N}[f] \tag{3.35}$$

is thereby obtained.

If we prove that

$$W_{21} \leq K \operatorname*{l.u.b.}_N |f|, \tag{3.36}$$

then in conjunction with (3.35), (3.34), it follows that

$$W_2 \leq Kd^\alpha H_{P,N}[f] + K \operatorname*{l.u.b.}_N |f|. \tag{3.37}$$

Proof of (3.36). Consider first the special case where $\tau' = \tau$. By the mean value theorem, the last factor in W_{21} is bounded by

$$K|\xi - \xi'| \int_{t^0}^{\tau} \int_{\partial B} |D_{\bar{\xi}}^2 G(x - \bar{\xi}, t - \tau)| \, dS_x \, dt \tag{3.38}$$

where $\bar{\xi}$ lies in the interval (ξ, ξ'). Using the inequality $|x - \bar{\xi}| \geq 3d/4$ $(x \in \partial B)$, we can estimate the integral in the same manner that we have estimated the integral in (3.20), and then, using $|\xi - \xi'| = d(P, Q) \leq d/4$, we get the inequality (3.36).

Consider next the special case where $\xi = \xi'$. Then

$$W_{21} \leq \frac{d^\alpha}{d(P, Q)^\alpha} (\operatorname*{l.u.b.}_N |f|) \int_{\tau'}^{\tau} d\sigma \left[\int_{t^0}^{\tau'} dt \int_{\partial B} |D_\xi D_\sigma G(x - \xi, t - \sigma)| \, dS_x\right]. \tag{3.39}$$

The integral in the brackets can be estimated in the same manner by which the integrals in (3.38), (3.20) were estimated. We obtain the bound K/d^2. Substituting this into (3.39) and using $\tau - \tau' = d(P, Q)^2 \leq d^2/16$, we again obtain the inequality (3.36).

Finally, the proof of (3.36) in the general case follows by writing the last factor of W_{21} in the form

$$\left[D_\xi \int_{t^0}^{\tau'} \int_B D_\xi G(x - \xi, t - \tau) \, dx \, dt - D_\xi \int_{t^0}^{\tau'} \int_B D_\xi G(x - \xi, t - \tau') \, dx \, dt\right]$$
$$+ \left[D_\xi \int_{t^0}^{\tau'} \int_B D_\xi G(x - \xi, t - \tau') \, dx \, dt - D_{\xi'} \int_{t^0}^{\tau'} D_{\xi'} G(x - \xi', t - \tau') \, dx \, dt\right]$$

and using the results of the previous two special cases.

To complete the estimation of H' it remains to consider W_1. This will involve more lengthy considerations than for the previous W_i, but once we finish with estimating W_1, the proof of Theorem 2 is easily completed.

Estimation of W_1. We split W_1 into four terms and, using (3.17), get

$$W_1 \leq V_1 + V_2 + V_3 + V_4, \tag{3.40}$$

where

$$\begin{aligned} V_1 = K \frac{d^\alpha}{d(P, Q)^\alpha} \int_{t^0}^{\tau' - \eta} \int_B &|D_\xi^2 G(x - \xi, t - \tau) \\ &- D_{\xi'}^2 G(x - \xi', t - \tau')| \\ &\cdot \left[(|x - \xi|^\alpha + (\tau - t)^{\alpha/2}) H_{P,N}[f]\right. \\ &+ \left.\left(\frac{|x - \xi|}{d} + \frac{\tau - t}{d^2}\right) \operatorname*{l.u.b.}_N |f|\right] dx \, dt, \end{aligned}$$

$$V_2 = K \frac{d^\alpha}{d(P, Q)^\alpha} \left| \int_{t^0}^{\tau'-\eta} \int_B D_{\xi'}^2 G(x - \xi', t - \tau')\, dx\, dt \right|$$

$$\cdot \left[(|\xi - \xi'|^\alpha + (\tau - \tau')^{\alpha/2}) H_{P,N}[f] \right.$$

$$\left. + \left(\frac{|\xi - \xi'|}{d} + \frac{\tau - \tau'}{d^2} \right) \operatorname*{l.u.b.}_N |f| \right],$$

$$V_3 = K \frac{d^\alpha}{d(P, Q)^\alpha} \int_{\tau'-\eta}^{\tau'} \int_B |D_\xi^2 G(x - \xi, t - \tau)|$$

$$\cdot \left[(|x - \xi|^\alpha + (\tau - t)^{\alpha/2}) H_{P,N}[f] \right.$$

$$\left. + \left(\frac{|x - \xi|}{d} + \frac{\tau - t}{d^2} \right) \operatorname*{l.u.b.}_N |f| \right] dx\, dt,$$

$$V_4 = K \frac{d^\alpha}{d(P, Q)^\alpha} \int_{\tau'-\eta}^{\tau'} \int_B |D_{\xi'}^2 G(x - \xi', t - \tau')|$$

$$\cdot \left[(|x - \xi'|^\alpha + (\tau' - t)^{\alpha/2}) H_{Q,N}[f] \right.$$

$$\left. + \left(\frac{|x - \xi'|}{d} + \frac{\tau' - t}{d^2} \right) \operatorname*{l.u.b.}_N |f| \right] dx\, dt.$$

Here $\eta = d(P, Q)^2/4$.

Using (3.11) and substituting $|x - \xi| = (\tau - t)^{1/2}\rho$ we find that the coefficient of $H_{P,N}[f]$ in V_3 is bounded by

$$K \frac{d^\alpha}{d(P, Q)^\alpha} \int_{\tau'-\eta}^{\tau'} \int_0^\infty \frac{(\tau - t)^{\alpha/2}\rho^\alpha + (\tau - t)^{\alpha/2}}{(\tau - t)^{(n+2)/2}} \exp\left[-K\rho\right](\tau - t)^{n/2}\rho^{n-1}\, d\rho\, dt$$

$$\leq K \frac{d^\alpha}{d(P, Q)^\alpha} \int_{\tau'-\eta}^{\tau'} \frac{dt}{(\tau - t)^{1-\alpha/2}} \leq K \frac{d^\alpha}{d(P, Q)^\alpha} \eta^{\alpha/2} = K d^\alpha.$$

Similarly we find that the coefficients of $\operatorname*{l.u.b.}_N |f|$ in V_3 are bounded by K. Hence

$$V_3 \leq K d^\alpha H_{P,N}[f] + K \operatorname*{l.u.b.}_N |f|. \tag{3.41}$$

In the same way

$$V_4 \leq K d^\alpha H_{Q,N}[f] + K \operatorname*{l.u.b.}_N |f|. \tag{3.42}$$

Using the divergence theorem and calculations similar to those in (3.21), we find that the integral in V_2 is bounded by K. Hence,

$$V_2 \leq K d^\alpha H_{P,N}[f] + K \operatorname*{l.u.b.}_N |f|. \tag{3.43}$$

It remains to estimate V_1. For simplicity, consider first the special case where $\tau = \tau'$. Let $V_1 = V_{11} + V_{12}$ where V_{11} is the coefficient of $H_{P,N}[f]$

in V_1 and V_{12} is the coefficient of $\underset{N}{\text{l.u.b.}}\,|f|$ in V_1. If ζ is a variable point in the interval (ξ, ξ'), then

$$V_{11} \leq K \frac{d^\alpha}{d(P, Q)^\alpha} \left| \int_\xi^{\xi'} d\zeta \int_{t^0}^{\tau'-\eta} dt \int_B |D_\zeta^3 G(x - \zeta, t - \tau)| \cdot [|x - \zeta|^\alpha + (\tau - t)^{\alpha/2} + |\zeta - \xi|^\alpha]\, dx \right| \cdot$$

Using (3.11) and substituting $|x - \zeta| = (\tau - t)^{1/2}\rho$, we obtain

$$V_{11} \leq K \frac{d^\alpha}{d(P, Q)^\alpha} \left| \int_\xi^{\xi'} \left(\frac{1}{\eta^{(1-\alpha)/2}} + \frac{|\zeta - \xi|^\alpha}{\eta^{1/2}} \right) d\zeta \right| \leq K d^\alpha. \tag{3.44}$$

V_{12} can be evaluated in a similar way. We thus get

$$V_1 \leq K d^\alpha H_{P,N}[f] + K \underset{N}{\text{l.u.b.}}\,|f|. \tag{3.45}$$

Consider next the special case where $\xi = \xi'$ and decompose V_1 into $V_{11} + V_{12}$ as before. Writing

$$V_{11} \leq K \frac{d^\alpha}{d(P, Q)^\alpha} \int_{\tau'}^{\tau} d\sigma \int_{t^0}^{\tau'-\eta} dt \int_B |D_\xi^2 D_\sigma G(x - \xi, t - \sigma)| [|x - \xi|^\alpha + (\sigma - t)^{\alpha/2} + (\tau - \sigma)^{\alpha/2}]\, dx,$$

we can proceed similarly to the previous case by using (3.11) and substituting $|x - \xi| = (t - \sigma)^{1/2}\rho$. We get $V_{11} \leq K d^\alpha$. V_{12} is treated in the same way and (3.45) thus follows.

In the general case, decomposing V_1 into $V_{11} + V_{12}$ we have,

$$\begin{aligned} V_{11} \leq{} & K \frac{d^\alpha}{d(P, Q)^\alpha} \int_{t^0}^{\tau'-\eta} \int_B |D_\xi^2 G(x - \xi, t - \tau) - D_\xi^2 G(x - \xi, t - \tau')| \\ & \cdot [|x - \xi|^\alpha + (\tau - t)^{\alpha/2}]\, dx\, dt \\ & + K \frac{d^\alpha}{d(P, Q)^\alpha} \int_{t^0}^{\tau'-\eta} \int_B |D_\xi^2 G(x - \xi, t - \tau') - D_{\xi'}^2 G(x - \xi', t - \tau')| \\ & \cdot [|x - \xi'|^\alpha + (\tau' - t)^{\alpha/2}]\, dx\, dt \\ & + K \frac{d^\alpha}{d(P, Q)^\alpha} \int_{t^0}^{\tau'-\eta} \int_B |D_\xi^2 G(x - \xi, t - \tau') - D_{\xi'}^2 (x - \xi', t - \tau')|dx\, dt \\ & \cdot [|\xi - \xi'|^\alpha + (\tau - \tau')^{\alpha/2}]. \end{aligned}$$

The first two terms are estimated by the previous two special cases. The last term is bounded by

$$K d^\alpha \left| \int_\xi^{\xi'} d\zeta \int_{t^0}^{\tau'-\eta} dt \int_B \left| D_\zeta^3 G(x - \zeta, t - \tau') \right| dx \right|.$$

Using (3.11) and substituting $|x - \zeta| = (\tau - t)^{1/2}\rho$ we find that the last expression is bounded by

$$K d^\alpha \left| \int_\xi^{\xi'} \frac{d\zeta}{\eta^{1/2}} \right| \leq K d^\alpha \frac{|\xi' - \xi|}{d(P, Q)} \leq K d^\alpha.$$

We have thus proved that $V_{11} \leq Kd^\alpha H_{P,N}[f]$. Similarly one estimates V_{12} and thus establishes (3.45).

Combining (3.45) with (3.41)–(3.43) and using (3.40), we get

$$W_1 \leq Kd^\alpha H_{P,N}[f] + Kd^\alpha H_{Q,N}[f] + K \operatorname*{l.u.b.}_N |f|. \tag{3.46}$$

We next combine (3.46) with (3.37), (3.33), (3.31) and find an estimate for H' (see (3.30)) which, when combined with (3.29), (3.23), completes the proof of (3.3) for $i = 2$.

We have thus completed the proof of (3.2), (3.3) in the case $i = 2$. To prove these inequalities for $i = 0, 1$ we apply D_ξ^i to both sides of (3.7) and then proceed step by step along the previous proof for $i = 2$. The details may be left to the reader.

We conclude this section by giving an extension of Theorem 2 to equations

$$\sum_{i,j=1}^{n} a_{ij} \frac{\partial^2 u}{\partial x_i \, \partial x_j} - \frac{\partial u}{\partial t} = f, \tag{3.47}$$

where a_{ij} are constants, $|a_{ij}| \leq k_1$, and for any real vector ξ,

$$\sum_{i,j=1}^{n} a_{ij}\xi_i\xi_j \geq k_2|\xi|^2 \qquad (k_2 > 0). \tag{3.48}$$

Theorem 2′. *Theorem 2 remains true if the equation* (3.1) *is replaced by the equation* (3.47), *but the constant* K *in the inequalities* (3.2), (3.3) *now depends on* k_1, k_2, n, α.

The proof is similar to that of Theorem 2 provided $G(x - \xi, t - \tau)$ is now defined as

$$\frac{(\det (a^{ij}))^{1/2}}{(2\sqrt{\pi})^n} (\tau - t)^{-n/2} \exp \left[- \frac{\sum a^{ij}(x_i - \xi_i)(x_j - \xi_j)}{4(\tau - t)} \right],$$

where (a^{ij}) is the inverse matrix to (a_{ij}). The inequalities (3.11) hold with 1/5 (in the exponent) replaced by some K. All the other calculations are the same provided Δ_x is replaced by $\sum a_{ij}\partial^2/\partial x_i \, \partial x_j$.

4. Derivation of the Interior Estimates

In this section we shall prove Theorem 1, Sec. 1. From Lemma 2, Sec. 2, it follows that it suffices to prove that

$$M \equiv M_{0,2}[u] \leq K(\operatorname*{l.u.b.}_D |u| + |f|_{2,\alpha}), \tag{4.1}$$

$$M_i' \equiv M_{0,i+\alpha}[u] \leq K(\operatorname*{l.u.b.}_D |u| + |f|_{2,\alpha}) \qquad (i = 0, 1, 2). \tag{4.2}$$

Set $M_{0,i} = M_{0,i}[u]$, $M_{0,i+\alpha} = M_{0,i+\alpha}[u]$. Without loss of generality we may assume that the $M_{0,i}$, $M_{0,i+\alpha}$ are all finite since otherwise we first establish the inequalities (4.1), (4.2) in D_ϵ and then take $\epsilon \to 0$, using the same argument as in the first paragraph of the proof of Lemma 2.

Since M is finite, there exists a point $P \in D$ and a derivative D_x^2 such that

$$\frac{1}{2} M \leq d_P^2 |D_x^2 u(P)|. \tag{4.3}$$

Let N be a semicube with top P and edge λd_P where $\lambda < 1/2(n+1)^{1/2}$ and is still to be determined. Writing (1.1) in the form

$$\begin{aligned} \Sigma\, a_{ij}(P) \frac{\partial^2 u}{\partial x_i\, \partial x_j} - \frac{\partial u}{\partial t} &= \Sigma\, [a_{ij}(P) - a_{ij}(Q)] \frac{\partial^2 u}{\partial x_i\, \partial x_j} \\ &\quad - \Sigma\, b_i(Q) \frac{\partial u}{\partial x_i} - c(Q)u + f(Q) \\ &\equiv F(Q) \qquad (Q = (x, t)) \end{aligned} \tag{4.4}$$

and applying Theorem 2′, we get

$$|D_x^2 u(P)| \leq K d^{-2} \operatorname*{l.u.b.}_N |u| + K \operatorname*{l.u.b.}_N |F| + K d^\alpha H_{P,N}[F] \equiv KI. \tag{4.5}$$

In the sequel we shall make use of the inequality

$$H_{P,N}[gh] \leq |g(P)| H_{P,N}(h) + (\operatorname*{l.u.b.}_N |h|) H_{P,N}[g]. \tag{4.6}$$

To estimate I we begin by noting that

$$d^{-2} \operatorname*{l.u.b.}_N |u| \leq \lambda^{-2} d_P^{-2} \operatorname*{l.u.b.}_D |u|. \tag{4.7}$$

Using the definitions (1.4)–(1.7) and the assumption (1.8), and noting that $d_Q \geq [1 - \lambda(n+1)^{1/2}] d_P > d_P/2$ if $Q \in N$, we get

$$\begin{aligned} &\operatorname*{l.u.b.}_N |F| \\ &\leq \operatorname*{l.u.b.}_N \left\{ |f| + \Sigma\, |b_i| \left| \frac{\partial u}{\partial x_i} \right| + |c||u| + \Sigma\, |a_{ij}(P) - a_{ij}(Q)| \left| \frac{\partial^2 u}{\partial x_i\, \partial x_j} \right| \right\} \\ &\leq K d_P^{-2} (|f|_{2,\alpha} + |u|_{0,1}) + K \frac{(\lambda d_P)^\alpha}{(d_P/2)^\alpha} d_P^{-2} M \\ &\leq K d_P^{-2} \{ |f|_{2,\alpha} + |u|_{0,1} + \lambda^\alpha M \}. \end{aligned} \tag{4.8}$$

Using (1.4)–(1.8) and the rule (4.6) we find that

$$\begin{aligned} d^\alpha H_{P,N}[F] &\leq K d^\alpha d_P^{-2-\alpha} |f|_{2,\alpha} + K d^\alpha \{ \Sigma\, |b_i(P)| H_{P,N}[D_x u] \\ &\quad + |c(P)| H_{P,N}[u] \} + K d^\alpha \{ \Sigma\, H_{P,N}[a_{ij}] (\operatorname*{l.u.b.}_N |D_x^2 u|) \\ &\quad + \Sigma\, H_{P,N}[b_i] (\operatorname*{l.u.b.}_N |D_x u|) + H_{P,N}[c] (\operatorname*{l.u.b.}_N |u|) \} \\ &\leq K \lambda^\alpha d_P^{-2} |f|_{2,\alpha} + K \lambda^\alpha d_P^\alpha \sum_{i=0}^{1} d_P^{i-2} H_{P,N}[D_x^i u] \\ &\quad + K \lambda^\alpha d_P^{-2} |u|_{0,2}. \end{aligned} \tag{4.9}$$

Since $H_{P,N}[D_x^i u] \le K d_P^{-i-\alpha} M_{0,i+\alpha}$, we get

$$d^\alpha H_{P,N}[F] \le K\lambda^\alpha d_P^{-2} \left\{ |f|_{2,\alpha} + |u|_{0,2} + \sum_{i=0}^{1} M_{0,i+\alpha} \right\}. \tag{4.10}$$

Substituting the inequalities (4.7), (4.8), (4.10) into the right-hand side of (4.5), we get

$$d_P^2 I \le K\lambda^{-2}|u|_0 + K|f|_{2,\alpha} + K|u|_{0,1} + K\lambda^\alpha(M + M_0' + M_1'). \tag{4.11}$$

Since, by (2.3), for any $\epsilon > 0$

$$|u|_{0,1} \le \epsilon M + \frac{C}{\epsilon}|u|_0 \tag{4.12}$$

where C is independent of ϵ, taking $\epsilon = \lambda^\alpha$ we obtain from (4.11) and (4.5),

$$d_P^2|D_x^2 u(P)| \le d_P^2 I \le K\lambda^{-2}|u|_0 + K|f|_{2,\alpha} + K\lambda^\alpha(M + M_0' + M_1'). \tag{4.13}$$

Substituting this into (4.3) we obtain

$$M \le K\lambda^{-2}|u|_0 + K|f|_{2,\alpha} + K\lambda^\alpha(M + M_0' + M_1'). \tag{4.14}$$

We shall now derive bounds for the M_i'. We begin with M_2'. Since M_2' is finite, there exist two points P, Q in D and a derivative D_x^2 such that

$$\frac{1}{2} M_2' \le d_{PQ}^{2+\alpha} \frac{|D_x^2 u(P) - D_x^2 u(Q)|}{d(P, Q)^\alpha}. \tag{4.15}$$

We may assume that the t-coordinate of P is greater or equal to the t-coordinate of Q.

There are two cases:

(a) $d(P, Q) \ge \lambda d_{PQ}/4(n+1)^{1/2}$. Then,

$$M_2' \le K d_P^2 \lambda^{-\alpha}|D_x^2 u(P)| + K d_Q^2 \lambda^{-\alpha}|D_x^2 u(Q)| \le K\lambda^{-\alpha} M. \tag{4.16}$$

(b) $d(P, Q) < \lambda d_{PQ}/4(n+1)^{1/2}$. Then $d(P, Q) < \lambda d_P/4(n+1)^{1/2}$. Let N be the semicube with top P and edge λd_P. Applying Theorem 2′, Sec. 3, with $d = \lambda d_P/(n+1)^{1/2}$ to the equation (4.4) we get

$$d^\alpha \frac{|D_x^2 u(P) - D_x^2 u(Q)|}{d(P, Q)^\alpha} \le KI + K d^\alpha H_{Q,N}[F]. \tag{4.17}$$

Since I was already estimated in (4.13), it remains to estimate $d^\alpha H_{Q,N}[F]$. Analogously to (4.9), (4.10) we obtain

$$d^\alpha H_{Q,N}[F] \le K\lambda^\alpha d_P^{-2} \left\{ |f|_{2,\alpha} + |u|_{0,2} + \sum_{i=0}^{1} M_{0,i+\alpha} \right\} + K d^\alpha \sum |a_{ij}(P) - a_{ij}(Q)| H_{Q,N}[D_x^2 u]. \tag{4.18}$$

The last term on the right-hand side is bounded by

$$K d^\alpha d_{PQ}^{-\alpha} d(P, Q)^\alpha H_{Q,N}[D_x^2 u] \le K\lambda^\alpha H_{Q,N}[D_x^2 u] \le K\lambda^{2\alpha} d_P^{-2} M_2'.$$

Substituting this into (4.18) we get a bound on $d^\alpha H_{Q,N}[F]$. Using this bound and (4.13), we get from (4.15), (4.17)

$$(4.19)\quad M_2' \le K\lambda^{-\alpha}d_P^2 I + K\lambda^{-\alpha}d_P^2 d^\alpha H_{Q,N}[F]$$
$$\le K\lambda^{-2-\alpha}|u|_0 + K\lambda^{-\alpha}|f|_{2,\alpha} + K(M + M_0' + M_1') + K\lambda^\alpha M_2'.$$

By the same method by which we have derived (4.16), (4.19) one can also show that, if $i = 0, 1$, then either (case (a))

$$(4.20)\qquad M_i' \le K\lambda^{-\alpha}M_{0,i},$$

or (case (b))

$$(4.21)\quad M_i' \le K\lambda^{-i-\alpha}|u|_0 + K\lambda^{2-i-\alpha}|f|_{2,\alpha}$$
$$+ K\lambda^{2-i}(M + M_0' + M_1') + K\lambda^{2-i+\alpha}M_2'.$$

Hence M_i' is bounded by the sum of the right-hand sides of (4.20) and (4.21). Taking next λ sufficiently small (depending only on K) we get

$$(4.22)\quad M_0' + M_1' \le K\lambda^{-1-\alpha}|u|_0 + K\lambda^{1-\alpha}|f|_{2,\alpha} + K\lambda M$$
$$+ K\lambda^{1+\alpha}M_2' + K\lambda^{-\alpha}M_{0,1}.$$

Substituting this into (4.19) we get, upon taking λ to be sufficiently small (depending only on K),

$$(4.23)\qquad M_2' \le K\lambda^{-2-\alpha}|u|_0 + K\lambda^{-\alpha}|f|_{2,\alpha} + KM + K\lambda^{-\alpha}M_{0,1}.$$

We can now easily complete the proof of the theorem. Indeed, in case (a) we add (4.14), (4.22) and evaluate the M_2' which appears on the right-hand side of (4.22) by (4.16). We get

$$(4.24)\quad M + M_0' + M_1' \le K\lambda^{-2}|u|_0 + K|f|_{2,\alpha}$$
$$+ K\lambda^\alpha(M + M_0' + M_1') + K\lambda^{-\alpha}M_{0,1}.$$

Taking λ such that $K\lambda^\alpha < 1/4$ and then using Lemma 2 to conclude that (with λ now being fixed)

$$K\lambda^{-\alpha}M_{0,1} < \frac{1}{4}M + C|u|_0 \qquad (C = C(K)),$$

we obtain from (4.24)

$$(4.25)\qquad M + M_0' + M_1' \le K|u|_0 + K|f|_{2,\alpha}.$$

Since, by (4.16), a similar bound also holds for M_2', the proof of the theorem is completed in case (a).

In case (b), substituting (4.23) into (4.22) we get

$$M_0' + M_1' \le K\lambda^{-1-\alpha}|u|_0 + K\lambda^{1-\alpha}|f|_{2,\alpha} + K\lambda M + K\lambda^{-\alpha}M_{0,1}.$$

Combining this inequality with (4.14), we get

$$M + M_0' + M_1' \le K\lambda^{-2}|u|_0 + K|f|_{2,\alpha} + K\lambda^\alpha(M + M_0' + M_1') + K\lambda^{-\alpha}M_{0,1}.$$

Since this inequality coincides with (4.24), the inequality (4.25) follows as above. From (4.23) we also obtain the desired bound on M_2'. The proof of Theorem 1, Sec. 1, is thus completed.

5. A Fundamental Lemma

Calculations similar to those used in Secs. 3, 4 in deriving the interior estimates will be used in Secs. 6, 7 in deriving the boundary estimates. Additionally we shall need an auxiliary lemma to which the present section is devoted.

Let R denote the interval $-1 < x < 1$, let S_τ ($S_1 = S$) denote the rectangle $-1 < x < 1$, $0 < t < \tau$, and let

$$V_j(x, t; \xi, \tau) = \frac{(\tau - t)^{-j/2}}{2\sqrt{\pi}} \exp\left[-\frac{(x-\xi)^2}{4(\tau - t)}\right], \qquad V_1 = V. \tag{5.1}$$

Clearly, for some constant C,

$$\left|\frac{\partial}{\partial \xi} V(x, t; \xi, \tau)\right| \leq C|x - \xi| V_3(x, t; \xi, \tau), \tag{5.2}$$

$$\left|\frac{\partial^2}{\partial \xi^2} V(x, t; \xi, \tau)\right| \leq C|x - \xi|^2 V_5(x, t; \xi, \tau) + C V_3(x, t; \xi, \tau). \tag{5.3}$$

We also have,

$$\left|\frac{\partial^k}{\partial \tau^k}\frac{\partial^j}{\partial \xi^j} V(x, t; \xi, \tau)\right| \leq C' V_{j+2k+1}(x, 2t; \xi, 2\tau) \tag{5.4}$$

where C' depends only on k, j.

Let f be a Hölder continuous (exponent α) function in the closure $\overline{S}$ of S, i.e.,

$$\frac{|f(x, t) - f(x', t')|}{[(x - x')^2 + |t - t'|]^{\alpha/2}} \leq H[f] < \infty \qquad ((x, t) \in \overline{S}, (x', t') \in \overline{S}) \tag{5.5}$$

where $H[f]$ is the Hölder coefficient of f.

In this section various positive constants which depend only on α will be denoted by A.

The main concern of the present section is the establishing of the following lemma.

Lemma 6. *If $f(\pm 1, 0) = 0$ then the function*

$$v(\xi, \tau) = \int_0^\tau \int_R V(x, t; \xi, \tau) f(x, t)\, dx\, dt \tag{5.6}$$

satisfies the Hölder condition

$$\frac{|v_\tau(\xi, \tau) - v_{\tau'}(\xi', \tau')|}{[(\xi - \xi')^2 + |\tau - \tau'|]^{\alpha/2}} \leq AH[f] \tag{5.7}$$

for all $(\xi, \tau) \in S$, $(\xi', \tau') \in S$, where $v_\tau = \partial v/\partial \tau$.

Proof. Consider first the special case where $\tau = \tau'$. We may assume that $\xi' < \xi$. It suffices to prove (5.7) in case

$$0 \leq \xi' < \xi < 1. \tag{5.8}$$

Indeed, the case $-1 < \xi' < \xi \leq 0$ can be reduced to the previous one by making a transformation $\xi \to -\xi$, $x \to -x$ in (5.6) and changing the roles of ξ' and ξ, whereas the case $-1 < \xi' < 0 < \xi < 1$ follows by writing

$$v_\tau(\xi, \tau) - v_\tau(\xi', \tau) = [v_\tau(\xi, \tau) - v_\tau(0, \tau)] + [v_\tau(0, \tau) - v_\tau(\xi', 0)]$$

and using the previous two cases.

From the assertion (1.3.27) of Theorem 5, Chap. 1, we get, upon specializing to $a_{ij} = \delta_{ij}$, $n = 1$,

$$v_\tau(\xi, \tau) = -f(\xi, \tau) - w_1(\xi, \tau) + w_2(\xi, \tau) \tag{5.9}$$

where

$$w_1(\xi, \tau) = \Sigma \pm \int_0^\tau \left[\frac{\partial}{\partial \xi} V(\pm 1, t; \xi, \tau)\right] f(\xi, t)\, dt, \tag{5.10}$$

$$w_2(\xi, \tau) = \iint_{S_T} \left[\frac{\partial^2}{\partial \xi^2} V(x, t; \xi, \tau)\right] [f(x, t) - f(\xi, t)]\, dx\, dt. \tag{5.11}$$

Consider first w_2. Setting $\gamma = (\xi - \xi')^2$, $B_\gamma = S_\tau - S_{\tau-\gamma}$ (if $\tau - \gamma < 0$ then $B_\gamma = S_\tau$), we can write

$$w_2(\xi, \tau) - w_2(\xi', \tau) = J_1 - J_2 + J_3 - J_4, \tag{5.12}$$

where

$$J_1 = J_1(\xi, \tau) = \iint_{B\gamma} \left[\frac{\partial^2}{\partial \xi^2} V(x, t; \xi, \tau)\right] [f(x, t) - f(\xi, t)]\, dx\, dt,$$

$$J_2 = J_1(\xi', \tau),$$

$$J_3 = \iint_{S_{T-\gamma}} \left[\frac{\partial^2}{\partial \xi^2} V(x, t; \xi, \tau) - \frac{\partial^2}{\partial \xi'^2} V(x, t; \xi', \tau)\right] [f(x, t) - f(\xi, t)]\, dx\, dt,$$

$$J_4 = \iint_{S_{T-\gamma}} \left[\frac{\partial^2}{\partial \xi'^2} V(x, t; \xi', \tau)\right] [f(\xi, t) - f(\xi', t)]\, dx\, dt.$$

Using (5.4) and substituting $|x - \xi| = (\tau - t)^{1/2}\rho$, we get

$$|J_1| \leq AH[f] \int_{\tau-\gamma}^\tau \int_R V_3(x, 2t; \xi, 2\tau)|x - \xi|^\alpha\, dx\, dt$$

$$\leq AH[f] \int_{\tau-\gamma}^\tau \frac{dt}{(\tau - t)^{1-\alpha/2}} \leq AH[f]\gamma^{\alpha/2}.$$

By the definition of γ it then follows that

$$|J_1| = AH[f](\xi - \xi')^\alpha. \tag{5.13}$$

Similarly,

$$|J_2| \leq AH[f](\xi - \xi')^\alpha. \tag{5.14}$$

To estimate J_3 we use the relation

$$g(\xi) - g(\xi') = \int_{\xi'}^\xi g'(\zeta)\, d\zeta \tag{5.15}$$

with $g = \partial^2 V/\partial\xi^2$, and (5.4). We get

$$|J_3| \leq AH[f] \int_{\xi'} d\zeta \int_0^{\tau-\gamma} V_4(x, 2t; \zeta, 2\tau)|x - \xi|^\alpha \, dx \, dt.$$

Using the inequality $|x - \xi|^\alpha \leq |x - \zeta|^\alpha + |\zeta - \xi|^\alpha$ and then substituting $|x - \zeta| = (\tau - t)^{1/2}\rho$, we get

$$\begin{aligned}|J_3| &\leq AH[f] \int_{\xi'}^{\xi} d\zeta \int_0^{\tau-\gamma} \frac{dt}{(\tau - t)^{(3-\alpha)/2}} \\ &+ AH[f] \int_{\xi'}^{\xi} (\xi - \zeta)^\alpha \, d\zeta \int_0^{\tau-\gamma} \frac{d\tau}{(\tau - t)^{3/2}} \\ &\leq AH[f] \int_{\xi'}^{\xi} \frac{d\zeta}{\gamma^{(1-\alpha)/2}} + AH[f] \int_{\xi'}^{\xi} \frac{(\xi - \zeta)^\alpha}{\gamma^{1/2}} \, d\zeta.\end{aligned}$$

Hence,

$$|J_3| \leq AH[f](\xi - \xi')^\alpha. \tag{5.16}$$

As will be seen shortly, there is no need to estimate J_4.

Turning to w_1, we can write

$$w_1(\xi, \tau) - w_1(\xi', \tau) = K_1 + K_2 + K_3, \tag{5.17}$$

where

$$K_1 = \Sigma \pm \int_0^\tau \left[\frac{\partial}{\partial\xi} V(\pm 1, t; \xi, \tau) - \frac{\partial}{\partial\xi'} V(\pm 1, t; \xi', \tau)\right] f(\xi, t) \, dt,$$

$$K_2 = \Sigma \pm \int_{\tau-\gamma}^{\tau} \left[\frac{\partial}{\partial\xi'} V(\pm 1, t; \xi', \tau)\right] [f(\xi, t) - f(\xi', t)] \, dt$$

$$K_3 = \Sigma \pm \int_0^{\tau-\gamma} \left[\frac{\partial}{\partial\xi'} V(\pm 1, t; \xi', \tau)\right] [f(\xi, t) - f(\xi', t)] \, dt.$$

Noting that in J_4 $\partial/\partial\xi'$ may be replaced by $-\partial/\partial x$, and then integrating with respect to x, we obtain the relation

$$K_3 + J_4 = 0. \tag{5.18}$$

It remains to estimate K_1, K_2. By (5.2),

$$|K_2| \leq AH[f] \, \Sigma \int_{\tau-\gamma}^{\tau} |\pm 1 - \xi'| V_3(\pm 1, t; \xi', \tau)(\xi - \xi')^\alpha \, dt$$

Substituting $z = (\pm 1 - \xi')^2/(\tau - t)$ we get

$$|K_2| \leq AH[f](\xi - \xi')^\alpha \int_0^\infty z^{-1/2} \exp\left[-\frac{z}{4}\right] dz.$$

Hence

$$|K_2| \leq AH[f](\xi - \xi')^\alpha. \tag{5.19}$$

To estimate K_1 write

$$K_1 = L_1 + L_2 \tag{5.20}$$

where

$$L_1 = \Sigma \pm \int_0^\tau \left[\frac{\partial}{\partial \xi} V(\pm 1, t; \xi, \tau) - \frac{\partial}{\partial \xi'} V(\pm 1, t; \xi', \tau)\right] [f(\xi, t) - f(\pm 1, t)]\, dt,$$

$$L_2 = \Sigma \pm \int_0^\tau \left[\frac{\partial}{\partial \xi} V(\pm 1, t; \xi, \tau) - \frac{\partial}{\partial \xi'} V(\pm 1, t; \xi', \tau)\right] f(\pm 1, t)\, dt.$$

Using (5.15) with $g = \partial V/\partial \xi$ and (5.4), we obtain

$$|L_1| \leq AH[f]\, \Sigma \int_{\xi'} d\zeta \int_0^\tau V_3(\pm 1, 2t; \zeta, 2\tau) |\pm 1 - \xi|^\alpha\, dt.$$

Using the inequality $|\pm 1 - \xi|^\alpha \leq |\pm 1 - \zeta|^\alpha + (\xi - \zeta)^\alpha$ and then substituting $z = (\pm 1 - \zeta)^2/(\tau - t)$, the inequality

$$|L_1| \leq AH[f]\, \Sigma \int_{\xi'}^{\xi} \left[\frac{1}{|\pm 1 - \zeta|^{1-\alpha}} + \frac{(\xi - \zeta)^\alpha}{|\pm 1 - \zeta|}\right] d\zeta$$

is obtained. Noting that $\xi - \zeta \leq |\pm 1 - \zeta|$ (here we make use of (5.8)), we get

$$|L_1| \leq AH[f](\xi - \xi')^\alpha. \tag{5.21}$$

To evaluate L_2 introduce the functions

$$g_\pm(t) = \begin{cases} f(\pm 1, t) & \text{if } 0 \leq t < 1 \\ 0 & \text{if } -\infty < t < 0. \end{cases}$$

Then

$$\begin{aligned} L_2 = \Sigma \pm \int_{-\infty}^\tau \left[\frac{\partial}{\partial \xi} V(\pm 1, t; \xi, \tau) - \frac{\partial}{\partial \xi'} V(\pm 1, t; \xi', \tau)\right] [g_\pm(t) - g_\pm(\tau)]\, dt \\ + \Sigma \pm g_\pm(\tau) \int_{-\infty}^\tau \left[\frac{\partial}{\partial \xi} V(\pm 1, t; \xi, \tau) - \frac{\partial}{\partial \xi'} V(\pm 1, t; \xi', \tau)\right] dt \\ \equiv L_{21} + L_{22}. \end{aligned} \tag{5.22}$$

It is easily verified that

$$L_{22} = 0. \tag{5.23}$$

Indeed, the integral of L_{22} is equal to

$$\begin{aligned} \int_{\xi'}^{\xi} d\zeta \int_{-\infty}^{\tau} \frac{\partial^2}{\partial \zeta^2} V(\pm 1, t; \zeta, \tau)\, dt &= - \int_{\xi'}^{\xi} d\zeta \int_{-\infty}^{\tau} \frac{\partial}{\partial t} V(\pm 1, t; \zeta, \tau)\, dt \\ &= - \int_{\xi'}^{\xi} \lim_{\substack{\epsilon \to 0 \\ B \to -\infty}} [V(\pm 1, t; \zeta, \tau)]_B^{\tau - \epsilon} = 0. \end{aligned}$$

To estimate L_{21} we make use of the assumption $f(\pm 1, 0) = 0$, from which it follows that $g_\pm(t)$ are not only continuous functions but also Hölder continuous (exponent α) for $-\infty < t \leq 1$. Using this fact and (5.4), one obtains

$$|L_{21}| \leq AH[f] \Sigma \int_{\xi'}^{\xi} d\zeta \int_{-\infty}^{\tau} V_3(\pm 1, 2t; \zeta, 2\tau)(\tau - t)^{\alpha/2} \, dt.$$

Substituting $z = (\pm 1 - \zeta)^2/(\tau - t)$ we get

$$|L_{21}| \leq AH[f] \Sigma \int_{\xi'}^{\xi} |\pm 1 - \zeta|^{-1+\alpha} \, d\zeta \leq AH[f](\xi - \xi')^{\alpha}.$$

Combining this with (5.23), (5.22), we get

$$|L_2| \leq AH[f](\xi - \xi')^{\alpha}. \tag{5.24}$$

From (5.24), (5.21), (5.20), and (5.19) it follows that

$$|K_1 + K_2| \leq AH[f](\xi - \xi')^{\alpha}.$$

Combining this inequality with (5.18), (5.17), (5.16), (5.14), (5.13), (5.12), and (5.9), the proof of (5.7), in case $\tau' = \tau$, is completed.

If $\xi = \xi'$, $\tau < \tau'$ then we use for v_τ the representation

$$\begin{aligned} v_\tau(\xi, \tau) = &-2f(\xi, \tau) + \int_R V(x, 0; \xi, \tau) f(x, \tau) \, dx \\ &+ \iint_{S_\tau} \left[\frac{\partial}{\partial \tau} V(x, t; \xi, \tau) \right] [f(x, t) - f(x, \tau)] \, dx \, dt \end{aligned} \tag{5.25}$$

which can be deduced from (1.3.27). We then proceed to estimate $v_\tau(\xi, \tau) - v_\tau(\xi, \tau')$ in the following manner. Setting

$$w_3(\xi, \tau) = \iint_{S_\tau} \left[\frac{\partial}{\partial \tau} V(x, t; \xi, \tau) \right] [f(x, t) - f(x, \tau)] \, dx \, dt,$$

$$w_3(\xi, \tau') = \iint_{S_{\tau'}} = \iint_{S_\tau} + \iint_{S_{\tau'} - S_\tau} = w_{31}(\xi, \tau') + w_{32}(\xi, \tau'),$$

we split $w_3(\xi, \tau) - w_{31}(\xi, \tau')$ into four terms analogously to (5.12). Similarly we split

$$\int_R V(x, 0; \xi, \tau) f(x, \tau) \, dx - \int_R V(x, 0; \xi, \tau') f(x, \tau') \, dx$$

into two differences (analogously to (5.17) with $\gamma = 0$) and proceed similarly to the previous proof. w_{32} is estimated directly by

$$AH[f](\tau' - \tau)^{\alpha/2}.$$

The details may be left to the reader.

Having proved (5.7) in the two special cases of $\tau = \tau'$ and $\xi = \xi'$, the proof of the general case follows immediately.

Lemma 6 will now be generalized to n dimensions. Let R be a rectangle

$$-\beta_i < x_i < \beta_i \qquad (i = 1, \ldots, n)$$

and set $S_\tau = R \times (0, \tau)$, $S = S_1$. Denote by ∂R_n the set

$$\{(x, t); \quad x_n = \pm\beta_n, \quad t = 0, \quad -\beta_i < x_i < \beta_i \quad (i = 1, \ldots, n - 1)\}.$$

Assume that $f(x, t)$ is Hölder continuous (exponent α) in $\overline{S}$ and set

(5.26) $$G(x, t; \xi, \tau) = \frac{(\tau - t)^{n/2}}{(2\sqrt{\pi})^n} \exp\left[-\frac{|x - \xi|^2}{4(\tau - t)} \right] \qquad (t < \tau).$$

Lemma 6′. *Assume that $f(x, t) = 0$ on ∂R_n and consider the function*

(5.27) $$v(\xi, \tau) = \int_0^\tau \int_R G(x, t; \xi, \tau) f(x, t)\, dx\, dt.$$

Let N be any subdomain of S such that each of its limit points lies either in S, or on one of the open (n-dimensional) faces of S, or on ∂R_n, or on $\partial R_n \times \{t = 1\}$. Then there exists a constant A' depending only on n, α, N and the β_i, such that

(5.28) $$\frac{|v_\tau(\xi, \tau) - v_{\tau'}(\xi', \tau')|}{(|\xi - \xi'|^2 + |\tau - \tau'|)^{\alpha/2}} \le A' H[f]$$

for all $(\xi, \tau) \in N$, $(\xi', \tau') \in N$.

Proof. It will be enough to consider the case where $\tau' = \tau$. A representation similar to (5.9) holds. w_2 now is the integral

$$w_2 = \int_{S_T} \int [\Delta_\xi G(x, t; \xi, \tau)][f(x, t) - f(\xi, t)]\, dx\, dt$$

where Δ_ξ is the Laplace operator. $w_2(\xi, \tau) - w_2(\xi', \tau)$ can be treated by splitting it into four terms as before. w_1 consists of $2n$ boundary integrals taken over the $2n$ faces of the lateral boundary of S_τ. We can split $w_1(\xi, \tau) - w_1(\xi', \tau)$ into three terms as in (5.17), and then the relation (5.18) is valid. K_2 can be treated also as above. Proceeding with K_1, let us first note that for any boundary integral in K_1 taken on a face $x_i = \pm\beta_i$ with $i < n$, the desired bounds follow immediately from the fact that the points on this face are bounded away from the points of N. Note that the bounds obtained involve l.u.b. $|f|$, but since $f = 0$ on ∂R_n, l.u.b. $|f| \le (\Sigma\, \beta_i^2 + 1)^{\alpha/2} H[f]$. It remains to consider the boundary integrals over the faces $x_n = \pm\beta_n$ which appear in K_1. Denote by K_{11} the sum of the two terms in K_1 which involve these integrals.

Decomposing K_{11} analogously to (5.20), we can treat L_1 as before. Decomposing L_2 analogously to (5.22), we can estimate L_{21} as before upon making use of $f = 0$ on ∂R_n.

To consider L_{22}, suppose first that $f \equiv 1$. We may assume that $\xi_i' \ne \xi_i$ only for one i, and that $\xi_i' < \xi_i$. If $i = n$ then L_{22} is the sum $\Sigma \pm I_\pm$ where

$$I_\pm = \int \left[\int_{\xi_n'}^{\xi_n} d\zeta_n \int_{-\infty}^{\tau} \frac{\partial^2}{\partial \zeta_n^2} G(x_1, \ldots, x_{n-1}, \pm\beta_n, t; \xi_1, \ldots, \xi_{n-1}, \zeta_n, \tau)\, dt \right] dx_1 \cdots dx_{n-1}$$

$$= \int \left[\int_{\xi_n'}^{\xi_n} d\zeta_n \int_{-\infty}^{\tau} \left(-\frac{\partial G}{\partial t} - \frac{\partial^2 G}{\partial x_1^2} - \cdots - \frac{\partial^2 G}{\partial x_{n-1}^2} \right) dt \right] dx_1 \cdots dx_{n-1}.$$

The integral corresponding to $\partial G/\partial t$ vanishes (as in the case $n = 1$) whereas each integral corresponding to $\partial^2 G/\partial x_j^2$ $(1 \le j \le n - 1)$ can be rewritten as a difference of boundary integrals over $x_j = \pm\beta_j$ and, therefore, the desired bounds immediately follow.

If $\xi_i' \neq \xi_i$ for just one i with $i < n$, then $L_{22} = \Sigma \pm I_\pm$ where

$$I_\pm = \int \left[\int_{\xi_i'}^{\xi_i} d\zeta_i \int_{-\infty}^{\tau} \frac{\partial^2 G}{\partial \xi_n \partial \zeta_i} dt \right] dx_1 \cdots dx_{n-1}.$$

Integrating with respect to x_i (using $\partial G/\partial \zeta_i = -\partial G/\partial x_i$) we transform the integral into a difference of boundary integrals over the faces $x_i = \pm\beta_i$, and the desired bounds then follow.

If $f \not\equiv 1$ then we write the f which appears in L_{22} in the form

$$[f(\bar{x}, \pm\beta_n, \tau) - f(\bar{\xi}, \pm\beta_n, \tau)] + f(\bar{\xi}, \pm\beta_n, \tau) \tag{5.29}$$

where $\bar{x} = (x_1, \ldots, x_{n-1})$, $\bar{\xi} = (\xi_1, \ldots, \xi_{n-1})$. The terms of L_{22} which involve $f(\bar{\xi}, \pm\beta_n, \tau)$ can be estimated by the previous special case. It thus remains to estimate the integrals corresponding to the expression in brackets in (5.29). We break the t-integral into

$$\int_{-\infty}^{\tau-\gamma} + \int_{\tau-\gamma}^{\tau} \qquad \text{where } \gamma = |\xi' - \xi|^2.$$

Using the mean value theorem we find that the $(\bar{x}, t)$-integral corresponding to $\int_{-\infty}^{\tau-\gamma}$ is bounded by

$$A_1 H[f] |\xi' - \xi| \int_{-\infty}^{\tau-\gamma} \int (\tau - t)^{-(n+2)/2} \exp\left[-\frac{|\bar{x} - \bar{\zeta}|^2}{5(\tau - t)} \right] |\bar{x} - \bar{\xi}|^\alpha \, d\bar{x}\, dt,$$

where A_i are used to denote constants depending only on n, α and the β_j, $\bar{\zeta}$ is some point in the interval $(\bar{\xi}, \bar{\xi}')$ and $d\bar{x} = dx_1 \cdots dx_n$. Using the inequality $|\bar{x} - \bar{\xi}| \leq |\bar{x} - \bar{\zeta}| + |\xi' - \xi|$ one easily finds that the last expression is bounded by $A_2 H[f] |\xi' - \xi|^\alpha$.

As for the $(\bar{x}, t)$-integral corresponding to $\int_{\tau-\gamma}^{\tau}$, it suffices to estimate

$$\int_{\tau-\gamma}^{\tau} \int \frac{\partial}{\partial \xi_n} G(\bar{x}, \pm\beta_n, t; \xi, \tau) \cdot [f(\bar{x}, \pm\beta_n, \tau) - f(\bar{\xi}, \pm\beta_n, \tau)] \, d\bar{x}\, dt$$

and

$$\int_{\tau-\gamma}^{\tau} \int \frac{\partial}{\partial \xi_n'} G(\bar{x}, \pm\beta_n, t; \xi', \tau) \cdot [f(\bar{x}, \pm\beta_n, \tau) - f(\bar{\xi}, \pm\beta_n, \tau)] \, d\bar{x}\, dt \tag{5.30}$$

separately. Now the first integral is easily seen to be bounded by

$$A_3 H[f] \int_{\tau-\gamma}^{\tau} \frac{dt}{(\tau - t)^{(2-\alpha)/2}} = A_4 H[f] |\xi' - \xi|^\alpha. \tag{5.31}$$

If in the brackets of the second integral we replace $\bar{\xi}$ by $\bar{\xi}'$, then we obtain the same bound as in (5.31). It therefore remains to estimate (5.30) with the expression in the brackets replaced by $f(\bar{\xi}', \pm\beta_n, \tau) - f(\bar{\xi}, \pm\beta_n, \tau)$. For this integral we get the bound

$$A_5 H[f] \left\{ \int_{\tau-\gamma}^{\tau} \int |\pm\beta_n - \xi_n'| (\tau - t)^{-(n+2)/2} \right.$$

$$\left. \exp\left[- \frac{|\bar{x} - \bar{\xi}'| + (\pm\beta_n - \xi_n')^2}{4(\tau - t)} \right] d\bar{x}\, dt \right\} |\xi' - \xi|^\alpha$$

$$\leq A_6 H[f] \left\{ \int_{\tau-\gamma}^{\tau} |\pm\beta_n - \xi_n'| (\tau - t)^{-3/2} \exp\left[- \frac{(\pm\beta_n - \xi_n')^2}{4(\tau - t)} \right] dt \right\} |\xi' - \xi|^\alpha.$$

Substituting $z = (\pm\beta_n - \xi_n')^2/(\tau - t)$ we get the bound $A_7 H[f]\, |\xi' - \xi|^\alpha$. The proof that $|L_{22}|$ is bounded by $A_8 H[f] |\xi' - \xi|^\alpha$ is thereby completed.

For the sake of later reference we state the following corollary of Lemma 6′.

Lemma 6″. *Denote by ∂R the union of the open $(n - 1)$ – dimensional faces of the lower base of S and assume that f is Hölder continuous (exponent α) in $\bar{S}$ and that $f = 0$ on ∂R. Let N be any subdomain of S such that each of its limit points lies either in S, or on one of the n-dimensional open faces of S, or on ∂R, or on $\partial R \times \{t = 1\}$. Then the function $v(\xi, \tau)$ defined by (5.27) satisfies (5.28) for all $(\xi, \tau) \in N$, $(\xi', \tau') \in N$, where A' is a constant depending only on n, α, N, and* min β_i, max β_i.

The proof is obtained by dividing N into several sets for which Lemma 6′ can be applied (with the role of x_n being given to appropriate x_i's). The fact that A' depends on the β_i only through its dependence on their maximum and on their minimum follows from the proof of Lemma 6′.

6. An Auxiliary Theorem for the Boundary Estimates

Let $\xi^0 = (\xi_1^0, \ldots, \xi_n^0)$ be a fixed point and denote by B the n-dimensional rectangle

$$-d \leq x_i - \xi_i^0 \leq d \qquad (i = 1, \ldots, n - 1), \qquad -d \leq x_n - \xi_n^0 \leq \bar{d},$$

and by N_τ the set $B \times \{0 \leq t \leq \tau\}$. We shall always take

$$\bar{d} \leq d, \quad \tau \leq \tau^0 \leq d^2$$

and set $N = N_{\tau^0}$. The following theorem, which is of the same nature as Theorem 2, will play a decisive role in deriving the boundary estimates in the next section.

Theorem 3. *Let $f(x, t)$ be Hölder continuous (exponent α) in N, let u, $D_x u$, $D_x^2 u$, $D_t u$ be Hölder continuous (exponent α) functions in N, and assume that* (3.1) *holds in N and that u vanishes on the two faces $x_n = \bar{d}$ and $t = 0$ of N. There exists a constant K depending only on n, α such that, for $i = 0$,* 1, 2,

$$(6.1) \quad |D_x^i u(P)| \leq d^{-2} K \operatorname*{l.u.b.}_N |u| + d^{2-i} K \operatorname*{l.u.b.}_N |f| + d^{2-i+\alpha} K H_{P,N}[f] \equiv KI,$$

$$(6.2)\quad d^\alpha \frac{|D_x^i u(P) - D_x^i u(Q)|}{d(P, Q)^\alpha} \le KI + Kd^{2-i+\alpha}H[f]$$

$$\text{if } d(P, Q) \le \frac{d}{4}, Q \in N,$$

where $P = (\xi^0, \tau^0)$, *and* $H[f]$ *is the Hölder coefficient (exponent* α*) of* f *in* N.

Note that the right-hand side of (6.2) is not the same as the right-hand side of (3.3) since instead of $H_{Q,N}[f]$ we now have $H[f]$.

Proof. The notation of Sec. 3 will be freely used. Using Green's identity (1.8.4) with the functions $u(x, t)$ and $\varphi(x, t)\overline{G}(x, t; \xi, \tau)$, where

$$(6.3)\qquad \overline{G}(x, t; \xi, \tau) = G(x, t; \xi, \tau) - G(x, t; \xi', \tau),$$

G defined in (5.26) and $\xi' = (\xi_1', \ldots, \xi_n')$ defined by

$$\xi_i' = \xi_i \qquad (1 \le i \le n - 1), \qquad \xi_n' = 2\overline{d} - \xi_n,$$

and noting that $\overline{G}(x, t; \xi, \tau)$ vanishes on $x_n = \overline{d}$, we obtain, upon recalling that u vanishes on $x_n = \overline{d}$ and on $t = 0$,

$$\begin{aligned}(6.4)\quad u(\xi, \tau) &= -\int_0^\tau \int_B \varphi(x, t)f(x, t)\overline{G}(x, t; \xi, \tau)\, dx\, dt \\ &\quad + \int_0^\tau \int_B u(x, t)L_0^*[\varphi(x, t)\overline{G}(x, t; \xi, \tau)]\, dx\, dt \\ &\equiv -H_0 + J_0.\end{aligned}$$

H_0, J_0 are similar to the H_0, J_0 of Sec. 3.

Note that $\overline{G}$ and its derivatives satisfy the same inequalities (3.11) that G and its derivatives satisfy. This fact follows by observing that $|x - \xi'| \ge |x - \xi|$ if $x \in B$. For the sake of future references we write down these inequalities:

$$(6.5)\quad |D_t^i D_x^j D_\tau^k D_\xi^m \overline{G}(x, t; \xi, \tau)| \le C(\tau - t)^{-(2i+j+2k+m+n)/2} \exp\left[-\frac{|x - \xi|^2}{5(\tau - t)}\right]$$

where C depends only on i, j, k, m, n.

Later on we shall need the fact that

$$(6.6)\quad \frac{\partial^2}{\partial \xi_i\, \partial \xi_j} \int_0^\tau \int_B \overline{G}(x, t; \xi, \tau)\, dx\, dt \quad \text{is bounded,} \quad \text{if } i \ne n.$$

The proof is obtained by noting that $\partial \overline{G}/\partial \xi_i = -\partial \overline{G}/\partial x_i$ and integrating with respect to x_i (using the divergence theorem), then using (6.5) and substituting $z = d^2/(\tau - t)^{1/2}$.

With the aid of (6.5), (6.6) we can now proceed to prove (6.1) for all the derivatives $D_x^i u$ which are different from $\partial^2 u/\partial x_n^2$. The proof is almost word by word as the proof of (3.2) and the details are therefore omitted. As for $\partial^2 u/\partial x_n^2$, we first estimate $\partial u/\partial t$ and then get a bound for $\partial^2 u/\partial x_n^2$ by using the differential equation (3.1). The derivation of a bound on $\partial u/\partial t$ is obtained by differentiating (6.4) with respect to τ thus obtaining

$$(6.7)\quad \frac{\partial u(\xi,\tau)}{\partial \tau} = -\int_0^\tau \int_B [D_\tau \overline{G}(x,t;\xi,\tau)][\varphi(x,t)f(x,t) - \varphi(\xi,\tau)f(\xi,\tau)]\,dx\,dt$$
$$- [\varphi(\xi,\tau)f(\xi,\tau)] \lim_{\epsilon\to 0} \int_0^{\tau-\epsilon} \int_B D_\tau \overline{G}(x,t;\xi,\tau)\,dx\,dt$$
$$+ \int_0^\tau \int_B \varphi(x,t)f(x,t)(\Delta_x + D_t)D_\tau \overline{G}(x,t;\xi,\tau)\,dx\,dt.$$

The first and third integrals can be estimated by the calculations of Sec. 3 once we notice that the bounds on $D_\tau \overline{G}$ and its derivatives are (by (6.5)) the same as the bounds on $D_\xi^2 G$ and its derivatives (by (3.11)). Finally, the second integral on the right-hand side of (6.7) is bounded since, by writing $D_\tau \overline{G} = -D_t \overline{G}$ and integrating, we get

$$\int_B \overline{G}(x,0;\xi,\tau)\,dx - \lim_{\epsilon\to 0}\int_B \overline{G}(x,\tau-\epsilon;\xi,\tau)\,dx = \int_B \overline{G}(x,0;\xi,\tau)\,dx - 1.$$

The desired bound on $\partial u/\partial \tau$ is thus obtained.

As for (6.2), if $i = 0, 1$ then the inequality can be proved in the same manner as for the corresponding inequalities (3.3). For $i = 2$, if $D_x^2 \neq \partial^2/\partial x_n^2$ then the proof of (6.2) is still very similar to the proof of the analogous inequality in Sec. 3. It therefore remains to prove (6.2) for $D_x^2 = \partial^2/\partial x_n^2$. Since u satisfies (3.1), it suffices to prove that

$$(6.8)\quad d^\alpha \frac{|D_t u(P) - D_t u(Q)|}{d(P,Q)^\alpha} \le KI + Kd^\alpha H[f] \qquad \left(d(P,Q) \le \frac{1}{4}d\right).$$

To prove (6.8) we write

$$D_\tau u(\xi,\tau) = -H + J$$

where $-H$ denotes the sum of the first two terms on the right-hand side of (6.7) and J denotes the last integral on the right-hand side of (6.7). The considerations for J are as in Sec. 3, and we get

$$(6.9)\qquad d^\alpha \frac{|J(P) - J(Q)|}{d(P,Q)^\alpha} \le Kd^{-2}\,\underset{N}{\text{l.u.b.}}\,|u|.$$

Denoting $\varphi(Q)f(Q)$ by $g(Q)$ and noting that

$$H[g] \le (\underset{N}{\text{l.u.b.}}\,|\varphi|)H[f] + Kd^{-\alpha}\,\underset{N}{\text{l.u.b.}}\,|f|$$

it follows that if

$$(6.10)\qquad d^\alpha \frac{|H(P) - H(Q)|}{d(P,Q)^\alpha} \le Kd^\alpha H[g]$$

then the proof of (6.2) is completed.

To prove (6.10) write H in the form

$$(6.11) \quad H = \frac{\partial}{\partial \tau} \int_0^\tau \int_B g(x, t) G(x, t; \xi, \tau)\, dx\, dt - \frac{\partial}{\partial \tau'} \int_0^{\tau'} \int_B g(x, t) G(x, t; \xi', \tau')\, dx\, dt.$$

The transformation $(x, t) \to (\bar{x}, \bar{t})$ where $x = d\bar{x}$, $t = d^2\bar{t}$ maps B into $\overline{B}$, N into $\overline{N}$, $g(x, t)$ into $\bar{g}(\bar{x}, \bar{t})$ (i.e., $\bar{g}(\bar{x}, \bar{t}) = g(x, t)$), $H(Q)$ into $\overline{H}(\overline{Q})$, and $G(x, t; \xi, \tau)$ into $G(\bar{x}, \bar{t}; \bar{\xi}, \bar{\tau})/d^n$. We wish to apply Lemma 6″, Sec. 5. We therefore have to show that $\bar{g}$ vanishes on the boundary of the lower base of $\overline{N}$. Since φ vanishes on all the lateral faces of N with the possible exception of $x_n = \bar{d}$, it suffices to show that f vanishes on the intersection of the faces $x_n = \bar{d}$ and $t = 0$. This follows from the equation $L_0 u = f$ since, as $u = 0$ on $x_n = \bar{d}$ and on $t = 0$, $L_0 u = 0$ on the intersection of these two faces.

Lemma 6″ now can be applied. (Observe that Lemma 6″ remains true if $G(x, t; \xi, \tau)$ is replaced by $G(x, t; \xi', \tau)$.) We get

$$(6.12) \quad \frac{|\overline{H}(\overline{P}) - \overline{H}(\overline{Q})|}{d(\overline{P}, \overline{Q})^\alpha} \leq KH[\bar{g}]$$

in the closure of any subdomain E satisfying the restrictions made on N in Lemma 6″. Observing that $\bar{g} = 0$ in a neighborhood of some $\overline{N} - E$ and that the size of this neighborhood is independent of d (the property (3.4) of φ is being used), it follows from the form of H that (6.12) holds throughout $\overline{N}$. Since the edges of $\overline{B}$ are of lengths between 1 and 2, it also follows from Lemma 6″ that the constant K in (6.12) depends only on n, α.

Returning to the original coordinates, the inequality (6.10) is obtained from (6.12).

We conclude this section by extending Theorem 3 to equations of the form (3.47).

Theorem 3′. *Theorem* 3 *remains true for the parabolic equation* (3.47). *The constant* K *in* (6.1), (6.2) *depends only on* n, α, k_1, k_2 *where* k_2 *is a constant for which* (3.48) *holds and* $k_1 \geq |a_{ij}|$.

Proof. Note first that Theorem 3 remains true if the n-dimensional rectangle B is replaced by an n-dimensional rectangle B_0 of the form

$$-d\alpha_i \leq x_i - \xi_i^0 \leq d\alpha_i \quad (i = 1, \ldots, n-1), \quad -d\alpha_n \leq x_n - \xi_n \leq \bar{d}\alpha_n,$$

where α_i are arbitrary numbers. Note also that if we rotate B_0 and thus obtain a new rectangle B^*, then Theorem 3 remains true with respect to B^*. This follows by observing that orthogonal transformations (in the x-space) leave both the equation (3.1) and the euclidean distance invariant.

The proof of Theorem 3′ can now be easily completed. An appropriate linear transformation

(6.13) $$x_i' = \sum_{j=1}^{n} g_{ij}x_j \qquad (g_{ij} \text{ constants})$$

transforms the equation (3.47) into the equation (3.1) (in the variables $x_1', \ldots, x_n', t$). B is transformed into B'. B' is an n-dimensional rectangle of the type B^* considered above. From the previous remarks it follows that we can apply Theorem 3 to the transformed domain B'. If we use the obvious inequalities

$$\text{const.}\ |x - y| \leq |x' - y'| \leq \text{const.}\ |x - y|$$

where y' is the image of y under (6.13) and where the constants are positive, then the proof of Theorem 3′ is easily completed by going back to the original coordinates.

7. Derivation of the Boundary Estimates

In this section we shall prove a theorem more general than Theorem 6, Chap. 3. We first introduce some notation.

Let R be a fixed region on the manifold $\overline{B} + S$. For any point $P = (\xi, \tau)$ in D, define $\bar{d}_P$ to be the distance from P to $(\overline{B} + S_\tau) - R$. For any two points P, Q in D, set $\bar{d}_{PQ} = \min(\bar{d}_P, \bar{d}_Q)$. Analogously to (1.4)–(1.7), we set

(7.1) $$|g|_{p,m}^{R,D} = \sum_{j=0}^{m} M_{p,j}^{R,D}[g],$$

(7.2) $$|g|_{p,m+\alpha}^{R,D} = |g|_{p,m}^{R,D} + \sum_{j=0}^{m} M_{p,j+\alpha}^{R,D}[g],$$

where

(7.3) $$M_{p,j}^{R,D}[g] = \operatorname*{l.u.b.}_{P \in D} \bar{d}_P^{p+j} |D_x^j g(P)|,$$

(7.4) $$M_{p,j+\alpha}^{R,D}[g] = \operatorname*{l.u.b.}_{P,Q \in D} \bar{d}_{PQ}^{p+j+\alpha} \frac{|D_x^j g(P) - D_x^j g(Q)|}{d(P, Q)^\alpha}.$$

When there is no confusion, we abbreviate $|g|_{p,m}^{R,D}$, $|g|_{p,m+\alpha}^{R,D}$, etc., by $|g|_{p,m}^{R}$, $|g|_{p,m+\alpha}^{R}$, etc. Lemmas 1, 2, 3 of Sec. 2 remain true for the present norms.

Let ψ be a function defined on R and assume that there exists a function Ψ defined on $D + B_T + R$ such that

$$\Psi = \psi \quad \text{on } R, \qquad |\Psi|^* \equiv |\Psi|_{0,2+\alpha}^{R} + \left|\frac{\partial \Psi}{\partial t}\right|_{2,\alpha}^{R} < \infty.$$

We then define

(7.5) $$|\psi|_{2+\alpha}^{R} = \operatorname{g.l.b.} |\Psi|^*$$

where the g.l.b. is taken over all such extensions Ψ of ψ.

The following assumptions will be needed:

($\mathrm{A^R}$) $$|a_{ij}|_{0,\alpha}^{R,D} \leq \bar{K}_1, \quad |b_i|_{1,\alpha}^{R,D} \leq \bar{K}_1, \quad |c|_{2,\alpha}^{R,D} \leq \bar{K}_1;$$

(B) For any real vector ξ and for every $(x, t) \in D$,

$$\sum a_{ij}(x, t)\xi_i\xi_j \geq K_2|\xi|^2 \qquad (K_2 > 0);$$

(C^{R})
$$|f|_{2,\alpha}^{R,D} < \infty.$$

If the local representations (3.2.17) are only assumed to hold in neighborhoods of points of $R \cap \overline{S}$, and if h, $D_x h$, $D_x^2 h$, $D_t h$ are Hölder continuous (exponent α) then we say that D has the property (E^{R}).

The main result of the present section is the following theorem.

Theorem 4. *Let u be a solution of* (1.1) *in D and assume that* (A^{R}), (B), (C^{R}) *hold, that D has the property* (E^{R}), *and that u, $D_x u$, $D_x^2 u$, $D_t u$ are uniformly Hölder continuous (exponent α) in every open set whose closure lies in $D + B_T + R$. Assume further that* l.u.b. $|u| < \infty$ *and that $u = \psi$ on R, where $|\psi|_{2+\alpha}^R < \infty$. Let R_0 be any region in $\overline{B} + S$ which is contained in R and whose distance to the complement of R (in $\overline{B} + S$) is positive. Then there exists a constant $\overline{K}$, depending only on $\overline{K}_1$, K_2, R_0, R and D, such that*

$$|u|_{0,2+\alpha}^{R_0,D} \leq \overline{K}(|u|_0 + |\psi|_{2+\alpha}^R + |f|_{2,\alpha}^{R,D}). \tag{7.6}$$

Before proving the theorem, let us establish the following corollary.

Corollary. *Theorem* 6 *of Chap.* 3 *is a consequence of Theorem* 4.

Proof. If we prove that

$$\overline{|u|}_\alpha + \sum \overline{|D_x u|}_\alpha + \sum \overline{|D_x^2 u|}_\alpha \leq \overline{K}(|u|_0 + \overline{|\psi|}_{2+\alpha} + \overline{|f|}_\alpha), \tag{7.7}$$

then, using (1.1), we also get a similar bound on $\overline{|D_t u|}_\alpha$. Since by the maximum principle (see (2.3.12))

$$|u|_0 \leq \text{const. (l.u.b. } |\psi| + |f|_0),$$

the proof of Theorem 6, Chap. 3, thus follows from (7.7). To prove (7.7), take any two disjoint points P_1, P_2 on $\overline{B} + S$ and let R_1, R_2 be the complements in $\overline{B} + S$ of P_1, P_2, respectively. Let R_{01}, R_{02} be related to R_1, R_2, respectively, as R_0 is related to R (in Theorem 4), and such that for each point $Q \in \overline{B} + S$ there exists an $(n + 1)$-dimensional neighborhood V such that

$$V \cap (\overline{B} + S) = V \cap R_{0i} \qquad \text{for some } i.$$

Applying Theorem 4 to the pairs R_j, R_{0j} for $j = 1, 2$, the proof of (7.7) then follows.

Proof of Theorem 4. Since for every domain D_0 of D whose closure does not intersect R_0, the estimates of Theorem 1, Sec. 1 yield the necessary bounds on u, $D_x u$, $D_x^2 u$ and on their Hölder coefficients, it suffices to prove (7.6) not in the whole domain D, but only in domains Ω which are intersections of D with some balls V with centers P on R_0 and radii r sufficiently small.

There are three possibilities: either P lies on the base B of D, or P lies on the boundary S of D, or P lies in the intersection of $\bar{B}$ with $\bar{S}$. The first two cases may be treated as special cases of the third case. Hence it suffices to consider the case where $P \in \bar{B} \cap \bar{S}$.

We may take r sufficiently small so that $R \cap V = (\bar{B} + S) \cap V$ and so that the representation (3.2.17) is valid for $\bar{S} \cap V$. Assume for simplicity that $i = n$ in (3.2.17). The transformation

$$t' = t, \qquad x_i' = x_i \qquad (i = 1, \ldots, n-1) \tag{7.8}$$
$$x_n' = x_n - h(x_1, \ldots, x_{n-1}, t)$$

maps $S \cap V$ onto a set which lies on $x_n' = 0$ and, as we may assume, it maps Ω onto a domain Ω' for which $x_n' > 0$; also $t' > 0$. The mapping (7.8) is one-to-one. Moreover,

$$\text{const. } d(P', Q') \leq d(P, Q) \leq \text{const. } d(P', Q') \tag{7.9}$$

(the constants are positive) since $\partial h/\partial x_i$ and $\partial h/\partial t$ are bounded functions.

Define $v(x', t') = u(x, t)$, $\varphi(x', t') = \psi(x, t)$, $\Phi(x', t') = \Psi(x, t)$. Then v assumes the boundary values of Φ on the boundaries $x_n' = 0$ and $t' = 0$ of Ω'. Equation (1.1) is transformed into the equation

$$L'v \equiv \Sigma\, a_{ij}'(x', t') \frac{\partial^2 v}{\partial x_i'\, \partial x_j'} + \Sigma\, b_i'(x', t') \frac{\partial v}{\partial x_i'} + c(x', t')v - \frac{\partial v}{\partial t'} \tag{7.10}$$
$$= f'(x', t')$$

where $c'(x', t') = c(x, t)$, $f'(x', t') = f(x, t)$, $a_{\lambda\mu}'(x', t')$ is easily expressed in terms of $a_{ij}(x, t)$, $\partial h/\partial x_i$, and $b_\lambda'(x', t')$ is easily expressed in terms of $b_i(x, t)$, $\partial h/\partial x_i$, $\partial^2 h/\partial x_i\, \partial x_j$, $\partial h/\partial t$. Using the differentiability properties of h and (7.9) we conclude that the coefficients of L' satisfy the assumption $(\bar{\text{A}})$ of Chap. 3, Sec. 2, with $\bar{K}_1$ which depends on h and on the $\bar{K}_1$ in the condition (A^{R}). Clearly also (B) is valid with a different constant K_2 (depending also on h). Finally, since $f'(x', t') = f(x, t)$, f' is uniformly Hölder continuous (exponent α) in Ω'.

Since the points of Ω are bounded away from $(\bar{B} + S) - R$ (and thus $\bar{d}_Q \geq \text{const.} > 0$ if $Q \in \Omega$), if we prove that

$$\overline{|v|}_\alpha^{\Omega'} + \Sigma\, \overline{|D_{x'}v|}_\alpha^{\Omega'} + \Sigma\, \overline{|D_{x'}^2 v|}_\alpha^{\Omega'} \leq K'(|v|_0^{\Omega'} + \overline{|\Phi|}_{2+\alpha}^{\Omega'} + \overline{|f'|}_\alpha^{\Omega'}) \tag{7.11}$$

then the desired bounds would follow for u, and the proof of Theorem 4 would thereby be completed.

Introducing

$$w(x', t') = v(x', t') - \Phi(x', t') \tag{7.12}$$

we find that

$$L'w = f'' \qquad \text{in } \Omega' + [\bar{\Omega}' \cap \{t' = 0\}] + [\bar{\Omega}' \cap \{x_n' = 0\}], \tag{7.13}$$

where $f'' = f' - L'\Phi$, and

(7.14) $\quad w(x', t') = 0 \qquad$ on $[\overline{\Omega'} \cap \{t' = 0\}] + [\overline{\Omega'} \cap \{x'_n = 0\}]$.

We have

$$(7.15) \qquad \overline{|f''|}_\alpha^{\Omega'} \leq \text{const.}\ \overline{|\Phi|}_{2+\alpha}^{\Omega'} + \overline{|f|}_\alpha^{\Omega'}$$

the constant depending only on $\bar{K}_1$, R, R_0, D, and h.

To the system (7.13), (7.14) we apply the proof of Theorem 1, with $|\ \ |_{j,\alpha}$, etc. replaced by $\overline{|\ \ |}_{j,\alpha}$, etc.; instead of Theorem 2′, Sec. 3, we now use Theorem 3′, Sec. 6. Since that proof can be translated to the present situation almost word by word, the details are omitted. We note, however, that Theorem 3′ is slightly weaker than Theorem 2′ since in place of $H_{Q,N}[f]$ in (3.3) stands $H[f]$ in (6.2). This makes no difference in the proof of our theorem since what actually enters in the calculations of Sec. 4 is a bound not on $H_{Q,N}[f]$ but on l.u.b. $H_{Q,N}[f] \equiv H[f]$. We thus derive the inequality

$$(7.16) \qquad \overline{|w|}_\alpha^{\Omega'} + \Sigma\, \overline{|D_{x'}w|}_\alpha^{\Omega'} + \Sigma\, \overline{|D_{x'}^2 w|}_\alpha^{\Omega'} \leq K^*(|w|_0^{\Omega'} + \overline{|f''|}_\alpha^{\Omega'})$$

where K^* depends on the $\bar{K}_1$, K_2 in the conditions $(\bar{\text{A}})$, (B) for L'. Using (7.15), (7.16), the inequality (7.11) follows.

From the proof of Theorem 4 one gets more precise information on the constant $\bar{K}$ appearing in Theorem 6, Chap. 3, namely:

Let H be a bound on all the functions h in the local representations (3.2.17), on $D_x h$, $D_x^2 h$, $D_t h$ and on their Hölder coefficients (exponent α). Let a neighborhood D^* of $\bar{S}$ be covered by, say, m balls V_i with centers on $\bar{S}$ such that each portion $V_i \cap \bar{S}$ can be globally represented in the form (3.2.17). Suppose that for every $P \in D^* \cap D$ there exists a ball V_P of center P and radius $\geq r$ such that $V_P \cap (D^* \cap D)$ is contained in some $V_i \cap (D^* \cap D)$. Finally let the diameter of D be $\leq d$. Then the constant $\bar{K}$ in (3.2.21) depends only on $\bar{K}_1$, K_2, H, m, r, d, n, α.

8. Existence Theorems for the Heat Equation

In this section we shall prove the statements (a) made in the proofs of Theorems 7, 9 of Chap. 3. The proofs of Theorems 7, 9 and the proofs of Theorems 10–12 of Chap. 3 which were based, for $p = 0$, on Theorems 7, 9 would then be completed.

We begin with a basic existence theorem for the heat equation.

Theorem 5. *Assume that S has local barriers. Then for any continuous ψ on $\bar{B} + S$ there exists a unique solution of*

$$(8.1) \qquad \begin{aligned} L_0 u &\equiv \sum_{i=1}^{n} \frac{\partial^2 u}{\partial x_i^2} - \frac{\partial u}{\partial t} = 0 \qquad && \text{in } D + B_T, \\ u &= \psi && \text{on } \bar{B} + S. \end{aligned}$$

The proof is given in Problems 1–13 at the end of this chapter.

Using Theorem 5 we next prove:

Theorem 5′. *Assume that S has local barriers and that f is Hölder continuous (exponent α) in D. Then there exists a unique solution of*

$$\begin{aligned} L_0 u &= f \qquad \text{in } D + B_T, \\ u &= 0 \qquad \text{on } \overline{B} + S. \end{aligned} \tag{8.2}$$

Proof. Consider the function

$$v(x, t) = -\iint_{D_t} \frac{(t-\tau)^{-n/2}}{(2\sqrt{\pi})^n} \exp\left[-\frac{|x-\xi|^2}{4(t-\tau)}\right] f(\xi, \tau)\, d\xi\, d\tau. \tag{8.3}$$

By the results of Sec. 3, Chap. 1, it follows that v is a solution of the equation $L_0 v = f$ in $D + B_T$.

Since, by Theorem 5, there exists a unique solution w of

$$\begin{aligned} L_0 w &= 0 \qquad \text{in } D + B_T, \\ w &= -v \qquad \text{on } \overline{B} + S, \end{aligned} \tag{8.4}$$

it follows that there exists a solution u of (8.2), namely, $u = v + w$. The uniqueness follows the maximum principle.

We shall now establish the statement (a) made in the proof of Theorem 9, Chap. 3, i.e.,

Theorem 6. *Assume that S has local barriers and let $|f|_\alpha < \infty$. Then there exists a unique solution of* (8.2) *and it belongs to $C_{2+\alpha}$.*

Proof. By the proof of Theorem 5′, u exists and can be written in the form $v + w$ where v is given by (8.3) and w satisfies (8.4). We shall prove that w is an infinitely differentiable function in $D + B_T$. It suffices to prove this for any cylinder $D_1 = B_1 \times (t_0, t_1]$ lying in $D + B_T$, where B_1 is a ball. Using Green's formula (1.8.4) with w and G, where G is given by (5.26) we get, upon integration

$$w(x, t) = \int_{B_1} G(x, t; t_0, \tau) w(\xi, t_0)\, d\xi + \int_{t_0}^{t} \int_{\partial B_1} I(x, t; \xi, \tau)\, dS_\xi\, d\tau$$

where ∂B_1 is the boundary of B_1. The integrand I and all its derivatives with respect to x, t are continuous functions of $(x, t; \xi, \tau)$ provided $(x, t) \neq (\xi, \tau)$. The assertion concerning the infinite differentiability of $w(x, t)$ thus follows.

Consider next the function $v(x, t)$ in any closed subdomain D_0 of $D + B_T$. Let D_1 be a closed subdomain of $D + B_T$ such that $D_0 \subset D_1$ and such that the distance from D_0 to the complenent of D_1, in $0 \leq t \leq T$, is positive. Multiplying f by a continuously differentiable function h, in $0 \leq t \leq T$, which equals 1 in some neighborhood of D_0 and equals zero in the complement of D_1 with respect to $0 \leq t \leq T$ (see Problem 1, Chap.

1) and denoting the product by F, and setting $F = 0$ outside $D + B_T$, we form the function

$$V(x, t) = -\int_0 \int_{B_2} \frac{(t-\tau)^{-n/2}}{(2\sqrt{\pi})^n} \exp\left[-\frac{|x-\xi|^2}{4(t-\tau)}\right] F(\xi, \tau)\, d\xi\, d\tau$$

where B_2 is any cube containing $\{x; (x, t) \in \overline{D}$ for some $t\}$. Clearly $V(x, t) - v(x, t)$ is an infinitely differentiable function in D_0. By Lemma 6′, Sec. 5, V_t is Hölder continuous (exponent α) in D_0. The proof that $\partial^2 V/\partial x_i\, \partial x_j$ is Hölder continuous (exponent α) is much simpler than the proof of Lemma 6′, provided $i < n$, and is left to the reader. Since $\partial^2 V/\partial x_n^2$ can be expressed (from $L_0 V = F$) in terms of $\partial V/\partial t$ and the

$$\partial^2 V/\partial x_i^2 \ (1 \le i \le n-1),$$

it follows that $\partial v/\partial t$ and all the derivatives $D_x^2 v$ are Hölder continuous (exponent α) in D_0. The Hölder continuity of $D_x v$ follows also from the analogous result for V.

Combining the differentiability properties of v and w it follows that u, $D_x u$, $D_x^2 u$, $D_t u$ are Hölder continuous functions in $D + B_T$. Since also $|f|_\alpha < \infty$ and $\underset{D}{\text{l.u.b.}}\, |u| \le (\text{const.})\, \underset{D}{\text{l.u.b.}}\, |f| < \infty$, Theorem 1 implies that $u \in C_{2+\alpha}(D)$; this completes the proof of Theorem 6.

In proving the statement (a) made in the proof of Theorem 7, Chap. 3, we shall need the following lemma.

Lemma 7. *Let* $D = B \times (0, T)$ *where*

$$B = \{x; -\beta_i < x_i < \beta_i \ (i = 1, \ldots, n)\},$$

and denote by π *the open face of* D *lying in* $x_n = 0$. *Let* R *be an open set of* $\bar{\pi} + \overline{B}$ *and let* R_0 *be an open subset of* R, $\overline{R}_0 \subset R$. *If* ψ *is a continuous function on* $\overline{S} - \pi$ *vanishing on* $(\overline{S} - \pi) \cap (\bar{\pi} + \overline{B})$, *and if* $|f|_{2,\alpha}^R < \infty$ *and* $f = 0$ *on* $\bar{\pi} \cap \overline{B}$, *then there exists a unique solution* u *of*

$$\begin{aligned} L_0 u &= f && \text{in } D + B_T. \\ u &= 0 && \text{on } \bar{\pi} + \overline{B}, \\ u &= \psi && \text{on } S - \pi, \end{aligned} \tag{8.5}$$

and

$$|u|_{0,2+\alpha}^{R_0} \le K(\underset{S-\pi}{\text{l.u.b.}}\, |\psi| + |f|_{2,\alpha}^R). \tag{8.6}$$

Proof. Consider the function

$$u_1(x, t) = -\iint_{D_t} \overline{G}(\xi, \tau; x, t) f(\xi, \tau)\, d\xi\, d\tau$$

where $\overline{G}(x, t; \xi, \tau)$ is defined by (6.3), ξ' being the reflection of ξ with respect to the hyperplane $x_n = \beta_n$. u_1 satisfies

$$(8.7)\qquad \begin{aligned} L_0 u_1 &= f \quad && \text{in } D + B_T, \\ u_1 &= 0 \quad && \text{on } \bar{\pi} + \overline{B}. \end{aligned}$$

Let D_0 be any open cylinder whose closure lies in $D + R$. By modifying the definition of f outside D_0 (as in the considerations for v in the proof of Theorem 6) we conclude that the functions

$$u_1, \quad D_x u_1, \quad D_x^2 u_1, \quad D_t u_1$$

are uniformly Hölder continuous (exponent α) in D_0.

In view of (8.7), u is a solution of (8.5) if and only if $u = u_1 + u_2$ where u_2 is a solution of

$$(8.8)\qquad \begin{aligned} L_0 u_2 &= 0 \quad && \text{in } D + B_T, \\ u_2 &= 0 \quad && \text{on } \bar{\pi} + \overline{B}, \\ u_2 &= \psi - u_1 \quad && \text{on } S - \pi. \end{aligned}$$

It can be verified by direct calculations that the function

$$u_2(x, t) = \int_{S-\pi} \frac{\partial \hat{G}(x, t; \xi, \tau)}{\partial \nu(\xi, \tau)} [\psi(\xi, \tau) - u_1(\xi, \tau)]\, d\sigma(\xi, \tau)$$

is a solution of (8.8). Here $d\sigma(\xi, \tau)$ is the surface element on $S - \pi$, $\nu(\xi, \tau)$ is the inward normal to $S - \pi$ at (ξ, τ) and $\hat{G}(x, t; \xi, \tau)$ is Green's function for $L_0 u = 0$ in the rectangle D. In verifying that $u_2(x, t)$ satisfies (8.8), one uses the explicit formula for $\hat{G}$ (see (3.7.8), (3.7.9)).

From the explicit formula of $\hat{G}$ it also follows that $u_2(x, t)$ is infinitely differentiable in any closed domain contained in $D + R$. Combining this with the differentiability properties which were previously established for u_1, it follows (using Theorem 1, Sec. 1) that $u = u_1 + u_2$ has a finite norm $|u|_{0,2+\alpha}^{R_0}$. (8.6) now follows by Theorem 4, Sec. 7.

We shall extend Lemma 7 to general parabolic equations. Instead of (8.5) we consider

$$(8.9)\qquad \begin{aligned} Lu &= f \quad && \text{in } D + B_T, \\ u &= 0 \quad && \text{on } \bar{\pi} + \overline{B}, \\ u &= \psi \quad && \text{on } S - \pi. \end{aligned}$$

Lemma 8. *Let D, R, R_0, ψ, f be as in Lemma* 7 *and assume that L satisfies* ($\mathrm{A^R}$), (B) *of Sec.* 7. *Then there exists a unique solution of* (8.9) *and it satisfies* (8.6).

Proof. The proof is based on the method of continuity and is similar to the proofs of Theorems 7, 9 of Chap. 3. The analogue of the statement (a) states that for any f as in Lemma 7 there exists a unique solution of (8.5) having a finite norm $|u|_{0,2+\alpha}^{R_0}$ satisfying (8.6). The proof of the statement (b) is based on the boundary estimates of Theorem 4 and the details may be left to the reader. To prove (c), i.e., that if $\lambda_m \in \Sigma$ and $\lambda_m \to \sigma$

then $\sigma \in \Sigma$, we first show (as in the proofs of Theorem 7, 9) that $\lim u_m = u$ satisfies (8.6). (For simplicity, a convergent subsequence of $\{u_m\}$ is again denoted by $\{u_m\}$.) Finally, it remains to prove that u is continuous also up to the boundary $\bar{S} - \pi$ and $u = \psi$ on $S - \pi$, i.e.,

$$u(P) \to \psi(Q) \qquad \text{as } P \to Q,\, P \in D + B_T. \tag{8.10}$$

The proof is somewhat different from the proof of the analogous statement in Theorem 9, Chap. 3, since in that proof $\psi \equiv 0$. We may assume in what follows that $c \leq 0$.

Extend ψ to a continuous function on $\bar{B} + S$ by setting $\psi = 0$ on $\bar{\pi} + \bar{B}$. At each point $Q \in \bar{S} - \pi$ there exists a barrier w_Q with respect to all the L_λ (given by (3.4.3)). It is easily seen that given any $\epsilon > 0$ there is a k sufficiently large such that

$$|\psi - \psi(Q)| < \epsilon + kw_Q \qquad \text{on } \bar{B} + S.$$

The function $W = \epsilon + k_1 w_Q$, where $k_1 = \max\ (k, \text{l.u.b.}\ |f - c\psi(Q)|)$, satisfies

$$|\psi - \psi(Q)| < W \qquad \text{on } \bar{B} + S. \tag{8.11}$$

Next,

$$L_{\lambda_m} W = c\epsilon + k_1 L_{\lambda_m} w_Q \leq -\text{l.u.b.}\ |f - c\psi(Q)| \qquad \text{in } D + B_T \tag{8.12}$$

since $c \leq 0$. Introduce the function $\tilde{W} = W \pm [u_m - \psi(Q)]$. By (8.11), $\tilde{W} > 0$ on $\bar{B} + S$, and, by (8.12), $L\tilde{W} \leq 0$ in $D + B_T$. Hence, by the maximum principle, $\tilde{W} > 0$ in $\bar{D}$, i.e.,

$$|u_m(P) - \psi(Q)| < \epsilon + k_1 w_Q(P) \qquad \text{for all } P \in D + B_T.$$

Taking $m \to \infty$ we get

$$|u(P) - \psi(Q)| \leq \epsilon + k_1 w_Q(P).$$

If we next let $P \to Q$ and recall that ϵ is an arbitrary positive number, then (8.10) follows.

We can now establish the statement (a) made in the proof of Theorem 7, Chap. 3, namely,

Theorem 7. *Let D have the property $(\bar{\bar{\text{E}}})$ and let $\overline{|f|}_\alpha < \infty$, $f = 0$ on ∂B. Then there exists a unique solution of (8.2) and it belongs to $\bar{C}_{2+\alpha}$.*

Proof. By Theorem 8′, Sec. 4, Chap. 3, S has local barriers. Hence, by Theorem 6, there exists a unique solution u and $u \in C_{2+\alpha}$. It thus remains to show that the functions

$$u, \quad D_x u, \quad D_x^2 u, \quad D_t u \tag{8.13}$$

are uniformly Hölder continuous (exponent α) in some $(D + B_T)$-neighborhood of $\bar{B} + S$, i.e., for any $P \in \bar{B} + S$ there exists an $(n + 1)$-dimensional neighborhood V of P with diameter sufficiently small so that the functions

(8.13) are uniformly Hölder continuous (exponent α) in $(D + B_T) \cap V$. It suffices to consider the case where $P \in \bar{B} \cap \bar{S}$.

Without loss of generality we may assume that, for some $(n+1)$-dimensional neighborhood W of P, $D \cap W$ is a cylinder with a rectangular base and that $S \cap W$ is one lateral face whereas $B \cap W$ is the lower base. Indeed, if this is not the case then we first perform a transformation of coordinates in a neighborhood of P (compare the proof of Theorem 4 and, in particular, (7.8)–(7.10)) and then restrict ourselves, in the image of that neighborhood, to a cylinder with a rectangular base.

Since u vanishes on $S \cap W$ and on $\bar{B} \cap W$, Lemma 8 can be applied thus implying that the functions (8.13) are uniformly Hölder continuous (exponent α) in $D \cap W$. The proof of Theorem 7 is thereby completed.

We conclude this section with some additional results which readily follow from some theorems and arguments given in this and in the preceding chapter.

Theorem 8. *Corollary* 1 *of Theorem* 12, *Chap.* 3, *remains true even if the assumption that* $u \in \bar{C}_{2+\alpha}(D_1)$ *is omitted.*

Indeed, we have only to prove the assertion for $p = 0$ in $(D + B_T) \cap V$ where V is a ball with center P and sufficiently small radius, and P is any point on S_0. The assertion in this case follows by performing a transformation (compare the proof of Theorem 7) which leads to a situation where Lemma 8 can be applied.

Theorem 9. *Let* (A^R), (B), (C^R), (E^R) *be satisfied, let* R, R_0, ψ *be as in Theorem* 4, *and let* $L\psi = f$ *on* $R \cap \partial B$. *If* S *has local barriers then the first initial-boundary value problem*

$$(8.14) \qquad \begin{aligned} Lu &= f \qquad \text{in } D + B_T, \\ u &= \psi \qquad \text{on } \bar{B} + S \end{aligned}$$

has a unique solution u, *and* $|u|_{0,2+\alpha}^{R_0,D} < \infty$.

Proof. The existence of a solution u follows from Theorem 9, Chap. 3. To prove that $|u|_{0,2+\alpha}^{R_0,D} < \infty$ we consider u in a $(D + B_T)$-neighborhood of P, $P \in R_0$, perform a local transformation and then apply Lemma 8.

Another way of proving Theorem 8 is by the method of continuity.

Having proved Theorem 1, Sec. 1, Theorem 4, Sec. 7, the existence theorems 7, 9 of Chap. 3, and the differentiability theorems 12–14 of Chap. 3, as well as their corollaries, one can now state an obvious improvement of Theorem 1, namely, *Theorem* 1 *remains true even if the assumptions concerning the Hölder continuity of* u, $D_x u$, $D_x^2 u$, $D_t u$ *are omitted.* A similar extension of Theorem 4 is also valid provided D has the property $(\bar{\bar{E}})$.

9. Elliptic Equations

The proof of the Schauder estimates for elliptic equations can be given along the same lines as for parabolic equations. The role of the fundamental solution of the heat equation is now given to the function

$$\Gamma(x, y) = \begin{cases} \dfrac{1}{\omega_n} |x - y|^{2-n} & \text{if } n > 2 \\ \dfrac{1}{2\pi} \log \dfrac{1}{|x - y|} & \text{if } n = 2, \end{cases}$$

where ω_n is the area of the unit hypersphere in R^n.

The solutions of $\Delta u = f$ (where Δ is the Laplace operator $\Sigma\, \partial^2/\partial x_i^2$) can be represented in the form

$$u(x) = -\int f(y)\varphi(y)\Gamma(x - y)\, dy + \int u(y)\, \Delta[\varphi(y)\Gamma(x - y)]\, dy.$$

The actual calculations in deriving the estimates are technically different from those for parabolic equations; see [19].

In the derivation of the boundary estimates, the role of $\overline{G}$ in (6.3) is given to

$$\Gamma(x, y) = \frac{1}{\omega_n} (|x - y|^{2-n} - |x - y'|^{2-n})$$

where y' is the reflection of y with respect to some hyperplane x_n = const. In proving the analogue of Theorem 3, Sec. 6, the only "dangerous" derivative is $\partial^2 u/\partial x_n^2$ and it can be estimated from the equation $\Delta u = f$. The situation is therefore entirely different from that which occurs for parabolic equations, where we have to deal with two "dangerous" terms at the same time, namely, $\partial u/\partial t$ and $\partial^2 u/\partial x_n^2$. Thus, in the elliptic case there is no need to prove an analogue of Lemma 6′, Sec. 5.

Consider next the question of proving the statements (a) which are needed in the proofs of Theorems 18, 19 of Chap. 3. These statements can be established by considerations analogous to those of Sec. 8. One needs therefore an analogue of Lemma 6′; namely, one has to show that the function

$$v(x, t) = \int_R \Gamma(x - y) f(y)\, dy,$$

where $R = \{x;\ -\beta_i < x_i < \beta_i\ (i = 1, \ldots, n - 1),\ -\beta_n < x_n < 0\}$, belongs to $\overline{C}_{2+\alpha}(R_0)$ for every domain R_0 whose closure lies in $R + \pi$, where $\pi = \{x;\ -\beta_i < x_i < \beta_i\ (i = 1, \ldots, n - 1),\ x_n = 0\}$, provided f belongs to $\overline{C}_{2+\alpha}(R)$. Since it can be shown (see [63]) that $\Delta v = -f$, the Hölder continuity of the only "dangerous" term $\partial^2 v/\partial x_n^2$ follows from the Hölder continuity of the $\partial^2 v/\partial x_i^2$ $(i = 1, 2, \ldots, n - 1)$.

A more direct (but also more involved) method of establishing (a), without employing the Schauder estimates, is by proving directly the following:

Kellogg's Lemma. *Let ∂D belong to $C^{2+\alpha}$ and let $\psi \in C^{2+\alpha}$. Then there exists a unique solution of the Dirichlet problem*

$$\Delta u = 0 \quad \text{in } D,$$

$$u = \psi \quad \text{on } \partial D,$$

and $u \in \overline{C}_{2+\alpha}(D)$.

For proof, see [62].

PROBLEMS

1. By a rectangle R we shall understand a set $\{(x, t); a_i < x_i < b_i \ (i = 1, \ldots, n), a_0 < t < b_0\}$. Its normal boundary will be denoted by ∂R. L_0 is the heat operator. A function $v(x, t)$ in $D + B_T$ is said to be *supercaloric* if
   (i) $v(x, t)$ is piecewise continuous in $D + B_T$, its discontinuities occur on a finite number of hyperplanes $t = \tau$ where $v(x, \tau + 0) \geq v(x, \tau - 0)$ (we then define $v(x, \tau) = v(x, \tau - 0)$) and $v(x, \tau)$ is a continuous function of x, and
   (ii) for every $(\xi, \tau) \in D + B_T$ there exists a $\delta > 0$ such that for any rectangle R with (ξ, τ) at the center of its upper base and with diameter less than δ, $v \geq u$ in R for any function u satisfying $L_0 u = 0$ in R, $u \leq v$ on ∂R.

   In (ii), u is not required to be continuous on ∂R. It is allowed to have discontinuities on a finite number of sets $\partial R \cap \{t = \tau\}$ (and then $u(x, \tau) = u(x, \tau - 0)$), and $u(x, \tau)$ is assumed to be a continuous function of x. Prove that if $G^*(x, t; \xi, \tau)$ is Green's function of $L_0^* u = 0$ with respect to R and if we set $R_0 = \overline{R} \cap \{t = a_0\}$, $R_{1i} = R_0 \cap \{x_i = a_i\}$, $R_{2i} = R_0 \cap \{x_i = b_i\}$, then the solution of $L_0 u = 0$ with boundary values u on ∂R exists and is given by

$$(\bigstar) \qquad \begin{aligned} u(y, \sigma) = &\int_{R_0} u(x, a_0) G^*(x, a_0; y, \sigma)\, dx \\ &+ \sum_{i=1}^{n} \int_{a_0}^{\sigma} \int_{R_{1i}} \left[u \frac{\partial G^*}{\partial x_i} \right]_{x_i = a_i} \frac{dx}{dx_i}\, dt \\ &- \sum_{i=1}^{n} \int_{a_0}^{\sigma} \int_{R_{2i}} \left[u \frac{\partial G^*}{\partial x_i} \right]_{x_i = b_i} \frac{dx}{dx_i}\, dt \end{aligned}$$

where $dx/dx_i = dx_1 \ldots dx_{i-1}\, dx_{i+1} \ldots dx_n$. Hence

$$(\bigstar\bigstar) \quad \begin{aligned} v(\xi, \tau) \geq &\int_{R_0} u(x, a_0) G^*(x, a_0; \xi, \tau)\, dx + \sum_{i=1}^{n} \int_{a_0}^{\tau} \int_{R_{1i}} \left[u \frac{\partial G^*}{\partial x_i} \right]_{x_i = a_i} \frac{dx}{dx_i}\, dt \\ &- \sum_{i=1}^{n} \int_{a_0}^{\tau} \int_{R_{2i}} \left[u \frac{\partial G^*}{\partial x_i} \right]_{x_i = b_i} \frac{dx}{dx_i}\, dt. \end{aligned}$$

[*Hint:* Use the form of G at the end of Sec. 7, Chap. 3, to verify that the right-hand side of ($\bigstar$), call it w, satisfies $L_0 w = 0$ in R, $w = u$ on ∂R (except for a finite number of sets $\partial R \cap \{t = \tau\}$). Prove that $w = u$ in R by applying the maximum principle to $u_\epsilon \pm (w - u)$ where u_ϵ is defined by the right-hand side of (★) with u replaced by a function f which equals $\underset{R}{\text{l.u.b.}} (|u| + |w|)$ in an ϵ-neighborhood of the discontinuities of u, w and which vanishes elsewhere.]

2. Prove that if the inequality (★★) holds with the boundary values $u = v$ (on ∂R), then it also holds for any boundary values u which are $\leq v$ (on ∂R).
[*Hint:* Show that in (★★)

$$(\bigstar\bigstar\bigstar) \qquad G^* \geq 0, \quad \left.\frac{\partial G^*}{\partial x_i}\right|_{x_i = a_i} \geq 0, \quad \left.\frac{\partial G^*}{\partial x_i}\right|_{x_i = b_i} \leq 0.]$$

3. Use the representation for u given by the right-hand side of (★) to prove that if $\{u_m\}$ is a sequence of solutions of $L_0 u = 0$ in $D + B_T$ which is uniformly bounded, then for any i, j, $\{D_x^i D_t^j u_m\}$ is a uniformly bounded sequence in closed subsets of $D + B_T$ (for a proof based on the a priori estimates, see Chap. 3. Sec. 6).

4. Prove that if v is supercaloric in $D + B_T$ and $L_0 u = 0$ in $D + B_T$, then $v + u$ is supercaloric in $D + B_T$.

5. Prove that if v is supercaloric in $D + B_T$ and if $L_0 u = 0$ in $D + B_T$, $\liminf_{P \to Q} (v - u) \geq 0$ for all $Q \in \bar{B} + S$, then $v \geq u$ in $D + B_T$.
[*Hint:* If $w \equiv v - u$ takes a negative minimum $-M$ then for some point $P \in D + B_T$ and some rectangle R, with P at the center of its upper base, $w(P) = -M$, $w > -M$ at some points of ∂R. Use (★★), (★★★) to derive $-M > -M$.]

6. Prove that if v satisfies (*i*) and if (★) holds for $u = v$ on the right-hand side and for all (ξ, τ), R as in (*ii*), then (*ii*) holds and hence v is supercaloric.
[*Hint:* Use the proof of Problem 5.]

7. Let v be a supercaloric function in $D + B_T$ and let D_0 be a rectangle contained in D. Define

$$w = \begin{cases} v & \text{in } D - \bar{D}_0 \\ u & \text{in } \bar{D}_0 \end{cases}$$

where $L_0 u = 0$, $u = v$ on ∂D_0 (the normal boundary of D_0). Prove that w is supercaloric.
[*Hint:* If (ξ, τ) does not belong to ∂D_0, (★★) holds if the diameter of R is sufficiently small. If $(\xi, \tau) \in \partial D_0$ use $v \geq u$ in $\bar{D}_0$ and (★★★) to establish (★★). Now use Problem 6.]

8. Let v be a continuous supercaloric function in $D + B_T$ and let $\{D_m\}$ be a sequence of rectangles, $\bar{D}_m \subset D + B_T$. Define a sequence of functions $\{v_m\}$ inductively by

$$v_m(x, t) = \begin{cases} v_{m-1}(x, t) & \text{in } D - \bar{D}_m \\ u(x, t) & \text{in } \bar{D}_m, \end{cases}$$

where $L_0 u = 0$ in D_m, $u = v_{m-1}$ on ∂D_m. Prove that $\{v_m(x, t)\}$ is a monotone decreasing sequence of supercaloric functions, and that v_m has at most m hyperplanes $t = \tau$ of discontinuity.

9. Prove that any function satisfying $L_0 v \leq 0$ in $D + B_T$ is supercaloric.
[*Hint:* Use Green's formula in a rectangle R for v, G^*.]

10. A *weak barrier* at Q $(Q \in \bar{B} + S)$ is a function w_Q which is continuous in $\bar{D}$, zero at Q, positive in $\bar{D} - Q$, and supercaloric in $D + B_T$. By Problem 9, barriers are weak barriers. Prove that if weak barriers exist at all the points of S then there exists a solution of

$$L_0 u = 0 \qquad \text{in } D + B_T,$$
$$u = \psi \qquad \text{on } \bar{B} + S,$$

for any continuous function ψ on $\bar{B} + S$, which can be extended into a continuous function v in $\bar{D}$, supercaloric in $D + B_T$.
[*Hint:* Construct v_m as in Problem 8, with D_m such that $\bigcup_m \bar{D}_m = D + B_T$ and such that each D_j is repeated an infinite number of times in the sequence $\{D_m\}$. By Problems 8, 3, $u = \lim u_m$ exists and satisfies $L_0 u = 0$. Next, if $Q \in \bar{B} + S$, $v < \psi(Q) + \epsilon + kw_Q$ in $\bar{D}$. Hence, $v_m < \psi(Q) + \epsilon + kw_Q$, $u < \psi(Q) + \epsilon + kw_Q$. Similarly $u > \psi(Q) - \epsilon - kw_Q$.]

11. Extend the result of Problem 10 to the case where ψ is a polynomial.
[*Hint:* Write $\psi = (\psi + Ct) - Ct$ and use Problem 9.]

12. Extend the result of Problem 11 to the case where ψ is any continuous function.
[*Hint:* Approximate ψ by polynomials.]

13. Prove Theorem 5, Sec. 8.
[*Hint:* Use Problem 12 step by step on t-intervals.]

CHAPTER 5

THE SECOND INITIAL-BOUNDARY VALUE PROBLEM

Introduction. In Chap. 3 we have solved the first initial-boundary value problem by the method of continuity, using a priori estimates. In the present chapter we shall solve the second initial-boundary value problem by reducing it, with the aid of potentials, to an integral equation. To give a brief description of this method, consider the case of the heat equation $L_0 u = 0$ in a cylinder $D \times (0, T]$. The function

$$U(x, t) = \int_0^t \int_S \Gamma(x, t; \xi, \tau)\varphi(\xi, \tau)\, dS_\xi\, d\tau, \tag{0.1}$$

where S is the boundary of D and dS_ξ is the surface element of S, is called a *single-layer potential* with *density* φ (with respect to Γ); Γ is the fundamental solution of the heat equation. $U(x, t)$ is defined and is continuous in $\bar{D} \times [0, T]$, and as x in D tends to a point x^0 on S along nontangential directions,

$$\frac{\partial}{\partial \nu} U(x, t) \to -\frac{1}{2}\varphi(x^0, t) + \frac{\partial}{\partial \nu} U(x^0, t) \tag{0.2}$$

where ν is the inward normal to S at x^0. This *jump relation* is the most significant feature of the single-layer potentials.

The problem

$$\begin{aligned} L_0 u &= 0 \quad && \text{in } D \times (0, T], \\ u(x, 0) &= 0 && \text{on } \bar{D}, \\ \frac{\partial u}{\partial \nu} + \beta u &= g && \text{on } S \times (0, T], \end{aligned} \tag{0.3}$$

can be solved by taking u to be the function in (0.1) with φ satisfying the equation

$$\frac{1}{2}\varphi(x^0, t) = \int_0^t \int_S \left[\frac{\partial}{\partial \nu}\Gamma(x^0, t; \xi, \tau) + \beta(x^0, t)\Gamma(x^0, t; \xi, \tau)\right]\varphi(\xi, \tau)\, dS_\xi\, d\tau - g(x^0, t);$$

this is an integral equation of Volterra type. The general second initial-boundary value problem where $L_0 u = f$, $u(x, 0) = \psi(x)$ can be solved in a similar manner.

In the last section of the present chapter we shall extend the concepts of fundamental solution and potentials to elliptic equations and consider the Neumann problem.

1. Summary of Results on Fundamental Solutions

In this chapter the notation and results of Chap. 1 will be freely used. However, for the convenience of the reader we sum up some of the main facts in a form which will be useful in the following sections.

Let

$$(1.1)\qquad Lu = \sum_{i,j=1}^{n} a_{ij}(x, t) \frac{\partial^2 u}{\partial x_i\, \partial x_j} + \sum_{i=1}^{n} b_i(x, t) \frac{\partial u}{\partial x_i} + c(x, t)u - \frac{\partial u}{\partial t}$$

be an operator defined in a bounded closed domain $\Omega = \overline{D} \times [0, T]$ where D is a domain in R^n, and assume:

(A_1) L is parabolic in Ω,

(A_2) the coefficients of L satisfy the Hölder conditions (1.1.4), (1.1.5), (1.1.6) in Ω.

Using the italicized statement following Theorem 2, Sec. 1, Chap. 3, we extend the coefficients of L to a larger set $\Omega_0 = \overline{D}_0 \times [0, T]$ where D_0 is a bounded domain containing $\overline{D}$, so that (A_1), (A_2) hold with respect to Ω_0.

The fundamental solution $\Gamma(x, t; \xi, \tau)$ can now be constructed in Ω_0. It has the form

$$(1.2)\qquad \Gamma(x, t; \xi, \tau) = Z(x, t; \xi, \tau) + \int_\tau^t \int_{D_0} Z(x, t; \eta, \sigma)\Phi(\eta, \sigma; \xi, \tau)\, d\eta\, d\sigma,$$

where Z is given by (1.2.5), (1.2.6), (1.2.4) and Φ is the solution of the integral equation (1.4.1) with D replaced by D_0, and it can be expanded into a series (1.4.4). Φ satisfies the inequality (1.4.8).

Γ can also be written in the form

$$(1.3)\qquad \Gamma(x, t; \xi, \tau) = Z(x, t; \xi, \tau) + Z_0(x, t; \xi, \tau),$$

where Z_0 is the integral on the right-hand side of (1.2), and the following inequalities hold for x, ξ in D_0 and $0 \leq \tau < t \leq T$:

$$(1.4)\qquad |\Gamma(x, t; \xi, \tau)| \leq \frac{\text{const.}}{(t-\tau)^\mu} \frac{1}{|x-\xi|^{n-2\mu}} \qquad (0 < \mu < 1),$$

$$(1.5)\qquad |D_x Z(x, t; \xi, \tau)| \leq \frac{\text{const.}}{(t-\tau)^\mu} \frac{1}{|x-\xi|^{n+1-2\mu}} \qquad \left(\frac{1}{2} < \mu < 1\right),$$

$$(1.6)\qquad |D_x Z_0(x, t; \xi, \tau)| \leq \frac{\text{const.}}{(t-\tau)^\mu} \frac{1}{|x-\xi|^{n+1-2\mu-\alpha}} \qquad \left(1 - \frac{\alpha}{2} < \mu < 1\right).$$

(1.6) follows from (1.5), (1.4.8), upon using Lemma 2, Sec. 4, Chap. 1.

The reason for constructing $\Gamma(x, t; \xi, \tau)$ in $\overline{D}_0 \times [0, T]$ (instead of $\overline{D} \times [0, T]$) is that we want $D_x\Gamma(x, t; \xi, \tau)$ to be defined and continuous not only for x, ξ in D but also for x, ξ in a neighborhood of $\overline{D}$.

We shall denote the boundary of D by S. We say that S is of class $C^{1+\lambda}$ $(0 < \lambda < 1)$ if, locally, S can be represented in the form (3.8.6) with h in class $C^{1+\lambda}$. From now on it will always be assumed (in this chapter) that S is of class $C^{1+\lambda}$.

Let $x^0 \in S$ and denote by π the tangent hyperplane to S at x^0. Consider the closed ball B_δ with center x^0 and radius δ. It intersects S in a portion S_δ and, if δ is sufficiently small, the orthogonal projection of S_δ on π defines a one-to-one mapping between the points ξ of S_δ and the points ξ' of some set S'_δ (ξ' is the projection of ξ). The functions $\xi = \xi(\xi')$, $\xi' = \xi'(\xi)$ are Hölder continuous (exponent λ). S'_δ is a closed domain in π and x^0 lies in its interior.

Let x be a point on the inward normal N_{x^0} to S at x^0 and let $|x - x^0|$ be sufficiently small, say $|x - x^0| < \delta_0$. Since S is of class $C^{1+\lambda}$, for any $\xi \in S_\delta$,

$$|\xi - \xi'| \leq \text{const.}\, |x^0 - \xi|^{1+\lambda}. \tag{1.7}$$

We shall now prove that

$$0 < \text{const.} \leq \frac{|x - \xi|}{|x - \xi'|} \leq \text{const.} \tag{1.8}$$

If $|x^0 - \xi| > 2|x - x^0|$ then, by (1.7),

$$|x - \xi'| \leq |x - x^0| + |x^0 - \xi| + |\xi - \xi'| \leq \text{const.}\, |x^0 - \xi|.$$

Since also

$$|x - \xi| \geq |x^0 - \xi| - |x - x^0| > |x^0 - \xi|/2,$$

the inequality

$$\frac{|x - \xi'|}{|x - \xi|} \leq \text{const.} \tag{1.9}$$

follows.

If $|x^0 - \xi| \leq 2|x - x^0|$, then, using (1.7) and the fact that x^0 is the nearest point on S to x (if $|x - x^0|$ is sufficiently small), we get

$$\begin{aligned} |x - \xi'| &\leq |x - \xi| + |\xi - \xi'| \leq |x - \xi| + \text{const.}\, |x^0 - \xi|^{1+\lambda} \\ &\leq |x - \xi| + \text{const.}\, |x - \xi|^{1+\lambda} \leq \text{const.}\, |x - \xi| \end{aligned}$$

and (1.9) is thus completely proved. The proof for the right-hand inequality in (1.8) is similar.

Note that δ, δ_0, and the constants in (1.7), (1.8) can be taken to be independent of x^0 and x.

2. The Jump Relation for Single-layer Potentials

Let $\varphi(x, t)$ be a continuous function on $S \times [0, T]$ and introduce (in $D_0 \times (0, T]$)

$$U(x, t) = \int_0^t \int_S \Gamma(x, t; \xi, \tau)\varphi(\xi, \tau)\, dS_\xi\, d\tau, \tag{2.1}$$

where dS_ξ is the surface element on S. U is called a *single-layer potential* of the *density* φ (with respect to Γ).

If $x \in S$ then the integral in (2.1) is improper and is defined as $\lim_{\epsilon \to 0} \int_0^{t-\epsilon} \int_S$. From (1.4) it follows that this improper integral is absolutely convergent. Furthermore, using arguments similar to those used in the proof of Lemma 1, Sec. 3, Chap. 1, we conclude that $U(x, t)$ is a continuous function not only for $x \in \bar{D}_0 - S$ but also for x in $\bar{D}_0$. It is easily seen that $U(x, t)$ is continuous in the closed domain Ω_0 when defined to be zero on $t = 0$.

By (1.3)

$$U(x, t) = V(x, t) + W(x, t) \tag{2.2}$$

where

$$V(x, t) = \int_0^t \int_S Z(x, t; \xi, \tau)\varphi(\xi, \tau)\, dS_\xi\, d\tau, \tag{2.3}$$

$$W(x, t) = \int_0^t \int_S Z_0(x, t; \xi, \tau)\varphi(\xi, \tau)\, dS_\xi\, d\tau. \tag{2.4}$$

Clearly, if $x \in D_0 - S$,

$$D_x W(x, t) = \int_0^t \int_S D_x Z_0(x, t; \xi, \tau)\varphi(\xi, \tau)\, dS_\xi\, d\tau. \tag{2.5}$$

Using (1.6) and arguments similar to those used in the proof of Lemma 1, Sec. 3, Chap. 1, one can show that the right-hand side of (2.5), denote it by $W_1(x, t)$, exists also when $x \in S$ (as an improper, absolutely convergent integral) and $W_1(x, t)$ is a continuous function also for x in a neighborhood of S. Thus, in particular, for any x^0 on S,

$$\lim_{x \to x^0} D_x W(x, t) = \int_0^t \int_S D_x Z_0\, (x^0, t; \xi, \tau)\, \varphi(\xi, \tau)\, dS_\xi\, d\tau \tag{2.6}$$

Clearly, if $x \in D$ then

$$D_x V(x, t) = \int_0^t \int_S D_x Z(x, t; \xi, \tau)\varphi(\xi, \tau)\, dS_\xi\, d\tau, \tag{2.7}$$

but the behavior of $D_x V(x, t)$ as x tends to a point x^0 on S is entirely different from the behavior of $D_x W(x, t)$. We shall be interested in the behavior of the first derivative of U, or V, primarily in one particular direction, namely, the direction of the inward conormal at x^0. Before stating the main result, we introduce some notation.

Let $x^0 \in S$ and denote by $\nu(x^0, t^0)$ the inward conormal at (x^0, t^0), i.e., the components $\nu^i(x^0, t^0)$ of $\nu(x^0, t^0)$ are given by

$$\nu^i(x^0, t^0) = \sum_{j=1}^{n} a_{ij}(x^0, t^0)N^j_{x^0} \qquad (1 \leq i \leq n), \qquad \nu^{n+1}(x^0, t^0) = 0, \tag{2.8}$$

where $N_{x^0} = (N^1_{x^0}, \ldots, N^n_{x^0})$ is the inward normal to S at x^0. Since (a_{ij}) is a positive definite matrix, $\nu(x^0, t^0)$ points into the interior of $\Omega \cap \{t = t^0\}$. The derivative of $U(x, t)$ in the conormal direction (at (x^0, t)) is given by

$$\frac{\partial U(x, t)}{\partial \nu(x^0, t)} = \sum_{i,j=1}^{n} a_{ij}(x^0, t) \cos(N_{x^0}, x_j) \frac{\partial U(x, t)}{\partial x_i}, \tag{2.9}$$

where $\cos(N_{x^0}, x_j)$ is cosine of the angle between the direction of N_{x^0} and the positive x_j-axis.

Let $K = K(x^0)$ be any finite closed cone in R^n with vertex x^0, such that $K \subset D + \{x^0\}$. Thus, every direction from x^0 into a point x in K points into the interior of D.

We can now state the main result of this section.

Theorem 1. *Let the assumptions* (A_1), (A_2) *hold, let* S *be of class* $C^{1+\lambda}$ *for some* $0 < \lambda < 1$, *and let* φ *be a continuous function on* $S \times [0, T]$. *Then, for any* $x^0 \in S$, $0 < t \leq T$, *the function* $U(x, t)$ *satisfies the relation*

$$\lim_{\substack{x \to x^0 \\ x \in K}} \frac{\partial U(x, t)}{\partial \nu(x^0, t)} = -\frac{1}{2}\varphi(x^0, t) + \int_0^t \int_S \frac{\partial \Gamma(x^0, t; \xi, \tau)}{\partial \nu(x^0, t)} \varphi(\xi, \tau)\, dS_\xi\, d\tau. \tag{2.10}$$

Thus the single-layer potentials experience a jump across the lateral boundary of Ω on which the density φ is situated. The relation (2.10) is called a *jump relation.*

Proof. We shall first show that the improper integral on the right-hand side of (2.10) exists and is, in fact, absolutely convergent. We shall need an elementary lemma similar to Lemma 2, Sec. 4, Chap. 1 (for proof, see Problem 1):

Lemma 1. *If* $0 \leq a < n - 1$, $0 \leq b < n - 1$ *then*

$$\int_S \frac{dy}{|x - y|^a |y - \xi|^b} \leq \begin{cases} \text{const.}\ |x - \xi|^{n-1-a-b} & \text{if } a + b > n - 1, \\ \text{const.} & \text{if } a + b < n - 1. \end{cases} \tag{2.11}$$

If we prove that

$$\left| \frac{\partial \Gamma(x^0, t; \xi, \tau)}{\partial \nu(x^0, t)} \right| \leq \frac{\text{const.}}{(t - \tau)^\mu} \frac{1}{|x^0 - \xi|^{n+1-2\mu-\beta}} \qquad (\beta = \min(\alpha, \lambda)), \tag{2.12}$$

for any $(1 - \beta/2) < \mu < 1$, then the absolute convergence of the integral on the right-hand side of (2.10) follows by applying Lemma 1 with $a = n + 1 - 2\mu - \beta$, $b = 0$.

To prove (2.12) we first consider $\partial Z/\partial \nu$:

$$\begin{aligned}(2.13)\quad \frac{\partial Z(x, t; \xi, \tau)}{\partial \nu(x^0, t^0)} &= -\frac{1}{2} C(\xi, \tau)(t-\tau)^{-1-n/2} \exp\left[-\frac{\vartheta^{\xi,\tau}(x, \xi)}{4(t-\tau)}\right] \\ &\quad \cdot \sum_{i,j,k} a_{ij}(x^0, t) a^{ik}(\xi, \tau)(x_k - \xi_k) \cos(N_{x^0}, x_j) \\ &= F_0(x, t; \xi, \tau) + F_1(x, t; \xi, \tau),\end{aligned}$$

where

$$\begin{aligned}(2.14)\quad & F_0(x, t; \xi, \tau) \\ &= \frac{1}{2} C(\xi, \tau)(t-\tau)^{-1-n/2} \exp\left[-\frac{\vartheta^{\xi,\tau}(x, \xi)}{4(t-\tau)}\right] |x - \xi| \cos(N_{x^0}, \overrightarrow{x\xi}),\end{aligned}$$

$$\begin{aligned}(2.15)\quad & F_1(x, t; \xi, \tau) \\ &= -\frac{1}{2} C(\xi, \tau)(t-\tau)^{-1-n/2} \exp\left[-\frac{\vartheta^{\xi,\tau}(x, t)}{4(t-\tau)}\right] \\ &\quad \cdot \sum_{i,j,k} [a_{ij}(x^0, t) - a_{ij}(\xi, \tau)] a^{ik}(\xi, \tau)(x_k - \xi_k) \cos(N_{x^0}, x_j);\end{aligned}$$

$\overrightarrow{x\xi}$ denotes the vector which connects x to ξ.

Since $|\cos(N_{x^0}, \overrightarrow{x^0\xi})| \leq \text{const.}\ |x^0 - \xi|^\lambda$,

$$(2.16)\qquad |F_0(x^0, t; \xi, \tau)| \leq \frac{\text{const.}}{(t-\tau)^\mu} \frac{1}{|x^0 - \xi|^{n+1-2\mu-\lambda}}.$$

Next, using (1.1.4) we find that

$$\begin{aligned}(2.17)\quad |F_1(x^0, t; \xi, \tau)| &\leq \frac{\text{const.}}{(t-\tau)^{\mu_0}} \frac{1}{|x^0 - \xi|^{n+1-2\mu_0-\alpha}} \\ &\quad + \frac{\text{const.}}{(t-\tau)^{\mu_1 - \alpha/2}} \frac{1}{|x^0 - \xi|^{n+1-2\mu_1}}\end{aligned}$$

for any $(1 - \alpha/2) < \mu_0 < 1$, $1 < \mu_1 < 1 + \alpha/2$. Hence,

$$(2.18)\quad |F_1(x^0, t; \xi, \tau)| \leq \frac{\text{const.}}{(t-\tau)^\mu} \frac{1}{|x^0 - \xi|^{n+1-2\mu-\alpha}} \qquad \left(1 - \frac{\alpha}{2} < \mu < 1\right).$$

Combining (2.17) we get a bound on $\partial Z/\partial \nu$ which, when combined with (1.6) for $x = x^0$, yields the desired inequality (2.12).

We next prove the jump relation (2.10) in case that $x \to x^0$ and x lies on the normal N_{x^0}. Since then $|x^0 - \xi| \leq \text{const.}\ |x - \xi|$,

$$|a_{ij}(x^0, t) - a_{ij}(\xi, t)| \leq \text{const.}\ |x - \xi|^\alpha.$$

One can therefore proceed to estimate $F_1(x, t; \xi, \tau)$ similarly to the previous estimation of $F_1(x^0, t; \xi, \tau)$ in (2.17), (2.18) and thus obtain

$$(2.19)\qquad |F_1(x, t; \xi, \tau)| \leq \frac{\text{const.}}{(t-\tau)^\mu} \frac{1}{|x - \xi|^{n+1-2\mu-\alpha}}.$$

Thus, $F_1(x, t; \xi, \tau)$ is bounded by the same bound as $D_x Z_0(x, t; \xi, \tau)$. We conclude, as in the case of $D_x Z_0$, that the integral

$$V_1(x, t) = \int_0^t \int_S F_1(x, t; \xi, \tau)\varphi(\xi, \tau)\, dS_\xi\, d\tau$$

satisfies the relation

$$V_1(x, t) \to V_1(x^0, t) \qquad \text{as } x \to x^0.$$

Combining this fact with (2.6), it thus remains to show that the integral

$$V_0(x, t) = \int_0^t \int_S F_0(x, t; \xi, \tau)\varphi(\xi, \tau)\, dS_\xi\, d\tau \tag{2.20}$$

satisfies the jump relation

$$\lim_{x \to x^0} V_0(x, t) = -\frac{1}{2}\varphi(x^0, t) + V_0(x^0, t). \tag{2.21}$$

Write

$$V_0(x, t) = I_\delta(x, t) + J_\delta(x, t) \tag{2.22}$$

where (recall (2.14))

$$I_\delta(x, t) = \frac{1}{2}\int_0^t \int_{S_\delta} \frac{C(\xi, \tau)\varphi(\xi, \tau)}{(t - \tau)^{1+n/2}} \exp\left[-\frac{\vartheta^{\xi,\tau}(x, \xi)}{4(t - \tau)}\right] \cdot |x - \xi| \cos(N_{x^0}, \overrightarrow{x\xi})\, dS_\xi\, d\tau \tag{2.23}$$

and J_δ is the complementary part. (For the notation of π, S_δ, S'_δ, ξ', see Sec. 1.)

We shall later compare I_δ with

$$I'_\delta(x, t) = \frac{1}{2}\int_0^t \int_{S'_\delta} \frac{C(x^0, \tau)\varphi(x^0, \tau)}{(t - \tau)^{1+n/2}} \exp\left[-\frac{\vartheta^{x^0,t}(x, \xi')}{4(t - \tau)}\right] |x - \xi'| \cdot \cos(N_{x^0}, \overrightarrow{x\xi'})\, dS'_\xi\, d\tau \tag{2.24}$$

where dS'_ξ is the surface element (at ξ') on the hyperplane π.

As for the behavior of I'_δ, we shall establish the relation

$$\lim_{x \to x^0} I'_\delta(x, t) = -\frac{1}{2}\varphi(x^0, t). \tag{2.25}$$

Integrating with respect to τ and substituting $t - \tau = \vartheta^{x^0,t}(x, \xi')/4\rho$, we find that

$$I'_\delta(x, t) = 2^{n-1}\int_{S'_\delta} \frac{|x - \xi'| \cos(N_{x^0}, \overrightarrow{x\xi'})}{[\vartheta^{x^0,t}(x, \xi')]^{n/2}} \psi(x, \xi', t)\, dS'_\xi \tag{2.26}$$

where

$$\psi(x, \xi', t) = \int_{\vartheta^{x^0,t}(x,\xi')/4t}^{\infty} \rho^{n/2-1} e^{-\rho} C\left(x^0, t - \frac{\vartheta^{x^0,t}(x, \xi')}{4\rho}\right) \cdot \varphi\left(x^0, t - \frac{\vartheta^{x^0,t}(x, \xi')}{4\rho}\right) d\rho. \tag{2.27}$$

$\psi(x, \xi', t)$ is a continuous function of (x, ξ') for all x on N_{x^0} (and sufficiently close to x^0) and for all $\xi' \in S'_\delta$. In particular, $\psi(x^0, x^0, t) = \lim \psi(x, \xi', t)$ (as $x \to x^0$, $\xi' \to x^0$), and

$$\psi(x^0, x^0, t) = C(x^0, t)\varphi(x^0, t) \int_0^\infty \rho^{n/2-1}e^{-\rho}\, d\rho.$$

Since the last integral is equal to $\Gamma(n/2) = 2\pi^{n/2}/\omega_n$ where ω_n is the area of the unit hypersphere in R^n, we get

$$\psi(x^0, x^0, t) = C(x^0, t)\varphi(x^0, t)2\pi^{n/2}/\omega_n. \tag{2.28}$$

Let Λ be the unit hypersphere with center x in R^n and denote by ξ'' the intersection of Λ with the ray with origin x and direction $\overrightarrow{x\xi'}$. Introduce coordinates $(\xi''_1, \ldots, \xi''_n)$ for ξ'', which are proportional to

$$(\xi'_1 - x_1, \ldots, \xi'_n - x_n)$$

so that $\Sigma\, (\xi''_i)^2 = 1$, and denote the surface element of Λ (at ξ'') by $d\omega(\xi'')$. Divide S'_δ into two regions, $S'_{1\delta}$ which contains x^0 in its interior and its complement (in S'_δ) $S'_{2\delta}$. As ξ' varies in $S'_{1\delta}$, ξ'' varies in $S''_{1\delta}$.

Recalling that x lies on N_{x^0} we find that

$$\begin{aligned} I'_\delta(x, t) = & -2^{n-1}\psi(x^0, x^0, t) \int_{S''_{1\delta}} \frac{d\omega(\xi'')}{[\Sigma\, a^{ij}(x^0, t)\xi''_i\xi''_j]^{n/2}} \\ & - 2^{n-1} \int_{S''_{1\delta}} \frac{[\psi(x, \xi', t) - \psi(x^0, x^0, t)]}{[\Sigma\, a^{ij}(x^0, t)\xi''_i\xi''_j]^{n/2}}\, d\omega(\xi'') \\ & + 2^{n-1} \int_{S'_{2\delta}} \frac{|x - \xi'| \cos (N_{x^0}, \overrightarrow{x\xi'})}{[\vartheta^{x_0,t}(x, \xi')]^{n/2}}\, \psi(x, \xi', t)\, dS_{\xi'}. \end{aligned} \tag{2.29}$$

Since $\psi(x, \xi', t)$ is a continuous function, given any $\epsilon > 0$ we can find an $\eta = \eta(\epsilon)$ such that if the diameter of $S_{1\delta}$ is less than η then the second term on the right-hand side of (2.29) is bounded by ϵ. η will be fixed from now on. Since $\cos (N_{x^0}, \overrightarrow{x\xi'}) \to 0$ if $x \to x^0$, and since $|x - \xi'|$ is bounded away from zero if $\xi' \in S'_{2\delta}$, we conclude that if $|x - x^0| < \eta_0$ (for some η_0 depending on η, ϵ) then the third term on the right-hand side of (2.29) is bounded by ϵ.

As for the first term on the right-hand side of (2.29), in view of (2.28) it is equal to

$$-\frac{2^n\pi^{n/2}}{\omega_n} C(x^0, t)\varphi(x^0, t) \int_{S''_{1\delta}} \frac{d\omega(\xi'')}{[\Sigma\, a^{ij}(x^0, t)\xi''_i\xi''_j]^{n/2}}\cdot \tag{2.30}$$

In order to evaluate the integral in (2.30) for $x \to x^0$, we consider the function

$$I(t) = \int_{|\xi|<1} t^{-n/2} \exp\left[-\frac{\Sigma\, a^{ij}\xi_i\xi_j}{4t}\right] d\xi$$

where (a^{ij}) is a positive definite matrix and a^{ij} are constants. Substituting first $\xi_i = \rho\xi_i''$ and then $\sigma = \rho^2\vartheta/4t$ where $\vartheta = \sum a^{ij}\xi_i''\xi_j''$, we get

$$I(t) = 2^{n-1}\int_\Lambda \vartheta^{-n/2}\left[\int_0^{\vartheta/4t} \sigma^{n/2-1}e^{-\sigma}\,d\sigma\right]d\omega(\xi'').$$

Hence,

$$\lim_{t\to 0} I(t) = 2^{n-1}\Gamma\left(\frac{n}{2}\right)\int_\Lambda \vartheta^{-n/2}\,d\omega(\xi'').$$

Since on the other hand we have, by Theorem 1, Sec. 2, Chap. 1,

$$\lim_{t\to 0} I(t) = [\det(a^{ij})]^{-1/2}(2\sqrt{\pi})^n,$$

we conclude that

$$\int_\Lambda \frac{d\omega(\xi'')}{[\sum a^{ij}\xi_i''\xi_j'']^{n/2}} = [\det(a^{ij})]^{-1/2}\omega_n.$$

If we decompose the integral $\int_\Lambda$ on the left-hand side into two integrals $\int_{\Lambda_1} + \int_{\Lambda_2}$ where Λ_1, Λ_2 are complementary hemispheres on Λ, then $\int_{\Lambda_1} = \int_{\Lambda_2}$ since the points of Λ_1, Λ_2 correspond to each other in a one-to-one way by the mapping $\xi'' \to -\xi''$ which preserves $d\omega(\xi'')$ and $[\sum a^{ij}\xi_i''\xi_j'']^{n/2}$. We therefore have:

$$\int_{\Lambda_1} \frac{d\omega(\xi'')}{[\sum a^{ij}\xi_i''\xi_j'']^{n/2}} = [\det(a^{ij})]^{-1/2}\omega_n/2.$$

Noting now that, as $x \to x^0$, $S_{1\delta}''$ tends to a hemisphere on Λ, and using the last equality with $a^{ij} = a^{ij}(x^0, t)$, we find that the integral in (2.30) tends to

$$[\det(a^{ij}(x^0, t))]^{-1/2}\omega_n/2$$

as $x \to x^0$.

Recalling the definition of $C(x^0, t)$ in (1.2.6) it follows that the expression in (2.30), and hence the first term on the right-hand side of (2.29), converges to $-\varphi(x^0, t)/2$ as $x \to x^0$. Since each of the other two terms remains bounded by any given ϵ, the relation (2.25) holds.

Recalling the definition of J_δ in (2.22) we see that for the variable of integration ξ in the integral of J_δ, $|x - \xi| \geq \text{const.} > 0$ for all x sufficiently close to x^0. Hence the integral is a continuous function in some small neighborhood of x^0. It follows that

$$\lim_{x\to x^0} J_\delta(x, t) = J_\delta(x^0, t). \tag{2.31}$$

Combining (2.31), (2.25) with (2.22), and noting that $I_\delta'(x^0, t) = 0$, we conclude that the jump relation (2.21) would follow if we prove that

$$I_\delta(x, t) - I_\delta'(x, t) \to I_\delta(x^0, t) \qquad \text{as } x \to x^0. \tag{2.32}$$

To prove (2.32), take $\delta_1 < \delta$ and write

$$
\begin{aligned}
I_\delta(x,t) &= I_{\delta_1}(x,t) + \bar{I}_{\delta_1}(x,t),\\
I_\delta(x^0,t) &= I_{\delta_1}(x^0,t) + \bar{I}_{\delta_1}(x^0,t),\\
I'_\delta(x,t) &= I'_{\delta_1}(x,t) + \bar{I}'_{\delta_1}(x,t),
\end{aligned}
\tag{2.33}
$$

where $\bar{I}_{\delta_1}(x,t)$ is the complementary part to $I_{\delta_1}(x,t)$ (i.e., the ξ-integration is taken over the set $S_\delta - S_{\delta_1}$), etc.

We first prove that for any $\epsilon > 0$,

$$
|I_{\delta_1}(x,t) - I'_{\delta_1}(x,t)| < \epsilon \tag{2.34}
$$

provided δ_1 is sufficiently small, independently of x. Using (1.7), (1.8) we get

$$
\begin{aligned}
||x-\xi|\cos(N_{x^0},\overrightarrow{x\xi}) - |x-\xi'|\cos(N_{x^0},\overrightarrow{x\xi'})| &= |\xi-\xi'| \leq \text{const. } |x^0-\xi|^{1+\lambda}\\
&\leq \text{const. } |x-\xi|^{1+\lambda},
\end{aligned}
$$

$$
\begin{aligned}
&\left|\exp\left[-\frac{\vartheta^{\xi,\tau}(x,\xi)}{4(t-\tau)}\right] - \exp\left[-\frac{\vartheta^{x^0,t}(x,\xi')}{4(t-\tau)}\right]\right|\\
&\qquad \leq \exp\left[-\frac{K|x-\xi|^2}{t-\tau}\right]\frac{|\vartheta^{\xi,\tau}(x,\xi) - \vartheta^{x^0,t}(x,\xi')|}{4(t-\tau)} \qquad (K \text{ a constant}),
\end{aligned}
$$

$$
\begin{aligned}
&|\vartheta^{\xi,\tau}(x,\xi) - \vartheta^{x^0,t}(x,\xi')|\\
&\qquad \leq \text{const. } [(|x^0-\xi|^\alpha + |t-\tau|^{\alpha/2})|x-\xi|^2 + |\xi-\xi'|\,|x-\xi|]\\
&\qquad \leq \text{const. } [(|x-\xi|^\alpha + |t-\tau|^{\alpha/2})|x-\xi|^2 + |x-\xi|^{2+\lambda}].
\end{aligned}
$$

With the aid of these inequalities it follows that

$$
\begin{aligned}
&\left|\frac{|x-\xi|\cos(N_{x^0},\overrightarrow{x\xi})}{2(t-\tau)^{1+n/2}}\exp\left[-\frac{\vartheta^{\xi,\tau}(x,\xi)}{4(t-\tau)}\right] - \frac{|x-\xi'|\cos(N_{x^0},\overrightarrow{x\xi'})}{2(t-\tau)^{1+n/2}}\right.\\
&\qquad \left.\cdot\exp\left[-\frac{\vartheta^{x^0,t}(x,\xi')}{4(t-\tau)}\right]\right| \leq \frac{\text{const.}}{(t-\tau)^\mu}\frac{1}{|x-\xi|^{n+1-2\mu-\beta}} \qquad (\beta = \min(\alpha,\lambda)).
\end{aligned}
$$

Hence,

$$
\begin{aligned}
&|I_{\delta_1}(x,t) - I'_{\delta_1}(x,t)| \\
&\leq \text{const. } \int_0^t\int_{S_{\delta_1}}\frac{1}{(t-\tau)^\mu}\frac{1}{|x-\xi|^{n+1-2\mu-\beta}}\,dS_\xi\,d\tau\\
&\quad + \text{l.u.b.}\left|\frac{C(\xi,\tau)\varphi(\xi,\tau)}{\cos\gamma(\xi)} - C(x^0,\tau)\varphi(x^0,\tau)\right|\left|\int_0^t\int_{S_{\delta_1}}\frac{|x-\xi'|\cos(N_{x^0},\overrightarrow{x\xi'})}{2(t-\tau)^{1+n/2}}\right.\\
&\quad \left.\cdot\exp\left[-\frac{\vartheta^{x^0,t}(x,\xi')}{4(t-\tau)}\right]dS_{\xi'}\,d\tau\right|,
\end{aligned}
\tag{2.35}
$$

where $\gamma(\xi)$ is the angle between the normals N_{x^0} and N_ξ. The first integral on the right-hand side of (2.35) is absolutely convergent and can be made arbitrarily small if δ_1 is sufficiently small. The second integral on the right-hand side of (2.35) is bounded independently of δ_1 since it coincides with $I'_\delta(x,t)$ (which was proved to be bounded) when $\delta = \delta_1$ and

$$C(x^0, \tau)\varphi(x^0, \tau) \equiv 1.$$

Since the expression outside the second integral tends to zero with δ_1 (because C, φ are continuous and $\cos \gamma(\xi) \to 1$), the proof of (2.34) is completed.

Since the integrand of $I_{\delta_1}(x^0, t)$ is bounded by a constant times $|F_0(x^0, t; \xi, \tau)|$, using (2.16) we find that

$$|I_{\delta_1}(x^0, t)| < \epsilon \tag{2.36}$$

if δ_1 is sufficiently small.

Since $|x - \xi|$, $|x^0 - \xi|$, $|x - \xi'|$ in the integrands of $\bar{I}_{\delta_1}(x, t)$, $\bar{I}_{\delta_1}(x^0, t)$, $\bar{I}'_{\delta_1}(x, t)$, respectively, are bounded away from zero, if δ_1 is now fixed, and since $\cos(N_{x^0}, \overrightarrow{x\xi'}) \to 0$ if $x \to x^0$,

$$|\bar{I}_{\delta_1}(x, t) - \bar{I}_{\delta_1}(x^0, t)| < \epsilon, \qquad |\bar{I}'_{\delta_1}(x, t)| < \epsilon \tag{2.37}$$

if x is sufficiently close to x^0.

Combining (2.37), (2.36), (2.34) with (2.33), the proof of (2.32) is completed. This also completes the proof of (2.10) in the case where x tends to x^0 along the normal N_{x^0}.

Consider now the general case where x approaches x^0 and $x \in K$. Let $\bar{x}$ be the nearest point to x on S. We first evaluate

$$\frac{\partial U(x, t)}{\partial \nu(x^0, t)} - \frac{\partial U(x, t)}{\partial \nu(\bar{x}, t)} = \sum_{i,j=1}^{n} [a_{ij}(x^0, t) \cos(N_{x^0}, x_j) - a_{ij}(\bar{x}, t) \cos(N_{\bar{x}}, x_j)] \frac{\partial U(x, t)}{\partial x_i}. \tag{2.38}$$

By (1.5), (1.6),

$$\left| \frac{\partial U(x, t)}{\partial x_i} \right| \leq \text{const.} \int_S \frac{dS_\xi}{|x - \xi|^{n+1-2\mu}} \leq \frac{\text{const.}}{|x - \bar{x}|^\epsilon} \int_S \frac{dS_\xi}{|x - \xi|^{n+1-2\mu-\epsilon}} \leq \frac{\text{const.}}{|x - \bar{x}|^\epsilon} \tag{2.39}$$

for any $\epsilon > 0$, since μ is any positive number < 1 and $|x - \bar{x}| \leq |x - \xi|$. Since N_{x^0} is Hölder continuous (exponent λ) on S, the expression in brackets on the right-hand side of (2.38) is bounded by $0(|x^0 - \bar{x}|^\beta)$. Hence, if ϵ in (2.39) is taken to be less than β then

$$\left| \frac{\partial U(x, t)}{\partial \nu(x^0, t)} - \frac{\partial U(x, t)}{\partial \nu(\bar{x}, t)} \right| \leq \text{const.} \frac{|x^0 - \bar{x}|^\beta}{|x - \bar{x}|^\epsilon} \to 0 \qquad \text{as } x \to x^0; \tag{2.40}$$

here we have used the fact that $x \in K$, which implies that

$$|x^0 - \bar{x}| \leq \text{const.}\, |x - \bar{x}|.$$

On the other hand we have already proved that

$$\left| \frac{\partial U(x, t)}{\partial \nu(\bar{x}, t)} + \frac{1}{2} \varphi(\bar{x}, t) - \int_0^t \int_S \frac{\partial \Gamma(\bar{x}, t; \xi, \tau)}{\partial \nu(\bar{x}, t)} \varphi(\xi, \tau)\, dS_\xi\, d\tau \right| < \epsilon$$

if $|x - \overline{x}| < \delta$ and, as follows from the proof, δ can be taken to be independent of $\overline{x}$. Since $\varphi(\overline{x}, t) \to \varphi(x^0, t)$ as $\overline{x} \to x^0$, the proof of (2.10) would follow if we show that the integral

$$M(\overline{x}, t) = \int_0^t \int_S \frac{\partial \Gamma(\overline{x}, t; \xi, \tau)}{\partial \nu(\overline{x}, t)} \varphi(\xi, \tau)\, dS_\xi\, d\tau$$

converges to $M(x^0, t)$ when $\overline{x} \to x^0$. The continuity of $M(\overline{x}, t)$, however, follows by using the inequality (2.12) and arguments similar to those used in proving Lemma 1, Sec. 3, Chap. 1.

3. Solution of the Second Initial-boundary Value Problem

With the aid of Theorem 1, Sec. 2, we shall solve in this section the second initial-boundary value problem. With Lu defined by (1.1) and with a given continuous function $\beta(x, t)$ on $S \times [0, T]$, the problem in question is to find a solution u of

$$Lu(x, t) = f(x, t) \quad \text{in } D \times (0, T], \tag{3.1}$$

$$u(x, 0) = \psi(x) \quad \text{on } \overline{D}, \tag{3.2}$$

$$\frac{\partial u(x, t)}{\partial \nu(x, t)} + \beta(x, t) u(x, t) = g(x, t) \quad \text{on } S \times (0, T], \tag{3.3}$$

where f, ψ, g are any given functions, and the conormal derivative in (3.3) is defined by

$$\frac{\partial u(x, t)}{\partial \nu(x, t)} = \lim_{\substack{y \to x \\ y \in K}} \frac{\partial u(y, t)}{\partial \nu(x, t)}, \tag{3.4}$$

where K is any finite closed cone with vertex x, which is contained in $D + \{x\}$. The definition (3.4) is more restrictive than the definition in (2.5.8).

Theorem 2. *Let L satisfy* (A_1), (A_2) *and let S belong to $C^{1+\lambda}$. If f is Hölder continuous (exponent α) in x, uniformly in $\Omega = \overline{D} \times [0, T]$, if ψ is continuous in $\overline{D}$ and vanishes in some D-neighborhood of the boundary ∂D of D, and if g is continuous on $S \times [0, T]$, then there exists a unique solution of the second initial-boundary value problem* (3.1)–(3.3).

Proof. We first prove the existence of a solution $u(x, t)$. We try to find it in the form

$$u(x, t) = \int_0^t \int_S \Gamma(x, t; \xi, \tau) \varphi(\xi, \tau)\, dS_\xi\, d\tau + \int_D \Gamma(x, t; \xi, 0) \psi(\xi)\, d\xi - \int_0^t \int_D \Gamma(x, t; \xi, \tau) f(\xi, \tau)\, d\xi\, d\tau \tag{3.5}$$

where φ is to be determined.

The proof of uniqueness follows from the following more general fact.

Lemma 2. *If u is a solution of* (3.1)–(3.3) *and if* (A_1), (A_2) *hold and S is of class $C^{1+\lambda}$, then for all $(x, t) \in \Omega$,*

$$|u(x, t)| \leq K(\operatorname*{l.u.b.}_{\Omega} |f| + \operatorname*{l.u.b.}_{S\times[0,T]} |g| + \operatorname*{l.u.b.}_{D} |\psi|) \tag{3.11}$$

where K is a constant depending only on L, β, Ω.

Proof. By the proof of existence in Theorem 2 applied to $u = v - 1$, there exists a function $v(x, t)$ satisfying

$$\begin{aligned} Lv(x, t) &= -1 \qquad \text{in } D \times (0, T], \\ v(x, 0) &= 1 \qquad \text{on } \overline{D}, \\ \frac{\partial v(x, t)}{\partial \nu(x, t)} + \beta(x, t)v(x, t) &= -1 \qquad \text{on } S \times (0, T]. \end{aligned}$$

Applying Theorem 17, Sec. 6, Chap. 2, to the functions u and $\pm Av$, where

$$A = \text{l.u.b.}\, |f| + \text{l.u.b.}\, |g| + \text{l.u.b.}\, |\psi|,$$

we find that $|u(x, t)| \leq Av(x, t)$, from which (3.11) follows.

We shall now consider the case where the assumption that $\psi = 0$ in some D-neighborhood of ∂D is omitted. If no assumptions are made in addition to those of Theorem 2, then instead of (3.7) we now have

$$|F(x, t)| \leq \frac{\text{const.}}{t^{\epsilon}} \qquad \text{for any } \frac{1}{2} < \epsilon < 1, \tag{3.12}$$

and from (3.10) we get,

$$|\varphi(x, t)| \leq \frac{\text{const.}}{t^{\epsilon}} \qquad \text{for any } \frac{1}{2} < \epsilon < 1. \tag{3.13}$$

Observe now that Theorem 1, Sec. 2, can be applied even if $\varphi(x, t)$ is not assumed to be bounded, but instead is assumed to satisfy (3.13). Indeed, this follows by dividing the single-layer potential into two parts $\int_0^\delta \int_S + \int_\delta^t \int_S$; the first integral is clearly continuously differentiable for $x \in \overline{D}$, $t > \delta$, whereas the proof of Theorem 1, Sec. 2, applies to the second integral (if $t > \delta$) with trivial modifications. We conclude that if the integral equation (3.8) has a continuous solution $\varphi(x, t)$ for $x \in S$, $0 < t \leq T$ and if the solution satisfies (3.13), then $u(x, t)$ satisfies the boundary condition (3.3). Hence if φ and u are defined by (3.10) and (3.5) respectively, then u satisfies (3.3).

$u(x, t)$ also satisfies (3.1), and

$$u(x, 0) = \psi(x) \qquad \text{in } D. \tag{3.14}$$

We sum up:

Corollary 1. *If in Theorem 2 we omit the assumption that $\psi = 0$ in some D-neighborhood of ∂D, then there exists a solution of* (3.1), (3.14), (3.3).

Consider the function

$$(3.6)\quad F(x, t) = \int_D \frac{\partial\Gamma(x, t; \xi, 0)}{\partial\nu(x, t)} \psi(\xi)\, d\xi - \int_0^t \int_D \frac{\partial\Gamma(x, t; \xi, \tau)}{\partial\nu(x, t)} f($$

$$+ \beta(x, t) \int_D \Gamma(x, t; \xi, 0)\psi(\xi)\, d\xi$$

$$- \beta(x, t) \int_0^t \int_D \Gamma(x, t; \xi, \tau) f(\xi, \tau)\, d\xi\, d$$

on $S \times (0, T]$. Using (1.4)–(1.6) one finds that $F(x, t)$ is a c function satisfying

$$(3.7)\quad |F(x, t)| \leq \text{const.}$$

If $\varphi(x, t)$ is a continuous function on $S \times [0, T]$ then, by usi orem 1, Sec. 2, and the definition of F one readily sees that the b condition (3.3) reduces to the condition

$$(3.8)\quad \varphi(x, t) = 2 \int_0^t \int_S \left[\frac{\partial\Gamma(x, t; \xi, \tau)}{\partial\nu(x, t)} + \beta(x, t)\Gamma(x, t; \xi, \tau)\right]$$

$$\cdot \varphi(\xi, \tau)\, dS_\xi\, d\tau + 2$$

Setting

$$M(x, t; \xi, \tau) = 2\frac{\partial\Gamma(x, t; \xi, \tau)}{\partial\nu(x, t)} + 2\beta(x, t)\Gamma(x, t; \xi, \tau)$$

and using (1.4), (2.12), we get

$$(3.9)\quad |M(x, t; \xi, \tau)| \leq \frac{\text{const.}}{(t - \tau)^\mu} \frac{1}{|x - \xi|^{n+1-2\mu-\beta}}.$$

It is easy to show that there exists a continuous bounded solution φ to integral equation (3.8), expressed in the form

$$(3.10)\quad \varphi(x, t) = 2F(x, t) + 2 \sum_{\nu=1}^{\infty} \int_0^t \int_S M_\nu(x, t; \xi, \tau) F(\xi, \tau)\, dS_\xi\, d\tau$$

where $M_1 = M$ and

$$M_{\nu+1}(x, t; \xi, \tau) = \int_0^t \int_S M(x, t; y, \sigma) M_\nu(y, \sigma; \xi, \tau)\, dS_y\, d\sigma.$$

Let us just establish the convergence of the series on the right-hand side of (3.10). This follows by calculations similar to those in Sec. 4, Chap. 1, concerning the integral equation (1.4.1), provided we make use of (3.9) and of Lemma 1, Sec. 2.

Having proved that u satisfies (3.3), we proceed to prove (3.1), (3.2). (3.1) follows from Theorem 9, Sec. 5, Chap. 1, and from $L\Gamma = 0$. As for (3.2), using (1.4) one finds that, for all $x \in \overline{D}$, the first and third integrals on the right-hand side of (3.5) tend to zero as $t \to 0$. Since the second integral converges to $\psi(x)$, by the property (1.1.7) of fundamental solutions, (3.2) follows.

Note that the solution is not asserted to be continuous or even bounded in Ω, i.e., it may be unbounded near $\partial D \cap \{t = 0\}$. Under additional assumptions on ψ and the a_{ij}, we may assert that the solution is a continuous function in Ω, namely:

Corollary 2. *Suppose that in Theorem 2 we omit the assumption that $\psi = 0$ in some D-neighborhood of ∂D and instead assume that in some neighborhood D' of ∂D the functions $\psi(x)$, $a_{ij}(x, 0)$ are defined and are continuously differentiable. Then the assertion of Theorem 2 remains true.*

Proof. We proceed as in the proof of Theorem 1, but replace the second integral on the right-hand side of (3.5) by

$$A(x, t) \equiv \int_{D^*} \Gamma(x, t; \xi, 0)\psi(\xi)\, d\xi$$

for some domain $D^* \supset \bar{D}$, $\bar{D}^* \subset D + D'$. If we show that

$$|D_x A(x, t)| \leq \frac{\text{const.}}{t^\epsilon} \qquad \text{for some } \epsilon < \frac{1}{2}, \tag{3.15}$$

then (3.12), (3.13) hold for some $\epsilon < \frac{1}{2}$, and the proof of Corollary 2 then follows by combining the proof of Theorem 2 and the proof of its first corollary.

It suffices to prove (3.15) for the function

$$\tilde{A}(x, t) = \int_{D_0} \Gamma(x, t; \xi, 0)\psi(\xi)\, d\xi$$

where D_0 is some neighborhood of ∂D contained in D'. Consider first the case of the heat equation. Then

$$\begin{aligned} D_x\tilde{A}(x, t) &= \int_{D_0} D_x\Gamma(x, t; \xi, 0)\psi(\xi)\, d\xi = -\int_{D_0} [D_\xi\Gamma(x, t; \xi, 0)]\psi(\xi)\, d\xi \\ &= \int_{D_0} \Gamma(x, t; \xi, 0)D_\xi\psi(\xi)\, d\xi + \text{boundary integrals,} \end{aligned}$$

and (3.15) follows with $\epsilon = 0$.

In the general case, we write $\Gamma = Z + \int Z\Phi$ (see (1.2.8)) and decompose $\tilde{A}$ accordingly. The first integral is treated similarly to the case of the heat equation (making use of the differentiability of $a_{ij}(x, 0)$). The derivative of the second integral is estimated directly (using (1.5), (1.4.8)); the bound obtained being const./t^ϵ for any $\epsilon > (1 - \alpha)/2$.

We conclude this section by proving a useful lemma similar to Lemma 2, which assumes more on β but less on L, S. It is analogous to (d), Sec. 3, Chap. 2.

Lemma 3. *Let u be a solution of (3.1)–(3.3) and assume that L is parabolic in $\Omega = \bar{D} \times [0, T]$ with continuous coefficients, that $\beta(x, t) \leq -\mu_0 < 0$ and that S is of class C'. Then*

$$(3.16) \qquad |u(x, t)| \le K(\underset{\Omega}{\text{l.u.b.}} |f| + \underset{S\times[0,T]}{\text{l.u.b.}} |g| + \underset{D}{\text{l.u.b.}} |\psi|)$$

where K is a constant depending only on L, β, Ω.

Proof. Without loss of generality we may assume that $c(x, t) \le 0$. Consider the function (compare (2.3.11))

$$h(x) = A - \exp [\lambda(x_1 - x_1^0)] \qquad (A > 1).$$

Take λ sufficiently large so that $Lh < 0$. Next take A sufficiently large so that

$$\frac{\partial h(x)}{\partial \nu(x, t)} + \beta(x, t)h(x) \le \lambda \exp [\lambda(x_1 - x_1^0)] - \mu_0\{A - \exp [\lambda(x_1 - x_1^0)]\} < 0$$

on S. We can now apply Theorem 17, Sec. 6, Chap. 2 to $\pm u$ and

$$w(x) \equiv A'(\text{l.u.b.} |f| + \text{l.u.b.} |g| + \text{l.u.b.} |\psi|)h(x)$$

where A' is some constant depending only on h, and thus conclude that $|u(x, t)| \le w(x)$, from which (3.16) follows.

4. Further Results on Single-layer Potentials

We shall state without proof some additional results concerning single-layer potentials (under the assumptions of Theorem 1).

Theorem 3. *The function $U(x, t)$, defined in* (2.1), *is Hölder continuous (exponent ϑ) in $\Omega = \bar{D} \times [0, T]$ for any $0 < \vartheta < 1$, i.e.,*

$$\frac{|U(x, t) - U(x^0, t^0)|}{|x - x^0|^\vartheta + |t - t^0|^{\vartheta/2}} \le \text{const.} < \infty \qquad \text{for all } (x, t) \in \Omega,\ (x^0, t^0) \in \Omega.$$

Theorem 4. *The function*

$$H(x, t) = \int_0^t \int_S \frac{\partial \Gamma(x, t; \xi, \tau)}{\partial \nu(x, t)} \varphi(\xi, \tau)\, dS_\xi\, d\tau$$

is Hölder continuous on $S \times [0, T]$ with any exponent less than $2\beta/3$, where $\beta = \min(\alpha, \lambda)$.

Theorem 5. *If $\varphi(x, t)$ is Hölder continuous in x, uniformly with respect to $x \in S$, $\epsilon \le t \le T$ (for any $\epsilon > 0$), then $\partial U(x, t)/\partial x_i$ $(i = 1, \ldots, n)$ are uniformly continuous in $D \times [\epsilon, T]$, for any $\epsilon > 0$.*

We also state one result concerning volume potentials.

Theorem 6. *If $f(x, t)$ is a continuous function in Ω then the volume potential*

$$M(x, t) = \int_0^t \int_D \Gamma(x, t; \xi, \tau) f(\xi, \tau)\, d\xi\, d\tau$$

and its first derivatives $\partial M/\partial x_i$ are Hölder continuous (exponent ϑ) in Ω, for any $\vartheta < 1$.

The proofs of Theorems 3, 4, 6 are based on considerations of the same type that have been used in studying the properties of potentials in Sec. 2 and in Chap. 1; for details, see [98]. The proof of Theorem 5 (for details, see [100]) is somewhat more delicate, and in the course of the proof it is also shown that if $x^0 \in S$ and D_{x^0} is a derivative in a tangential direction to S at x^0, then

$$\lim_{\substack{x \to x^0 \\ x \in D}} D_{x^0} U(x, t) = \int_0^t \int_S D_{x^0} \Gamma(x^0, t; \xi, \tau) \varphi(\xi, \tau) \, dS_\xi \, d\tau. \tag{4.1}$$

The improper integral on the right-hand side of (4.1) is not absolutely integrable in general; it converges in the sense of the principal value of Cauchy.

(4.1) shows that in contrast to the jump of the conormal derivatives of single-layer potentials, the tangential derivatives vary continuously across the boundary S.

From Theorem 6 it follows that the function $F(x, t)$ defined in (3.6) is Hölder continuous on $S \times (0, T]$. Using Theorems 3, 4 one can establish a Hölder condition for the series $\Sigma\, M_\nu(x, t; \xi, \tau)$ of (3.10). From (3.10) one then derives the Hölder continuity of φ. Hence Theorem 5 can be applied to deduce that the first integral on the right-hand side of (3.5) is uniformly continuously x-differentiable in $\bar{D} \times [\epsilon, T]$ for any $\epsilon > 0$. The same conclusion is valid for the third integral (by Theorem 6) and also, obviously, for the second integral. We have thus arrived at the following corollary.

Corollary. *The derivatives $\partial u/\partial x_i$ of the solution u of Theorem 2 are uniformly continuous in $D \times (\epsilon, T]$ for any $\epsilon > 0$.*

5. Integral Equations

The theory of linear integral equations is well developed, and the fundamental facts of this theory have been known for a long time. It is therefore natural that one tries to reduce problems of differential equations to problems of integral equations. Thus, in Chap. 1, Sec. 4, and in the present chapter (Sec. 3) we have reduced problems of existence of solutions to problems of solving integral equations of Volterra type. In the next section and also in some of the following chapters we shall encounter more general integral equations. We shall state in this section some of the main results about integral equations that will be needed in the future.

Let x and y vary in a bounded, closed (and measurable) set G of R^n and let $K(x, y)$ be a continuous complex-valued function defined on $G \times G$. Given any continuous complex-valued function $f(x)$ and any complex number λ, consider the equation

$$\varphi(x) - \lambda \int_G K(x, y)\varphi(y)\, dy = f(x). \tag{5.1}$$

Every equation, for φ, of the form (5.1) is called a *linear integral equation of Fredholm type* (of the *first kind*), or briefly, an *integral equation.* (If $\varphi(x)$ is missing on the left-hand side then the equation is of the *second kind.*) $K(x, y)$ is called the *kernel* of the integral equation. In what follows, unless the contrary is explicitly stated, we consider only continuous solutions.

Together with (5.1) consider the corresponding homogeneous equation

$$\varphi(x) - \lambda \int_G K(x,y)\varphi(y)\, dy = 0. \tag{5.2}$$

If, for some λ, there exists a nontrivial solution of (5.2) (i.e., a solution $\varphi \not\equiv 0$) then λ is called an *eigenvalue* of (5.2), or of K, and the solution φ is called an *eigenfunction.* The linear space consisting of all the eigenfunctions which correspond to some λ is called the *eigenspace* corresponding to λ. Denote its dimension by $N(\lambda)$.

Together with (5.1), (5.2) we consider the integral equations

$$\psi(x) - \bar{\lambda} \int_G \overline{K(y, x)}\psi(y)\, dy = g(x), \tag{5.3}$$

$$\psi(x) - \bar{\lambda} \int_G \overline{K(y, x)}\psi(y)\, dy = 0, \tag{5.4}$$

whose kernel $K^*(x, y) \equiv \overline{K(y, x)}$ is called the *adjoint* of the kernel $K(x, y)$. We call (5.3), (5.4) the *adjoint equations* of (5.1), (5.2) respectively.

Fredholm alternative states the following:

(a) If λ is not an eigenvalue for (5.2) then $\bar{\lambda}$ is not an eigenvalue for (5.4), and for every continuous functions f, g there exist unique solutions φ, ψ of (5.1), (5.3) respectively.

(b) If λ is an eigenvalue for (5.2) then $\bar{\lambda}$ is an eigenvalue for (5.4). The dimension $N(\lambda)$ is finite and equals the dimension $N^*(\bar{\lambda})$ of the eigenspace of solutions of (5.4). Finally, for any continuous function f (g) there exists a solution of (5.1) ((5.3)) if and only if f (g) is orthogonal to all the eigenfunctions of (5.4) ((5.2)), i.e., $\int_G f(x)\overline{\psi(x)}\, dx = 0$ for all the ψ which satisfy (5.4) ((5.2)).

Clearly, if λ is an eigenvalue and if (5.1) has a solution, the solution is not unique; the difference of two solutions is an eigenfunction.

From (a), (b) we deduce the following useful fact.

(c) If the homogeneous equation has no nontrivial solutions then the nonhomogeneous equation has a unique solution for any continuous right-hand side.

In the future we shall need the following extension of (a), (b):

(d) If $K(x, y)$ is continuous for $x \in G$, $y \in G$, $x \neq y$, and if

$$(5.5) \qquad |K(x, y)| \leq \frac{\text{const.}}{|x - y|^{n-\epsilon}} \qquad \text{for some } \epsilon > 0,$$

then the Fredholm alternative (i.e., (a), (b)) remains valid.

A kernel $K(x, y)$ which becomes unbounded as $x \to y$ is called a *singular* kernel. A singular kernel satisfying (5.5) is said to be *integrable.*

(e) All the previous results remain true if the euclidean measure dx is replaced by a general measure $d\mu(x)$. If G is a compact domain on a continuously differentiable n-dimensional manifold then the statements (a)–(d) remain valid when dx is replaced by the surface element of G.

(f) All the previous results remain true if $K(x, y)$ is piecewise continuous in $G \times G$, and if f, φ and g, ψ are taken to be piecewise continuous.

(g) There exists a positive number λ_0 such that each λ with $|\lambda| < \lambda_0$ is not an eigenvalue, and the solution of (5.1) can be represented in the form

$$(5.6) \qquad \varphi(x) = f(x) + \lambda \int_G R(x, y; \lambda) f(y)\, dy,$$

where

$$(5.7) \qquad R(x, y; \lambda) = \sum_{j=1}^{\infty} \lambda^{j-1} K_j(x, y)$$

and $K_1(x, y) = K(x, y)$,

$$K_{j+1}(x, y) = \int_G K_j(x, \xi) K(\xi, y)\, d\xi.$$

The function $R(x, y; \lambda)$ is called the *resolvent* of K.

If $G = \{(z, t); z \in B_t, a \leq t \leq b\}$ where the B_t are uniformly bounded, closed (and measurable) domains, then (5.1) can be written in the form

$$(5.8) \qquad \varphi(z, t) - \lambda \int_a^b \int_{B_T} K(z, t; \zeta, \tau) \varphi(\zeta, \tau)\, d\zeta\, d\tau = f(z, t).$$

Suppose that

$$(5.9) \qquad K(z, t; \zeta, \tau) = 0 \qquad \text{if } \tau > t.$$

Then (5.8) reduces to

$$(5.10) \qquad \varphi(z, t) - \lambda \int_a^t \int_{B_T} K(z, t; \zeta, \tau) \varphi(\zeta, \tau)\, d\zeta\, d\tau = f(z, t).$$

This integral equation is said to be of *Volterra type*. Any equation of Volterra type (5.10) can be written as an equation of Fredholm type (5.8) if we extend K by (5.9). Since the extended K will only have discontinuities of the first kind at $t = \tau$, all the results (a)–(g) remain true for (5.10). In addition:

(h) Equations of Volterra type have no eigenvalues, and the solution can be represented in the form (5.6), (5.7).

6. Elliptic Equations

Let

$$Lu \equiv \sum_{i,j=1}^{n} a_{ij}(x) \frac{\partial^2 u}{\partial x_i \partial x_j} + \sum_{i=1}^{n} b_i(x) \frac{\partial u}{\partial x_i} + c(x)u \tag{6.1}$$

be an elliptic operator in a bounded domain D of R^n. For any $y \in D$ consider the elliptic operator with constant coefficients

$$L_0 u \equiv \sum_{i,j=1}^{n} a_{ij}(y) \frac{\partial^2 u}{\partial x_i \partial x_j} \cdot \tag{6.2}$$

Let $(a^{ij}(y))$ be the inverse matrix to $(a_{ij}(y))$ and introduce

$$\Gamma_0(x, y) = \begin{cases} \dfrac{[\Sigma\, a^{ij}(y)(x_i - y_i)(x_j - y_j)]^{(2-n)/2}}{(n-2)\omega_n[\det (a^{ij}(y))]^{1/2}} & \text{if } n > 2, \\[2ex] \dfrac{\log\, [\Sigma\, a^{ij}(y)(x_i - y_i)(x_j - y_j)]^{-1/2}}{2\pi[\det (a^{ij}(y))]^{1/2}} & \text{if } n = 2. \end{cases} \tag{6.3}$$

It is easy to verify that $L_0\Gamma_0(x, y) = 0$ and that

$$\begin{aligned} \Gamma_0(x, y) &= 0(|x - y|^{2-n}), \\ D_x\Gamma_0(x, y) &= 0(|x - y|^{1-n}), \\ D_x^2\Gamma_0(x, y) &= 0(|x - y|^{-n}). \end{aligned}$$

Definition. A function $\Gamma(x, y)$ defined for all x, y in $\bar{D}$, $x \neq y$, and satisfying, for some $\lambda > 0$ and all x, y in D, $x \neq y$,

$$D_x^i[\Gamma(x, y) - \Gamma_0(x, y)] = 0(|x - y|^{2-i-n+\lambda}) \qquad (i = 0, 1, 2), \tag{6.4}$$

is called a *fundamental solution* of $Lu = 0$ if, for each $y \in D$, as a function of x Γ satisfies the equation

$$L\Gamma(x, y) = 0 \qquad \text{for } x \in D,\ x \neq y. \tag{6.5}$$

If, in addition, $\Gamma(x, y) = 0$ for all $y \in D$ and $x \in S$ (where S is the boundary of D), then Γ is called a *Green's function.* Clearly, $\Gamma_0(x, y)$ is a fundamental solution of $L_0 u = 0$.

Using the analogue of Green's identity (1.8.4) for elliptic operators, and assuming that u is smooth in $\bar{D}$, that $v = \Gamma^*(x, y)$ (where Γ^* is the fundamental solution of the adjoint equation $L^*v = 0$ of $Lv = 0$) is smooth in $\bar{D}$, and that the boundary S is smooth, we get, upon integration,

$$\begin{aligned} u(y) = -\int_D \Gamma^*(x, y)Lu(x)\, dx - \sum_{i=1}^{n} \int_S \Bigg\{ \sum_{j=1}^{n} \Bigg[\Gamma^*(x, y)a_{ij}(x) \frac{\partial u(x)}{\partial x_j} \\ -u(x)a_{ij}(x) \frac{\partial \Gamma^*(x, y)}{\partial x_j} - u(x)\Gamma^*(x, y) \frac{\partial a_{ij}(x)}{\partial x_j} \Bigg] \\ + b_i(x)\Gamma^*(x, y)u(x) \Bigg\} \cos (N_x, x_i)\, dS_x \end{aligned} \tag{6.6}$$

where N_x is the inward normal to S at x. This is a useful representation formula.

Fundamental solutions can be constructed by the parametrix method, in the form

$$\Gamma(x, y) = \Gamma_0(x, y) + \int_D \Gamma_0(x, \xi)\Phi(\xi, y)\, d\xi + \sum_i \alpha_i(x)\beta_i(y), \tag{6.7}$$

under the assumptions that L is elliptic in $\bar{D}$ and that its coefficients are Hölder continuous in $\bar{D}$. In the process of constructing Γ, *volume potentials*

$$V(x) = \int_D \Gamma_0(x, \xi) f(\xi)\, d\xi$$

are introduced. It is shown that if f is Hölder continuous then $LV(x) = -f(x)$. The integral equation for Φ is of Fredholm type. The functions α_i, β_i are chosen so as to ensure that, in case the alternative (b) occurs (see Sec. 5), the integral equation for Φ still has a solution.

Single-layer potentials are functions of the form

$$U(x) = \int_S \Gamma(x, y)\varphi(y)\, dS_y. \tag{6.8}$$

If S is of class $C^{1+\lambda}$ then it can be shown that for any $x^0 \in S$,

$$\lim_{\substack{x \to x^0 \\ x \in K}} \frac{\partial U(x^0)}{\partial \nu(x^0)} = -\frac{1}{2}\varphi(x^0) + \int_S \frac{\partial \Gamma(x^0, y)}{\partial \nu(x^0)} \varphi(y)\, dS_y \tag{6.9}$$

where $\nu(x^0)$ is the inward conormal at x^0 and K is any finite closed cone with vertex x^0, which is contained in $D + \{x^0\}$. This jump relation can be used to reduce the Neumann problem to a problem of solving an integral equation of the form

$$\varphi(x) = 2 \int_S \left[\frac{\partial \Gamma(x, y)}{\partial \nu(x)} + \beta(x)\Gamma(x, y)\right] \varphi(y)\, dS_y + F(x). \tag{6.10}$$

This equation is of Fredholm type. From Sec. 5 it follows that if $F \equiv 0$ implies $\varphi \equiv 0$ then for any continuous F there exists a unique continuous solution φ of (6.10). Thus, if $\beta \leq 0$, $\beta \not\equiv 0$, $c \leq 0$ then by a uniqueness result mentioned at the end of Sec. 7, Chap. 2, it follows that, if S is of class C^2, there exists a unique solution of (6.10). Also, if $\beta(x) < 0$ for all x then it suffices to assume that S is of class $C^{1+\lambda}$. We conclude that in either case there exists a unique solution of the Neumann problem

$$\begin{aligned} Lu &= f \qquad \text{in } D, \\ \frac{\partial u}{\partial \nu} + \beta u &= g \qquad \text{on } S. \end{aligned} \tag{6.11}$$

The solution has uniformly continuous first derivatives in D (compare the corollary in Sec. 4).

We conclude this section by proving a useful lemma analogous to Lemma 3, Sec. 3.

Lemma 4. *Let L be an elliptic operator with continuous coefficients in $\overline{D}$, let $c(x) \leq 0$, $\beta(x) \leq -\mu_0 < 0$, and let S be of class C'. If u is a solution of* (6.11) *then*

$$|u(x)| \leq K(\underset{D}{\text{l.u.b.}}\ |f| + \underset{S}{\text{l.u.b.}}\ |g|) \tag{6.12}$$

where K is a constant depending only on L, D, β.

The inequality (6.12) is the analogue of (2.7.5) for solutions of the Neumann problem.

Proof. Let $h(x)$ be the function introduced in the proof of Lemma 3, Sec. 3, and let

$$v(x) = A'(\text{l.u.b. } |f| + \text{l.u.b. } |g|)h(x) \qquad (A' > 0).$$

For some A' depending only on h, the function $\tilde{u} = v \pm u$ satisfies $L\tilde{u} < 0$ in D, $\partial\tilde{u}/\partial\nu + \beta\tilde{u} < 0$ on S. If $\tilde{u}$ is negative at some points of $\overline{D}$, then $\tilde{u}$ attains its negative minimum at some point $y \in S$. Since then $\partial\tilde{u}/\partial\nu \geq 0$, $\beta\tilde{u} > 0$ at y, we obtain a contradiction. Hence, $\tilde{u} \geq 0$ in $\overline{D}$, i.e., $|u| \leq v$, and the proof is completed.

PROBLEMS

1. Prove Lemma 1, Sec. 2.
   [*Hint:* For $|x - \xi|$ sufficiently small compare the integral over S_δ with an integral over S'_δ.]
2. Assume that (A_1), (A_2) hold and that S is of class $C^{1+\lambda}$. Extend L into $0 \leq t \leq T$ by Lemma 1, Sec. 7, Chap. 3, and construct Γ as in Chap. 1. Let $(\xi, \tau) \in D \times [0, T)$. Prove that there exists a continuous function $\varphi(\eta, \sigma; \xi, \tau)$, for $\sigma > \tau$, satisfying

   $$|\varphi(\eta, \sigma; \xi, \tau)| \leq \text{const.}\ (\sigma - \tau)^{-\mu}|\eta - \xi|^{-n-1+2\mu+\beta} \qquad (\beta = \min(\alpha, \lambda))$$

   such that the function

   $$H(x, t; \xi, \tau) = \int_\tau^t \int_S \Gamma(x, t; \eta, \sigma)\varphi(\eta, \sigma; \xi, \tau) dS_\eta\, d\sigma$$

   satisfies the condition

   $$\frac{\partial H(x, t; \xi, \tau)}{\partial\nu(x, t)} - \text{H}(x, t; \xi, \tau) = \frac{\partial\Gamma(x, t; \xi, \tau)}{\partial\nu(x, t)} - \Gamma(x, t; \xi, \tau),$$

   where $\partial/\partial\nu$ is defined by (2.9) with N_{x^0} now being the outward normal to S at x^0.
   [*Hint:* Use Theorem 1, (1.4), (2.12) and Lemma 1.]
3. Let L be a parabolic operator with continuous coefficients in the strip $0 \leq t \leq T$. Prove: if $Lu = 0$ in the complement $\hat{\Omega}$ of Ω (in $0 \leq t \leq T$), if $u(x, 0) = 0$

in the complement $\tilde{D}$ of D (in R^n), if $\partial u/\partial \nu - u = 0$ on $S \times (0, T]$ and if $\text{l. u. b.}_{0 \le t \le T} |u(x, t)| \to 0$ as $|x| \to \infty$, then $u \equiv 0$ in $\hat{\Omega}$. ($\partial/\partial \nu$ is defined as in Problem 2.)
[*Hint:* Use the maximum principle.]

4. Under the same assumptions as in Problem 2, show that $G(x, t; \xi, \tau) = H(x, t; \xi, \tau) - \Gamma(x, t; \xi, \tau)$ is Green's function for $Lu = 0$ in D. (Compare this result with Theorem 16, Sec. 7, Chap. 3.)
[*Hint:* Use Problems 2, 3.]

5. *Neumann's function,* in a cylinder, is defined analogously to Green's function in Sec. 7, Chap. 3, except that the boundary condition (3.7.4) is replaced by

$$\frac{\partial u(x, t)}{\partial \nu(x, t)} + \beta(x, t)u(x, t) = 0.$$

Prove the existence of a Neumann's function.
[*Hint:* Proceed similarly to the proof of Theorem 16, Sec. 7, Chap. 3, and make use of Lemma 2, Sec. 3 (instead of the maximum principle).]

6. Let S belong to the class C^3. Construct a function $\zeta(x)$ which is twice continuously differentiable in $\bar{D}$ and satisfies: $\partial\zeta(x^0)/\partial N_{x^0} = -1$, $\partial\zeta(x^0)/\partial T_{x^0} = 0$ for all $x^0 \in S$, where N_{x^0} is the inward normal to S at x^0 and T_{x^0} is any tangential direction to S at x^0.
[*Hint:* Let $\sigma(x)$ be the distance from x to S. For some $\epsilon_0 > 0$, take $\zeta(x) = (\epsilon_0 - \sigma(x))^3/3\epsilon_0^2$ if $\sigma(x) < \epsilon_0$, $\zeta(x) = 0$ otherwise.]

7. Let L be a parabolic operator with continuous coefficients in $\bar{D} \times [0, T]$, let S be of class C^3, and let $\tau(x)$ be a continuous direction which varies on S and points into the interior of D. Prove that if u satisfies (3.1), (3.2) and $\partial u/\partial \tau + \beta u = g$ on $S \times (0, T]$, where β is any continuous function, then the inequality (3.11) holds where K depends only on L, β, Ω and on the function τ.
[*Hint:* Compare u with $v = \pm A e^{at+b\zeta}$ where A, a, b are constants and ζ is as in Problem 6, and use Theorem 17, Chap. 2.]

CHAPTER 6

ASYMPTOTIC BEHAVIOR OF SOLUTIONS

Introduction. Let $u(x, t)$ be a solution of the heat equation $u_{xx} - u_t = 0$ in the half strip $0 \leq x \leq 1$, $0 \leq t < \infty$ and assume that $u(x, 0) = f(x)$ $(0 \leq x \leq 1)$, $u(0, t) = u(1, t) = 0$ $(0 < t < \infty)$. If $f(x)$ is continuously differentiable and $f(0) = f(1) = 0$, then $f(x)$ can be expanded into a uniformly convergent sine-series $\sum_{n=1}^{\infty} a_n \sin n\pi x$. The solution $u(x, t)$ is then given explicitly by

$$u(x, t) = \sum_{n=1}^{\infty} a_n e^{-n^2\pi^2 t} \sin n\pi x. \tag{0.1}$$

Hence, $u(x, t) \to 0$ as $t \to \infty$. Moreover, $u(x, t)$ tends to zero exponentially, i.e.,

$$|u(x, t)| \leq \text{const. } e^{-\pi^2 t}. \tag{0.2}$$

In this chapter we prove theorems concerning the convergence, as $t \to \infty$, of solutions of the first and second initial-boundary value problems. We also derive asymptotic expansions for the solutions. The domain where the solution is defined is not necessarily cylindrical.

From (0.1) we also see that if $u(x, t) \to 0$, as $t \to \infty$, faster than any exponential $e^{-\lambda t}$ $(\lambda > 0)$, then $u \equiv 0$. In Sec. 8 we shall prove a similar result for general parabolic equations. The methods used can also be employed to prove uniqueness theorems for "backward" parabolic equations. Such uniqueness theorems are proved in Sec. 7 and they are used in Sec. 8.

1. Convergence of Solutions of the First Initial-boundary Value Problem

Let D be a domain in the $(n + 1)$-dimensional space of the variables $(x, t) = (x_1, \ldots, x_n, t)$, bounded by a domain B on $t = 0$ and by a manifold S in the half-space $0 < t < \infty$. Assume that for every $\tau > 0$ the set $B_\tau = D \cap \{t = \tau\}$ is a nonempty bounded domain, and set

$$D_\tau = D \cap \{0 < t < \tau\}, \qquad S_\tau = S \cap \{0 < t \leq \tau\}.$$

Let C denote a bounded n-dimensional domain in the x-space. Denote the boundary of C, B_t by ∂C, ∂B_t respectively.

We shall later assume that $B_t \to C$ in the following sense:

(D_1) As $t \to \infty$,

$$\underset{(x,t)\in B_t}{\text{l.u.b.}}\ \underset{y\in C}{\text{g.l.b.}}\ |x - y| \to 0, \qquad \underset{y\in C}{\text{l.u.b.}}\ \underset{(x,t)\in B_t}{\text{g.l.b.}}\ |x - y| \to 0;$$

(D_2) there is a one-to-one continuous correspondence $x \leftrightarrow x_t$ between the points x of ∂C and the points x_t for which $(x_t, t) \in \partial B_t$, and as $t \to \infty$,

$$\underset{x\in\partial C}{\text{l.u.b.}}\ |x_t - x| \to 0.$$

(D_1) is equivalent to the statement that to every $\epsilon > 0$ there exists an A_ϵ such that, if $t > A_\epsilon$, for any $y \in C$ and $(\bar{x}, t)$ in B_t there exist points (x, t) in B_t and $\bar{y} \in C$ such that $|x - y| < \epsilon$, $|\bar{x} - \bar{y}| < \epsilon$.

Definitions. We say that $w(x, t) \to v(y)$ ($(x, t) \in \bar{D}$, $y \in \bar{C}$) as $x \to y$, $t \to \infty$, and write

$$\lim_{\substack{x\to y\\ t\to\infty}} w(x, t) = v(y),$$

if for every $\epsilon > 0$ there exist $\delta > 0$, $t_0 > 0$ such that $|w(x, t) - v(y)| < \epsilon$ whenever $|x - y| < \delta$, $t > t_0$. If δ and t_0 are independent of y, then we say that the convergence is *uniform* in $\bar{D}$. We say that $w(x, t) \to 0$ as $t \to \infty$ (or, also, $\lim_{t\to\infty} w(x, t) = 0$) *uniformly* in $\bar{D}$ (S) if for any $\epsilon > 0$ there exists a t_0 such that $|w(x, t)| < \epsilon$ whenever $t > t_0$, $(x, t) \in \bar{D}$ ($t > t_0$, $(x, t) \in S$). Similar definitions hold for $\lim\sup w(x, t)$, $\lim\inf w(x, t)$. Thus, if in the last definition the inequality $|w(x, t)| < \epsilon$ is replaced by $w(x, t) < \epsilon$, then we say that $\limsup_{t\to\infty} w(x, t) \le 0$ uniformly in $\bar{D}$ (S).

Consider the equations

$$(1.1)\quad Lu \equiv \sum_{i,j=1}^{n} a_{ij}(x, t) \frac{\partial^2 u}{\partial x_i\, \partial x_j} + \sum_{i=1}^{n} b_i(x, t) \frac{\partial u}{\partial x_i} + c(x, t)u - \frac{\partial u}{\partial t} = f(x, t) \qquad \text{in } D,$$

$$(1.2)\quad L_0 v \equiv \sum_{i,j=1}^{n} a_{ij}(x) \frac{\partial^2 v}{\partial x_i\, \partial x_j} + \sum_{i=1}^{n} b_i(x) \frac{\partial v}{\partial x_i} + c(x)v = f(x) \qquad \text{in } C,$$

where $u = u(x, t)$, $v = v(x)$ satisfy the boundary conditions

$$(1.3)\qquad u(x, t) = h(x, t) \qquad \text{on } S,$$

$$(1.4)\qquad v(x) = h(x) \qquad \text{on } \partial C.$$

We shall need the following assumptions:

(A) There exists a constant $M' > 0$ such that $\sum a_{ij}(x, t)\xi_i\xi_j \ge M'|\xi|^2$ for all $(x, t) \in \bar{D}$ and for all real vectors ξ, and the coefficients of L

are continuous functions in $\overline{D}$ and they are bounded, i.e., $|a_{ij}| \leq M$, $|b_i| \leq M$, $|c| \leq M$.

(B) $a_{ij}(x, t) \to a_{ij}(y)$, $b_i(x, t) \to b_i(y)$, $c(x, t) \to c(y)$ as $x \to y$, $t \to \infty$, uniformly in $\overline{D}$, and $a_{ij}(x)$, $b_i(x)$, $c(x)$ are Hölder continuous (exponent α) in $\overline{C}$.

(C) $f(x, t)$ is continuous in $\overline{D}$ and $f(x, t) \to f(y)$ as $x \to y$, $t \to \infty$, uniformly in $\overline{D}$; furthermore, $f(x)$ is Hölder continuous (exponent α) in $\overline{C}$.

With the notation of the assumption (D_2) we next state an assumption on h:

(E) $h(x, t)$ is continuous on S and $h(x_t, t) \to h(x)$ as $t \to \infty$, uniformly with respect to $x \in \partial C$.

Note that (E) implies that $h(x)$ is a continuous function on ∂C.

We can now state theorems concerning the convergence of solutions of (1.1), (1.3) to solutions of (1.2), (1.4).

Theorem 1. *Let u be a solution of* (1.1), (1.3) *where f is a continuous function in $\overline{D}$ and h is a continuous function on S and let* (A) *be satisfied. Assume further that the projections on R^n of the domains B_t $(0 \leq t < \infty)$ are uniformly bounded and that*

$$\lim_{t\to\infty} f(x, t) = 0, \quad \lim_{t\to\infty} h(x, t) = 0, \quad \limsup_{t\to\infty} c(x, t) \leq 0 \tag{1.5}$$

uniformly in $\overline{D}$, S, and $\overline{D}$ respectively. Then

$$\lim_{t\to\infty} u(x, t) = 0 \qquad \text{uniformly in } \overline{D}. \tag{1.6}$$

Theorem 2. *Assume that* (A), (B), (C), (D_1), (D_2), (E) *hold, that $c(x) \leq 0$, and that ∂C is of class $C^{3+\alpha}$. If $u(x, t)$ is a solution of* (1.1), (1.3) *then*

$$\lim_{\substack{x\to y\\ t\to\infty}} u(x, t) = v(y) \qquad \text{uniformly in } \overline{D}, \tag{1.7}$$

where $v(x)$ is the unique solution of (1.2), (1.4).

The proofs of Theorems 1, 2 are given in Secs. 2 and 3 respectively.

2. Proof of Theorem 1

Consider the function

$$\varphi(x) = e^{\lambda R} - e^{\lambda x_1} \tag{2.1}$$

where R is any positive number satisfying $R \geq 2x_1$ for all $(x, t) \in \overline{D}$ and λ is a positive constant to be determined later. $\varphi(x)$ satisfies

$$L\varphi(x) = -a_{11}(x, t)\lambda^2 e^{\lambda x_1} - b_1(x, t)\lambda e^{\lambda x_1} + c(x, t)(e^{\lambda R} - e^{\lambda x_1}) \qquad \text{in } D.$$

Using (A) we can take λ sufficiently large so that

$$L\varphi(x) < -2e^{\lambda x_1} + c(x, t)(e^{\lambda R} - e^{\lambda x_1}) \quad \text{in } D. \tag{2.2}$$

λ is now fixed. From the inequality in (1.5) it follows that for some $\bar{\sigma}$ sufficiently large,

$$c(x, t)(e^{\lambda R} - e^{\lambda x_1}) < e^{\lambda x_1} \quad \text{in } \bar{D} - D_{\bar{\sigma}}. \tag{2.3}$$

Inserting this inequality into (2.2) we get

$$L\varphi(x) < -\delta \quad \text{in } \bar{D} - D_{\bar{\sigma}} \tag{2.4}$$

where $\delta = \underset{(x,t) \in D}{\text{g.l.b.}}\ e^{\lambda x_1}$.

Set

$$\delta_0 = \underset{(x,t) \in D}{\text{g.l.b.}}\ \varphi(x), \qquad \delta_1 = \underset{(x,t) \in D}{\text{l.u.b.}}\ \varphi(x) \tag{2.5}$$

and consider the function

$$\psi(x, t) = \epsilon\frac{\varphi(x)}{\delta} + \epsilon\frac{\varphi(x)}{\delta_0} + A\frac{\varphi(x)}{\delta_0}e^{-\gamma(t-\sigma)} \quad \text{in } D - D_\sigma, \tag{2.6}$$

where ϵ, A, γ are any positive constants and $\sigma \geq \bar{\sigma}$. By (2.4),

$$L\psi(x, t) < -\epsilon - \delta\frac{\epsilon}{\delta_0} - \delta\frac{A}{\delta_0}e^{-\gamma(t-\sigma)} + \gamma\frac{A\,\delta_1}{\delta_0}e^{-\gamma(t-\sigma)}. \tag{2.7}$$

Taking

$$\gamma = \delta/\delta_1, \tag{2.8}$$

it follows that

$$L\psi(x, t) < -\epsilon \quad \text{in } \bar{D} - D_\sigma. \tag{2.9}$$

Clearly,

$$\psi(x, \sigma) > A \quad \text{for } (x, \sigma) \in B_\sigma, \tag{2.10}$$

$$\psi(x, t) > \epsilon \quad \text{for } (x, t) \in S - S_\sigma. \tag{2.11}$$

Now, from (1.5) it follows that for any $\epsilon > 0$ there exists a $\sigma = \sigma(\epsilon)$ such that

$$|f(x, t)| < \epsilon \quad \text{in } \bar{D} - D_\sigma, \tag{2.12}$$

$$|h(x, t)| < \epsilon \quad \text{on } S - S_\sigma. \tag{2.13}$$

We may assume that $\sigma(\epsilon) \geq \bar{\sigma}$. Taking $A = \underset{B_\sigma}{\text{l.u.b.}}\ |u(x, \sigma)|$ and using the inequalities (2.9)–(2.13), we can apply Theorem 16, Sec. 6, Chap. 2, to $\pm u$ and ψ in $D - D_\sigma$. We then conclude that $|u(x, t)| < \psi(x, t)$, i.e.,

$$|u(x, t)| \leq A_1\epsilon + A_2e^{-\gamma(t-\sigma)} \leq 2A_1\epsilon \quad \text{if } t \geq \sigma - \frac{1}{\gamma}\log\frac{A_1\epsilon}{A_2},$$

where A_1, A_2 are positive constants, A_1 depending only on φ. This completes the proof of Theorem 1.

From the previous proof we also derive the following corollaries.

Corollary 1. *If $t > \sigma > \bar{\sigma}$ then*

$$(2.14)\quad |u(x,t)| \le A_0 \{\underset{D_t - D_\sigma}{\text{l.u.b.}} |f| + \underset{S_t - S_\sigma}{\text{l.u.b.}} |h| + [\underset{B_\sigma}{\text{l.u.b.}} |u(x,\sigma)|] e^{-\gamma(t-\sigma)}\},$$

where A_0 is a constant depending only on φ.

Corollary 2. *If (1.5) is replaced by*

$$(2.15)\quad \limsup_{t\to\infty} |f(x,t)| \le \epsilon, \quad \limsup_{t\to\infty} |h(x,t)| \le \epsilon, \quad \limsup_{t\to\infty} c(x,t) \le 0$$

uniformly in $\bar{D}$, S, and $\bar{D}$ respectively, where $\epsilon > 0$, then

$$(2.16)\qquad \limsup_{t\to\infty} |u(x,t)| \le A_0\epsilon \qquad \text{uniformly in } \bar{D},$$

where A_0 is a constant depending only on φ.

Corollary 3. *If (1.5) is replaced by*

$$(2.17)\quad \liminf_{t\to\infty} f(x,t) \ge -\epsilon, \quad \limsup_{t\to\infty} h(x,t) \le \epsilon, \quad \limsup_{t\to\infty} c(x,t) \le 0$$

uniformly in $\bar{D}$, S, and $\bar{D}$ respectively, where $\epsilon > 0$, then

$$(2.18)\qquad \limsup_{t\to\infty} u(x,t) \le A_0\epsilon \qquad \text{uniformly in } \bar{D},$$

where A_0 is a constant depending only on φ.

The inequality (2.14) can be used to give explicit bounds on the solution. We shall give two examples.

Corollary 4. *If (1.5) is replaced by*

$$(2.19)\quad |f(x,t)| < N(1+t)^{-\mu}, \quad |h(x,t)| < N(1+t)^{-\mu}, \quad c(x,t) \le 0 \quad (N > 0,\ \mu > 0),$$

then

$$(2.20)\qquad |u(x,t)| \le N'N(1+t)^{-\mu} \qquad \text{in } D$$

where N' is a constant depending only on φ.

Proof. Define $M(t) = \underset{B_t}{\text{l.u.b.}} |u(x,t)|$. We can apply (2.14) with $\bar{\sigma} = 0$ (see (2.3)) and thus get

$$(2.21)\qquad M(t) \le \frac{2A_0N}{(1+\sigma)^\mu} + A_0M(\sigma)e^{-\gamma(t-\sigma)}.$$

Let a be a number (depending only on A_0, γ) such that

$$(2.22)\qquad \frac{A_0e^{-\gamma a}}{(1+\sigma)^\mu} < \frac{1}{2(1+\sigma+a)^\mu} \qquad \text{for all } \sigma \ge 0.$$

Applying the maximum principle to $\pm u +$ const. φ we find that (2.20) holds for $0 \le t \le a$ if $N' \ge N''$, where N'' depends only on φ. If we prove that whenever (2.20) holds in B_t for some $t = \sigma$ it also holds in B_t for $t = \sigma + a$, then the proof of the corollary is completed.

Substituting (2.20) with $t = \sigma$ into (2.21) we get, for $t = \sigma + a$,

$$M(t) \leq 2A_0 N \frac{(1 + \sigma + a)^\mu}{(1 + \sigma)^\mu} \frac{1}{(1 + t)^\mu} + \frac{A_0 e^{-\gamma a}}{(1 + \sigma)^\mu} N'N.$$

Using (2.22) and taking $N' \geq 4A_0(1 + a)^\mu$, we get (2.20) in $B_t = B_{\sigma+a}$.

Corollary 5. *If* (1.5) *is replaced by*

$$(2.23) \qquad |f(x, t)| < Ne^{-\mu t}, \quad |h(x, t)| < Ne^{-\mu t}, \quad c(x, t) \leq 0 \quad (N > 0, \mu > 0),$$

then

$$(2.24) \qquad |u(x, t)| \leq N'Ne^{-\nu t} \qquad \text{in } D,$$

where $\nu = \min(\gamma', \mu)$ *for any* $\gamma' < \gamma$, *and* N' *is a constant depending only on* φ.

The proof is similar to the proof of Corollary 4 and is left to the reader.

3. Proof of Theorem 2

We first prove Theorem 2 in case $h(x)$ is a polynomial.

By Theorem 18, Sec. 8, Chap. 3, there exists a unique solution $v(x)$ of (1.2), (1.4) and $v \in \overline{C}_{2+\alpha}(C)$. We want to prove that for any $\epsilon > 0$ there exist $\beta > 0$, $\rho > 0$ such that

$$(3.1) \quad |u(x, t) - v(y)| \leq A\epsilon \qquad \text{for all } (x, t) \in D - D_\rho, y \in C, |x - y| < \beta,$$

where A is a constant independent of ϵ, β, ρ. One is tempted to consider the function $w = u - v$ and apply to it Theorem 1. This, however, cannot be done since D is not necessarily cylindrical. We shall therefore modify this idea as follows: we first construct solutions v^δ in domains C_δ which contain $\overline{C}$, such that $v^\delta \to v$ as $\delta \to 0$, and then estimate $u - v^\delta$ by applying Theorem 1 (or rather, its corollary 2).

Extend the coefficients of L_0 into some neighborhood $N(C)$ of $\overline{C}$ (by Theorem 2, Sec. 1, Chap. 3) so that the extended L_0 is still elliptic, its coefficients are Hölder continuous (exponent α) and (the extended) $c(x)$ is ≤ 0. For any $\delta > 0$ sufficiently small construct a manifold ∂C_δ whose points lie on the outward normals to ∂C at a distance δ from ∂C (for more details, see Sec. 8, Chap. 3). ∂C_δ can be covered by a finite number m_0 of neighborhoods V_i such that the portion ∂C_δ^i lying in the ith neighborhood can be globally represented in the form (3.8.6) with h in $C^{2+\alpha}$. m_0 and the V_i can be taken to be independent of δ. Since ∂C is of class $C^{3+\alpha}$, we can take the $(2 + \alpha)$-norm of the h's to be bounded by a constant H_0 independent of δ.

Denote by C_δ the interior of ∂C_δ. By Theorem 18, Sec. 8, Chap. 3, for any Hölder continuous function $g(x)$ in $\overline{C}^\delta$ and for any function $\varphi(x)$ in $\overline{C}_{2+\alpha}(C_\delta)$ there exists a unique solution $w(x)$ of

$$(3.2)\qquad \begin{aligned} L_0 w(x) &= g(x) \qquad \text{in } C_\delta,\\ w(x) &= \varphi(x) \qquad \text{on } \partial C_\delta, \end{aligned}$$

and $w \in \overline{C}_{2+\alpha}(C_\delta)$. Furthermore, by Schauder's boundary estimates,

$$(3.3)\qquad \overline{|w|}_{2+\alpha}^{C_\delta} \le H(\overline{|\varphi|}_{2+\alpha} + \overline{|g|}_{\alpha}^{C_\delta}).$$

From the proof of the boundary estimates (see the remark at the end of Sec. 7, Chap. 4) it follows that the constant H is *independent* of δ.

Denote by x^δ the point of ∂C_δ which lies on the outward normal to ∂C at x. The correspondence $x \leftrightarrow x^\delta$ is one-to-one and $x^\delta \to x$ as $\delta \to 0$, uniformly with respect to $x \in \partial C$.

Extend $f(x)$ to $N(C)$ so that the extension (which is again denoted by f) is Hölder continuous (exponent α) and let v^δ be the unique solution of

$$(3.4)\qquad \begin{aligned} L_0 v^\delta(x) &= f(x) \qquad \text{in } C_\delta,\\ v^\delta(x) &= h(x) \qquad \text{on } \partial C_\delta. \end{aligned}$$

By the previous remarks concerning (3.2), (3.3) we infer that

$$(3.5)\qquad \overline{|v^\delta|}_{2+\alpha}^{C_\delta} \le H' \qquad (H' \text{ independent of } \delta).$$

It follows that

$$(3.6)\qquad \underset{x \in \partial C}{\text{l.u.b.}}\ |v^\delta(x) - v^\delta(x^\delta)| \to 0 \qquad \text{as } \delta \to 0.$$

Using (3.4), (3.6), we find that

$$(3.7)\qquad |v^\delta(x) - h(x)| < \epsilon \qquad (x \in \partial C)$$

if $\delta \le \delta_0$, where $\delta_0 = \delta_0(\epsilon)$ is independent of x, Since also $L_0(v^\delta - v) = 0$, we can use the maximum principle and get

$$(3.8)\qquad \underset{y \in C}{\text{l.u.b.}}\ |v^\delta(y) - v(y)| \le \epsilon.$$

From now on we take δ to be a fixed positive number $\le \delta_0$.

Consider the function

$$(3.9)\qquad w(x, t) = u(x, t) - v^\delta(x) \qquad \text{for } (x, t) \in D - D_\sigma.$$

By (D_1), if σ is sufficiently large then the projections of all the B_t ($t > \sigma$) lie in C_δ (i.e., $(x, t) \in B_t$ implies $x \in C_\delta$) and, therefore, $w(x, t)$ is well defined. By (3.5) and the assumption (B) it follows that

$$(3.10)\qquad (L - L_0)v^\delta(x) \to 0 \qquad \text{as } t \to \infty, \text{ uniformly in } D - D_\sigma.$$

Writing

$$Lw = Lu - (L - L_0)v^\delta - L_0 v^\delta = [f(x, t) - f(x)] + (L - L_0)v^\delta$$

and using (3.10) and the assumption (C), we conclude that

$$(3.11) \qquad |Lw(x, t)| < \epsilon \qquad \text{in } D - D_\sigma$$

if σ is sufficiently large, depending on ϵ.

Next,

$$w(x_t, t) = h(x_t, t) - v^\delta(x_t)$$
$$= [h(x_t, t) - h(x)] + [h(x) - v^\delta(x)] + [v^\delta(x) - v^\delta(x_t)].$$

Using the assumption (E), (3.7), and the continuity of the function $v^\delta(x)$ in conjunction with the assumption (D_2), we then get

$$(3.12) \qquad |w(x_t, t)| < 2\epsilon \qquad \text{on } S - S_\sigma$$

if σ is sufficiently large, depending on ϵ.

Applying Corollary 2 of Theorem 1 to $w(x, t)$ and recalling the definition of w, we obtain

$$(3.13) \qquad |u(x, t) - v^\delta(x)| < A_1\epsilon \qquad \text{in } D - D_\rho,$$

if ρ is sufficiently large ($\rho > \sigma$), where A_1 is a constant independent of ϵ, ρ.

Since v^δ is a continuous function, there exists a positive number β such that

$$(3.14) \qquad |v^\delta(x) - v^\delta(y)| < \epsilon \qquad \text{if } |x - y| < \beta,\ x \in C_\delta,\ y \in C.$$

Combining the inequalities (3.8), (3.13), (3.14), we obtain

$$|u(x, t) - v(y)| < (A_1 + 2)\epsilon,$$

which completes the proof of the theorem in case $h(x)$ is a polynomial.

If $h(x)$ is not a polynomial then the existence of $v(x)$ follows by Theorem 19, Sec. 8, Chap. 3. The previous proof fails because (3.5) does not hold in general. We therefore approximate $h(x)$ by a polynomial $\hat{h}(x)$ so that

$$(3.15) \qquad \underset{x \in \partial C}{\text{l.u.b.}}\ |\hat{h}(x) - h(x)| < \epsilon.$$

By the maximum principle we get

$$(3.16) \qquad \underset{x \in C}{\text{l.u.b.}}\ |\hat{v}(x) - v(x)| < \epsilon,$$

where $\hat{v}$ is the solution of (1.2), (1.4) with h replaced by $\hat{h}$.

All the arguments used in the proof when h is a polynomial can be extended with trivial changes to the present case of the pair $u(x, t)$, $\hat{v}(x)$. We thus obtain

$$|u(x, t) - \hat{v}(y)| < A_2\epsilon$$

for all $t > \rho$, $|x - y| < \beta$. Combining this inequality with (3.16), the proof of Theorem 2 is completed.

We conclude this section with two corollaries.

Corollary 1. *If D is a cylinder then the assertion of Theorem 2 remains valid if ∂C is only assumed to be of class $C^{2+\alpha}$.*

Corollary 2. *If the coefficients of L are independent of t and D is a cylinder, then the assertion of Theorem 2 remains valid if the assumption that ∂C is of class $C^{2+\alpha}$ is replaced by the assumption that there exist barriers at all the points of ∂C.*

The proof of both corollaries is obtained by applying Theorem 1 to $u - v$. Note that in the case of Corollary 2, the existence of $v(x)$ follows by Theorem 19, Sec. 8, Chap. 3.

4. Asymptotic Expansions of Solutions

Consider the case where D is a cylinder, and assume that

$$a_{ij}(x, t) = \sum_{k=0}^{m} a_{ij}^k(x)t^{-k} + o(t^{-m}),$$

$$b_i(x, t) = \sum_{k=0}^{m} b_i^k(x)t^{-k} + o(t^{-m}),$$

$$c(x, t) = \sum_{k=0}^{m} c^k(x)t^{-k} + o(t^{-m}),$$

$$f(x, t) = \sum_{k=0}^{m} f^k(x)t^{-k} + o(t^{-m}),$$

$$h(x, t) = \sum_{k=0}^{m} h^k(x)t^{-k} + o(t^{-m}),$$

where throughout this section, for any $m \geq 0$, $t^m o(t^{-m}) \to 0$ as $t \to \infty$, uniformly with respect to x.

Assume further that all the functions $a_{ij}^k(x)$, $b_i^k(x)$, $c^k(x)$, $f^k(x)$ are Hölder continuous (exponent α) in $\overline{B}$ (B is the base of D) and that the $h^k(x)$ belong to $\overline{C}_{2+\alpha}(B)$. Assume, finally, that ∂B is of class $C^{2+\alpha}$, that $c(x) \leq 0$ and that L satisfies the assumption (A).

Setting

$$L_k = \sum_{i,j=1}^{n} a_{ij}^k(x) \frac{\partial^2}{\partial x_i \, \partial x_j} + \sum_{i=1}^{n} b_i^k(x) \frac{\partial}{\partial x_i} + c^k(x),$$

we can now state the result of this section.

Theorem 3. *Under the foregoing assumptions, if $u(x, t)$ is a solution of* (1.1), (1.3) *then*

$$u(x, t) = \sum_{k=0}^{m} u^k(x)t^{-k} + o(t^{-m}) \tag{4.1}$$

where the $u^k(x)$ are defined successively as the unique solutions of the Dirichlet problems

$$(4.2)\quad L_0u^k(x) = f^k(x) - (k-1)u^{k-1}(x) - \sum_{i=1}^{k} L_iu^{k-i}(x) \qquad (x \in B),$$

$$(4.3)\qquad u^k(x) = h^k(x) \qquad (x \in \partial B)$$

(If $k = 0$ then the right-hand side of (4.2) is understood to be $f^0(x)$.).

Proof. From Corollary 1 in Sec. 3 it follows that $u(x, t) = u^0(x) + o(1)$ where u^0 is determined by (4.2), (4.3) for $k = 0$.

We shall proceed by induction. Assuming the assertion to hold for $m - 1$ we shall prove it for m. More precisely, we assume that $u^1, \ldots, u^{m-1}$ belong to $\bar{C}_{2+\alpha}(B)$ and that (4.1)–(4.3) hold for $k = 0, \ldots, m - 1$, and we shall prove that (4.1) holds with a u^m in $\bar{C}_{2+\alpha}(B)$ and satisfying (4.2), (4.3) for $k = m$.

Writing

$$(4.4)\qquad u(x, t) = \sum_{k=0}^{m-1} u^k(x)t^{-k} + t^{-m}v(x, t)$$

and substituting into (4.1) we get, after making use of (4.2),

$$\left(L_0 - \frac{\partial}{\partial t}\right) v + \sum \epsilon_{ij} \frac{\partial^2 v}{\partial x_i\, \partial x_j} + \sum \epsilon_i \frac{\partial v}{\partial x_i} + \epsilon v$$

$$= f^m - (m-1)u^{m-1} - \sum_{i=1}^{m} L_iu^{m-i} + \epsilon',$$

where $\epsilon_{ij} = o(1)$, $\epsilon_i = o(1)$, $\epsilon = o(1)$, $\epsilon' = o(1)$; here we have made use of the boundedness in B of the derivatives D_xu^k, $D_x^2u^k$.

Since also

$$v(x, t) = h^m(x) + o(1) \qquad \text{on } S,$$

we can apply Corollary 1, Sec. 3, and thus conclude that

$$(4.5)\qquad v(x, t) = u^m(x) + o(1)$$

where u^m satisfies (4.2), (4.3) for $k = m$. By Theorem 18, Sec. 8, Chap. 3, u^m belongs to $\bar{C}_{2+\alpha}(B)$. Substituting (4.5) into (4.4), the proof is completed.

By approximating h^m by polynomials, as in the proof of Theorem 2, one finds that *Theorem* 3 *remains true also if $h^m(x)$ is only assumed to be a continuous function.*

5. Convergence of Solutions of the Second Initial-boundary Value Problem

We now turn to solutions satisfying the second boundary condition and prove results analogous to Theorems 1, 2. $u(x, t)$ satisfies the differential equation (1.1) and the boundary condition

$$(5.1)\qquad \frac{\partial u(x, t)}{\partial \nu(x, t)} + g(x, t, u(x, t)) = h(x, t) \qquad \text{on } S,$$

whereas $v(x)$ satisfies the differential equation (1.2) and the boundary condition

$$(5.2)\qquad \frac{\partial v(x)}{\partial \nu(x)} + g(x)v(x) = h(x) \qquad \text{on } \partial C.$$

Here $\partial/\partial\nu(x, t)$, $\partial/\partial\nu(x)$ are the inward conormal derivatives, i.e.,

$$(5.3)\qquad \frac{\partial u(x, t)}{\partial \nu(x, t)} = \lim_{\substack{y\to x \\ y\in K(x,t)}} \sum_{i,j=1}^{n} a_{ij}(x, t) \cos (N_{x,t}, x_j) \frac{\partial u(y, t)}{\partial y_i},$$

$$(5.4)\qquad \frac{\partial v(x)}{\partial \nu(x)} = \lim_{\substack{y\to x \\ y\in K(x)}} \sum_{i,j=1}^{n} a_{ij}(x) \cos (N_x, x_j) \frac{\partial v(y)}{\partial y_i},$$

where $N_{x,t}$, N_x are the inward normals to ∂B_t (at (x, t)) and ∂C (at x) respectively, and $K(x, t)$, $K(x)$ are finite closed cones with vertices (x, t), x, contained in $B_t + \{(x, t)\}$ and $C + \{x\}$ respectively. If D is a cylinder then the definition (5.3) coincides with the definition of conormal derivatives given in Sec. 5, Chap. 2. The existence of solutions of (1.1), (5.1) was established in that case in Chap. 5, provided $g(x, t, u)$ is a linear function of u, i.e., $g(x, t, u) = g_0(x, t)u$.

We shall need stronger assumptions on D than the assumptions (D_1), (D_2) of Sec. 1.

$(D_1)'$ The projections on R^n of the domains B_t $(0 \le t < \infty)$ are uniformly bounded, and the boundaries ∂B_t belong to class C^1.

$(D_2)'$ ∂C is of class $C^{2+\alpha}$; (D_1), (D_2) hold, and the direction cosines of $N_{x_t,t}$ tend to the direction cosines of N_x as $t \to \infty$, uniformly with respect to $x \in \partial C$.

Remark. From (B), $(D_2)'$ it follows that if $w(x)$ is a continuously differentiable function in some neighborhood of ∂C then, as $t \to \infty$,

$$(5.5)\qquad \frac{\partial w(x_t)}{\partial \nu(x_t, t)} \to \frac{\partial w(x)}{\partial \nu(x)} \qquad \text{uniformly in } x \in \partial C.$$

We next formulate two assumptions on $g(x, t, u)$.

(F_1) $g(x, t, u)$ is a continuous function for (x, t) on S and for $-\infty < u < \infty$, and, for some constant $\mu_1 > 0$,

$$(5.6)\qquad \frac{g(x, t, u)}{-u} > \mu_1 \qquad (x, t) \in S,\ 0 < |u| < \infty).$$

(F_2) As $t \to \infty$, $g(x_t, t, u) \to g(x)u$ uniformly in $x \in \partial C$ and u in bounded sets.

Note that (F_1) implies that $g(x, t, 0) \equiv 0$, and (F_2) implies that $g(x)$ is a continuous function on ∂C and

$$g(x) \leq -\mu_1 < 0. \tag{5.7}$$

We can now state analogs of Theorems 1, 2 with (1.3), (1.4) replaced by (5.1), (5.2).

Theorem 4. *Let u be a solution of* (1.1), (5.1) *where $f(x, t)$ is a continuous function in $\overline{D}$, $h(x, t)$ is a continuous function on S, and let* (A), $(D_1)'$, (F_1) *hold. If*

$$\lim_{t\to\infty} f(x, t) = 0, \qquad \lim_{t\to\infty} h(x, t) = 0, \qquad \limsup_{t\to\infty} c(x, t) \leq 0 \tag{5.8}$$

uniformly in $\overline{D}$, S, and $\overline{D}$ respectively, then

$$\lim_{t\to\infty} u(x, t) = 0 \qquad \text{uniformly in } \overline{D}. \tag{5.9}$$

Theorem 5. *Assume that* (A), (B), (C), $(D_1)'$, $(D_2)'$, (E), (F_1), (F_2) *hold and that $c(x) \leq 0$. If $u(x, t)$ is a solution of* (1.1), (5.1), *then*

$$\lim_{\substack{x\to y\\ t\to\infty}} u(x, t) = v(y) \tag{5.10}$$

uniformly in $\overline{D}$, where $v(x)$ is the unique solution of (1.2), (5.2).

The proof of Theorem 5 will be given in the next section.

Proof of Theorem 4. Introducing the function $\varphi(x)$ defined in (2.1) and using (5.6), we find that for any function $H = H(t) > 0$,

$$\frac{\partial(H\varphi(x))}{\partial\nu(x, t)} + g(x, t, H\varphi(x)) \leq \left\{-\lambda e^{\lambda x_1}\frac{\partial x_1}{\partial\nu(x, t)} - \mu_1(e^{\lambda R} - e^{\lambda x_1})\right\} H.$$

With λ fixed so that (2.2) holds, take R sufficiently large such that, for some constant $\mu_2 > 0$,

$$\lambda e^{\lambda x_1} - \mu_1(e^{\lambda R} - e^{\lambda x_1}) < -\mu_2 \qquad \text{on } S. \tag{5.11}$$

We then get the inequality

$$\frac{\partial(H\varphi(x))}{\partial\nu(x, t)} + g(x, t, H\varphi(x)) < -H\mu_2 \qquad \text{on } S \tag{5.12}$$

(for any $H = H(t) > 0$).

With λ, R now fixed, choose $\bar{\sigma}$ such that (2.3) and, consequently, (2.4) hold.

Introducing

$$\psi(x, t) = \epsilon\frac{\varphi(x)}{\delta} + \epsilon\frac{\varphi(x)}{\mu_2} + A\frac{\varphi(x)}{\delta_0}e^{-\gamma(t-\sigma)}$$

where $A = \text{l.u.b.}_{x\in B_\sigma} |u(x, \sigma)|$ we find that (2.9), (2.10) hold, and

$$\frac{\partial\psi(x, t)}{\partial\nu(x, t)} + g(x, t, \psi(x, t)) < -\epsilon. \tag{5.13}$$

We can now proceed similarly to the proof of Theorem 1, making use of Theorem 17, Sec. 6, Chap. 2.

It is clear from the above proof that Corollaries 1–5 of Sec. 2 extend to the present case.

6. Proof of Theorem 5

We first give the proof in case $h(x)$ is a polynomial. The notation of Sec. 3 concerning C_δ, ∂C_δ, x^δ will be maintained in the sequel.

We extend the functions $b_i(x)$, $c(x)$, $f(x)$, $g(x)$ to some neighborhood $N(C)$ of $\overline{C}$ so that they remain Hölder continuous (exponent α) and so that $c(x) \leq 0$ in $N(C)$. The coefficients $a_{ij}(x)$, however, are extended in the following special way: if $x \in N(C)$, $x \notin \overline{C}$, let x^0 be the nearest point to x on ∂C and set $a_{ij}(x) = a_{ij}(x^0)$. Since ∂C is of class $C^{2+\alpha}$, the extended a_{ij} are certainly Hölder continuous (exponent α) in $N(C)$.

Denoting by $\nu_\delta(x)$ the inward conormal to ∂C_δ at x ($x \in \partial C_\delta$), it is easily seen that $\nu_\delta(x^\delta)$ and $\nu(x)$ are in the same direction, for any $x \in \partial C$.

By the results of Sec. 6, Chap. 5, there exists a unique solution $v(x)$ of the Neumann problem (1.2), (5.2). There also exists a unique solution $v^\delta(x)$ of

$$L_0 v^\delta(x) = f(x) \qquad \text{in } C_\delta, \tag{6.1}$$

$$\frac{\partial v^\delta(x)}{\partial \nu_\delta(x)} + g(x)v^\delta(x) = h(x) \qquad \text{on } \partial C_\delta. \tag{6.2}$$

The following lemma will be needed.

Lemma 1. *Let x vary on ∂C and let x^δ be the point on ∂C_δ which lies on the outward normal to ∂C at x. Then*

$$\operatorname*{l.u.b.}_{y \in C} |v(y) - v^\delta(y)| \to 0 \qquad \text{as } \delta \to 0, \tag{6.3}$$

$$\operatorname*{l.u.b.}_{x \in \partial C} |v^\delta(x) - v^\delta(x^\delta)| \to 0 \qquad \text{as } \delta \to 0, \tag{6.4}$$

$$\operatorname*{l.u.b.}_{x \in \partial C} \left| \frac{\partial v^\delta(x)}{\partial \nu(x)} - \frac{\partial v^\delta(x^\delta)}{\partial \nu_\delta(x^\delta)} \right| \to 0 \qquad \text{as } \delta \to 0. \tag{6.5}$$

We shall first complete the proof of Theorem 5 using Lemma 1, and then prove Lemma 1.

By Lemma 1 and the continuity of $g(x)$, $h(x)$,

$$\operatorname*{l.u.b.}_{x \in C} |v(y) - v^\delta(y)| < \epsilon, \tag{6.6}$$

$$\operatorname*{l.u.b.}_{x \in \partial C} \left| \frac{\partial v^\delta(x)}{\partial \nu(x)} - \frac{\partial v^\delta(x^\delta)}{\partial \nu_\delta(x^\delta)} \right| < \epsilon, \tag{6.7}$$

$$\operatorname*{l.u.b.}_{x \in \partial C} |g(x)v^\delta(x) - g(x^\delta)v^\delta(x^\delta)| < \epsilon, \tag{6.8}$$

$$\text{(6.9)} \qquad \operatorname*{l.u.b.}_{x \in C} |h(x) - h(x^\delta)| < \epsilon,$$

if δ is sufficiently small. From now on δ is a fixed positive number such that (6.6)–(6.9) hold.

Consider the function

$$\text{(6.10)} \qquad w(x, t) = u(x, t) - v^\delta(x) \qquad \text{in } D - D_\sigma$$

where σ is sufficiently large such that whenever $(x, t) \in B_t$ and $t \geq \sigma$ then $x \in C_\delta$. $w(x, t)$ is then well defined in $D - D_\sigma$ and satisfies

$$\text{(6.11)} \quad Lw = Lu - (L - L_0)v^\delta - L_0 v^\delta = [f(x, t) - f(x)] - (L - L_0)v^\delta.$$

We may next assume that σ is such that there exists a closed set C_0 contained in C_δ, such that whenever $(x, t) \in B_t$ and $t \geq \sigma$ then $x \in C_0$. From the interior estimates of Schauder (see Sec. 8, Chap. 3) it follows that

$$\text{(6.12)} \qquad \overline{|v^\delta|}^{C_0}_{2+\alpha} < \infty.$$

Hence, using the assumption (B) we conclude that $(L - L_0)v^\delta \to 0$ as $t \to \infty$, uniformly in $D - D_\sigma$. From (6.11) it then follows that

$$\text{(6.13)} \qquad |Lw(x, t)| < \epsilon \qquad \text{in } D - D_\sigma,$$

if σ is sufficiently large.

We turn to the boundary condition for w. First we notice that since (2.14) holds for u (see the remark at the end of Sec. 5),

$$\text{(6.14)} \qquad |u(x, t)| \leq \text{const.} < \infty \qquad \text{in } D.$$

By $(D_2)'$ and the remark following it,

$$\text{(6.15)} \qquad \operatorname*{l.u.b.}_{x \in \partial C} |g(x_t)v^\delta(x_t) - g(x)v^\delta(x)| \to 0 \qquad \text{as } t \to \infty,$$

$$\text{(6.16)} \qquad \operatorname*{l.u.b.}_{x \in \partial C} \left| \frac{\partial v^\delta(x_t)}{\partial \nu(x_t, t)} - \frac{\partial v^\delta(x)}{\partial \nu(x)} \right| \to 0 \qquad \text{as } t \to \infty.$$

Now, on ∂B_t we have

$$\begin{aligned}\text{(6.17)} \quad \frac{\partial w(x_t, t)}{\partial \nu(x_t, t)} + g(x_t)w(x_t, t) &= [g(x_t)u - g(x_t, t, u)] + [h(x_t, t) - h(x^\delta)] \\ &\quad + \left[\frac{\partial v^\delta(x^\delta)}{\partial \nu_\delta(x^\delta)} - \frac{\partial v^\delta(x_t)}{\partial \nu(x_t, t)}\right] \\ &\quad + [g(x^\delta)v^\delta(x^\delta) - g(x_t)v^\delta(x_t)] \\ &\equiv I_1 + I_2 + I_3 + I_4.\end{aligned}$$

As $t \to \infty$, $I_1 \to 0$ by (F_2) (here we make use of (6.14)); $|I_2|$ becomes smaller than 2ϵ by (E) and (6.9); $|I_3|$ becomes smaller than 2ϵ by (6.7), (6.16), and $|I_4|$ becomes smaller than 2ϵ because of (6.8), (6.15). The above statements hold uniformly with respect to $x \in \partial C$. It follows that

$$(6.18) \qquad \left| \frac{\partial w(x, t)}{\partial \nu(x, t)} + g(x)w(x, t) \right| < 7\epsilon \qquad \text{for } (x, t) \in S - S_\sigma$$

provided σ is sufficiently large.

Since Corollary 2 of Sec. 2 holds also for the case of the second initial-boundary value problem (see the remark at the end of Sec. 5), it follows from (6.13), (6.18) and from the definition of $w(x, t)$ that

$$(6.19) \qquad |u(x, t) - v^\delta(x)| < A_1\epsilon \qquad \text{in } D - D_\rho$$

if ρ is sufficiently large, where A_1 is a constant independent of ϵ, ρ.

Since v^δ is a continuous function in C_δ, there exists a positive number $\beta > 0$ such that

$$(6.20) \qquad |v^\delta(x) - v^\delta(y)| < \epsilon \qquad \text{if } |x - y| \le \beta,\ y \in \overline{C},\ x \in C_\delta.$$

Combining (6.19), (6.20), and (6.6), we get

$$(6.21) \qquad |u(x, t) - v(y)| < (A_1 + 2)\epsilon \qquad \text{if } |x - y| \le \beta,$$
$$y \in \overline{C},\ (x, t) \in D - D_\rho.$$

Since ϵ is arbitrary, the proof of Theorem 5, in case h is a polynomial, is completed.

If $h(x)$ is not a polynomial, let $\hat{h}(x)$ be a polynomial satisfying (3.15) and let $\hat{v}(x)$ be the solution of (1.2), (5.2) with h replaced by $\hat{h}$. By Lemma 4, Sec. 6, Chap. 5,

$$(6.22) \qquad |\hat{v}(x) - v(x)| \le A_2\epsilon \qquad \text{for all } x \in C,$$

where A_2 is a constant independent of ϵ. Defining v^δ to be the solution of (6.1), (6.2) with $h(x)$ replaced by $\hat{h}(x)$ and proceeding as before, we arrive at the inequality (compare (6.21))

$$|u(x, t) - \hat{v}(y)| < (A_1 + 2)\epsilon$$

provided (x, t), y are as in (6.21). Combining this inequality with (6.22), the proof of Theorem 5 is thereby completed.

It remains to prove Lemma 1.

Proof of Lemma 1. We shall first give a proof in case the $a_{ij}(x)$ are uniformly continuously differentiable in a C-neighborhood of ∂C. Denote by $v^\delta(x)$ the solution of

$$(6.23) \qquad \begin{aligned} L_0 v^\delta(x) - \frac{\partial v^\delta(x)}{\partial t} &= f(x) && \text{for } x \in C_\delta,\ 0 < t \le 1, \\ v^\delta(x) &= v^\delta(x) && \text{for } x \in C_\delta, \\ \frac{\partial v^\delta(x)}{\partial \nu_\delta(x)} + g(x)v^\delta &= h(x) && \text{for } x \in \partial C_\delta,\ 0 < t \le 1. \end{aligned}$$

Since $v^\delta(x)$ is uniformly continuously differentiable in C_δ (see Sec. 6, Chap. 5), we can extend it, as well as the $a_{ij}(x)$, so that the extended functions are continuously differentiable in a neighborhood of ∂C_δ. Corollary 2 of Sec. 3, Chap. 5 can then be employed. We conclude that

$$(6.24)\quad v^\delta(x) = \int_0^t \int_{\partial C_\delta} \Gamma(x, t; \xi, \tau)\varphi_\delta(\xi, \tau)\, dS_\xi^\delta\, d\tau \\ + \int_{C_\delta} \Gamma(x, t; \xi, 0)v^\delta(\xi)\, d\xi - \int_0^t \int_{C_\delta} \Gamma(x, t; \xi, \tau)f(\xi)\, d\xi\, d\tau,$$

where dS_ξ^δ is the surface element of ∂C_δ, and where φ_δ is the solution of the integral equation

$$(6.25)\quad \varphi_\delta(x, t) = 2\int_0^t \int_{\partial C_\delta} \left[\frac{\partial \Gamma(x, t; \xi, \tau)}{\partial \nu_\delta(x)} \right. \\ \left. + g(x)\Gamma(x, t; \xi, \tau)\right] \varphi_\delta(\xi, \tau)\, dS_\xi^\delta\, d\tau + 2F_\delta(x, t);$$

F_δ has a form similar to (5.3.6) so that it can be estimated by

$$(6.26)\quad \operatorname*{l.u.b.}_{x} |F_\delta(x, t)| \le \frac{A}{t^\eta} \operatorname{l.u.b.} |v^\delta(x)| + A$$

for any $\eta > \frac{1}{2}$, where A is a constant independent of δ.

By Lemma 4, Sec. 6, Chap. 5, it follows that $v^\delta(x)$, as a solution of (6.1), (6.2), satisfies

$$(6.27)\quad \operatorname*{l.u.b.}_{x \in C_\delta} |v^\delta(x)| \le A_0$$

where A_0 is independent of δ (notice that the constant K in (5.6.12) depends on D only through its dependence on the diameter of D; this follows from the proof of (5.6.12)).

Substituting (6.27) into (6.26) we obtain a bound on F_δ. If we solve (6.25) by iteration, making use of the inequality (compare (5.2.12))

$$(6.28)\quad \left|\frac{\partial \Gamma(x, t; \xi, \tau)}{\partial \nu_\delta(x, t)}\right| \le \frac{A_1}{(t - \tau)^\mu} \frac{1}{|x - \xi|^{n-2\mu-\alpha}} \qquad (x \in \partial C_\delta,\ \xi \in \partial C_\delta)$$

where A_1 is independent of δ (since ∂C is of class $C^{2+\alpha}$), and then use the bound obtained for F_δ, we get

$$(6.29)\quad \operatorname*{l.u.b.}_{x \in \partial C_\delta} |\varphi_\delta(x, t)| \le \frac{A_2}{t^\eta},$$

where A_2 is independent of δ.

From the form of $F_\delta(x, t)$ it follows (using (6.27)) that this function is continuous in (x, t) $(x \in \partial C_\delta,\ t_0 \le t \le 1$ for any $t_0 > 0)$ uniformly with respect to δ. Since the same can be proved (using (6.28), (6.29)) for the integral on the right-hand side of (6.25), it follows that $\varphi_\delta(x, t)$ is a continuous function of (x, t) $(x \in \partial C_\delta,\ t_0 \le t \le 1$ for any $t_0 > 0)$, uniformly with respect to δ.

Using (6.27), (6.29), and the proof of (5.2.10) when $x \to x^0$ along the normal N_{x^0}, we then obtain from (6.24)

$$\lim_{\bar{x}\to x^\delta} \left| \frac{\partial v^\delta(\bar{x})}{\partial \nu_\delta(x^\delta)} - \frac{\partial v^\delta(x^\delta)}{\partial \nu_\delta(x^\delta)} \right| \to 0 \tag{6.30}$$

uniformly with respect to δ and $x^\delta \in \partial C_\delta$, where $\bar{x}$ lies on the inward normal to ∂C_δ at x^δ. From the manner by which we have extended the coefficients $a_{ij}(x)$ outside C and from the fact that the normals to ∂C at x and to ∂C_δ at x^δ are in the same direction, it follows that $\partial w(\bar{x})/\partial \nu_\delta(x^\delta) = \partial w(\bar{x})/\partial \nu(x)$ if $x \in \partial C$. Hence, (6.5) is a consequence of (6.30).

(6.4) follows by using (6.24) and the inequalities (6.27), (6.29).

Finally, to prove (6.3) we observe that by (6.4), (6.5), and (6.2),

$$\underset{x\in\partial C}{\text{l.u.b.}} \left| \left[\frac{\partial v^\delta(x)}{\partial \nu(x)} + g(x)v^\delta(x) \right] - \left[\frac{\partial v(x)}{\partial \nu(x)} + g(x)v(x) \right] \right| \to 0 \qquad \text{as } \delta \to 0.$$

Applying Lemma 4, Sec. 6, Chap. 5, to $v^\delta(x) - v(x)$ in C, (6.3) follows.

The proof of Lemma 1 can be given also by considering v^δ as a solution of (6.1), (6.2) instead of (6.23), and representing it in terms of potentials of the fundamental solution of $L_0 u = 0$. In this proof, which is similar to the previous one, the assumption that the $a_{ij}(x)$ are uniformly continuously differentiable in a C-neighborhood of ∂C and the fact that $v^\delta(x)$ has uniformly continuous first derivatives in C_δ are not needed. The details of this proof are omitted.

7. Uniqueness for Backward Parabolic Equations

If in a parabolic equation we substitute $-t$ for t then we obtain an equation of the form

$$\sum_{i,j=1}^{n} a_{ij}(x, t) \frac{\partial^2 u}{\partial x_i \, \partial x_j} + \sum_{i=1}^{n} b_i(x, t) \frac{\partial u}{\partial x_i} + c(x, t)u + \frac{\partial u}{\partial t} = f(x, t). \tag{7.1}$$

This equation is called a *backward parabolic equation*. The first initial-boundary value problem for such equations consists of finding a solution $u(x, t)$ for (7.1) in some domain $D_T + B_T$ satisfying the initial and boundary conditions

$$u(x, 0) = \psi(x) \qquad \text{on } B, \tag{7.2}$$

$$u(x, t) = h(x, t) \qquad \text{on } S_T, \tag{7.3}$$

where the notation D_t, B_t, S_t, etc. is the same as in the previous sections.

Whereas, in general, there does not exist a solution to this problem, we shall prove in this section that the solution, if existing, is unique.

The reason for considering this question in the present chapter is that the results which we shall derive will be needed in the next section in studying the behavior of solutions of parabolic equations as $t \to \infty$. The methods of the present section are similar to those of the next section.

Throughout this section it will always be assumed that:

(G_T) D_T is a cylinder with bounded base B; ∂B is of class C^1; $a_{ij}(x, t)$ are continuously differentiable functions in $\overline{D}_T$, and

$$\nu_0|\xi|^2 \leq \sum_{i,j=1}^{n} a_{ij}(x, t)\xi_i\xi_j \leq \nu_1|\xi|^2 \qquad ((x, t) \in \overline{D}_T, \ \xi \text{ real}),$$

where ν_0, ν_1 are positive constants.

We shall be primarily concerned with functions $u(x, t)$ which satisfy the differential inequality

$$(L_0u)^2 \leq c_1u^2 + c_2|D_xu|^2 \qquad \text{in } D_T, \tag{7.4}$$

where

$$L_0u(x, t) \equiv \sum_{i,j=1}^{n} \frac{\partial}{\partial x_i}\left(a_{ij}(x, t)\,\frac{\partial u(x, t)}{\partial x_j}\right) - \frac{\partial u(x, t)}{\partial t} \tag{7.5}$$

and where c_1, c_2 are nonnegative constants. The functions $u(x, t)$ are always assumed, for simplicity, to be twice continuously differentiable in $\overline{D}_T$. In (7.4), the notation

$$|D_xu|^2 = \sum_{i=1}^{n} \left(\frac{\partial u}{\partial x_i}\right)^2$$

has been used.

Theorem 6. *Assume that* (G_T) *holds* $(T < \infty)$ *and that* $u(x, t)$ *satisfies* (7.4) *and the terminal and boundary conditions*

$$u(x, T) = 0 \qquad \text{on } B_T, \tag{7.6}$$

$$u(x, t) = 0 \qquad \text{on } S_T. \tag{7.7}$$

Then $u \equiv 0$ *in* D_T.

As an immediate consequence we conclude that there exists at most one solution to the first initial-boundary value problem for backward parabolic equations of the form (7.1).

Proof. We introduce the notation

$$(u, v) = \int_{D_T} u(x, t)v(x, t)\,dx\,dt,$$

$$||u|| = (u, u)^{1/2},$$

$$||u||_1 = \left[\sum_{i=1}^{n} \left\|\frac{\partial u}{\partial x_i}\right\|^2\right]^{1/2},$$

and denote by P_T the set of all functions v twice continuously differentiable in $\overline{D}_T$, which vanish on S_T, B, and B_T. Setting

$$\lambda(t) = t - T - \eta \qquad (\eta > 0),$$

we shall first prove two lemmas concerning functions $v \in P_T$.

Lemma 2. *For any* $v \in P_T$ *and for any positive integer* m,

$$||\lambda^{-m}L_0v||^2 \geq m||\lambda^{-m-1}v||^2 - A||\lambda^{-m}v||_1^2 \tag{7.8}$$

where A *depends only on bounds on* $\partial a_{ij}/\partial t$.

Proof. The function $z = \lambda^{-m}v$ belongs to P_T and satisfies

$$L_0v = \lambda^m Ez - \lambda^m \frac{\partial z}{\partial t} - m\lambda^{m-1}z,$$

where

$$Eu(x, t) \equiv \sum_{i,j=1}^{n} \frac{\partial}{\partial x_i}\left(a_{ij}(x, t)\, \frac{\partial u(x, t)}{\partial x_j}\right). \tag{7.9}$$

It follows that

$$\begin{aligned}||\lambda^{-m}L_0v||^2 = \left|\left| \frac{\partial z}{\partial t} \right|\right|^2 - 2\left(\frac{\partial z}{\partial t}, Ez\right)& \\ + 2m\left(\frac{z}{\lambda}, \frac{\partial z}{\partial t}\right) + ||Ez - m\lambda^{-1}z||^2&.\end{aligned} \tag{7.10}$$

We proceed to evaluate the terms on the right-hand side of (7.10). First, since z vanishes on B, B_T,

$$2m\left(\frac{z}{\lambda}, \frac{\partial z}{\partial t}\right) = m \int_{D_T} \frac{\partial}{\partial t}\left(\frac{z^2}{\lambda}\right) dx\, dt + m||\lambda^{-1}z||^2 = m||\lambda^{-m-1}v||^2. \tag{7.11}$$

Next, since z vanishes on S_T, B, B_T,

$$\begin{aligned}-2\left(\frac{\partial z}{\partial t}, Ez\right) &= 2\int_{D_T} \sum_{i,j} \frac{\partial}{\partial t}\left(\frac{\partial z}{\partial x_i}\right) a_{ij} \frac{\partial z}{\partial x_j}\, dx\, dt \\ &= \int_{D_T} \frac{\partial}{\partial t}\left(\sum_{i,j} a_{ij} \frac{\partial z}{\partial x_i}\frac{\partial z}{\partial x_j}\right) dx\, dt \\ &\quad - \int_{D_T} \sum_{i,j} \frac{\partial a_{ij}}{\partial t} \frac{\partial z}{\partial x_i}\frac{\partial z}{\partial x_j}\, dx\, dt \\ &\geq -A||z||_1^2 = -A||\lambda^{-m}v||_1^2.\end{aligned} \tag{7.12}$$

Substituting the results of (7.11), (7.12) into (7.10), and dropping out the first and last terms on the right-hand side of (7.10), (7.8) follows.

Lemma 3. *There exist positive constants* η_0, μ_0, m_1 *which depend only on* ν_0 *(in the condition* (G_T)*) and on bounds on* $\partial a_{ij}/\partial t$ *such that if* $0 < T \leq \mu$, $v \in P_T$ *and* m *is an integer* $\geq m_1$, *and if* $\mu \leq \mu_0$, $0 < \eta \leq \eta_0$, *then*

$$\rho||\lambda^{-m}L_0v||^2 \geq ||\lambda^{-m-1}v||^2 + \frac{1}{2}||\lambda^{-m}v||_1^2 \tag{7.13}$$

where

$$\rho = \frac{2}{\nu_0}\left[\mu + \eta + \frac{1 + \nu_0}{2m} + \frac{(\mu + \eta)^2}{2}\right]. \tag{7.14}$$

Proof. We begin with the identity

$$(7.15)\qquad -(\lambda^{-m-1}v, \lambda^{-m+1}L_0v) = \left(\lambda^{-2m}v, \frac{\partial v}{\partial t}\right) - (\lambda^{-2m}v, Ev)$$

and evaluate the right-hand side by means of integration by parts, making use of the fact that $v \in P_T$. First,

$$(7.16)\qquad -\left(\lambda^{-2m}v, \frac{\partial v}{\partial t}\right) = -\frac{1}{2}\int_{D_T}\frac{\partial}{\partial t}(\lambda^{-2m}v^2)\,dx\,dt - m\int_{D_T}\lambda^{-2m-1}v^2\,dx\,dt$$
$$\leq m(T+\eta)||\lambda^{-m-1}v||^2,$$

since $|\lambda(t)| \leq T + \eta$.

Next,

$$(7.17)\qquad -(\lambda^{-2m}v, Ev) = \int_{D_T}\lambda^{-2m}\sum_{i,j}a_{ij}\frac{\partial v}{\partial x_i}\frac{\partial v}{\partial x_j}\,dx\,dt \geq \nu_0||\lambda^{-m}v||_1^2.$$

Substituting (7.16), (7.17) into (7.15) and estimating the integrand on the left-hand side by Cauchy's inequality, we get

$$\nu_0||\lambda^{-m}v||_1^2 \leq \left[m(T+\eta)+\frac{1}{2}\right]||\lambda^{-m-1}v||^2 + \frac{1}{2}||\lambda^{-m+1}L_0v||^2.$$

Using the inequality $||\lambda^{-m+1}L_0v||^2 \leq (T+\eta)^2||\lambda^{-m}L_0v||^2$ and Lemma 2, we then obtain

$$\nu_0||\lambda^{-m}v||_1^2 \leq \left[\frac{m(T+\eta)+1/2}{m} + \frac{(T+\eta)^2}{2}\right]||\lambda^{-m}L_0v||^2$$
$$+\left(T+\eta+\frac{1}{2m}\right)A||\lambda^{-m}v||_1^2.$$

Choose μ_0, η_0, m_0 such that $2[\mu_0 + \eta_0 + 1/2m_0]A \leq \nu_0$. Then, if $T < \mu \leq \mu_0$, $0 < \eta \leq \eta_0$, $m \geq m_0$,

$$(7.18)\qquad ||\lambda^{-m}v||_1^2 \leq \frac{2}{\nu_0}\left[(\mu+\eta)+\frac{1}{2m}+\frac{(\mu+\eta)^2}{2}\right]||\lambda^{-m}L_0v||^2.$$

Combining (7.18) with (7.8) we get

$$||\lambda^{-m-1}v||^2 + \left(1-\frac{A}{m}\right)||\lambda^{-m}v||_1^2 \leq \frac{2}{\nu_0}\left[\mu+\eta+\frac{1+\nu_0}{2m}+\frac{(\mu+\eta)^2}{2}\right]$$
$$\cdot||\lambda^{-m}L_0v||^2.$$

Taking finally $m_1 \geq m_0$ such that $(1 - A/m) \geq 1/2$ if $m \geq m_1$, the proof of (7.13) is completed.

We now turn to the proof of Theorem 6. By decreasing μ, η and by increasing m_1 if necessary, we get

$$(7.19)\qquad 2(\mu+\eta)^2\rho c_1 \leq 1, \qquad 2\rho c_2 \leq 1.$$

It suffices to prove the theorem under the assumption that $T \leq \mu$ since otherwise we first prove that $u \equiv 0$ in $D_T - D_{T-\mu}$, then in $D_{T-\mu} - D_{T-2\mu}$, etc.

Let $0 < t_1 < t_2 < T$ and introduce a twice continuously differentiable function $\zeta(t)$ satisfying: $\zeta(t) = 0$ if $0 \le t < t_1$, $\zeta(t) = 1$ if $t_2 < t \le T$. The function $v = \zeta u$ belongs to P_T and therefore, by Lemma 3,

$$
\begin{aligned}
(7.20)\quad \rho \int_{t_1}^{t_2} \int_B (\lambda^{-m} L_0 v)^2\, dx\, dt + \rho \int_{t_2}^{T} \int_B (\lambda^{-m} L_0 u)^2\, dx\, dt \\
= \rho \|\lambda^{-m} L_0 v\|^2 \ge \|\lambda^{-m-1} v\|^2 + \frac{1}{2} \|\lambda^{-m} v\|_1^2 \\
\ge \int_{t_2}^{T} \int_B (\lambda^{-m-1} u)^2\, dx\, dt + \frac{1}{2} \int_{t_2}^{T} \int_B \lambda^{-2m} |D_x u|^2\, dx\, dt.
\end{aligned}
$$

In view of (7.4),

$$
\begin{aligned}
\rho \int_{t_2}^{T} \int_B (\lambda^{-m} L_0 u)^2\, dx\, dt \le \rho c_1 (\mu + \eta)^2 \int_{t_2}^{T} \int_B (\lambda^{-m-1} u)^2\, dx\, dt \\
+ \rho c_2 \int_{t_2}^{T} \int_B \lambda^{-2m} |D_x u|^2\, dx\, dt.
\end{aligned}
$$

Substituting this into (7.20) and using (7.19), we obtain

$$
2\rho \int_{t_1}^{t_2} \int_B (\lambda^{-m} L_0 v)^2\, dx\, dt \ge \int_{t_2}^{T} \int_B (\lambda^{-m-1} u)^2\, dx\, dt.
$$

This inequality implies, for any $t_2 < t_3 < T$,

$$
2\rho (T + \eta - t_2)^{-m} \left[\int_{t_1}^{t_2} \int_B (L_0 v)^2\, dx\, dt \right] \ge (T + \eta - t_3)^{-m-1} \left[\int_{t_3}^{T} \int_B u^2\, dx\, dt \right]
$$

which is impossible for sufficiently large m unless the integral on the right-hand side is zero, i.e., $u(x, t) = 0$ if $x \in B$, $t_3 < t < T$. Since t_3 can be taken arbitrarily small, $u \equiv 0$ in D_T.

From the proof of Theorem 6 the following corollary is obtained.

Corollary 1. *Theorem 6 remains true if the inequality (7.4) is replaced by the weaker inequality*

$$
(7.21)\quad \int_{B_t} (L_0 u)^2\, dx \le c_1 \int_{B_t} u^2\, dx + c_2 \int_{B_t} |D_x u|^2\, dx.
$$

Writing the parabolic operator

$$
(7.22)\quad L_1 u \equiv \sum_{i,j=1}^{n} a_{ij}(x, t) \frac{\partial^2 u}{\partial x_i\, \partial x_j} - \frac{\partial u}{\partial t}
$$

in the form

$$
(7.23)\quad L_1 u = L_0 u + \sum_{i=1}^{n} \tilde{b}_i(x, t) \frac{\partial u}{\partial x_i} \qquad \left(\tilde{b}_j = - \sum_{i=1}^{n} \frac{\partial a_{ij}}{\partial x_i} \right)
$$

and noting that u satisfies (7.21) if and only if

$$
(7.24)\quad \int_{B_t} (L_1 u)^2\, dx \le c_1 \int_{B_t} u^2\, dx + c_3 \int_{B_t} |D_x u|^2\, dx
$$

for some constant c_3, we conclude:

Corollary 2. *Theorem 6 and Corollary 1 remain true if the inequalities (7.4) and (7.21) respectively, are replaced by the inequality (7.24), where L_1 is defined by (7.22).*

Consider now the case where u satisfies the second boundary condition

$$\frac{\partial u(x, t)}{\partial \nu(x, t)} + g(x, t)u(x, t) = 0 \qquad \text{on } S_T \tag{7.25}$$

instead of (7.7). In the definition of P_T the condition $v = 0$ on S_T is replaced by $\partial v/\partial \nu + gv = 0$ on S_T. The previous calculations remain unchanged except for (7.12), (7.17). Integrating by parts on the left-hand side of (7.12) and making use of (7.25) for v, we get an additional term (on the right-hand side)

$$-\int_{S_T} g \frac{\partial (z^2)}{\partial t} \, dS = \int_{S_T} \frac{\partial g}{\partial t} z^2 \, dS, \tag{7.26}$$

and integrating by parts on the left-hand side of (7.17) we get the additional term

$$-\int_{S_T} \lambda^{-2m} g v^2 \, dS. \tag{7.27}$$

If $g \leq 0$, $\partial g/\partial t \geq 0$, the additional terms are nonnegative and can be dropped out. Hence:

Corollary 3. *Theorem 6 and Corollaries 1, 2 extend to the case where the condition (7.7) is replaced by the condition (7.25), provided g, $\partial g/\partial t$ are continuous functions and $g \leq 0$, $\partial g/\partial t \geq 0$.*

We shall show that the assumptions $g \leq 0$, $\partial g/\partial t \geq 0$ can be omitted if ∂B is of class C^2. First we prove an elementary lemma.

Lemma 4. *Let B be a bounded domain whose boundary ∂B is of class C^2, and let $w(x)$ be a continuously differentiable function in $\overline{B}$. Then, for any $\epsilon > 0$,*

$$\int_{\partial B} (w(x))^2 \, dS_x \leq \epsilon \int_B |D_x w(x)|^2 \, dx + \frac{A'}{\epsilon} \int_B (w(x))^2 \, dx, \tag{7.28}$$

where dS_x is the surface element on ∂B, and A' is a constant depending only on B.

Proof. Let x_δ be the point lying on the inward normal N_x to ∂B at x, at a distance δ from x. The manifold $\partial \tilde{B}_\delta$ generated by the x_δ, as x varies on ∂B, is of class C^1. Let dS_x^δ denote the surface element on $\partial \tilde{B}_\delta$, and let $\tilde{B}_\delta$ denote the region bounded by ∂B and $\partial \tilde{B}_\delta$.

Writing, for $x \in \partial B$,

$$w(x) = w(x_\delta) - \int_{x_\delta}^{x} \frac{\partial}{\partial N_x} w(\zeta) d\zeta$$

where ζ varies in the interval (x_δ, x), we get

$$|w(x)|^2 \leq 2(w(x_\delta))^2 + 2\left(\int_{x\delta}^{x} |D_x w| \, d\zeta\right)^2 \leq 2(w(x_\delta))^2 + 2\delta \int_{x\delta}^{x} |D_x w|^2 \, d\zeta.$$

Integrating with respect to $x \in \partial B$ and δ, we find

$$\delta \int_{\partial B} (w(x))^2 \, dS_x \leq 2C_1 \int_{\tilde{B}\delta} (w(x))^2 \, dx + 2C_1\delta^2 \int_{\tilde{B}\delta} |D_x w(x)|^2 \, dx$$

where C_1 depends only on B (in fact, $C_1 = \text{l.u.b. } dS_x/dS_x^\delta$). Replacing on the right-hand side the integration over $\tilde{B}_\delta$ by integration over the whole domain B, and taking $2C_1\delta = \epsilon$, (7.28) follows.

We now go back to the case where u satisfies (7.4), (7.6) and (7.25). On the right hand side of (7.12) there appears then the additional term of (7.26) which, by Lemma 4, is larger than

$$(7.29) \quad -\epsilon \int_{D_T} |D_x z|^2 \, dx \, dt - \frac{A''}{\epsilon} \int_{D_T} z^2 \, dx \, dt = -\epsilon ||\lambda^{-m} v||_1 - \frac{A''}{\epsilon} ||\lambda^{-m} v||^2$$

if $A'' > A'$ l.u.b. $|\partial g/\partial t|$. Substituting

$$(7.30) \quad -||\lambda^{-m} v||^2 \geq -(T + \eta)^2 ||\lambda^{-m-1} v||^2,$$

into the right-hand side of (7.29) and then substituting the result into the right-hand side of (7.12), we obtain, by proceeding as in the proof of Lemma 2,

$$||\lambda^{-m} L_0 v||^2 \geq \left[m - \frac{A''}{\epsilon}(T + \eta)^2\right] ||\lambda^{-m-1} v||^2 - (A + \epsilon)||\lambda^{-m} v||_1^2.$$

Hence, if $\epsilon = 1$ and m is sufficiently large, then

$$(7.31) \quad ||\lambda^{-m} L_0 v||^2 \geq \frac{1}{2} m ||\lambda^{-m-1} v||^2 - (A + 1)||\lambda^{-m} v||_1^2.$$

Consider next the term of (7.27) which appears on the right-hand side of (7.17) when (7.7) is replaced by (7.25). By Lemma 4 and (7.30), it is larger than

$$-\epsilon ||\lambda^{-m} v||_1^2 - \frac{A^*}{\epsilon}(T + \eta)^2 ||\lambda^{-m-1} v||^2$$

if $A^* > A'$ l.u.b. $|g|$. Taking $\epsilon = \nu_0/2$ we can proceed as in the proof of Lemma 3 and obtain an inequality of the form (7.13) with a different ρ, and $\rho \to 0$ if $\mu \to 0$, $\eta \to 0$, $m \to \infty$. The proof of Theorem 6 can now be extended, almost word by word, to the present case. We thus obtain the following result.

Theorem 7. *Assume that* (G_T) *holds* ($T < \infty$), *that ∂B is of class C^2 and that g, $\partial g/\partial t$ are continuous functions. If $u(x, t)$ satisfies* (7.6), (7.25) *and one of the inequalities* (7.4), (7.21), (7.24), *then $u \equiv 0$ in D_T.*

8. Lower Bounds on the Rate of Decay of Solutions

In this section we consider functions $u(x, t)$ satisfying a differential inequality of the form

$$(L_0u)^2 \leq c_1(t)u^2 + c_2(t)|D_xu|^2 \tag{8.1}$$

in the whole cylinder D_∞. The notation of Sec. 7 concerning L_0, L_1, D, S, etc. will be freely used.

Introducing the function

$$a(t) = \underset{B_t}{\text{l.u.b.}} \sum_{i,j} \left|\frac{\partial a_{ij}(x, t)}{\partial t}\right| \tag{8.2}$$

we shall prove:

Theorem 8. *Assume that* (G_∞) *holds and that* $u(x, t)$ *satisfies* (8.1) *and the boundary condition* (7.7) *with* $T = \infty$. *If, for some* $\eta > 1$,

$$c_1(t) = 0(t^{\eta-2}), \qquad c_2(t) = o(t^{-1}), \qquad a(t) = o(t^{-1}), \tag{8.3}$$

and

$$e^{\lambda t^\eta} \int_{B_t} |D_xu|^2 \, dx \to 0 \qquad \text{as } t \to \infty, \qquad \text{for any } \lambda > 0, \tag{8.4}$$

then $u \equiv 0$ *in* D_∞. *If*

$$c_1(t) = 0(t^{-2}), \qquad c_2(t) = o(t^{-2}), \qquad a(t) = o(t^{-2}) \tag{8.5}$$

and (8.4) *holds for* $\eta = 1$, *then* $u \equiv 0$ *in* D_∞.

Proof. For $\eta \geq 1$, $\lambda > 0$, β real, introduce

$$K(t; \beta, \lambda, \eta) = t^{2\beta} \exp [2\lambda t^\eta]. \tag{8.6}$$

We denote by Q_η the class of all twice continuously differentiable functions $v(x, t)$ in $\overline{D}_\infty$ satisfying

$$v \equiv 0 \qquad \text{on } S_\infty, \tag{8.7}$$

$$v \equiv 0 \qquad \text{in } D_\rho \tag{8.8}$$

for some $\rho > 0$ depending on v, and

$$K(t; \beta, \lambda, \eta) \int_{B_t} |D_xv(x, t)|^2 \, dx \to 0 \qquad \text{as } t \to \infty \tag{8.9}$$

for any $\lambda > 0$, β real. It then follows (see Problem 5) that

$$K(t; \beta, \lambda, \eta) \int_{B_t} (v(x, t))^2 \, dx \to 0 \qquad \text{as } t \to \infty. \tag{8.10}$$

In what follows various positive constants that depend only on η and ν_0 (of the condition (G_∞)) will be denoted by A.

Lemma 5. *If* $v \in Q_\eta$, $K = K(t; \beta, \lambda, \eta)$, *then*

$$(8.11) \quad \int_{D_\infty} [\lambda\eta(\eta - 1)t^{\eta-2} - \beta t^{-2}]Kv^2\, dx\, dt \leq \int_{D_\infty} K(L_0 v)^2\, dx\, dt + \int_{D_\infty} a(t)\, K|D_x v|^2\, dx\, dt.$$

Proof. The function $z(x, t) = K(t, \frac{1}{2}\beta, \frac{1}{2}\lambda, \eta)v(x, t)$ belongs to Q_η, and

$$K(L_0 v)^2 = \left[Ez - \frac{\partial z}{\partial t} + (\beta t^{-1} + \lambda\eta t^{\eta-1})z\right]^2.$$

Using the inequality $(a + b + c)^2 \geq 2b(a + c)$, we obtain

$$(8.12) \qquad K(L_0 v)^2 \geq -2\frac{\partial z}{\partial t} Ez - 2(\beta t^{-1} + \lambda\eta t^{\eta-1})z\frac{\partial z}{\partial t}.$$

Integrating over D_T we get

$$-2\int_{D_T} \frac{\partial z}{\partial t} \sum_{i,j} \frac{\partial}{\partial x_i}\left(a_{ij}\frac{\partial z}{\partial x_j}\right) dx\, dt - \int_{D_T} (\beta t^{-1} + \lambda\eta t^{\eta-1})\frac{\partial(z^2)}{\partial t}\, dx\, dt \leq \int_{D_T} K(L_0 v)^2\, dx\, dt.$$

Integrating by parts on the left-hand side and using (8.7), (8.8) we get (compare (7.11), (7.12))

$$(8.13) \quad \int_{D_T} [\lambda\eta(\eta - 1)t^{\eta-2} - \beta t^{-2}]z^2\, dx\, dt \leq \int_{D_T} K(L_0 v)^2\, dx\, dt + \int_{D_T} \sum_{i,j} \frac{\partial a_{ij}}{\partial t}\frac{\partial z}{\partial x_i}\frac{\partial z}{\partial x_j}\, dx\, dt + J_T$$

where J_T consists of integrals taken over B_T and, in view of (8.9), (8.10), $J_T \to 0$ as $T \to \infty$. Taking $T \to \infty$ in (8.13) and recalling (8.2), (8.11) follows.

Lemma 6. *If* $v \in Q_\eta$, $K = K(t; \beta, \lambda, \eta)$, $|\beta| \geq 1$, *then*

$$(8.14) \quad \int_{D_\infty} K|D_x v|^2\, dx\, dt \leq A\int_{D_\infty} tK(L_0 v)^2\, dx\, dt + A\int_{D_\infty} (\lambda\eta t^{\eta-1} + |\beta|t^{-1})Kv^2\, dx\, dt.$$

Proof. Consider the identity

$$-\int_{D_T} KvL_0 v\, dx\, dt = \frac{1}{2}\int_{D_T} K\frac{\partial(v^2)}{\partial t}\, dx\, dt - \int_{D_T} KvEv\, dx\, dt.$$

Integrating by parts on the right-hand side and using (8.7), (8.8), we find

$$(8.15) \quad -\int_{D_T} KvL_0 v\, dx\, dt = -\beta\int_{D_T} t^{-1}Kv^2\, dx\, dt - \lambda\eta\int_{D_T} t^{\eta-1}Kv^2\, dx\, dt + \int_{D_T} K\sum_{i,j} a_{ij}\frac{\partial v}{\partial x_i}\frac{\partial v}{\partial x_j}\, dx\, dt + I_T$$

where I_T is an integral over B_T which tends to zero as $T \to \infty$ (by (8.10)).

Hence

$$(8.16)\quad \int_{D_T} K|D_x v|^2\,dx\,dt \le A\int_{D_T} K|vL_0 v|\,dx\,dt + A|\beta|\int_{D_T} t^{-1}Kv^2\,dx\,dt + A\lambda\eta\int_{D_T} t^{\eta-1}Kv^2\,dx\,dt + A|I_T|.$$

Inserting the inequality

$$\int_{D_T} K|vL_0 v|\,dx\,dt \le 2\int_{D_T} tK(L_0 v)^2\,dx\,dt + 2\int_{D_T} t^{-1}Kv^2\,dx\,dt$$

into (8.16), and taking $T \to \infty$, (8.14) follows.

We next prove a lemma analogous to Lemma 3, Sec. 7, which will be the key lemma in the proof of Theorem 8.

Lemma 7. *If $v \in Q_\eta$, $K = K(t;\beta,\lambda,\eta)$, $\eta > 1$, $|\beta| > 2$, and $a(t) = o(t^{-1})$, and if $v \equiv 0$ in D_{T^*} where T^* is some constant (to be defined later) depending only on ν_0, η and the function $a(t)$, then*

$$(8.17)\quad \lambda\int_{D_\infty} t^{\eta-2}Kv^2\,dx\,dt + \int_{D_\infty} t^{-1}K|D_x v|^2\,dx\,dt \le A\int_{D_\infty} K(L_0 v)^2\,dx\,dt$$

for all λ sufficiently large (depending only on β, T^*, η).

Proof. The integrand on the left-hand side of (8.11) is larger than $A\lambda t^{\eta-2}Kv^2$ if $\eta > 1$ and λ is sufficiently large (depending on T^*, β). Hence,

$$(8.18)\quad \lambda\int_{D_\infty} t^{\eta-2}Kv^2\,dx\,dt \le A\int_{D_\infty} K(L_0 v)^2\,dx\,dt + A\int_{D_\infty} a(t)K|D_x v|^2\,dx\,dt.$$

Replacing β by $\beta + 1/2$ and by $\beta - (\eta - 1)/2$ in (8.18), we get

$$(8.19)\quad \lambda\int_{D_\infty} t^{\eta-1}Kv^2\,dx\,dt \le A\int_{D_\infty} tK(L_0 v)^2\,dx\,dt + A\int_{D_\infty} ta(t)K|D_x v|^2\,dx\,dt,$$

$$(8.20)\quad \lambda\int_{D_\infty} t^{-1}Kv^2\,dx\,dt \le A\int_{D_\infty} t^{1-\eta}K(L_0 v)^2\,dx\,dt + A\int_{D_\infty} t^{1-\eta}a(t)K|D_x v|^2\,dx\,dt.$$

Inserting the last two inequalities into the right-hand side of (8.14), we get, if $T^* \ge 1$,

$$(8.21)\quad \int_{D_\infty} K|D_x v|^2\,dx\,dt \le A\int_{D_\infty} tK(L_0 v)^2\,dx\,dt + A\int_{D_\infty} ta(t)K|D_x v|^2\,dx\,dt.$$

Replacing β by $\beta - 1/2$ in (8.21) and adding the result to (8.18), we find

$$(8.22)\quad \lambda\int_{D_\infty} t^{\eta-2}Kv^2\,dx\,dt + \int_{D_\infty} [t^{-1} - 2Aa(t)]K|D_x v|^2\,dx\,dt \le A\int_{D_\infty} K(L_0 v)^2\,dx\,dt.$$

Since $a(t) = o(t^{-1})$, for some T^* sufficiently large, $4Aa(t) < 1/t$ if $t \geq T^*$. Hence, if $v \equiv 0$ in D_{T^*},

$$[t^{-1} - 2Aa(t)]|D_xv|^2 \geq (2t)^{-1}|D_xv|^2 \qquad \text{in } D_\infty.$$

Substituting this into (8.22), we obtain (8.17).

We turn to the proof of Theorem 8 in case $\eta > 1$. Let $T_1 > T^*$ be such that $2Atc_2(t) < 1$ if $t > T_1$, where A is the constant appearing in (8.17). Let $\zeta(t)$ be a twice continuously differentiable function satisfying for some $T_2 > T_1$: $\zeta(t) = 0$ if $0 \leq t \leq T_1$, $\zeta(t) = 1$ if $T_2 \leq t < \infty$. In view of (7.7), (8.4), the function $v = \zeta u$ belongs to Q_η. Applying to it (8.17), we get

$$\lambda \int_{T_2}^\infty \int_B t^{\eta-2}Ku^2\,dx\,dt + \int_{T_2}^\infty \int_B t^{-1}K|D_xu|^2\,dx\,dt$$
$$\leq A \int_{T_1}^{T_2} \int_B K(L_0v)^2\,dx\,dt + A \int_{T_2}^\infty \int_B K(L_0u)^2\,dx\,dt.$$

Substituting $(L_0u)^2$ from (8.1), and using the assumption $c_1(t) = 0(t^{\eta-2})$ and the choice of T_1 we get, if λ is sufficiently large,

$$\lambda \int_{T_2}^\infty \int_B t^{\eta-2}Ku^2\,dx\,dt + \int_{T_2}^\infty \int_B t^{-1}K|D_xu|^2\,dx\,dt \tag{8.23}$$
$$\leq 2A \int_{T_1}^{T_2} \int_B K(L_0v)^2\,dx\,dt.$$

From (8.23) it follows that, for any $T > T_2$,

$$T^{2\beta} \exp\,[2\lambda T^\eta] \left\{\lambda \int_T^\infty \int_B [t^{\eta-2}u^2 + t^{-1}|D_xu|^2]\,dx\,dt\right\}$$
$$\leq 2AT_2^{2\beta} \exp\,[2\lambda T_2^\eta] \left\{\int_{T_1}^{T_2} \int_B (L_0v)^2\,dx\,dt\right\}.$$

This inequality cannot hold for sufficiently large λ unless the integral on the left-hand side is zero. We conclude that $u \equiv 0$ in $D_\infty - D_T$. If we now apply Theorem 6 then we find $u \equiv 0$ also in D_T. This completes the proof of the theorem in case $\eta > 1$.

To prove the theorem in case $\eta = 1$, we first establish an analogue of Lemma 7 to this case.

Lemma 8. *If $v \in Q_1$, $K = K(t; \beta, \lambda, 1)$, and $a(t) = o(t^{-2})$, and if $v \equiv 0$ in D_{T^*} where T^* depends only on ν_0 and on the function $a(t)$, then, for all $\beta < 0$, $\lambda < -A\beta$,*

$$|\beta| \int_{D_\infty} t^{-2}Kv^2\,dx\,dt + \int_{D_\infty} t^{-2}K|D_xv|^2\,dx\,dt \leq A \int_{D_\infty} K(L_0v)^2\,dx\,dt. \tag{8.24}$$

Proof. By Lemma 5,

$$-\beta \int_{D_\infty} t^{-2}Kv^2\,dx\,dt \leq \int_{D_\infty} K(L_0v)^2\,dx\,dt \tag{8.25}$$
$$+ A \int_{D_\infty} a(t)\,K|D_xv|^2\,dx\,dt.$$

By Lemma 6 with β replaced by $\beta - 1$,

$$(8.26)\quad \int_{D_\infty} t^{-2}K|D_x v|^2\,dx\,dt \le A\int_{D_\infty} t^{-1}K(L_0 v)^2\,dx\,dt + A\int_{D_\infty}(\lambda t^{-2} + |\beta|t^{-3})Kv^2\,dx\,dt.$$

Since $a(t) = o(t^{-2})$, $2Aa(t) < t^{-2}$ in (8.25) if $t \ge T^*$. Taking λ such that $4A\lambda < |\beta|$ in (8.26), and increasing T^* if necessary so that $4A|\beta|t^{-1} < |\beta|$ in (8.26) if $t \ge T^*$, and finally adding (8.25), (8.26), we obtain (8.24).

Using Lemma 8 instead of Lemma 7, we can now proceed to complete the proof of Theorem 8 in case $\eta = 1$ by the same reasoning as in the case $\eta > 1$.

From the proof of Theorem 8 one sees that Theorem 8 remains true if the differential inequality (8.1) is replaced by the weaker inequality

$$(8.27)\quad \int_{B_t}(L_0 u)^2\,dx \le c_1(t)\int_{B_t} u^2\,dx + c_2(t)\int_{B_t}|D_x u|^2\,dx$$

Writing the parabolic operator L_1, defined in (7.22), in the form (7.23) we obtain:

Corollary 1. *Theorem* 8 *remains valid if the inequality* (8.1) *is replaced by*

$$(8.28)\quad \int_{B_t}(L_1 u)^2\,dx \le c_1(t)\int_{B_t} u^2\,dx + c_2(t)\int_{B_t}|D_x u|^2\,dx$$

provided $\operatorname{l.u.b.}_x \sum_j (\tilde{b}_j(x, t))^2$ *satisfies the same restrictions as* $c_2(t)$.

Consider finally the situation where the first boundary condition (7.7) with $T = \infty$ is replaced by the second boundary condition (7.25) with $T = \infty$, and assume that u satisfies, in addition to (8.4), the condition:

$$(8.29)\quad e^{\lambda t\eta}\int_{B_t}|u|^2\,dx \to 0 \qquad \text{as } t \to \infty, \qquad \text{for any } \lambda > 0.$$

Modifying the definition of Q_η accordingly (requiring that (8.8), (8.9), (8.10) be satisfied and that $\partial v/\partial\nu + gv = 0$ on S_∞) we can proceed with the same calculations as before, except for two places where some differences occur. Firstly, on the left-hand side of (8.13) there appears the additional term

$$(8.30)\quad -\int_{S_T} g\,\frac{\partial(z^2)}{\partial t}\,dS = \int_{S_T}\frac{\partial g}{\partial t}\,z^2\,dS - \int_{\partial B_T} gz^2\,d\sigma$$

where $d\sigma$ is the surface element of ∂B_T, and, secondly, on the right-hand side of (8.15) there appears the additional term

$$(8.31)\quad -\int_{S_T} Kgv^2\,dS.$$

If $g \le 0$, $\partial g/\partial t \ge 0$ then the terms in (8.31) and on the right-hand side of (8.30) are all nonnegative and can therefore be dropped out. The proof of Theorem 8 thus extends to the present situation.

If

$$(8.32)\quad g(t) = 0(1), \quad \frac{\partial g}{\partial t} = 0(t^{\eta-2}) \qquad \text{when } \eta > 1,$$

$$(8.33) \qquad g(t) = 0(1), \quad \frac{\partial g}{\partial t} = 0(t^{-2}) \qquad \text{when } \eta = 1,$$

then the terms on the right-hand side of (8.30) and the term in (8.31) can be estimated (using Lemma 4 with an appropriate choice of ϵ; compare the proof of Theorem 7) so as to get results similar to Lemmas 5, 6. The extension of the proof of Theorem 8 to the present case is then obvious. The details may be left to the reader.

We sum up the results:

Theorem 9. *Assume that* (G_∞) *holds, that* g, $\partial g/\partial t$ *are continuous functions, that* ∂B *is of class* C^2 *and that* $u(x, t)$ *satisfies* (8.1) *or* (8.27), *and the boundary condition* (7.25) *with* $T = \infty$. *If for some* $\eta > 1$, (8.3), (8.4), (8.29) *hold, and if either* (8.32) *holds or* $g \leq 0$, $\partial g/\partial t \geq 0$, *then* $u \equiv 0$ *in* D_∞. *If* (8.5) *and* (8.4), (8.29), *with* $\eta = 1$, *hold, and if either* (8.33) *holds or* $g \leq 0$, $\partial g/\partial t \geq 0$, *then* $u = 0$ *in* D_∞.

The theorem extends to inequalities of the form (8.28).

PROBLEMS

1. Prove a theorem analogous to Theorem 3, Sec. 4, for solutions of the second initial-boundary value problem.

2. Extend Theorems 1, 4 to the case where (1.1) is replaced by $Lu(x, t) = f(x, t) + k(x, t, u)$, where $k(x, t, u)$ is a continuous function for $(x, t) \in \bar{D}$, $-\infty < u < \infty$, satisfying $|k(x, t, u)| \leq \mu_0|u|$ where μ_0 is sufficiently small (depending on M, M' and a uniform bound on the projections on R^n of the domains B_t $(0 \leq t < \infty)$). [*Hint:* Take $\psi = 2\epsilon\varphi/\delta + \epsilon\varphi/\delta_0 + A\varphi e^{-\gamma(t-\sigma)}/\delta_0$, $\gamma = \delta/2\delta_1$, $\delta_2 = \delta/2\delta_1$. Show that $L\psi + \epsilon + \delta_2\psi < 0$, $Lu + \epsilon + \mu_0|u| > 0$ and use Theorem 16, Sec. 6, Chap. 2.]

3. Let L be uniformly parabolic with bounded continuous coefficients in a domain D lying in a paraboloid $|x|^2 < \beta(t + 1)$ $(\beta > 0)$ and in $t > 0$. Assume that for every $\tau > 0$, $B_\tau = D \cap \{t = \tau\}$ is a nonempty domain and that
$$c \leq 0, \qquad \Sigma\,(a_{ii} + x_i b_i) \geq \lambda > 0 \qquad \text{if } |x| \text{ is sufficiently large.}$$
Prove that if $Lu = k(x, t, u)$ in D, $u = h$ on the boundary ∂D of D, and if for some positive constants A, δ and for a constant $\bar{A}$ sufficiently small (depending only on L)
$$|k(x, t, u)| \leq \frac{\bar{A}|u|}{t + 1}, \qquad |h(x, t)| \leq \frac{A}{(t + 1)^\delta},$$
then $u(x, t) \to 0$ as $t \to \infty$, uniformly in D.
[*Hint:* The function
$$\psi(x, t) = (t + 1)^{-\epsilon} \exp\left[-\frac{H|x|^2}{(t + 1)}\right]$$

satisfies $L\psi + A'(t+1)^{-1}\psi < 0$ if $\nu = \text{g.l.b.} \sum a_{ij}\xi_i\xi_j$ ($|\xi| = 1$, $(x, t) \in D$), $H = 1/4\nu$, $A' = H\lambda$, and $\epsilon \le H\lambda$. On ∂D, $\psi(x, t) \ge (t+1)^{-\epsilon} \exp[-H\beta]$. Take $\epsilon = \min(H\lambda, \delta)$ and apply Theorem 16, Sec. 6, Chap. 2, to $M\psi$ and $\pm u$ (M some constant).]

4. Prove a result similar to that of Problem 3 in case $k(x, t, u) \equiv f(x, t)$ and $|f(x, t)| \le A(t+1)^{-1-\epsilon}$.

5. Let $w(x)$ be a continuously differentiable function in a closed bounded domain G and let $w = 0$ on the boundary ∂G of G. Prove

$$\int_G (w(x))^2\, dx \le A \int_G |D_x w(x)|^2\, dx$$

where A depends only on G.
[*Hint:* Set $x' = (x_2, \ldots, x_n)$, $m(x') = \{x_1; (x_1, x') \in G\}$, G_1 = projection of G on the x'-space. Then

$$|w(x_1, x')|^2 \le \text{const.} \int_{m(x')} [\partial w(\xi_1, x')/\partial \xi_1]^2\, d\xi_1.$$

Integrate with respect to $x_1 \in m(x')$ and $x' \in G_1$. (G is assumed to be such that $\int_{G_1} \left[\int_{m(x_1)} k(x_1, x')\, dx_1\right] dx' = \int_G k(x)\, dx$ for any continuous function $k(x)$ in G.)]

6. Extend Theorems 8, 9 of Sec. 8 to $L + c$, $L_1 + c$ where $c(x, t)$, $\partial c(x, t)/\partial t$ are continuous functions and $\partial c/\partial t = 0(t^{\eta-2})$ if $\eta > 1$, $\partial c/\partial t = 0(t^{-2})$ if $\eta = 1$.

7. Prove that Theorem 8, Sec. 8 can be improved in case $\eta = 1$, by replacing (8.5) by

$$c_1(t) = 0(t^{-1-\epsilon}),\ c_2(t) = 0(t^{-1-\epsilon}),\ a(t) = o(c_2(t))$$

for some $\epsilon > 0$.
[*Hint:* Setting

$$g(t) = \exp\left[-\gamma \int_0^t c_2(s)ds\right] \qquad (\gamma > 1),$$

$$\sigma(t) = \lambda \int_0^t \frac{1}{g(\tau)} \left[\int_0^\tau g(s)c_1(s)ds\right] d\tau \qquad (\lambda > 0),$$

prove that if $v \in Q_1$ and $v(t) = 0$ for all $t < t_0$ (t_0 depends only on c_1, c_2, L), then

$$(*) \quad \int_{D_\infty} g(t)e^{2\sigma(t)}(L_0 v)^2\, dx\, dt \ge \lambda \int_{D_\infty} g(t)e^{2\sigma(t)}c_1(t)v^2\, dx\, dt$$

$$+ \tfrac{1}{2}\nu_0\gamma \int_{D_\infty} g(t)e^{2\sigma(t)}c_2(t)|D_x v|^2\, dx\, dt.$$

(*) plays a role similar to that of Lemmas 7, 8 in the proof of Theorem 8. To prove (*) set $z = e^{\sigma(t)}v$ and observe that the left-hand side is not less than

$$-2\int_{D_\infty} g(t)\frac{\partial z}{\partial t} Ez\, dx\, dt - 2\int_{D_\infty} g(t)\frac{\partial z}{\partial t} z \frac{d\sigma}{dt}\, dx\, dt \equiv J_1 + J_2.$$

Use $d[g(d\sigma/dt)]/dt = \lambda c_1 g$ to show that J_2 is equal to the first term on the right hand side of (*). Evaluate J_1 by using integration by parts.]

CHAPTER 7

SEMILINEAR EQUATIONS. NONLINEAR BOUNDARY CONDITIONS

Introduction. In this chapter we shall deal with two problems. The first consists of solving the equation

$$(0.1) \qquad Lu = f(x, t, u, \partial u/\partial x_1, \ldots, \partial u/\partial x_n)$$

under the first initial-boundary conditions, where L is a linear parabolic operator. The second problem consists of solving the equation $Lu = f(x, t)$ under the second initial-boundary condition, with the boundary condition being nonlinear, i.e.,

$$(0.2) \qquad \frac{\partial u}{\partial \nu} = g(x, t, u).$$

Equation (0.1) is the simplest type of a nonlinear equation, and is called *semilinear* equation. The procedure that will be used to solve it can serve as a standard model for solving more general nonlinear problems. It is the following:

Set $w = Tv$ if w is the solution of

$$Lw = f(x, t, v, \partial v/\partial x_1, \ldots, \partial v/\partial x_n)$$

satisfying the same initial-boundary conditions as u. Prove that when v is restricted to an appropriate set, Tv is defined and T has a fixed point.

In order to prove that Tv is defined, the existence theory for linear equations will be used. In order to establish the existence of a fixed point for T, we shall need some a priori inequalities and a certain fixed point theorem. The decisive and most difficult part of the problem is the derivation of the a priori estimates.

The second problem (see (0.2)) can also be solved by using the previous method, namely, we set $w = Tv$ if w is a solution of $Lw = f(x, t)$, satisfying the same initial condition as u, and

$$\frac{\partial w}{\partial \nu} = g(x, t, v(x, t)).$$

We then try to establish the existence of a fixed point for T.

The condition (0.2) arises naturally in physical problems. Thus, if u

is the temperature in the domain, this condition may represent the law of cooling of Newton; the nonlinearity of g in u occurs for very high or very low temperatures.

The first problem is treated in Sec. 4 and the second problem is treated in Sec. 5. A priori estimates needed in Sec. 4 are derived in Secs. 2, 3. In Sec. 1 we introduce the concept of quasi-linear equations and also state two fixed point theorems.

1. Nonlinear Equations. Fixed Point Theorems

The nonlinear equation

$$\Phi\left(x, t, u, \frac{\partial u}{\partial x_i}, \frac{\partial^2 u}{\partial x_i\,\partial x_j}, \frac{\partial u}{\partial t}\right) = 0 \qquad (i, j = 1, \ldots, n) \tag{1.1}$$

is said to be *parabolic* with respect to a solution $u(x, t)$ if

$$\frac{\partial \Phi}{\partial t} < 0, \qquad \left(\frac{\partial \Phi}{\partial(\partial^2 u/\partial x_i\,\partial x_j)}\right) \text{ is positive definite,} \tag{1.2}$$

where in the argument of Φ we substitute $u = u(x, t)$. If (1.1) is parabolic with respect to any solution then we simply say that (1.1) is *parabolic*.

Nonlinear parabolic equations of the form

$$\sum_{i,j=1}^{n} \Phi_{ij}(x, t, u, \nabla_x u)\,\frac{\partial^2 u}{\partial x_i\,\partial x_j} - \Phi_0(x, t, u, \nabla_x u)\,\frac{\partial u}{\partial t} = f(x, t, u, \nabla_x u), \tag{1.3}$$

where $\nabla_x u = \left(\frac{\partial u}{\partial x_1}, \ldots, \frac{\partial u}{\partial x_n}\right)$, are called *quasi-linear* parabolic equations. We may then assume that $\Phi_0 \equiv 1$ since otherwise we can divide the equation by Φ_0. In Sec. 4 we shall deal with a special class of quasi-linear parabolic equations, namely, equations of the form (1.3) for which $\Phi_0 \equiv 1$ and the Φ_{ij} are functions of (x, t) only. These equations are called *semi-linear* equations.

The concept of Banach spaces was introduced in Sec. 1, Chap. 3. We now give a few more definitions that will be needed in stating some fixed point theorems later on in this section.

A subset Y of a Banach space X is said to be *precompact* if for any sequence $\{x_m\}$ in Y there exists a subsequence $\{x_{m'}\}$ which is convergent to some element x of X. x need not belong to Y. If for any sequence $\{x_m\}$ in Y there exists a subsequence which converges to an element of Y then Y is called a *compact* set. Clearly, Y is compact if and only if Y is precompact and closed; Y is precompact if and only if Y is a subset of a compact set.

A subset Y of a Banach space X is said to be *bounded* if it is contained in some finite ball, i.e., if there exists a finite number R such that $\|x\| < R$ for all $x \in Y$. Compact sets are necessarily bounded and closed.

An important example of precompact sets is given in the following theorem.

Theorem 1. *Let D be a bounded domain in R^n and let $C_{p+\alpha}$, $\overline{C}_{p+\alpha}$ be defined with respect to D (see Sec. 8, Chap. 3). Then, for any $0 < \alpha < \beta < 1$, $0 \leq p \leq q$, the bounded subsets of the spaces $C_{q+\beta}$, $\overline{C}_{q+\beta}$ (defined with respect to D) are precompact subsets of $C_{p+\alpha}$, $\overline{C}_{p+\alpha}$ respectively.*

Proof. It is enough to prove the theorem for $\overline{C}_{q+\beta}$. We may clearly assume that $q = p$. Consider first the case $p = 0$. Let $\{u_m\}$ be a bounded sequence in $\overline{C}_\beta$, i.e.,

$$(1.4) \quad |u_m(x)| \leq A, \qquad \frac{|u_m(x) - u_m(y)|}{|x - y|^\beta} \leq A \qquad \text{(for all } x \in D,\ y \in D,\ x \neq y)$$

for some constant A. By the Ascoli-Arzela theorem there exists a subsequence which is uniformly convergent in D. Denote this subsequence, for simplicity, again by $\{u_m\}$ and set $u(x) = \lim u_m(x)$. Taking $m \to \infty$ in (1.4) we find that $u \in \overline{C}_\beta$. It remains to prove that $u_m \to u$ in the norm of $\overline{C}_\alpha$, i.e., for any $\epsilon > 0$ and for all $x \in D$, $y \in D$, $x \neq y$,

$$I_m(x, y) \equiv \frac{|[u_m(x) - u(x)] - [u_m(y) - u(y)]|}{|x - y|^\alpha} \leq \epsilon \qquad \text{if } m \geq m_0$$

where m_0 is independent of x, y.

Now, if $|x - y| < \delta$ then

$$|I_m(x, y)| \leq |x - y|^{\beta-\alpha} \frac{|u_m(x) - u_m(y)|}{|x - y|^\beta} + |x - y|^{\beta-\alpha} \frac{|u(x) - u(y)|}{|x - y|^\beta}$$
$$\leq 2A\delta^{\beta-\alpha} = \epsilon$$

if $\delta = (\epsilon/2A)^{1/(\beta-\alpha)}$. On the other hand, if $|x - y| > \delta$ then, since $u_m(x) \to u(x)$ uniformly in D,

$$|I_m(x, y)| \leq \frac{1}{\delta^\alpha} [|u_m(x) - u(x)| + |u_m(y) - u(y)|] < \epsilon \qquad \text{if } m \geq m_0$$

where m_0 is independent of x, y.

This completes the proof in case $p = 0$. The proof for $p > 0$ is similar.

Let T be an operator defined on a subset Y of a Banach space X, and suppose that T maps Y into X. T is said to be *continuous* at a point $x^0 \in Y$ if for any $\epsilon > 0$ there exists a $\delta > 0$ such that $||x - x^0|| < \delta$, $x \in Y$ imply $||Tx - Tx^0|| < \epsilon$. This is equivalent to saying that for any sequence $\{x_m\}$ in Y, $x_m \to x^0$ implies $Tx_m \to Tx^0$. If T is continuous at all the points of Y then we say that T is *continuous on Y*.

We denote by TY the set $\{Ty; y \in Y\}$.

A set $K \subset X$ is called a *convex* set if whenever $x \in K$, $y \in K$, also

$\vartheta x + (1 - \vartheta)y \in K$ for all $0 < \vartheta < 1$, i.e., if together with each pair of its points it also contains the interval connecting the pair.

We can now state a fixed point theorem due to Schauder.

Theorem 2 (Schauder). *Let Y be a closed convex subset of a Banach space X and let T be a continuous operator on Y such that TY is contained in Y and TY is precompact. Then T has a fixed point, i.e., there exists a point $y_0 \in Y$ such that $Ty_0 = y_0$.*

In contrast to Theorem 1, Sec. 1, Chap. 3, the present fixed point theorem does not assert uniqueness. However it is a far deeper theorem than Theorem 1 of Chap. 3. In the case where X is a euclidean space, Theorem 2 is known as Brouwer's fixed point theorem.

We shall now give another fixed point theorem due to Leray and Schauder.

A transformation T defined on a subset Y of X, with $TY \subset X$, is said to be *compact* if it maps bounded subsets of Y into compact subsets of X.

Consider a transformation

$$y = T(x, k) \tag{1.5}$$

where x, y belong to a Banach space X and k is a real parameter which varies in a bounded interval, say $a \leq k \leq b$.

Assume:

(a) $T(x, k)$ is defined for all $x \in X$, $a \leq k \leq b$.

(b) For any fixed k, $T(x, k)$ is continuous in X, i.e., for any $x^0 \in X$ and for any $\epsilon > 0$ there exists a $\delta > 0$ such that $||T(x, k) - T(x^0, k)|| < \epsilon$ if $||x - x^0|| < \delta$.

(c) For x in bounded sets of X, $T(x, k)$ is uniformly continuous in k, i.e., for any bounded set $X_0 \subset X$ and for any $\epsilon > 0$ there exists a $\delta > 0$ such that if $x \in X_0$, $|k_1 - k_2| < \delta$, $a \leq k_1, k_2 \leq b$, then $||T(x, k_1) - T(x, k_2)|| < \epsilon$.

(d) For any fixed k, $T(x, k)$ is a compact transformation, i.e., it maps bounded subsets of X into compact subsets of X.

(e) There exists a (finite) constant M such that every possible solution x of $x - T(x, k) = 0$ ($x \in X$, $k \in [a, b]$) satisfies: $||x|| \leq M$.

(f) The equation $x - T(x, a) = 0$ has a unique solution in X.

Theorem 3 (Leray-Schauder). *Under the assumptions* (a)–(f) *there exists a solution of the equation $x - T(x, b) = 0$.*

Theorem 3 remains true if $T(x, k)$ is defined only for $||x|| \leq M'$, $a \leq k \leq b$ for some $M' > M$, and if the assumptions (a)–(f) are modified accordingly.

In the application of Theorem 3, it is quite often the case that $T(x, k) = kT(x)$, $a = 0$, so that (f) is trivially satisfied.

2. A Priori Estimates of the Type $1 + \delta$

In Chap. 4 we have derived estimates on the $(2 + \alpha)$-norm of solutions. We shall now derive, by a different method, an estimate on the $(1 + \delta)$-norm of solutions, for any $0 < \delta < 1$. The notation of Chap. 3, Sec. 2, will be freely used. In addition, we introduce the following notation:

$$\overline{|v|}^D_{1+\delta} = \overline{|v|}^D_{\delta} + \Sigma\, \overline{|D_x v|}^D_{\delta}, \tag{2.1}$$

$$\overline{|v|}^D_{1-0} = \overline{|v|}^D_{0} + L^D[v], \tag{2.2}$$

$$\overline{|v|}^D_{2-0} = \overline{|v|}^D_{1-0} + \Sigma\, \overline{|D_x v|}^D_{1-0}, \tag{2.3}$$

where

$$L^D[v] = \underset{(x,t) \in D, (x',t') \in D}{\text{l.u.b.}} \frac{|v(x, t) - v(x', t')|}{|x - x'| + |t - t'|}.$$

$L^D[v] < \infty$ if and only if $v(x, t)$ is uniformly Lipschitz continuous in D.

We denote by $\overline{C}_{1+\delta}(D)$ the set of all functions v with $\overline{|v|}_{1+\delta} < \infty$ and introduce in this set the norm $\overline{|\quad|}^D_{1+\delta}$. $\overline{C}_{1+\delta}(D)$ is a Banach space. The spaces $\overline{C}_{1-0}(D)$, $\overline{C}_{2-0}(D)$ are defined in a similar manner. It should be noted that the spaces $\overline{C}_{1+\delta}$ $(0 < \delta < 1)$ are not contained in the spaces $\overline{C}_{2+\alpha}$ if $\delta > \alpha$.

When there is no confusion we abbreviate $\overline{|v|}^D_{1+\delta}$, $\overline{C}_{1+\delta}(D)$, etc. by $\overline{|v|}_{1+\delta}$, $\overline{C}_{1+\delta}$, etc.

Definitions. If $\overline{S}$ can locally be represented in the form (3.2.17) with h, $D_x h$, $D_x^2 h$, $D_t h$ Hölder continuous (exponent α), then we say that S is of class $\overline{C}_{2+\alpha}$. If the functions $D_x D_t h$ also exist, and are continuous, then we say that S is of class $\overline{C}_{2+\alpha}$. If $\overline{S}$ can locally be represented in the form (3.2.17) with h, $D_x h$ Lipschitz continuous (i.e., $L[h] < \infty$, $L[D_x h] < \infty$), then we say that S is of class $\overline{C}_{2-0}$. These definitions are quite different from the definition given in Sec. 8, Chap. 3 (the paragraph containing (3.8.6)).

If S is of class $\overline{C}_q$, then $\overline{S}$ can be covered by a finite number of balls V_j such that $S_j = \overline{S} \cap V_j$ can be represented in the form (3.2.17) with h in $\overline{C}_q$. Take such a finite covering which will be fixed from now on. Given a function v on $\overline{S}$, it can be written on each S_j as a function of $(x_1, \ldots, x_{i-1}, x_{i+1}, \ldots, x_n, t)$ in some region S_j^*, by using (3.2.17). We then define $|v|_p^S$ to be $\max_j |v|_p^{S^*_j}$, and say that v belongs to $\overline{C}_p$ on S if $|v|_p^S < \infty$.

Given a linear parabolic operator

$$Lu \equiv \sum_{i,j=1}^{n} a_{ij}(x, t) \frac{\partial^2 u}{\partial x_i\, \partial x_j} + \sum_{i=1}^{n} b_i(x, t) \frac{\partial u}{\partial x_i} + c(x, t)u - \frac{\partial u}{\partial t}, \tag{2.4}$$

consider the first initial-boundary value problem

$$Lu(x, t) = f(x, t) \qquad \text{in } D + B_T, \tag{2.5}$$

$$u = 0 \qquad \text{on } \overline{B} + S. \tag{2.6}$$

We shall need the following assumptions:

(A) L is parabolic in $\overline{D}$, i.e., there exists a positive constant H_0 such that, for each $(x, t) \in \overline{D}$ and for every real vector ξ,

$$\sum_{i,j=1}^{n} a_{ij}(x, t)\xi_i\xi_j \geq H_0 \sum_{i=1}^{n} \xi_i^2. \tag{2.7}$$

(B) a_{ij} are Hölder continuous (exponent α) in $\overline{D}$, b_i and c are continuous in $\overline{D}$ and, in addition, a_{ij} belong to $\overline{C}_{1-0}$ on S. Thus,

$$\sum \overline{|a_{ij}|}_\alpha^D + \sum |b_i|_0^D + |c|_0^D \leq H_1, \tag{2.8}$$

$$\sum |a_{ij}|_{1-0}^S \leq H_2. \tag{2.9}$$

We can now state the a priori $(1 + \delta)$-estimate.

Theorem 4. *Assume that S belongs both to $\overline{C}_{2+\alpha}$ and to $\overline{C}_{2-0}$, and that* (A), (B) *hold. Let $f(x, t)$ be a continuous function in $\overline{D}$ vanishing on ∂B, and let $u(x, t)$ be a solution of* (2.5), (2.6). *Then, for any positive $\delta < 1$ there exists a constant K depending only upon δ, H_0, H_1, H_2 and D such that*

$$\overline{|u|}_{1+\delta}^D \leq K|f|_0^D. \tag{2.10}$$

The proof is given in the remaining part of this section and in the next section. Use will be made of existence theorems for linear equations derived in Chaps. 3, 5. Since Theorem 4 will be applied to solve quasi-linear equations, making use of the existence theorems for linear equations derived in Chap. 3, it is justified to look upon (2.10) as an a priori estimate.

In what follows we denote by K each constant which (unless the contrary is explicitly stated) depends only on δ, D, and the H_i. We begin with the following lemma.

Lemma 1. *It suffices to prove Theorem 4 under the additional assumptions that f, c, and the b_i are Hölder continuous (exponent α) in $\overline{D}$.*

Proof. Suppose that Theorem 4 has already been proved under the additional assumptions stated in the lemma. We then have to prove Theorem 4 without making these additional assumptions. By the Weierstrass approximation theorem, there exist sequences $\{f^m\}$, $\{b_i^m\}$, $\{c^m\}$ of polynomials such that

$$\epsilon_m \equiv |f^m - f|_0^D \to 0, \quad |b_i^m - b_i|_0^D \to 0, \quad |c^m - c|_0^D \to 0 \quad \text{as } m \to \infty. \tag{2.11}$$

Denote by L_m the operator L when b_i, c are replaced by b_i^m, c^m respectively. We can modify the f^m so that they are still continuously differentiable and $\epsilon_m \to 0$, and so that $f^m = 0$ on ∂B. Indeed this can be done by multiplying each f^m by an appropriate smooth function which vanishes on ∂B and which is equal to 1 outside some neighborhood of ∂B (see Problem 1, Chap. 1). By Theorem 7, Sec. 3, Chap. 3, there exists a solution u_m of

$$Lu_m = f^m \quad \text{in } D + B_T, \qquad u_m = 0 \quad \text{on } B + S, \tag{2.12}$$

and by the assumption of the lemma,

$$\overline{|u_m|}_{1+\delta} \leq K|f^m|_0 \leq K(|f|_0 + \epsilon_m). \tag{2.13}$$

From (2.13) it follows, using the Ascoli-Arzela theorem, that there exists a subsequence of $\{u_m\}$, which we denote, for simplicity, again by $\{u_m\}$, such that

$$u_m \to v, \quad \nabla_x u_m \to \nabla_x v \quad \text{uniformly in } D, \quad \text{as } m \to \infty, \tag{2.14}$$

where $v(x) = \lim u_m(x)$ by definition. From (2.13), (2.14) we find (as in the proof of Theorem 3, Sec. 2, Chap. 3) that

$$\overline{|v|}_{1+\delta} \leq K \lim_{m\to\infty} (|f|_0 + \epsilon_m) = K|f|_0. \tag{2.15}$$

If we prove that

$$|u - u_m|_0^D \to 0 \qquad \text{as } m \to \infty \tag{2.16}$$

then it would follow that $u \equiv v$ and hence, by (2.15), the proof of (2.10) is completed; hence Lemma 1 follows.

To prove (2.16) we note that

$$L(u_m - u) = L_m u_m + (L - L_m)u_m - Lu = (L - L_m)u_m + f^m - f \equiv \tilde{f}_m.$$

From (2.11), (2.13) it follows $(L - L_m)u_m \to 0$, $f^m - f \to 0$, uniformly in D, if $m \to \infty$. Hence,

$$|\tilde{f}_m|_0^D \to 0 \qquad \text{as } m \to \infty.$$

By the maximum principle (i.e., by (2.3.12)) we then get,

$$|u_m - u|_0^D \leq K|\tilde{f}_m|_0^D \to 0 \qquad \text{as } m \to \infty,$$

and (2.16) is thereby proved.

In view of Lemma 1 we may assume in what follows that f, c, and the b_i are Hölder continuous (exponent α) in $\overline{D}$.

Extend the coefficients of L into a closed cylinder Ω_0 containing $\overline{D}$ and having its base E_0 on $t = 0$, in such a manner that the extended functions are Hölder continuous (exponent α) and such that (2.7), (2.8) hold in Ω_0 with H_0, H_1 possibly replaced by other constants depending only on H_0, H_1; see Lemma 1, Sec. 7, Chap. 3.

Let $\Gamma(x, t; \xi, \tau)$ be the fundamental solution in Ω_0 (for the extension of L) constructed in Chap. 1. The notation of (1.2.4)–(1.2.6) and of (1.2.8) with D and Ω replaced by E_0 and Ω_0, respectively, and some of the inequalities derived in Chap. 1 will be used in the sequel. We shall also need the additional inequalities:

$$|D_x Z(x, t; \xi, \tau) - D_{x'} Z(x', t; \xi, \tau)| \leq \frac{\text{const.}}{(t - \tau)^{\mu}} \frac{|x - x'|}{|x - \xi|^{n+2-2\mu}} \tag{2.17}$$

if $|x - \xi| > 2|x - x'|$, and

(2.18) $$|D_xZ(x, t; \xi, \tau) - D_xZ(x, t'; \xi, \tau)| \leq \frac{\text{const.}}{(t-\tau)^\mu} \frac{t'-t}{|x-\xi|^{n+3-2\mu}}$$

if $t' > t > \tau$, for any $0 < \mu < 1$. Here D_x is any partial derivative $\partial/\partial x_i$. The proof of (2.17) follows by using the mean value theorem and (1.3.18). The proof of (2.18) is similar.

Let E be any subdomain of B with $\overline{E} \subset B$, and let $\Omega = \overline{E} \times [0, \sigma]$ for some $\sigma \leq T$. We shall need the following lemma.

Lemma 2. *For any $0 < \delta < 1$ there exists a constant K depending only on E, δ, H_0, H_1 such that, for every continuous function $h(x, t)$ in Ω, the function*

(2.19) $$v(x, t) = \int_0^t \int_E \Gamma(x, t; \xi, \tau) h(\xi, \tau)\, d\xi\, d\tau$$

satisfies the inequality

(2.20) $$\overline{|v|}_{1+\delta}^{\Omega} \leq K\sigma^\gamma |h|_0^\Omega \qquad \left(\gamma = \frac{1-\delta}{2}\right).$$

The reader may note the similarity between this lemma and Theorem 6, Sec. 4, Chap. 5. The present lemma, however, gives a bound on the Hölder coefficients of $\partial v/\partial x_i$ which tends to zero with σ; this fact will be of fundamental importance in the applications of the lemma.

Proof. In the following proof, various constants that depend only on E, δ, H_0, H_1 will be denoted by K. By Sec. 5, Chap. 1 (see (1.5.1)–(1.5.4)),

(2.21) $$\begin{aligned} v(x, t) &= \int_0^t \int_E Z(x, t; \xi, \tau)\, h(\xi, \tau)\, d\xi\, d\tau \\ &\quad + \int_0^t \int_E Z(x, t; \xi, \tau) h_1(\xi, \tau)\, d\xi\, d\tau \\ &= v_1(\xi, \tau) + v_2(\xi, \tau), \end{aligned}$$

where

(2.22) $$h_1(x, t) = \int_0 \int_E \Phi(x, t; \xi, \tau) h(\xi, \tau)\, d\xi\, d\tau$$

Using (1.4.8) one obtains

(2.23) $$|h_1|_0 \leq K|h|_0.$$

To prove (2.20) it will be enough to prove that

(2.24) $$I_1 \equiv |D_xv_1(x, t) - D_{x'}v_1(x', t)| \leq K|x - x'|^\delta \sigma^\gamma |h|_0,$$

(2.25) $$I_2 \equiv |D_xv_1(x, t) - D_xv_1(x, t')| \leq K|t - t'|^{\delta/2} \sigma^\gamma |h|_0.$$

Indeed, using (2.21), (2.23) we then conclude that

(2.26) $$H_\delta[D_xv] \leq K\sigma^\gamma |h|_0.$$

Moreover, the method that will be used below to prove (2.24), (2.25) can easily be modified to yield similar bounds on D_xv_1, $H_\delta[v_1]$. Similar bounds are then obtained for $D_x v_2$, $H_\delta[v_2]$ (by using (2.23)) and, consequently, also for $D_x v$, $H_\delta[v]$. Combining these bounds with (2.26), the proof of (2.20) is completed.

To prove (2.24), recall that, by Theorem 3, Sec. 3, Chap. 1, D_x commutes with the integral of v_1. Letting E_1 be the set of points ξ of E satisfying $|\xi - x| > 2|x - x'|$ and setting $E_2 = E - E_1$ we then obtain, upon using (1.4.9), (2.17),

$$(2.27)\quad I_1 \le K|h|_0 \int_0^t \int_{E_1} \frac{1}{(t-\tau)^\mu} \frac{|x-x'|}{|x-\xi|^{n+2-2\mu}}\, d\xi\, d\tau$$
$$+ K|h|_0 \int_0^t \int_{E_2} \frac{1}{(t-\tau)^\mu} \left(\frac{1}{|x-\xi|^{n+1-2\mu}} + \frac{1}{|x'-\xi|^{n+1-2\mu}} \right) d\xi\, d\tau$$

provided $0 < \mu < 1$. Hence, if $2\mu > 1$,

$$I_1 \le K|h|_0\, \sigma^{1-\mu}(|x-x'|^{1-(2-2\mu)} + |x-x'|^{2\mu-1}) = 2K|h|_0 \sigma^{1-\mu}|x-x'|^{2\mu-1}.$$

Taking $2\mu - 1 = \delta$ we find

$$I_1 \le K|x-x'|^\delta |h|_0 \sigma^\gamma \qquad \text{if } \gamma = \frac{1-\delta}{2}.$$

It remains to prove (2.25). We may assume that $t' > t$. Let E_3 be the set of all points ξ in E satisfying $|\xi - x| > (t'-t)^{1/2}$, and set $E_4 = E - E_3$. Using (1.4.9), (2.18) we get, for any $0 < \lambda < 1$, $0 < \mu < 1$,

$$(2.28)\qquad I_2 \le K|h|_0 \int_0^t \int_{E_3} \frac{t'-t}{(t-\tau)^\mu} \frac{1}{|x-\xi|^{n+3-2\mu}}\, d\xi\, d\tau$$
$$+ K|h|_0 \int_t^{t'} \int_{E_3} \frac{1}{(t'-\tau)^\lambda} \frac{1}{|x-\xi|^{n+1-2\lambda}}\, d\xi\, d\tau$$
$$+ K|h|_0 \int_0^{t'} \int_{E_4} \frac{1}{(t'-\tau)^\mu} \frac{1}{|x-\xi|^{n+1-2\mu}}\, d\xi\, d\tau.$$

Hence, if $2\lambda < 1$, $2\mu > 1$,

$$I_2 \le K|h|_0\{(t'-t)\sigma^{1-\mu}(t'-t)^{(2\mu-3)/2} + (t'-t)^{1-\lambda}(t'-t)^{(2\lambda-1)/2}$$
$$+ \sigma^{1-\mu}(t'-t)^{(2\mu-1)/2}\}.$$

Taking $2\mu - 1 = \delta$ and noting that $(t'-t)^{1/2} \le (t'-t)^{\delta/2}\sigma^\gamma$ if $\gamma = (1-\delta)/2$, the inequality of (2.25) follows; the proof of Lemma 2 is thereby completed.

We conclude this section with the following remark.

Remark. Applying Lemma 2 to the second term on the right-hand side of (1.2.8) (with D replaced by E_0) it follows that if ξ varies on the boundary ∂E of E whereas x is restricted to a closed domain E_* lying in the interior of E, then Γ as a function of (x, t) (we write $\Gamma = \Gamma(\cdot, \cdot; \xi, \tau)$) satisfies

$$(2.29)\qquad \overline{|\Gamma(\cdot, \cdot; \xi, \tau)|}_{1+\delta}^{\Omega^*} \le K\sigma^\gamma \qquad \left(0 < \delta < 1, \gamma = \frac{1-\delta}{2}\right),$$

where $\Omega_* = E_* \times [0, \sigma]$, and K depends only on E_*, E, δ, H_0, H_1; here $\Gamma(x, t; \xi, \tau)$ is defined to be zero if $t < \tau$.

3. Completion of the Proof of Theorem 4

The remaining proof is divided into three parts.

3.1. Interior Estimates for Small t. We shall estimate the $(1 + \delta)$-norm of u in the interior of D_σ. Since f and the coefficients of L are Hölder continuous in $\bar{D}$, by Theorem 7, Sec. 3, Chap. 3, u belongs to $\bar{C}_{2+\alpha}(D)$. Let G, E be subdomains of B such that $G \subset \bar{G} \subset E \subset \bar{E} \subset B$, and let the boundary ∂E of E be of class $C^{1+\beta}$ for some $0 < \beta < 1$ (for definition of $C^{1+\beta}$, see Sec. 8, Chap. 3). If σ is sufficiently small then the cylinder $\Omega = \bar{E} \times [0, \sigma]$ is contained in $D + B$. Denote by $\overline{|u|}_{q,\sigma,A}$ the q-norm of u in the cylinder $A \times [0, \sigma]$, and by $\overline{|u|}_{q,\sigma}$ the q-norm of u in D_σ. We shall now estimate $\overline{|u|}_{1+\delta,\sigma,G}$.

From Theorem 2, Sec. 3, Chap. 5, and from the construction of the solution by (5.3.5), (5.3.6), (5.3.8) it follows that

$$u(x, t) = -\int_0^t \int_E \Gamma(x, t; \xi, \tau) f(\xi, \tau)\, d\xi\, d\tau + \int_0^t \int_{\partial E} \Gamma(x, t; \xi, \tau) k(\xi, \tau)\, d\Sigma\, d\tau, \tag{3.1}$$

where $d\Sigma$ is the surface element on ∂E and k is the solution of the integral equation ($y \in \partial E$, $0 < t \leq \sigma$)

$$k(y, t) = 2\int_0^t \int_{\partial E} \frac{\partial \Gamma(y, t; \xi, \tau)}{\partial \nu(y, t)} k(\xi, \tau)\, d\Sigma\, d\tau - 2\int_0^t \int_E \frac{\partial \Gamma(y, t; \xi, \tau)}{\partial \nu(y, t)} f(\xi, \tau)\, d\xi\, d\tau - 2\frac{\partial u(y, t)}{\partial \nu(y, t)}. \tag{3.2}$$

Expanding k into a series analogously to (5.3.10) we get

$$|k|_0 \leq K(|f|_{0,\sigma,E} + \Sigma\, |D_x u|_{0,\sigma,E}), \tag{3.3}$$

where the sum is taken over all the first-order x-derivatives. If we now apply Lemma 2 and the remark at the end of Sec. 2, we obtain the inequality

$$\overline{|u|}_{1+\delta,\sigma,G} \leq K\sigma^\gamma[|f|_{0,\sigma,E} + \Sigma\, |D_x u|_{0,\sigma,E}] \qquad \left(\gamma = \frac{1-\delta}{2}\right), \tag{3.4}$$

where K depends on G, E, δ, H_0, H_1

3.2. Boundary Estimates for Small t. There is a finite number of $(n + 1)$-dimensional open balls V_j ($j = 1, \ldots, j_0$) such that each portion $S_j = \bar{S} \cap V_j$ can be globally represented in the form (3.2.17) with h in $\bar{C}_{2+\alpha}$ and in $\bar{C}_{2-0}$. If we take G_0 to be a subdomain of B such that all the points of $B - G_0$ are sufficiently close to ∂B, then $\overline{D_\sigma - \Omega}$ is covered by the

V_j for all σ sufficiently small, where $\Omega = \overline{G}_0 \times [0, \sigma]$. We shall now estimate the $(1 + \delta)$-norm of u in subsets of $D_j = D_\sigma \cap V_j$.

Assume for simplicity that (3.2.17) holds, for S_j, with $i = n$. Perform in D_j the transformation

$$
\begin{aligned}
y_m &= x_m \qquad \text{for } m = 1, \ldots, n-1, \\
y_n &= x_n - h(x_1, \ldots, x_{n-1}, t)
\end{aligned}
\tag{3.5}
$$

which, as we may assume, maps D_j in a one-to-one manner onto a domain Δ lying in the half-space $y_n > 0$. Let Σ be the interior of the image of $S_\sigma \cap V_j$; it lies on $y_n = 0$. Putting

$$U(y, t) = u(x, t), \qquad F(y, t) = f(x, t), \qquad C(y, t) = c(x, t),$$

the equation $Lu = f$ becomes

$$\Lambda U \equiv \sum_{i,j=1}^{n} A_{ij}(y, t) \frac{\partial^2 U}{\partial y_i \, \partial y_j} + \sum_{i=1}^{n} B_i(y, t) \frac{\partial U}{\partial y_i} + C(y, t)U - \frac{\partial U}{\partial t} = F(y, t), \tag{3.6}$$

where

$$A_{ij} = \sum_{\lambda,\mu} a_{\lambda\mu} \frac{\partial y_i}{\partial x_\lambda} \frac{\partial y_j}{\partial x_\mu}, \qquad B_i = \sum_{\lambda,\mu} a_{\lambda\mu} \frac{\partial^2 y_i}{\partial x_\lambda \, \partial x_\mu} + \sum_{\lambda} b_\lambda \frac{\partial y_i}{\partial x_\lambda} - \frac{\partial y_i}{\partial t}. \tag{3.7}$$

By our assumptions it follows that A_{ij}, B_i, C are Hölder continuous (exponent α) in $\bar{\Delta}$ and that A_{ij} are of class C_{1-0} on Σ.

We wish to construct a "local" Green's function for Λ in Δ, that is, a fundamental solution of Λ in Δ which vanishes on Σ (compare (4.6.3)). We shall construct G in a form which generalizes (4.6.3). To accomplish this, we need the coefficients of $\partial^2/\partial y_i \, \partial y_n$ in Λ to vanish on Σ if $i < n$. Since this need not be the case, we perform another coordinate transformation such that the last property will hold for the transform of Λ. We try the transformation

$$
\begin{aligned}
z_i &= \varphi_i(\bar{y}, t) + y_n \psi_i(y, t) \qquad (i = 1, \ldots, n-1), \\
z_n &= y_n \psi_n(\bar{y}, t) \qquad \text{where } \bar{y} = (y_1, \ldots, y_{n-1}).
\end{aligned}
\tag{3.8}
$$

Denote by Δ', Σ' the images of Δ, Σ respectively. Putting

$$U'(z, t) = U(y, t), \qquad F'(z, t) = F(y, t), \qquad C'(z, t) = C(y, t) \tag{3.9}$$

we obtain for U' the equation

$$\Lambda' U' \equiv \sum_{i,j=1}^{n} A'_{ij}(z, t) \frac{\partial^2 U'}{\partial z_i \, \partial z_j} + \sum_{i=1}^{n} B'_i(z, t) \frac{\partial U'}{\partial z_i} + C'(z, t)U' - \frac{\partial U'}{\partial t} = F'(z, t) \tag{3.10}$$

where

$$A'_{ij} = \sum_{\lambda,\mu} A_{\lambda\mu} \frac{\partial z_i}{\partial y_\lambda} \frac{\partial z_j}{\partial y_\mu}, \qquad B'_i = \sum_{\lambda,\mu} A_{\lambda\mu} \frac{\partial^2 z_i}{\partial y_\lambda \, \partial y_\mu} + \sum_{\lambda} B_\lambda \frac{\partial z_i}{\partial y_\lambda} - \frac{\partial z_i}{\partial t}. \tag{3.11}$$

We shall now impose the conditions $A'_{in} = 0$ on Σ' (Σ' lies on $z_n = 0$), for $i = 1, \ldots, n-1$. Noting that $\partial z_n/\partial y_\mu = 0$ on Σ' if $\mu < n$, we obtain the equations

$$(3.12) \quad \sum_{\lambda=1}^{n-1} A_{\lambda n}(\bar{y}, 0, t) \frac{\partial \varphi_i(\bar{y}, t)}{\partial y_\lambda} + A_{nn}(\bar{y}, 0, t)\psi_i(\bar{y}, 0, t) = 0 \qquad (1 \le i \le n-1),$$

provided $\partial z_n/\partial y_n \neq 0$ on Σ'. These equations are satisfied if

$$(3.13) \quad \begin{aligned} &\psi_n = 1, \qquad \varphi_i = y_i \qquad (1 \le i \le n-1), \\ &\psi_i(\bar{y}, 0, t) = -\frac{A_{in}(\bar{y}, 0, t)}{A_{nn}(\bar{y}, 0, t)} \equiv \gamma_i(\bar{y}, t) \qquad (1 \le i \le n-1). \end{aligned}$$

Before extending ψ_i to $y_n > 0$ we need to introduce some notation. Let Σ'_0 be any closed n-rectangle of $\Sigma' + [\bar{\Sigma}' \cap \{t = 0\}]$ with one face lying on $t = 0$, let B'_0 be the open lower base of Δ', and let Δ'_0 be any open cylinder whose closure is contained in $\Delta' + B'_0 + \Sigma'_0$ having Σ'_0 as a part of its lateral boundary. Denote the base of Δ'_0 by E'_0.

There exists a three times continuously differentiable function $\rho(\xi, \tau)$ of $(\xi, \tau) = (\xi_1, \ldots, \xi_{n-1}, \tau)$ which vanishes outside a set $N = N_0 \times [0, \sigma]$ where N_0 is a neighborhood of the origin in the ξ-space, and which satisfies the conditions:

$$(3.14) \qquad |D_\xi^a D_\tau^b \rho(\xi, \tau)| \le \frac{\text{const.}}{\sigma^b} \qquad (0 \le a + b \le 3),$$

$$(3.15) \qquad \int_N \rho(\xi, \tau)\, d\xi\, d\tau = 1.$$

Furthermore, if N_0 is independent of σ, the constant in (3.14) is also independent of σ. Indeed, $\rho(\xi, \tau)$ can be constructed in the form $g_1(\xi)g_2(\tau)$ where g_1, g_2 are constructed by the method of Problem 1, Chap. 1.

Let $B_0, E_0, \Sigma_0, \Delta_0$ be the sets in the (y, t)-space whose images under (3.8) are $B'_0, E'_0, \Sigma'_0, \Delta'_0$ respectively. Choose N_0 so small that if $\xi \in N_0$, $(\bar{y}, t) \in \Sigma_0$ then $(\bar{y} + y_n\xi, t) \in \Sigma$ for all $(\bar{y}, y_n, t) \in \bar{\Delta}_0$. We now define $\psi_i(y, t)$ in $\bar{\Delta}_0$ by

$$(3.16) \qquad \psi_i(y, t) = \int_N \gamma_i(\bar{y} + y_n\xi, t + y_n\tau)\rho(\xi, \tau)\, d\xi\, d\tau.$$

Note that if $0 < t < \sigma$, $0 < t + y_n\tau \le K_0\sigma$ for some constant K_0 independent of σ. Since we could take $K_0\sigma$ instead of σ in all the previous considerations without changing anything else, we may assume that $\gamma_i(\bar{y}, t)$ is defined not only for $0 \le t \le \sigma$ but also for $0 \le t \le K_0\sigma$. Hence $\psi_i(y, t)$ is well defined in $\bar{\Delta}_0$.

From (3.15) it follows that $\psi_i(y, t)$ coincides with $\gamma_i(\bar{y}, t)$ for $y_n = 0$.

Substituting $\bar{y}_j + y_n\xi_j = \xi'_j$, $t + y_n\tau = \tau'$ in the integral of (3.16), we obtain

$$\psi_i(y, t) = \int \gamma_i(\xi', \tau')\rho\left(\frac{\xi' - \bar{y}}{y_n}, \frac{\tau' - t}{y_n}\right)\frac{d\xi'\,d\tau'}{y_n^n}, \tag{3.17}$$

from which we conclude that $\psi_i(y, t)$ is three times continuously differentiable for $y_n > 0$.

Using the fact that $\gamma_i(\xi, \tau)$ are Lipschitz continuous in (ξ, τ) it can further be shown that $D_y z$, $D_y^2 z$, $D_y D_t z$, $D_t z$ are bounded functions in Δ_0. Indeed, by (3.8), (3.13), it suffices to prove this statement for $y_n\psi_i(y, t)$. The boundedness of $D_y(y_n\psi_i)$ follows by using the formula (3.17). Next, if after differentiating with respect y (in (3.17)), we substitute back $\xi_j' = \bar{y}_j + y_n\xi_j$, $\tau' = t + y_n\tau$, and if $D_y = \partial/\partial y_k$, $k < n$, then we get

$$D_y(y_n\psi_i) = \int_N \gamma_i(\bar{y} + y_n\xi, t + y_n\tau)D_\xi\rho(\xi, \tau)\,d\xi\,d\tau. \tag{3.18}$$

Since γ_i is Lipschitz continuous, taking finite differences of $D_y(y_n\psi_i)$ with respect to some y_h or t and using (3.14) with $a = 1$, $b = 0$, we find that $D_y^2(y_n\psi_i)$ and $D_t D_y(y_n\psi_i)$ are bounded independently of σ. The case where the first D_y is $\partial/\partial y_n$ can be treated in the same manner. Finally, the boundedness of $D_t(y_n\psi_i)$ follows by taking finite differences with respect to t in (3.16).

The transformation (3.8) (with (3.13), (3.16)) is one-to-one and its Jacobian is different from zero if y_n is sufficiently small, independently of σ; we may clearly assume this to be the case.

Since the transformation (3.8) changes distances only by a factor bounded independently of σ, by the differentiability and boundedness properties proved above for the z_i (as functions of (y, t)) it follows that A'_{ij}, C' are Hölder continuous (exponent α) in $\overline{\Delta_0'}$ (their α-norm being independent of σ) and B'_i are Hölder continuous (exponent α) in closed subsets of $\overline{\Delta_0'}$ which are bounded away from $z_n = 0$ and, also B'_i are bounded in $\overline{\Delta_0'}$ independently of σ.

We can now proceed to construct a "local" Green's function in the variables (z, t). Denote by Δ_*' the reflection of Δ_0' with respect to the hyperplane $z_n = 0$ and put $\Delta_0'' = \Delta_0' \cup \Sigma_0' \cup \Delta_*'$. Without loss of generality we may assume that Δ_0'' is a cylinder with base E'' such that $\partial E''$ is of class $C^{1+\beta}$. Denoting by z^* the point $(z_1, \ldots, z_{n-1}, -z_n)$ which is the reflection of z with respect to the hyperplane $z_n = 0$, we extend the coefficients of Λ' into $\overline{\Delta_*'} - \Sigma_0'$ as follows:

$$A'_{ij}(z^*, t) = A'_{ij}(z, t) \qquad \text{if } 1 \le i \le n-1,\ 1 \le j \le n-1 \quad \text{or if } i = j = n,$$

$$A'_{in}(z^*, t) = A'_{ni}(z^*, t) = -A'_{in}(z, t) \qquad \text{if } 1 \le i < n,$$

$$B'_i(z^*, t) = B'_i(z, t) \qquad \text{if } 1 \le i \le n-1,\ B'_n(z^*, t) = -B_n(z, t),$$

$$C'(z^*, t) = C'(z, t).$$

Note that with the exception of B_n' all the extended coefficients are Hölder continuous (exponent α) in $\overline{\Delta_0''}$, and B_n' is bounded in $\overline{\Delta_0''}$ and Hölder continuous (exponent α) in closed subsets of $\overline{\Delta_0''}$ which are bounded away from $z_n = 0$.

We can now proceed by the method of Chap. 1 to construct a fundamental solution $\Gamma'(z, t; \xi, \tau)$ for the equation $\Lambda' u = 0$ in $\overline{\Delta_0''}$ (compare Problem 7, Chap. 1). We find that $\Gamma'(z, t; \xi, \tau)$ is a continuous function for $(z, t) \in \overline{\Delta_0''}$, $(\xi, \tau) \in \overline{\Delta_0''}$ provided $t > \tau$, and for every fixed $(\xi, \tau) \in \overline{\Delta_0''}$, Γ' satisfies the equation $\Lambda'\Gamma' = 0$ for $(z, t) \in \Delta_0' \cup \Delta_*'$ (but not necessarily for $(z, t) \in \Sigma_0'$, since B_n' is discontinuous on $z_n = 0$).

From the manner by which we have extended the coefficients of Λ' to Δ_*' it follows that $\Gamma'(z^*, t; \xi, \tau)$ is also a solution of $\Lambda'\Gamma' = 0$ as a function of (z, t) in $\Delta_0' \cup \Delta_*'$, for any fixed (ξ, τ) in $\overline{\Delta_0''}$. Hence the function

$$G'(z, t; \xi, \tau) = \Gamma'(z, t; \xi, \tau) - \Gamma'(z^*, t; \xi, \tau) \tag{3.19}$$

is a fundamental solution of $\Lambda' u = 0$ in $\overline{\Delta_0'}$. Also, $G'(z, t; \xi, \tau) = 0$ when $z_n = 0$. Thus, G' is the desired "local" Green's function for $\Lambda' u = 0$ in $\overline{\Delta_0'}$.

It is now easy to see, by slightly modifying the proof of Lemma 2, that Lemma 2 remains true if we replace Γ by G' and E by E_0' (the base of Δ_0').

We next proceed analogously to the proof of the interior estimates. First we wish to represent U' in the form

$$U'(z, t) = -\int_0^t \int_{E_0'} G'(z, t; \xi, \tau) F'(\xi, \tau)\, d\xi\, d\tau + \int_0^t \int_{S'} G'(z, t; \xi, \tau) k'(\xi, \tau)\, dS'\, d\tau$$

where S' is the part of the boundary of E_0' which does not lie on $z_n = 0$. Since the jump relation for G' is the same as that for Γ' we get for k' an integral equation analogous to (3.2).

Note next that the inequality (5.2.12) holds not only for $\Gamma'(z, t; \xi, \tau)$ $(z \in S', \xi \in S')$ but also for $\Gamma'(z^*, t; \xi, \tau)$ $(z \in S', \xi \in S')$. Indeed the proof for $\Gamma'(z^*, t; \xi, \tau)$ is the same as for $\Gamma'(z, t; \xi, \tau)$ (see (5.2.13)–(5.2.18)) if we assume, as we may, that all the normals to S' near $z_n = 0$ are orthogonal to the z_n axis (N_z and N_{z^*} are then in the same direction for $z \in S'$, z_n sufficiently small). The above representation for $U'(z, t)$ now follows by a theorem similar to Theorem 2, Sec. 3, Chap. 5, whose proof is also similar to that of Theorem 2, Chap. 5.

Using (5.2.12) for G', we derive for k' an inequality analogous to (3.3). We then get, if A_0' is any subset of $\overline{E_0'}$ which is bounded away from S',

$$\overline{|U'|}_{1+\delta,\sigma,A_0'} \leq K\sigma^{\gamma}(|F'|_{0,\sigma,E_0'} + \Sigma\, |D_z U'|_{0,\sigma,E_0'}). \tag{3.20}$$

Reversing the transformations (3.8), (3.5) and denoting by R_j the set in the (x, t)-space whose image in the (z, t)-space is the cylinder $A_0' \times [0, \sigma]$, we get (using the uniform boundedness of z_i, $D_y z_i$, $D_t z_i$, $D_y^2 z_i$, $D_y D_t z_i$ with respect to σ),

$$\overline{|u|}_{1+\delta}^{R_j} \leq K\sigma^\gamma(|f|_{0,\sigma} + \Sigma\, |D_x u|_{0,\sigma}). \tag{3.21}$$

It is clear that, if σ is sufficiently small, we can choose the Δ_0', A_0' in such a manner that the R_j $(j = 1, \ldots, j_0)$ will cover $D_\sigma - \Omega$ where Ω was defined at the beginning of 3.2. Taking G in (3.4) to be a subdomain of G_0 such that $[G \times (0, \sigma)] \cup \left(\bigcup_{j=1}^{j_0} R_j\right)$ covers D_σ and combining (3.4) (3.21), we get

$$\overline{|u|}_{1+\delta,\sigma} \leq K\sigma^\gamma(|f|_{0,\sigma} + \Sigma\, |D_x u|_{0,\sigma}). \tag{3.22}$$

Taking σ such that $2K\sigma^\gamma \leq 1$, we obtain the inequality

$$\overline{|u|}_{1+\delta,\sigma} \leq K\sigma^\gamma |f|_{0,\sigma}, \tag{3.23}$$

where K depends only on D, δ, H_0, H_1, H_2.

3.3. Completion of the proof. Having estimated in (3.23) the $(1 + \delta)$-norm of u in D_σ, we now proceed step by step to estimate the $(1 + \delta)$-norm of u in D. We first note that from the proof of (3.23) it follows that for any $0 \leq \rho \leq T - \sigma$ and for any solution u of $Lu = f$ in $D_{\rho+\sigma} - D_\rho$, which vanishes on $B_\rho = D \cap \{t = \rho\}$ and on $S_{\rho+\sigma} - S_\rho$, the inequality

$$\overline{|u|}_{1+\delta} \leq K\sigma^\gamma |f|_0 \qquad \text{in } D_{\rho+\sigma} - D_\rho \tag{3.24}$$

is valid provided σ is sufficiently small depending on D, δ, H_0, H_1, H_2, but not on ρ. From now on σ is a fixed positive number such that (3.24) holds for all $0 \leq \rho \leq T - \sigma$.

We introduce the domains

$$D^i = D \cap \left\{i\frac{\sigma}{2} < t < (i + 2)\frac{\sigma}{2}\right\}, \tag{3.25}$$

and denote by $\widehat{|v|}_{q,i}$ the q-norm of v in D^i.

Let $\zeta(t)$ be a twice continuously differentiable function of t which vanishes for $t \leq \frac{1}{2}$ and which equals 1 for $t \geq 1$. Consider the function

$$u^*(x, t) = \zeta\left(\frac{t}{\sigma}\right) u(x, t). \tag{3.26}$$

It vanishes on the lower base of D^1 and is equal to $u(x, t)$ for $t > \sigma$. Using (3.24) we get

$$\widehat{|u^*|}_{1+\delta,1} \leq K\sigma^\gamma \widehat{|Lu^*|}_{0,1}. \tag{3.27}$$

We proceed to estimate Lu^*. We have

(3.28) $$|Lu^*| \leq |f\zeta| + \left|u\,\frac{d\zeta}{dt}\right| \leq N|f| + \frac{N}{\sigma}\,|u|,$$

where N is a constant. To estimate $|u|$ we introduce the function $v = |f|_0^D t$. It satisfies

$$Lv = -|f|_0^D \quad \text{in } D, \qquad v \geq 0 \quad \text{on } B + S.$$

Applying the maximum principle to $v \pm u$ we get

(3.29) $$\underset{D_\rho}{\text{l.u.b.}}\ |u| \leq \rho|f|_0^D \qquad \text{for any } 0 < \rho \leq T.$$

Substituting (3.29) into (3.28) we get

$$\widehat{|Lu^*|}_{0,1} \leq (N + \tfrac{3}{2}N)|f|_0^D.$$

Substituting this inequality into (3.27) and recalling that $u^*(x, t) = u(x, t)$ if $\sigma < t < 3\sigma/2$ we get, in conjunction with (3.23),

(3.30) $$\overline{|u|}_{1+\delta,3\sigma/2} \leq [K\sigma^\gamma + K\sigma^\gamma(N + \tfrac{3}{2}N)]|f|_0^D \equiv K_1|f|_0^D.$$

The next step is to estimate the $(1 + \delta)$-norm of u in $D_{2\sigma}$. We use the previous method with D^1 replaced by D^2 and with $\zeta(t/\sigma)$ replaced by $\zeta((t + \frac{1}{2})/\sigma)$. We then obtain

(3.31) $$|u|_{1+\delta,2\sigma} \leq K_2|f|_0^D \qquad (K_2 \text{ constant}).$$

Note that the constants K_i depend on σ, but since σ is fixed and depends only on D, δ, and the H_j, the same is true of the K_i.

Proceeding in the above manner step by step, the proof of Theorem 4 is completed.

4. Existence Theorems for $Lu = f(x, t, u, \nabla u)$

In this section we consider the first initial-boundary value problem

(4.1) $$Lu = f(x, t, u, \nabla_x u) \qquad \text{in } D + B_T,$$

(4.2) $$u = \psi \qquad \text{on } \bar{B} + S.$$

We shall always assume that $f(x, t, u, w)$, where $w = (u_1, \ldots, u_n)$, is defined for

(4.3) $$(x, t) \in \overline{D}, \quad -\infty < u < \infty, \quad -\infty < u_i < \infty \quad (i = 1, \ldots, n).$$

We first prove some uniqueness theorems which require less differentiability assumptions on $f(x, t, u, w)$ than in Theorem 8, Sec. 3, Chap. 2.

Theorem 5. *Let L be parabolic with continuous bounded coefficients in $D + B_T$, and let $f(x, t, u, w)$ be monotone nondecreasing in u. Then there exists at most one solution of* (4.1), (4.2).

Proof. Consider first the special case where $f(x, t, u, w)$ is strictly monotone increasing in u and $c(x, t) \leq 0$, and suppose that u_1, u_2 are two solutions of (4.1), (4.2). If $u_1 \not\equiv u_2$ then we may suppose that $u_1 > u_2$ at some points of D. Hence the function $u = u_1 - u_2$ takes its positive maximum in $D + B_T$. Denoting by (x^0, t^0) a point where the maximum is obtained and using the facts that $\nabla_x u_1 = \nabla_x u_2$, $u_1 > u_2$ at (x^0, t^0), we get

$$Lu(x^0, t^0) = f(x^0, t^0, u_1(x^0, t^0), \nabla_x u_1(x^0, t^0)) \\ - f(x^0, t^0, u_2(x^0, t^0), \nabla_x u_1(x^0, t^0)) > 0.$$

Since on the other hand, by the proof of Lemma 1, Sec. 1, Chap. 2, $Lu(x^0, t^0) \leq 0$ at every point $(x^0, t^0) \in D + B_T$ where u takes a positive maximum, we have derived a contradiction.

To prove the theorem in the general case, we perform a transformation $v = e^{-\lambda t}u$ which carries $Lu = f$ into $(L - c)v = \tilde{f}$, where $\tilde{f} = fe^{-\lambda t} + (\lambda - c)v$. Taking $\lambda > c$, $\tilde{f}(x, t, v, w)$ becomes a strictly monotone function in v and the proof of the theorem then follows from the previous special case.

Theorem 6. *Let L be a parabolic operator with continuous bounded coefficients in $D + B_T$ and assume that $f(x, t, u, w)$ is Lipschitz continuous in u, uniformly with respect to (x, t, u, w) in bounded sets of* (4.3). *Then there exists at most one solution of* (4.1), (4.2).

Proof. Consider first the special case where $f(x, t, u, w)$ is Lipschitz continuous in u, uniformly in the set (4.3), i.e.,

$$|f(x, t, u, w) - f(x, t, \overline{u}, w)| \leq M|u - \overline{u}|. \tag{4.4}$$

The transformation $v = ue^{-\lambda t}$ takes (4.1) into $Lv = \tilde{f}$ where

$$\tilde{f}(x, t, v, \nabla_x v) = f(x, t, u, \nabla_x u)e^{-\lambda t} + \lambda v.$$

Taking $\lambda > M$, $\tilde{f}(x, t, v, w)$ becomes a monotone increasing function in v, and Theorem 5 can therefore be applied.

If (4.4) is not satisfied and if u_1, u_2 are two distinct solutions of (4.1), (4.2), then let A be any positive number satisfying

$$\operatorname*{l.u.b.}_{D} |u_i| < A, \qquad \operatorname*{l.u.b.}_{D} |\nabla_x u_i| < A \qquad \text{for } i = 1, 2,$$

where $|\nabla_x u| = [\Sigma\, (\partial u/\partial x_i)^2]^{1/2}$. Modify the definition of $f(x, t, u, w)$ outside the bounded set $\{(x, t, u, w);\ (x, t) \in \overline{D},\ |u| < A,\ |w| < A\}$ in such a manner that the modified function satisfies (4.4). The proof of the theorem then follows from the previous special case.

The following remarks will be useful later on.

Remark 1. Suppose we prove that any solution u of (4.1), (4.2) must be bounded by some constant K. Change then the definition of $f(x, t, u, w)$ for $|u| > K$, for instance by defining

$$\tilde{f}(x, t, u, w) = \begin{cases} f(x, t, u, w) & \text{for } |u| \leq K, \\ f(x, t, \pm K, w) & \text{for } \pm u > K. \end{cases} \tag{4.5}$$

If we then prove the existence of a solution u for the problem

$$Lu = \tilde{f} \quad \text{in } D + B_T, \qquad u = \psi \quad \text{on } \overline{B} + S, \tag{4.6}$$

then u is also a solution of the original problem (4.1), (4.2). The same is true of uniqueness. Note that if f is Hölder continuous in some of its variables, the same is true of the function $\tilde{f}$ defined in (4.5).

Remark 2. The following statement is a special case of Theorem 16, Sec. 6, Chap. 2.

Let f_1, f_2, φ_1, φ_2 be functions satisfying the inequalities

$$\begin{aligned} L\varphi_1 &< f_1(x, t, \varphi_1, \nabla_x\varphi_1) \text{ in } D, \\ Lu &\geq f_1(x, t, u, \nabla_x u) \text{ in } D, \\ \varphi_1 &> u \text{ on } \overline{B} + S, \end{aligned} \tag{4.7}$$

$$\begin{aligned} L\varphi_2 &> f_2(x, t, \varphi_2, \nabla_x\varphi_2) \text{ in } D, \\ Lu &\leq f_2(x, t, u, \nabla_x u) \text{ in } D, \\ \varphi_2 &< u \text{ on } \overline{B} + S. \end{aligned} \tag{4.8}$$

Then,

$$\varphi_2(x, t) < u(x, t) < \varphi_1(x, t) \qquad \text{in } D. \tag{4.9}$$

Using this remark we shall prove the following theorem.

Theorem 7. *Let $L \equiv \sum a_{ij}\partial^2/\partial x_i\, \partial x_j + \sum b_i\partial/\partial x_i - \partial/\partial t$ be a parabolic operator with continuous coefficients in D, and let $f(x, t, u, w)$ be a continuous function satisfying*

$$uf(x, t, u, 0) \leq A_1u^2 + A_2 \qquad (A_1 \geq 0, A_2 \geq 0) \tag{4.10}$$

for all $(x, t) \in D$, $-\infty < u < \infty$. Then, for any solution $u(x, t)$ of (4.1), (4.2),

$$|u(x, t)| \leq \left[\left(\frac{A_2}{b - A_1}\right)^{1/2} + \operatorname*{l.u.b.}_{B+S} |\psi|\right] e^{bt} \qquad \text{in } D, \tag{4.11}$$

for any $b > A_1$.

Proof. Denote by A the expression in brackets on the right-hand side of (4.11). Using (4.10) one can easily verify that, for any $\epsilon > 0$, the function $v = (A + \epsilon)e^{bt}$ satisfies

$$\begin{aligned} Lv &< f(x, t, v, \nabla_x v) && \text{in } D, \\ v &> \psi && \text{on } \overline{B} + S. \end{aligned}$$

Hence, by Remark 2, $u(x, t) < v(x, t)$. Similarly we get $u(x, t) > -v(x, t)$. Taking $\epsilon \to 0$ we obtain the inequality (4.11).

We shall now prove existence theorems for (4.1), (4.2). The following assumptions will be needed:

(B)′ a_{ij}, b_i, c are Hölder continuous (exponent α) in $\overline{D}$ and a_{ij} belong to C_{1-0} on S, i.e., (2.9) holds and, in addition,

$$\Sigma \, \overline{|a_{ij}|}_\alpha^D + \Sigma \, \overline{|b_i|}_\alpha^D + \overline{|c|}_\alpha^D \leq H_3. \tag{4.12}$$

(C) There exists a positive constant M_0 such that, for any $M \geq M_0$,

$$2K|f(x, t, u, \nabla_x u)| \leq M \qquad \text{in } D \tag{4.13}$$

for all functions $u = u(x, t)$ satisfying $\overline{|u|}_{1+\alpha}^D \leq M$; K is the constant appearing in (2.10) when $\delta = \alpha$.

Theorem 8. *Assume that S is of class $\overline{C}_{2-0}$ and of class $\overline{C}_{2+\alpha}$, that L satisfies the assumptions* (A), (B)′, *that $f(x, t, u, w)$ is Hölder continuous in bounded subsets of* (4.3), *and that* (C) *holds. If $\psi \in \overline{C}_{2+\delta}$ for some $\alpha < \delta < 1$, and if $L\psi(x, t) = f(x, t, \psi, \nabla_x\psi)$ on ∂B, then there exists a solution u of* (4.1), (4.2). *Furthermore, u belongs to $\overline{C}_{1+\delta}(D)$ and to $\overline{C}_{2+\gamma}$ for some $0 < \gamma < 1$.*

Proof. Consider the set C_M of functions $v(x, t)$ satisfying

$$\overline{|v|}_{1+\alpha}^D \leq M, \qquad v = \psi \qquad \text{on } \overline{B} + S,$$

and define a transformation $w = Zv$ on C_M as follows: w is the solution of

$$\begin{aligned} Lw &= f(x, t, v, \nabla_x v) && \text{in } D + B_T, \\ w &= \psi && \text{on } \overline{B} + S. \end{aligned} \tag{4.14}$$

Since $f(x, t, v(x, t), \nabla_x v(x, t)) \equiv F(x, t)$ is Hölder continuous with some positive exponent, and since $L\psi = F$ on ∂B, a unique solution w exists by Theorem 7, Sec. 3, Chap. 3. We shall prove that Z has a fixed point by employing Schauder's fixed point theorem (Theorem 2, Sec. 1).

First we prove that Z maps C_M into itself. The function $w_0 = w - \Psi$, where Ψ is some extension of ψ to D which belongs to $\overline{C}_{2+\alpha}(D)$, satisfies

$$\begin{aligned} Lw_0 &= f(x, t, v, \nabla_x v) - L\Psi && \text{in } D + B_T, \\ w_0 &= 0 && \text{on } \overline{B} + S. \end{aligned}$$

By Theorem 4,

$$\overline{|w_0|}_{1+\alpha} \leq K|f(x, t, v, \nabla_x v)|_0 + K|L\Psi|_0.$$

Taking $M > M_0$, $M > 4K|L\Psi|_0$ and applying the property (C), we conclude that

$$\overline{|w_0|}_{1+\alpha} \leq \tfrac{3}{4}M.$$

Hence, if $M > 4\overline{|\Psi|}_{1+\alpha}$,

$$\overline{|w|}_{1+\alpha} \leq M, \tag{4.15}$$

i.e., Z maps C_M into itself.

Similarly we obtain, for any $v \in C_M$,

$$\overline{|w|}_{1+\delta} \leq M' \qquad (M' \text{ depends on } M). \tag{4.16}$$

Now, from the proof of Theorem 1, Sec. 1 one sees that the set defined by (4.15), (4.16) is a compact subset of the set C_M (C_M is normed by $|\cdot|_{1+\alpha}^D$). Hence Z maps C_M into a compact subset of C_M.

To prove that Z is a continuous mapping we note that if $\overline{|u_m - u|}_{1+\alpha} \to 0$ then

$$\epsilon_m \equiv \underset{(x,t) \in D}{\text{l.u.b.}} \ |f(x, t, u, \nabla_x u) - f(x, t, u_m, \nabla_x u_m)| \to 0.$$

Since, by Theorem 4,

$$\overline{|Zu_m - Zu|}_{1+\alpha} \leq \text{const. } \epsilon_m \to 0,$$

$Zu_m \to Zu$ in the $(1 + \alpha)$-norms, i.e., Z is continuous.

Since, finally, C_M is a closed convex set of the Banach space $\overline{C}_{1+\alpha}(D)$, we can apply Theorem 2 and thus conclude that Z has a fixed point u. u is then a solution of (4.1), (4.2) and it belongs to $\overline{C}_{1+\delta}$. By Theorem 7, Sec. 3, Chap. 3, it follows that u also belongs to $\overline{C}_{2+\gamma}$, for some $0 < \gamma < 1$.

Using Theorem 7 we shall now derive existence theorems under more explicit conditions on f than the condition (C). First we consider the case where

$$|f(x, t, u, w)| \leq A(|u|) + \mu|w| \tag{4.17}$$

where $w = (u_1, \ldots, u_n)$, $|w| = (\Sigma\, v_i^2)^{1/2}$, μ is a positive constant, and $A(|u|)$ is any positive monotone increasing function of $|u|$.

Theorem 9. *Let L, S, ψ be as in Theorem 8 and assume that $f(x, t, u, w)$ is Hölder continuous in bounded subsets of (4.3), that (4.10) holds for some constants A_1, A_2 and that (4.17) holds for some positive monotone increasing function $A(|u|)$ and for some sufficiently small μ (depending only on D, L). If $L\psi = f(x, 0, \psi, \nabla_x\psi)$ on ∂B, then there exists a solution of (4.1), (4.2).*

Proof. From Theorem 7 it follows that $|u| \leq \lambda$ for some constant λ and for any possible solution of (4.1), (4.2). By Remark 1 and (4.17) we then conclude that, without loss of generality, we may assume that

$$|f(x, t, u, w)| \leq A' + \mu|w| \qquad (A' = A(\lambda)).$$

Taking μ such that $4n\mu K < 1$ it follows that if $|u| \leq M$, $\Sigma\, |D_x u| \leq M$ then

$$2K|f(x, t, u, \nabla_x u)| \leq M$$

provided $M > 4KA'$. The condition (C) is thus satisfied, and Theorem 9 then follows from Theorem 8.

Remark (a). Since the K in (4.13) depends only on δ, H_0, H_1, H_2 and D, the same is true of μ.

Remark (b). If $f(x, t, u, w)$ satisfies the inequality

$$|f(x, t, u, w)| \leq B_1 + B_1|u| + B_1|w|^\lambda + \tfrac{1}{2}\mu|w| \qquad (0 \leq \lambda < 1) \tag{4.18}$$

where B_1 is a constant, then f satisfies both (4.10) and (4.17), so that Theorem 9 can be applied.

Remark (c). The reader may verify that Theorem 9 remains true if (4.17) is replaced by the inequality

$$|f(x, t, u, w)| \leq A(|u|) + \tfrac{1}{2}\mu|w| + \epsilon p(|w|), \tag{4.19}$$

where $p(|w|)$ is any positive monotone increasing function and ϵ is sufficiently small.

Consider finally the case where $f(x, t, u, w)$ is not subjected to any growth conditions. If we take M sufficiently large (depending on Ψ, L, and D) and then apply the proof of Theorem 8 in a domain D_σ for some sufficiently small σ, making use of (3.23) instead of (2.10), we can establish the existence of a fixed point for Z. Hence,

Theorem 10. *Let L, S, ψ be as in Theorem 8 and assume that $f(x, t, u, w)$ is Hölder continuous in bounded subsets of* (4.3). *If $L\psi = f(x, 0, \psi, \nabla_x\psi)$ on ∂B, then there exists a solution of* (4.1), (4.2) *in D_σ, for some sufficiently small σ.*

Some of the results of the present section can be extended to solutions of the second initial-boundary value problem; see Problem 3.

5. Linear Equations with Nonlinear Boundary Conditions

In this section we consider the problem

$$Lu(x, t) = f(x, t) \quad \text{in } D, \tag{5.1}$$

$$u(x, 0) = \psi(x) \quad \text{on } B, \tag{5.2}$$

$$\frac{\partial u(x, t)}{\partial \nu(x, t)} = g(x, t, u(x, t), \varphi(t)) \quad \text{on } S, \tag{5.3}$$

where L is defined in (2.4), ν is the inward conormal derivative (for definition see Sec. 5, Chap. 2; also Sec. 2, Chap. 5), D is the infinite cylinder $B \times (0, \infty)$ with lateral boundary S, and $f(x, t)$, $g(x, t, u, v)$, $\varphi(t)$ are given functions.

We shall need the following assumptions:

(A) L is parabolic in $\bar{D}$.

(A)′ L is parabolic in $\bar{D}$ and $a_{11}(x, t) \geq K_1$, $b_1(x, t) \geq -K_2$ for some positive constants K_1, K_2.

(B) The coefficients a_{ij}, b_i, c of L are continuous in $\bar{D}$ and $c \leq 0$.

(B)′ The coefficients a_{ij}, b_i, c of L are Hölder continuous in $\bar{D}$ and $c \leq 0$.

Before stating conditions on g, consider the physical interpretation of the condition (5.3). If u is the temperature in D and $\varphi(t)$ is the temperature outside D then Newton's law of cooling states that $g(x, t, u, \varphi(t))$ is strictly monotone increasing in u, strictly monotone decreasing in $\varphi(t)$, and $g(x, t, u, \varphi(t)) = 0$ if $u = \varphi(t)$. We are thus led to assume:

(G_1) $g(x, t, u, v)$ is a continuous function for $(x, t) \in \bar{S}$, $-\infty < u < \infty$, $-\infty < v < \infty$.

(G_2) $g(x, t, u, v)$ is strictly increasing in u and strictly decreasing in v.

(G_3) $g(x, t, u, v) \to \pm\infty$ as $u \to \pm\infty$, uniformly with respect to (x, t) in $\bar{S}$ and v in bounded sets.

(G_4) $g(x, t, u, u) = 0$ for all $(x, t) \in \bar{S}$, $-\infty < u < \infty$.

(G_5) For any $K > 0$ there exists a positive constant η such that for all $(x, t) \in \bar{S}$, $|v| \leq 1 + \text{l.u.b.}\ |\varphi(t)|$, $|u'| \leq K$, $|u''| \leq K$, $u' < u''$,

$$g(x, t, u'', v) - g(x, t, u', v) > \eta(u'' - u'). \tag{5.4}$$

Note that (G_5) is satisfied if $\partial g/\partial u$ exists and is positive.

We shall prove in this section existence and uniqueness theorems. We shall also prove that if $f \to 0$, $\varphi \to A$ as $t \to \infty$, then $u \to A$ as $t \to \infty$. This statement is motivated from the physical interpretation of the condition (5.3).

Theorem 11. *Assume that the boundary ∂B of B is of class $C^{1+\beta}$ $(0 < \beta < 1)$, that L satisfies* (A), (B), *and that g satisfies* (G_2). *Then, for any $0 < \tau < \infty$, there exists at most one solution of* (5.1)–(5.3) *in D_τ.*

Proof. Suppose that u, v are two distinct solutions of (5.1)–(5.3) in D_τ. We may then assume that $u > v$ at some points of D. The function $w = u - v$ satisfies the equation $Lw = 0$ in D_τ. By the maximum principle, its positive maximum is attained at some points of $\bar{B} + S_\tau$. Since $w = 0$ on $\bar{B}$, w attains its positive maximum at some point $P = (x^0, t^0)$ of S_τ. Hence, $\partial w/\partial \nu \leq 0$ at P. Since, by (G_2),

$$\frac{\partial w}{\partial \nu} = g(x, t, u, \varphi(t)) - g(x, t, v, \varphi(t)) > 0 \qquad \text{at } P,$$

we have derived a contradiction.

Theorem 12. *Assume that ∂B belongs to $C^{1+\beta}$ $(0 < \beta < 1)$, that L satisfies* (A)$'$, (B), *and that g satisfies* (G_2), (G_3). *If u is a solution of* (5.1)–(5.3) *in some domain D_τ $(0 < \tau < \infty)$, then*

$$|u(x, t)| \leq H \qquad \text{in } D_\tau, \tag{5.5}$$

where H is a constant depending only on g, K_1, K_2, the diameter of B, and on upper bounds on $|f|$, $|\psi|$, $|\varphi|$.

Proof. The function $\zeta(x) = e^{\lambda R} - e^{\lambda x_1}$ satisfies

$$L\zeta < -1 \qquad \text{in } D_\tau,$$

$$\zeta > 1 \qquad \text{on } \bar{B},$$

provided λ, R are sufficiently large, depending on K_1, K_2, and the diameter of B (see Sec. 2, Chap. 6). Consider the function $w = H_0\zeta - u$ for any $H_0 \geq \text{l.u.b.}_{D_T} |f| + \text{l.u.b.}_B |\psi|$. It satisfies:

$$Lw < 0 \quad \text{in } D_\tau, \qquad w > 0 \quad \text{on } \overline{B}, \tag{5.6}$$

$$\frac{\partial w}{\partial \nu} = \tilde{g}(x, t, w) \qquad \text{on } S_\tau, \tag{5.7}$$

where $\tilde{g}(x, t, w) = H_0 \dfrac{\partial \zeta}{\partial \nu} - g(x, t, H_0\zeta - w, \varphi(t))$. By (G_3), $\tilde{g}(x, t, w) < 0$ on S_τ if $-w$ is sufficiently large (depending only on H_0 l.u.b. $|\zeta|$, an upper bound on $|\varphi|$, and on g), say $-w \geq H_1$.

Suppose now that w takes negative values in D_τ. Then, because of (5.6), its negative minimum, say m, must be attained at some point $P \in S_\tau$ and $\partial w/\partial \nu \geq 0$ at P. If $-m > H_1$ then $\tilde{g}(x, t, m) < 0$ and we obtain a contradiction to (5.7). Hence $-m \leq H_1$, i.e., $H_0\zeta - u \geq m \geq -H_1$, i.e., $u \leq H_0\zeta + H_1$. By considering $H_0\zeta + u$ we also obtain a lower bound on u, and the proof is completed.

Applying Theorem 12 to D_τ with τ's increasing to ∞ we get:

Corollary. *Let ∂B belong to $C^{1+\beta}$ $(0 < \beta < 1)$, let L satisfy* (A)′, (B) *and let g satisfy* (G_2), (G_3). *If u is a solution of* (5.1)–(5.3) *in D, and if* $\text{l.u.b.}_D |f| < \infty$, $\text{l.u.b.}_{0 \leq t < \infty} |\varphi(t)| < \infty$, *then*

$$|u(x, t)| \leq H \qquad \text{in } D \tag{5.8}$$

where H depends on the same quantities as in Theorem 12.

We shall now prove an existence theorem.

Theorem 13. *Assume that ∂B is of class $C^{1+\beta}$ $(0 < \beta < 1)$, that L satisfies* (A)′, (B)′, *that g satisfies* (G_1), (G_2), (G_3), *that $f(x, t)$ is Hölder continuous in x, uniformly in bounded subsets of $\overline{D}$, that $\varphi(t)$ is a continuous function for $0 \leq t < \infty$, and, finally, that $\psi(x)$ is a continuous function with compact support in B. Then there exists a unique solution of* (5.1)–(5.3).

Proof. We first prove existence in D_τ, for any $0 < \tau < \infty$. Let Z be the Banach space of all functions $v(x, t)$ which are continuous in $\overline{D}_\tau$, with the norm

$$||v|| = \text{l.u.b.}_{D_T} |v(x, t)|.$$

For any $R > 0$, denote by Z_R the set $\{v; v \in Z, ||v|| \leq R\}$. For every $v \in Z_R$ define $w = Tv$ as the solution of (5.1), (5.2) and

$$\frac{\partial w}{\partial \nu} = g(x, t, v(x, t), \varphi(t)).$$

By Theorem 2, Sec. 3, Chap. 5, w exists and is uniquely determined and, furthermore, it has the form

$$w(x, t) = \int_0^t \int_{\partial B} \Gamma(x, t; \xi, \tau)\rho(\xi, \tau)\, dB_\xi\, d\tau + G(x, t), \tag{5.9}$$

where dB_ξ is the surface element on ∂B,

$$G(x, t) = \int_B \Gamma(x, t; \xi, 0)\psi(\xi)\, d\xi - \int_0^t \int_B \Gamma(x, t; \xi, \tau) f(\xi, \tau)\, d\xi\, d\tau \tag{5.10}$$

and ρ is a solution of the integral equation

$$\rho(x, t) = 2 \int_0^t \int_{\partial B} \frac{\partial \Gamma(x, t; \xi, \tau)}{\partial \nu(x, t)} \rho(\xi, \tau)\, dB_\xi\, d\tau + 2\frac{\partial G(x, t)}{\partial \nu(x, t)} - 2g(x, t, v(x, t), \varphi(t)). \tag{5.11}$$

We shall prove that T has a fixed point.

Since, by Theorem 12, any possible solution of (5.1)–(5.3) is a priori bounded (see (5.5)), if we modify the definition of $g(x, t, u, v)$ for $|u| > H$ as in Remark 1, Sec. 4, then the solutions of the modified system coincide with the solutions of the original system. We conclude that, without loss of generality, we may assume that

$$|g(x, t, u, \varphi(t))| \leq \text{const.} < \infty \qquad \text{for all } (x, t) \in D_\tau,\ -\infty < u < \infty. \tag{5.12}$$

Using (5.12) and the inequality (compare (5.3.7))

$$\left| \frac{\partial G(x, t)}{\partial \nu(x, t)} \right| \leq \text{const.}$$

we find, from (5.11), that

$$|\rho(x, t)| \leq A \qquad \text{on } S_\tau,$$

where A is *independent* of $v \in Z$. (5.9) then yields

$$|w(x, t)| \leq A' \qquad \text{in } D_\tau$$

where A' is independent of $v \in Z$. Taking $R \geq A'$ we conclude that T maps Z_R into itself.

We shall now show that T is a continuous mapping. Let $v_m \in Z$ and let w_m, G_m, ρ_m be defined by (5.9), (5.10), (5.11) respectively, when $v = v_m$. We need to show that if $||v_m - v|| \to 0$ then $||w_m - w|| \to 0$. By (G_1),

$$\text{l.u.b.}_{S_T} |g(x, t, v_m(x, t), \varphi(t)) - g(x, t, v(x, t), \varphi(t))| \to 0 \qquad \text{as } m \to \infty.$$

Expanding the solutions ρ_m, ρ analogously to (5.3.10) we easily find that $\text{l.u.b.}_{S_T} |\rho_m - \rho| \to 0$ as $m \to \infty$. But then, by (5.9),

$$\text{l.u.b.}_{D_T} |w_m(x, t) - w(x, t)| \to 0 \qquad \text{as } m \to \infty,$$

i.e., $||w_m - w|| \to 0$, and the continuity of T is proved.

We shall next show that T maps Z_R into a compact subset of Z. The functions

$$\int_0^t \int_{\partial B} \Gamma(x, t; \xi, \tau)\rho(\xi, \tau)\, dB_\xi\, d\tau \qquad (\text{l.u.b. } |\rho| \leq A)$$

are easily seen to form an equicontinuous and uniformly bounded family of functions in $\overline{D}_\tau$. If we add to each of these functions the fixed function $G(x, t)$ defined by (5.10), the resulting family is again equicontinuous and uniformly bounded on $\overline{D}_\tau$. By the theorem of Ascoli-Arzela, this family is a precompact subset of Z. Since, by (5.9), this family contains the set $\{Tv; v \in Z_R\}$, T maps Z_R into a compact subset of Z.

Noting that Z_R is a closed convex set of Z we can then apply Schauder's fixed point theorem and thus conclude the existence of a function u in Z_R for which $Tu = u$; u is clearly a solution of (5.1)–(5.3) in D_τ. The uniqueness of the solution u was already proved in Theorem 12.

Having proved existence and uniqueness in D_τ, for arbitrary $0 < \tau < \infty$, we shall now denote that solution by u_τ. The function u defined by $u = u_\tau$ in D_τ, for every $0 < \tau < \infty$, is clearly the unique solution of (5.1)–(5.3) in D.

If instead of using Theorem 2, Sec. 3, Chap. 5, in the previous proof, we use Corollary 2 to that theorem, then we get:

Corollary. *Theorem* 13 *remains true if the assumption that* ψ *has a compact support in* B *is replaced by the assumptions that* $\psi(x)$ *and the* $a_{ij}(x, 0)$ *are defined and continuously differentiable in some neighborhood of the boundary* ∂B *of* B.

Remark. If the condition (G_3) is replaced by the weaker condition:

(G_3)′ $g(x, t, u, v) \to \pm\infty$ as $u \to \pm\infty$, uniformly with respect to (x, t) in S_τ and v in bounded sets, for any $0 < \tau < \infty$,

then Theorem 12 remains true except that H now depends also on τ. Using this weaker version of Theorem 12, we can prove Theorem 13 as before. Thus, *Theorem* 13 *and its corollary remain true if the assumption* (G_3) *is replaced by* (G_3)′.

We finally consider the asymptotic behavior of the solution.

Theorem 14. *Assume that* ∂B *is of class* $C^{1+\beta}$ $(0 < \beta < 1)$, *that* L *satisfies* (A)′, (B)′, *that* g *satisfies* (G_1)–(G_5), *and that*

$$\lim_{t\to\infty} f(x, t) = 0, \qquad \lim_{t\to\infty} c(x, t) = 0 \qquad \text{uniformly in } \overline{B}, \tag{5.13}$$

$$\varphi(\infty) = \lim_{t\to\infty} \varphi(t) \qquad \text{exists.} \tag{5.14}$$

If $u(x, t)$ *is a solution in* D *of* (5.1)–(5.3), *then*

$$\lim_{t\to\infty} u(x, t) = \varphi(\infty) \qquad \text{uniformly in } \overline{B}. \tag{5.15}$$

Proof. Since, by the corollary to Theorem 12, $u(x, t)$ is bounded in D, $c(x, t)u(x, t) \to 0$ uniformly in B, as $t \to \infty$. We may therefore rewrite (5.1) in the form $(L - c)u = \tilde{f}$ with $\tilde{f}(x, t) \to 0$ uniformly in B, as $t \to \infty$. It thus follows that, without loss of generality, we may assume in the sequel that $c(x, t) \equiv 0$.

Let ϵ be an arbitrary positive number and let σ be such that

$$|f(x, t)| < \epsilon \qquad \text{if } t \geq \sigma,\ x \in \overline{B}, \tag{5.16}$$

$$|\varphi(t) - \varphi(\infty)| < \epsilon \qquad \text{if } t \geq \sigma. \tag{5.17}$$

Let $|u(x, \sigma)| < C_1$ for $x \in \overline{B}$ and denote by v the solution if

$$\begin{aligned} Lv &= -\epsilon && \text{in } D - \overline{D}_\sigma, \\ v(x, \sigma) &= C_1 && \text{for } x \in \overline{B}, \\ \frac{\partial v}{\partial \nu} &= g(x, t, v, \varphi(\infty) + \epsilon) && \text{on } S - \overline{S}_\sigma \end{aligned} \tag{5.18}$$

Its existence follows from Theorem 13 applied to $u = v - C_1$. Arguing as in the proof of Theorem 17, Sec. 6, Chap. 2, we derive the inequality

$$u(x, t) < v(x, t) \qquad \text{in } D - D_\sigma. \tag{5.19}$$

To estimate v, introduce

$$w = v + \epsilon\vartheta \tag{5.20}$$

where $\vartheta(x) = e^{\lambda x_1}$ and choose λ such that $L\vartheta > 1$.

Because of (5.18), w satisfies:

$$\begin{aligned} Lw &> 0 && \text{in } D - \overline{D}_\sigma, \\ w(x, \sigma) &< C_2 && \text{for } x \in \overline{B}\ (C_2 = C_1 + 2\epsilon \text{ l.u.b. } \vartheta(x)), \\ \frac{\partial w}{\partial \nu} &= g_0(x, t, w) && \text{on } S - \overline{S}_\sigma, \end{aligned} \tag{5.21}$$

where $g_0(x, t, w) = g(x, t, w - \epsilon\vartheta, \varphi(\infty) + \epsilon) + \epsilon\partial\vartheta/\partial\nu$. Note next that, in view of the corollary to Theorem 12, the constant C_1 which bounds $|u(x, \sigma)|$ may be taken to be independent of σ. By the same corollary (replacing t by $t - \sigma$) we also conclude that v is bounded in $D - \overline{D}_\sigma$ independently of σ, ϵ $(0 < \epsilon < 1)$. The same therefore holds of w. Hence, if ϵ is sufficiently small then, by (G_5),

$$g_0(x, t, w) > g(x, t, w - N\epsilon, \varphi(\infty) + \epsilon) \tag{5.22}$$

for some $N > 0$ independent of ϵ.

Consider next the solution $s(x, t)$ of the system

$$\begin{aligned} Ls &= 0 && \text{in } D - \overline{D}_\sigma, \\ s(x, \sigma) &= C_2 && \text{for } x \in \overline{B}, \\ \frac{\partial s}{\partial \nu} &= g(x, t, s - N\epsilon, \varphi(\infty) + \epsilon) && \text{on } S - \overline{S}_\sigma. \end{aligned} \tag{5.23}$$

Its existence follows from Theorem 13 applied to $u = s - C_2$.

Using (5.22) and (5.21), (5.23), we see that Theorem 17, Sec. 6, Chap. 2, can be applied to yield

$$w(x, t) < s(x, t) \qquad \text{in } D - \overline{D}_\sigma. \tag{5.24}$$

It remains to estimate $s(x, t)$.

By (G_4), (G_5),

$$\begin{aligned} g(x, t, s - N\epsilon, \varphi(\infty) + \epsilon) & \\ &= g(x, t, s - N\epsilon, \varphi(\infty) + \epsilon) - g(x, t, \varphi(\infty) + \epsilon, \varphi(\infty) + \epsilon) \\ &> \eta(s - N\epsilon - \varphi(\infty) - \epsilon) \qquad (\eta > 0). \end{aligned}$$

It is important to observe that η is independent of s, since, by the corollary to Theorem 12 (applied to $u = s - N\epsilon$ and t replaced by $t - \sigma$) $s(x, t)$ is bounded in $D - \overline{D}_\sigma$, independently of σ, ϵ.

Setting $\tilde{s} = s - N\epsilon - \varphi(\infty) - \epsilon$, we find that

$$\begin{aligned} L\tilde{s} &= 0 && \text{in } D - \overline{D}_\sigma, \\ \tilde{s}(x, \sigma) &= C_2' && \text{for } x \in \overline{B} \qquad (C_2' = C_2 - N\epsilon - \varphi(\infty) - \epsilon), \\ \frac{\partial \tilde{s}}{\partial \nu} - \eta\tilde{s} &> 0 && \text{on } S - \overline{S}_\sigma. \end{aligned} \tag{5.25}$$

Similarly to the proof of Theorem 4, Sec. 5, Chap. 6, we find that for appropriate positive constants λ, R, μ, the function

$$z(x, t) = (e^{\lambda R} - e^{\lambda x_1})e^{-\mu(t-\sigma)}$$

satisfies·

$$\begin{aligned} Lz &< 0 && \text{in } D - \overline{D}_\sigma, \\ z(x, \sigma) &> C_2' && \text{for } x \in \overline{B}, \\ \frac{\partial z}{\partial \nu} - \eta z &< 0 && \text{on } S - \overline{S}_\sigma. \end{aligned} \tag{5.26}$$

Applying Theorem 17, Sec. 6, Chap. 2, to $\tilde{s}$, z, we get

$$\tilde{s}(x, t) < z(x, t) \text{ in } D - \overline{D}_\sigma.$$

Recalling the definitions of $\tilde{s}$ and z, we find that

$$s(x, t) - \varphi(\infty) - (N + 1)\epsilon < Ae^{-\mu(t-\sigma)} \tag{5.27}$$

where A is a constant independent of σ, ϵ.

Combining (5.19), (5.20), (5.24), (5.27), we obtain

$$u(x, t) < \varphi(\infty) + C_3\epsilon \tag{5.28}$$

provided $t \geq t_1$, where t_1 depends on ϵ and C_3 is a constant independent of ϵ. In a similar way one derives the inequality $u(x, t) > \varphi(\infty) - C_4\epsilon$ for all t sufficiently large, and the proof of Theorem 14 is thus completed.

PROBLEMS

1. Let $f(x, t, u, 0)$ be monotone nondecreasing in u and let $f(x, t, 0, 0) = 0$. Prove that $u = 0$ is the only solution of the equation $Lu = f(x, t, u, \nabla_x u)$ in $D + B_T$, which vanishes on $\overline{B} + S$ (L is parabolic with continuous coefficients).

2. Let L, S be as in Theorem 8, Sec. 4, and let $f(x, t, u, w)$ be Hölder continuous in bounded subsets of (4.3). Prove that if $f(x, 0, 0, 0) = 0$ on ∂B and $|f(x, t, u, w)| \le A + C|u|^\lambda$ where $\lambda > 1$, then there exists a solution of (4.1) which vanishes on $\overline{B} + S$, provided $2T < (A^{\lambda-1}C)^{-1/\lambda}$.
   [*Hint:* $\Phi = 2A(t + \epsilon)$ satisfies (for some $\epsilon > 0$) $L\Phi + A + C\Phi^\lambda < 0$ in D, $\Phi > 0$ in $\overline{D}$.]

3. From Theorem 2, Sec. 3, Chap. 5 (in particular (5.3.5)), and from the proof of Theorem 3, Sec. 4, Chap. 5, one derives the following analogue of Theorem 4:

   Theorem 4′. *Let D be a cylinder $B \times (0, T)$, let ∂B be of class $C^{1+\beta}$ $(0 < \beta < 1)$, let $f(x, t)$ be Hölder continuous in x, uniformly in D, and let L be defined by* (2.4) *with Hölder continuous coefficients satisfying* (2.7), (2.8). *If*

$$(\bigstar)\qquad \begin{aligned} Lu &= f &&\text{in } D + B_T,\\ u &= 0 &&\text{on } \overline{B},\\ \frac{\partial u}{\partial \nu} + g(x, t)u &= 0 &&\text{on } S, \end{aligned}$$

   where l.u.b. $|g| \le H'$, *then for any* $0 < \delta < 1$,

$$\overline{|u|}_\delta^D \le K|f|_0^D,$$

   where $K = K(D, \delta, H_0, H_1, H')$.

   Show that Theorem 7 extends to the system ($\bigstar$) if $g(x, t) \le -\mu_1 < 0$. Using this and Theorem 4′, prove an existence theorem for ($\bigstar$), analogous to Theorem 9, in case $g(x, t) \le -\mu_1 < 0$, $f = f(x, t, u)$, and $uf(x, t, u) \le A_1u^2 + A_2$, $|f(x, t, u)| \le A(|u|)$.

4. Let $f(u)$ be twice continuously differentiable and let $|f'(u)| \le \epsilon'$, $|f''(u)| \le \epsilon''$. Prove that for any functions u, v, and $w = u - v$,

$$d(P, Q)^{-\alpha}|[f(u(P)) - f(v(P))] - [f(u(Q)) - f(v(Q))]|$$
$$\le \epsilon' \frac{|w(P) - w(Q)|}{d(P, Q)^\alpha} + \epsilon''|w(Q)| \left\{ \frac{|u(P) - u(Q)|}{d(P, Q)^\alpha} + \frac{|v(P) - v(Q)|}{d(P, Q)^\alpha} \right\}.$$

   [*Hint:*

$$\begin{aligned} f(u(P)) - f(v(P)) &= \int_0^1 \frac{d}{d\vartheta} f(\vartheta u(P) + (1 - \vartheta)v(P))\, d\vartheta \\ &= [u(P) - v(P)] \int_0^1 f'(\vartheta u(P) + (1 - \vartheta)v(P))\, d\vartheta.] \end{aligned}$$

5. Prove: if the coefficients of L are Hölder continuous in $\overline{D}$, L parabolic in $\overline{D}$, S of class $\overline{C}_{2+\alpha}$, ψ of class $\overline{C}_{2+\alpha}$ and $f(x, t, u, D_xu, D_x^2u, D_tu)$ twice continuously differentiable in all its arguments, then there exists a unique solution of

$$Lu = \epsilon f(x, t, u, D_x u, D_x^2 u, D_t u) \qquad \text{in } D + B_T,$$
$$u = \psi \qquad \text{on } B + S,$$

provided ϵ is sufficiently small and provided $L\psi = \epsilon f(x, t, \psi, D_x\psi, D_x^2\psi, D_t\psi)$ on ∂B.

[*Hint:* Use the $(2 + \alpha)$-estimates of Theorem 6, Sec. 2, Chap. 3, Problem 4, extended to a function f of several variables, and Theorem 1, Sec. 1, Chap. 3.]

6. Assume that (A), (B) of Sec. 5 hold, that ∂B is of class $C^{1+\beta}$ $(0 < \beta < 1)$, that $g(x, t, u, v) \equiv g(x, t, u)$ is strictly increasing in u, and that $g(x, t, 1) \geq 0$, $g(x, t, 0) < 0$. Prove: if u is a solution of (5.1)–(5.3) in D_τ and $f \equiv 0$, then if $t > 0$, (a) $u > \min(0, \text{g.l.b. } \psi)$; (b) if l.u.b. $\psi \leq 1$ then $u \leq 1$; (c) if l.u.b. $\psi > 1$ then $u <$ l.u.b. ψ.

7. Assume that the coefficients of L are independent of t and satisfy (A), (B) of Sec. 5, and that ∂B is of class $C^{1+\beta}$ $(0 < \beta < 1)$. Let $g = g(x, t, u)$ be as in Problem 6 and, in addition, let $g(x, t, u)$ be monotone nonincreasing in t. Prove that if u is a solution of (5.1)–(5.3) with $\psi =$ const. < 1 and ≥ 0 and $f \equiv 0$, then $u(x, t)$ is monotone nondecreasing in t.

[*Hint:* Consider $v(x, t) = u(x, t + h) - u(x, t)$ and use Problem 6.]

8. Assume that (A)′, (B)′ of Sec. 5 hold, that $c \equiv 0$, that $g = g(x, t, u)$ is continuous $((x, t) \in \bar{S}, -\infty < u < \infty)$, that $g(x, t, 1) \equiv 0$, that $g(x, t, u)$ is strictly increasing in u and, finally, that for any $K > 0$ the inequality (5.4) holds for $g(x, t, u)$ $((x, t) \in \bar{S}, |u'| < K, |u''| < K, u' < u'')$ with η depending only on K. Prove that for any solution u of (5.1)–(5.3) with $f \equiv 0$,

$$u(x, t) = 1 + 0(e^{-\mu t}) \qquad \text{in } \bar{D},$$

for some $\mu > 0$.

[*Hint:* Let $C_1 = \min(0, \text{g.l.b. } \psi)$, $C_2 = \max(1, \text{l.u.b. } \psi)$ and let u_i be the solution of (5.1) – (5.3) with $f \equiv 0$, $\psi \equiv C_i$. Show that $u_1 \leq u \leq u_2$. Evaluate $1 - u_1$, $u_2 - 1$ by employing $z(x, t)$ of Sec. 5.]

CHAPTER 8

FREE BOUNDARY PROBLEMS

Introduction. In previous chapters we have dealt with solutions of the first and second initial-boundary value problems. In this chapter we shall deal with a new type of problem in which a part of the boundary is "free" (i.e., is not given) and has to be determined together with the solution of the differential system. An additional boundary condition is given on the free boundary.

Problems with a free boundary arise already in very simple physical situations. Consider, for instance, a thin block of ice occupying an interval $a \leq x < \infty$, and suppose that the temperature of the ice is everywhere 0°C and that at the point $x = a$ the temperature is kept at T°C, where $T > 0$. Then the ice will begin to melt and for every time $t > 0$ the water will occupy an interval $a \leq x < s(t)$. Denoting by u the water temperature we thus have:

$$(0.1)\qquad \begin{aligned} \alpha^2 u_{xx} - u_t &= 0 &&\text{for } a < x < s(t),\, t > 0,\\ u(a, t) &= T &&\text{for } t > 0,\\ u(s(t), t) &= 0 &&\text{for } t > 0, \end{aligned}$$

where α is some constant $\neq 0$.

$s(t)$ is the free boundary and is not given a priori. However, an additional condition is given on $x = s(t)$, namely, the law of conservation of energy. It has the form

$$(0.2)\qquad \frac{ds(t)}{dt} = -ku_x(s(t), t) \qquad \text{for } t > 0,$$

where k is some positive constant.

The problem (0.1), (0.2) is called a *Stefan problem.* Stefan problems are the free boundary problems which occur in processes of melting of solids (or crystallizing of liquids).

If the temperature v of the ice is not everywhere 0°C, then v satisfies

$$(0.3)\qquad \begin{aligned} \beta^2 v_{xx} - v_t &= 0 &&\text{for } s(t) < x < \infty,\, t > 0,\\ v(x, 0) &= \psi(x) &&\text{for } 0 < x < \infty,\\ v(s(t), t) &= 0 &&\text{for } t > 0 \end{aligned}$$

where $\beta \equiv$ const. $\neq 0$ and $\psi(x)$ is a given function ≤ 0. The condition (0.2) is replaced by

$$\frac{ds(t)}{dt} = -ku_x(s(t), t) + k_0 v_x(s(t), t) \qquad \text{for } t > 0, \tag{0.4}$$

where k_0 is some positive constant. The problem (0.1), (0.3), (0.4) is called a *two-phase* Stefan problem, whereas the problem (0.1), (0.2) is occasionally referred to as a *one-phase* Stefan problem.

In this chapter we shall be primarily concerned with Stefan problems in one dimension, and shall study questions of existence, uniqueness, and asymptotic behavior of solutions. In the final section we briefly mention other free boundary problems.

1. A Stefan Problem. Reduction to an Integral Equation

Consider the following problem: Find $s(t) > 0$ and $u(x, t)$ such that

$$u_{xx} = u_t \qquad \text{for } 0 < x < s(t),\ t > 0, \tag{1.1}$$

$$u(0, t) = f(t) \qquad \text{where } f(t) \geq 0 \text{ and } t > 0, \tag{1.2}$$

$$u(x, 0) = \varphi(x) \qquad \text{where } \varphi(x) \geq 0,\ 0 < x \leq b, \tag{1.3}$$

$$\text{and } \varphi(b) = 0,\ b > 0,$$

$$u(s(t), t) = 0 \qquad \text{for } t > 0 \text{ and } s(0) = b, \tag{1.4}$$

$$u_x(s(t), t) = -\frac{ds(t)}{dt} \qquad \text{for } t > 0. \tag{1.5}$$

$x = s(t)$ is the free boundary which is not given and is to be found together with $u(x, t)$. The conditions (1.2)–(1.4) form the first initial-boundary data, whereas (1.5) is a condition on the free boundary. The assumptions $f \geq 0$, $\varphi \geq 0$ are motivated by the physical background of the problem.

Definition. We say that $u(x, t)$, $s(t)$ form a solution of (1.1)–(1.5) for all $t < \sigma$ $(0 < \sigma \leq \infty)$ if (i) $\partial^2 u/\partial x^2$ and $\partial u/\partial t$ are continuous for $0 < x < s(t)$, $0 < t < \sigma$; (ii) u and $\partial u/\partial x$ are continuous for $0 \leq x \leq s(t)$, $0 < t < \sigma$; (iii) $u(x, t)$ is continuous also for $t = 0$, $0 < x \leq b$, and $0 \leq \liminf u(x, t) \leq \limsup u(x, t) < \infty$ as $t \to 0$, $x \to 0$ (if $\varphi(0) = f(0)$ then u is required to be continuous at $x = t = 0$); (iv) $s(t)$ is continuously differentiable for $0 \leq t < \sigma$, and (v) the equations (1.1)–(1.5) are satisfied.

Theorem 1. *Assume that $f(t)$ $(0 \leq t < \infty)$ and $\varphi(x)$ $(0 \leq x \leq b)$ are continuously differentiable functions. Then there exists a unique solution $u(x, t)$, $s(t)$ of the system (1.1)–(1.5) for all $t < \infty$. Furthermore, the function $x = s(t)$ is monotone nondecreasing in t.*

The proof is given in this section and in the next one. In this section we reduce the problem (1.1)–(1.5) to an equivalent problem of solving a nonlinear integral equation of Volterra type for $u_x(s(t), t)$. The procedure used applies to a large class of free boundary problems.

We first prove that if u, s form a solution of (1.1)–(1.5) for all $t < \sigma$, then $s(t)$ is monotone nondecreasing. By the maximum principle, $u(x, t) \geq 0$ for $0 < x < s(t)$, $0 \leq t \leq \sigma$. Since $u = 0$ on $x = s(t)$, $u_x \leq 0$ on $x = s(t)$. Hence, by (1.5), $ds/dt \geq 0$, i.e., $s(t)$ is monotone nondecreasing. We can actually prove more, namely:

If $\varphi(x) \not\equiv 0$ or if $f(t) \not\equiv 0$ in every interval $0 \leq t \leq \epsilon$ ($\epsilon > 0$) then $s(t)$ is strictly increasing.

Indeed, in the contrary case there exist two points t', t'' ($t' < t''$) such that $s(t') = s(t'')$. But then $s(t) = s(t')$ for all $t' < t < t''$ and, by (1.5), $u_x = 0$ on $x = s(t)$, $t' < t < t''$. Now, by the strong maximum principle and our assumptions on φ, f it follows that $u(x, t) > 0$ for $0 < x < s(t)$, $0 < t < \sigma$. Since $u(s(t), t) = 0$, we can apply Theorem 14, Sec. 5, Chap. 2, and conclude that $u_x < 0$ on $x = s(t)$, $t' < t < t''$, which is a contradiction

In reducing the problem (1.1)–(1.5) to a problem of solving an integral equation, we shall make use of the following lemma.

Lemma 1. *Let $\rho(t)$ $(0 \leq t \leq \sigma)$ be a continuous function and let $s(t)$ $(0 \leq t \leq \sigma)$ satisfy a Lipschitz condition. Then, for every $0 < t \leq \sigma$,*

$$\text{(1.6)} \quad \lim_{x \to s(t)-0} \frac{\partial}{\partial x} \int_0^t \rho(\tau) K(x, t; s(\tau), \tau)\, d\tau = \frac{1}{2}\rho(t) + \int_0^t \rho(\tau) \left[\frac{\partial}{\partial x} K(x, t; s(\tau), \tau)\right]_{x=s(t)} d\tau,$$

where

$$K(x, t; \xi, \tau) = \frac{1}{2\pi^{1/2}(t-\tau)^{1/2}} \exp\left\{-\frac{(x-\xi)^2}{4(t-\tau)}\right\}.$$

Thus, the lemma establishes a jump relation similar to that for single-layer potentials.

Proof. We shall first prove that for any fixed positive $\delta < t$, the integral

$$\text{(1.7)} \quad I \equiv \int_{t-\delta}^t \frac{x - s(\tau)}{2(t-\tau)} K(x, t; s(\tau), \tau)\, d\tau - \int_{t-\delta}^t \frac{s(t) - s(\tau)}{2(t-\tau)} K(s(t), t; s(\tau), \tau)\, d\tau$$

satisfies the relation

$$\text{(1.8)} \quad \limsup_{x \to s(t)-0} \left| I + \frac{1}{2} \right| \leq A\delta^{1/2};$$

here and in what follows, various constants which are independent of x, t, δ will be denoted by A (A may depend on σ).

Write $I = I_1 + I_2$ where

$$I_1 = \int_{t-\delta}^{t} \frac{x - s(t)}{2(t-\tau)} K(x, t; s(\tau), \tau)\, d\tau,$$

$$I_2 = \int_{t-\delta}^{t} \frac{s(t) - s(\tau)}{2(t-\tau)} [K(x, t; s(\tau), \tau) - K(s(t), t; s(\tau), \tau)]\, d\tau.$$

Since, by assumption, $|s(t) - s(\tau)| < A|t - \tau|$, we get

$$|I_2| \leq A \int_{t-\delta}^{t} \frac{d\tau}{(t-\tau)^{1/2}} \leq A\delta^{1/2}. \tag{1.9}$$

To evaluate I_1, introduce

$$J_1 = \int_{t-\delta}^{t} \frac{x - s(t)}{2(t-\tau)} K(x, t; s(t), \tau)\, d\tau. \tag{1.10}$$

Then

$$J_1 - I_1 = \int_{t-\delta}^{t} \frac{x - s(t)}{2(t-\tau)} K(x, t; s(t), \tau) \cdot \left\{1 - \exp\left[-\frac{(x - s(\tau))^2 - (x - s(t))^2}{4(t-\tau)}\right]\right\} d\tau. \tag{1.11}$$

The expression in brackets is bounded by

$$\frac{1}{4(t-\tau)} |s(t) - s(\tau)|(|x - s(t)| + |x - s(\tau)|) \leq A(|x - s(t)| + |s(t) - s(\tau)|).$$

Since it suffices to prove (1.8) for δ sufficiently small, and since x is going to approach $s(t)$, we may assume that the right-hand side of the last inequality is less than 1. Hence, the expression in braces in (1.11) is bounded by

$$A(|x - s(t)| + |s(t) - s(\tau)|).$$

Substituting this into (1.11) and using the elementary inequality $ye^{-y} \leq$ const. for $y \geq 0$, we find

$$|J_1 - I_1| \leq A \int_{t-\delta}^{t} \frac{d\tau}{(t-\tau)^{1/2}} + A \int_{t-\delta}^{t} d\tau \leq A\delta^{1/2}. \tag{1.12}$$

Now, as to J_1, substitute in the integral (1.10) $z = (t - \tau)/(x - s(t))^2$. Noting that $x - s(t) < 0$, we get

$$J_1 = -\frac{1}{4\pi^{1/2}} \int_0^{\delta'} z^{-3/2} \exp\left[-\frac{1}{4z}\right] dz \qquad \text{where } \delta' = \delta/(x - s(t))^2. \tag{1.13}$$

As $x \to s(t)$, $\delta' \to \infty$ and, consequently, $J_1 \to -\frac{1}{2}$. Combining this result with (1.12), (1.9) and recalling that $I = I_1 + I_2$, the relation (1.8) follows. From (1.12), (1.13) it also follows that

$$|I_1| \leq A. \tag{1.14}$$

Using the Lipschitz continuity of $s(t)$ we obtain

$$\int_{t-\delta}^{t} \frac{|s(t) - s(\tau)|}{2(t - \tau)} K(s(t), t; s(\tau), \tau)\, d\tau \leq A. \tag{1.15}$$

Finally, in addition to (1.8), (1.15), we shall need the inequality

$$\int_{t-\delta}^{t} \frac{|x - s(\tau)|}{2(t - \tau)} K(x, t; s(\tau), \tau)\, d\tau \leq A. \tag{1.16}$$

The proof follows by writing

$$\frac{|x - s(\tau)|}{2(t - \tau)} K \leq -\frac{x - s(t)}{2(t - \tau)} K + \frac{|s(t) - s(\tau)|}{2(t - \tau)} K$$

and using (1.14) and the Lipschitz continuity of s.

We shall now complete the proof of Lemma 1 with the aid of (1.8), (1.15), (1.16). Putting

$$\begin{aligned} L_1 = {} & \int_{t-\delta}^{t} \rho(\tau) \frac{x - s(\tau)}{2(t - \tau)} K(x, t; s(\tau), \tau)\, d\tau \\ & - \int_{t-\delta}^{t} \rho(\tau) \frac{s(t) - s(\tau)}{2(t - \tau)} K(s(t), t; s(\tau), \tau)\, d\tau, \end{aligned} \tag{1.17}$$

we claim that

$$\limsup_{x \to s(t)-0} \left|L_1 + \frac{1}{2}\rho(t)\right| \leq A\delta^{1/2} + A \operatorname*{l.u.b.}_{t-\delta \leq \tau \leq t} |\rho(t) - \rho(\tau)|. \tag{1.18}$$

Indeed this follows by writing, in (1.17), $\rho(\tau) = \rho(t) + [\rho(\tau) - \rho(t)]$ and using (1.8), (1.15), (1.16).

Observe next that the function

$$\begin{aligned} L_2 = {} & \int_{0}^{t-\delta} \rho(\tau) \frac{x - s(\tau)}{2(t - \tau)} K(x, t; s(\tau), \tau)\, d\tau \\ & - \int_{0}^{t-\delta} \rho(\tau) \frac{s(t) - s(\tau)}{2(t - \tau)} K(s(t), t; s(\tau), \tau)\, d\tau \qquad (0 < \delta < t) \end{aligned}$$

satisfies the relation

$$\lim_{x \to s(t)} L_2 = 0.$$

Combining this remark with (1.18) we get

$$\limsup_{x \to s(t)-0} \left|(L_1 + L_2) + \frac{1}{2}\rho(t)\right| \leq A\delta^{1/2} + \operatorname*{l.u.b.}_{t-\delta \leq \tau \leq t} |\rho(t) - \rho(\tau)|.$$

Since the left-hand side is independent of δ, and since the right-hand side can be made arbitrarily small if δ is sufficiently small, we get

$$\limsup_{x \to s(t)-0} \left|(L_1 + L_2) + \frac{1}{2}\rho(t)\right| = 0,$$

which is precisely the jump relation (1.6).

We shall now reduce the problem of solving (1.1)–(1.5) to a problem of solving an integral equation.

We introduce Green's function for the half-plane $x > 0$:

$$G(x, t; \xi, \tau) = K(x, t; \xi, \tau) - K(-x, t; \xi, \tau). \tag{1.19}$$

Suppose that u, s form a solution of (1.1)–(1.5). Integrating Green's identity

$$\frac{\partial}{\partial \xi}\left(G \frac{\partial u}{\partial \xi} - u \frac{\partial G}{\partial \xi}\right) - \frac{\partial}{\partial \tau}(Gu) = 0$$

over the domain $0 < \xi < s(\tau)$, $0 < \epsilon < \tau < t - \epsilon$ and letting $\epsilon \to 0$ we get, upon using (1.2)–(1.4),

$$\begin{aligned} u(x, t) &= \int_0^t u_\xi(s(\tau), \tau) G(x, t; s(\tau), \tau)\, d\tau + \int_0^t f(\tau)\, G_\xi(x, t; 0, \tau)\, d\tau \\ &\quad + \int_0^b \varphi(\xi) G(x, t; \xi, 0)\, d\xi \\ &\equiv M_1 + M_2 + M_3. \end{aligned} \tag{1.20}$$

Introducing

$$v(t) = u_x(s(t), t), \tag{1.21}$$

we proceed to differentiate both sides of (1.20) with respect to x and then take $x \to s(t) - 0$. Using Lemma 1, we find

$$\lim_{x \to s(t) - 0} \frac{\partial M_1}{\partial x} = \frac{1}{2} v(t) + \int_0^t v(\tau) G_x(s(t), t; s(\tau), \tau)\, d\tau. \tag{1.22}$$

Here we used the fact that the second term of G_x is a continuous function since $x + s(\tau) \geq b > 0$ (as $s(t)$ is monotone nondecreasing).

In order to evaluate $\lim \partial M_i / \partial x$ $(i = 2, 3)$ we introduce Neumann's function for the half-plane $x > 0$:

$$N(x, t; \xi, \tau) = K(x, t; \xi, \tau) + K(-x, t; \xi, \tau). \tag{1.23}$$

Using the relation $G_x = -N_\xi$ we get

$$\begin{aligned} \frac{\partial M_2}{\partial x} &= \int_0^t f(\tau) G_{x\xi}(x, t; 0, \tau)\, d\tau = \int_0^t f(\tau) N_\tau(x, t; 0, \tau)\, d\tau \\ &= -f(0) N(x, t; 0, 0) - \int_0^t f'(\tau) N(x, t; 0, \tau)\, d\tau. \end{aligned} \tag{1.24}$$

Similarly,

$$\begin{aligned} \frac{\partial M_3}{\partial x} &= \int_0^b \varphi(\xi) G_x(x, t; \xi, 0)\, d\xi = \varphi(0) N(x, t; 0, 0) \\ &\quad + \int_0^b \varphi'(\xi) N(x, t; \xi, 0)\, d\xi. \end{aligned} \tag{1.25}$$

Combining (1.22), (1.24), (1.25), we obtain from (1.20)

$$(1.26)\quad v(t) = 2[\varphi(0) - f(0)]N(s(t), t; 0, 0) + 2\int_0^b \varphi'(\xi)N(s(t), t; \xi, 0)\, d\xi$$
$$- 2\int_0^t f'(\tau)N(s(t), t; 0, \tau)\, d\tau + 2\int_0^t v(\tau)G_x(s(t), t; s(\tau), \tau)\, d\tau,$$

and, by (1.5), (1.21),

$$(1.27)\qquad s(t) = b - \int_0^t v(\tau)\, d\tau.$$

We have thus proved that for every solution u, s of the system (1.1)–(1.5) for all $t < \sigma$, the function $v(t)$, defined by (1.21), satisfies the nonlinear integral equation of Volterra type (1.26) (for $0 < t < \sigma$), where $s(t)$ is given by (1.27); $v(t)$ is continuous for $0 \leq t < \sigma$ and, therefore, by continuity, (1.26) is satisfied also at $t = 0$.

Suppose conversely that for some $\sigma > 0$, $v(t)$ is a continuous solution of the integral equation (1.26) for $0 \leq t < \sigma$, with $s(t)$ given by (1.27). Suppose further that $s(t) > 0$ for $0 \leq t < \sigma$. We shall prove that $u(x, t)$, $s(t)$ then form a solution of (1.1)–(1.5) for all $t < \sigma$, where $u(x, t)$ is defined by (1.20) with $u_\xi(s(\tau), \tau)$ replaced by $v(\tau)$.

First of all one can easily verify (1.1)–(1.3). We next differentiate $u(x, t)$ with respect to x and take $x \to s(t) - 0$. Using Lemma 1, the previous evaluations of $\partial M_i/\partial x$ for $i = 2, 3$, and the integral equation (1.26), we find that $u_x(s(t), t) = v(t)$. Since, by (1.27), $v(t) = -ds(t)/dt$, (1.5) follows. Note that u, s satisfy all the differentiability properties which are required in the definition of a solution. Thus it remains to prove that $u(s(t), t) \equiv 0$.

Integrate Green's identity (with G and u) in the domain $0 < \xi < s(\tau)$, $0 < \epsilon < \tau < t - \epsilon$ and let $\epsilon \to 0$. Comparing the integral representation thus obtained for $u(x, t)$ with the original definition of $u(x, t)$ by (1.20) (with $u_\xi(s(\tau), \tau) = v(\tau)$) we conclude that

$$(1.28)\quad \int_0^t u(s(\tau), \tau)G_\xi(x, t; s(\tau), \tau)\, d\tau = 0 \qquad \text{if } 0 < x < s(t),\ 0 < t < \sigma.$$

Taking $x \to s(t) - 0$ and using Lemma 1, we find that the function $\psi(t) = u(s(t), t)$ satisfies the integral equation

$$(1.29)\qquad \psi(t) = 2\int_0^t \psi(\tau)G_\xi(s(t), t; s(\tau), \tau)\, d\tau.$$

Since

$$|G_\xi(s(t), t; s(\tau), \tau)| \leq \frac{\text{const.}}{(t - \tau)^{1/2}},$$

the kernel of the integral equation is integrable and, consequently, $\psi(t) \equiv 0$, i.e., $u(s(t), t) \equiv 0$. We have thus proved:

Lemma 2. *The problem* (1.1)–(1.5) *for $t < \sigma$ is equivalent to the problem of finding a continuous solution $v(t)$ for the integral equation* (1.26) (*for $0 \leq t < \sigma$*) *where $s(t)$ is given by* (1.27) *and is positive.*

2. Existence and Uniqueness for Stefan Problems

In this section we shall complete the proof of Theorem 1. Denote by C_σ the set of all continuous functions $v(t)$ in the interval $0 \le t \le \sigma$ normed by $||v|| = \underset{0 \le t \le \sigma}{\text{l.u.b.}} |v(t)|$, and let $C_{\sigma,M}$ denote the subset $\{v; v \in C_\sigma, ||v|| \le M\}$ of C_σ. Denote by Tv the right-hand side of (1.26) where $s(t)$ is given by (1.27). For any $M > 0$, if σ is sufficiently small, say $2\sigma M < b$, then for every $v \in C_{\sigma,M}$ the function $s(t)$ (given by (1.27)) is positive (in fact, $s(t) > b/2$) and, consequently, Tv is well defined.

By elementary and straightforward calculation one can show that if

$$M = 1 + 4[\underset{0 \le x \le b}{\text{l.u.b.}} |\varphi'(x)|] \tag{2.1}$$

and if

$$AM\sigma^{1/2} \le 1 \tag{2.2}$$

for some constant A depending only on upper bounds on $|f'(t)|$, $|f(0) - \varphi(0)|$, b, $1/b$, then T maps $C_{\sigma,M}$ into itself and is a contraction; for details see [38]. Employing Theorem 1, Sec. 1, Chap. 3, we conclude that T has a unique fixed point v in $C_{\sigma,M}$. v is then a solution of (1.26). In view of Lemma 2, we have thus proved the existence of a solution of (1.1)–(1.5) for all $t < \sigma$, for some σ sufficiently small (restricted only by (2.2)).

To prove uniqueness for $t < \sigma$, suppose that u_0, s_0 is another solution of (1.1)–(1.5) for $t < \sigma$ and let v_0 be the corresponding solution of (1.26) for $0 \le t < \sigma$. It suffices to prove uniqueness for $t \le \sigma'$, for any $\sigma' < \sigma$. Let

$$\overline{M} = \max \{M, \underset{0 \le t \le \sigma'}{\text{l.u.b.}} |v_0(t)|\}$$

and let $\bar{\sigma}$ be any positive number satisfying $A\overline{M}\bar{\sigma}^{1/2} \le 1$, where A is the constant appearing in (2.2). Then by the same calculations which were used to prove that T maps $C_{\sigma,M}$ into itself and is a contraction one shows that T maps $C_{\bar{\sigma},\overline{M}}$ into itself and is a contraction. Hence, there exists at most one fixed point of T in $C_{\bar{\sigma},\overline{M}}$. It follows that $v(t) = v_0(t)$ for $0 \le t \le \bar{\sigma}$. Hence also $s(t) = s_0(t)$, $u(x, t) = u_0(x, t)$ if $0 \le t \le \bar{\sigma}$, $0 \le x \le s(t)$.

We next consider the system (1.1)–(1.5) for $t > \bar{\sigma}$, i.e., (1.1), (1.2), (1.4), (1.5) are considered for $t \ge \bar{\sigma}$ (instead of $t \ge 0$) whereas (1.3) is replaced by

$$u(x, \bar{\sigma}) = u(x, \bar{\sigma}) \qquad \text{for } 0 < x \le s(\bar{\sigma}).$$

This problem can again be transformed into an integral equation. All the considerations for the equation (1.26) extend to the present integral equation, provided M is replaced by M_0 where

$$M_0 = 1 + 4[\underset{0 \le x \le s(\bar{\sigma})}{\text{l.u.b.}} |u_x(x, \bar{\sigma})|]. \tag{2.3}$$

Similarly we reduce the problem (1.1)–(1.5) for u_0, s_0 in the interval $\bar{\sigma} \leq t < \sigma$ to an integral equation. Since $u(x, \bar{\sigma}) = u_0(x, \bar{\sigma})$, $s(\bar{\sigma}) = s_0(\bar{\sigma})$, the integral equations for $v(t)$ and $v_0(t)$ coincide. Repeating now the same argument as before we conclude that $v(t) = v_0(t)$ for $\bar{\sigma} \leq t \leq \tilde{\sigma}$ for any $\tilde{\sigma}$ satisfying

$$A\overline{M}_0(\tilde{\sigma} - \bar{\sigma})^{1/2} \leq 1 \qquad (\overline{M}_0 = \max \{M_0, \operatorname*{l.u.b.}_{0 \leq t \leq \sigma'} |v_0(t)|\}).$$

We can now proceed in the same manner as before step by step, noting that in each step the time interval can be taken to be $\geq \epsilon$ where ϵ satisfies

$$A \max \{1 + 4[\operatorname*{l.u.b.}_{0 \leq x \leq s(t), \bar{\sigma} \leq t \leq \sigma'} |u_x(x, t)|], \operatorname*{l.u.b.}_{0 \leq t \leq \sigma'} |v_0(t)|\}\epsilon^{1/2} = 1.$$

Having proved existence and uniqueness for all $t < \sigma$, where σ is any positive number satisfying (2.2), let us stress that the previous proof (see (2.1), (2.2)) shows also the following:

If instead of (1.1)–(1.5) for $t > 0$, we consider (1.1)–(1.5) for $t > \lambda$, i.e., (1.1), (1.2), (1.4), (1.5) hold for $t > \lambda$ and (1.3) replaced by $u(x, \lambda) = u(x, \lambda)$ for $0 < x \leq s(\lambda)$, and if

$$|u_x(x, \lambda)|, \qquad s(\lambda), \qquad 1/s(\lambda) \tag{2.4}$$

are bounded independently of λ, then there exists a unique solution for the problem in an interval $\lambda \leq t \leq \lambda + \epsilon$ where ϵ is some positive number independent of λ. For every λ, of course, the integral equation for $v(t)$ is different.

Since for any solution of (1.1)–(1.5) the function $s(t)$ is monotone nondecreasing, $1/s(\lambda) \leq 1/b$.

To complete the proof of Theorem 1 it suffices to prove the following statement: *For every $t_0 > 0$ there exists an $\epsilon > 0$ such that if the system (1.1)–(1.5) has a unique solution for all $t < t_0$ then it also has a unique solution for all $t < t_0 + \epsilon$.* In view of the previous remarks it suffices to show: If $u(x, t)$, $s(t)$ is a solution of (1.1)–(1.5) for all $t < t_0$ then, for all $\eta > 0$ sufficiently small, the functions

$$\operatorname*{l.u.b.}_{0 < x < s(t_0 - \eta)} |u_x(x, t_0 - \eta)|, \qquad s(t_0 - \eta) \tag{2.5}$$

are bounded independently of η. If we prove that

$$\operatorname*{l.u.b.}_{0 < t < t_0} |v(t)| < \infty \tag{2.6}$$

then from (1.27) follows the boundedness of $s(t)$ for $t < t_0$. Next, upon differentiating (1.20) with respect to x and using (2.6), (1.16) we also derive the boundedness of the first function in (2.5). Therefore, if we prove (2.6) then the proof of Theorem 1 is completed.

Proof of (2.6). We use for $v(t)$ the integral equation which corresponds to the system (1.1)–(1.5) in the interval $t_0 - \mu < t < t_0$ (μ sufficiently small). Since $u(0, t_0 - \mu) = f(t_0 - \mu)$, the equation is

$$(2.7)\qquad v(t) = 2\int_0^{s(t_0-\mu)} u_\xi(\xi, t_0-\mu)N(s(t), t; \xi, t_0-\mu)\,d\xi$$
$$- 2\int_{t_0-\mu}^{t} f'(\tau)N(s(t), t; 0, \tau)\,d\tau$$
$$+ 2\int_{t_0-\mu}^{t} v(\tau)G_x(s(t), t; s(\tau), \tau)\,d\tau$$
$$\equiv T_1 + T_2 + T_3.$$

Since $v(t) \le 0$ we only have to find a lower bound on $v(t)$. Introducing

$$(2.8)\qquad \psi(t) = \underset{t_0-\mu<\tau<t}{\text{g.l.b.}}\; v(\tau),$$

we proceed to evaluate T_3.

$$(2.9)\qquad T_3 = -\int_{t_0-\mu}^{t} v(\tau)\,\frac{s(t)-s(\tau)}{t-\tau}\,K(s(t), t; s(\tau), \tau)\,d\tau$$
$$+\int_{t_0-\mu}^{t} v(\tau)\,\frac{s(t)+s(\tau)}{t-\tau}\,K(-s(t), t; s(\tau), \tau)\,d\tau$$
$$\equiv T_3' + T_3''.$$

Since $s(t) - s(\tau) \ge 0$ and $v(\tau) \le 0$ we have

$$(2.10)\qquad T_3' \ge 0.$$

Since $s(t) + s(\tau) \ge 2b$ we have

$$(2.11)\qquad |T_3''| \le B_0|\psi(t)| \int_{t_0-\mu}^{t} \frac{1}{t-\tau} \exp\left[-\frac{b^2}{t-\tau}\right] d\tau$$
$$\le B_1|\psi(t)|\mu \le \frac{1}{2}\,|\psi(t)|,$$

where B_0, B_1 are constants depending only on b, and μ is such that $2B_1\mu \le 1$, μ is fixed from now on.

It is clear that

$$(2.12)\qquad |T_1 + T_2| \le B' \qquad \text{if } t_0 - \mu \le t < t_0$$

where B' is some constant depending on μ but not on t. Combining (2.7), (2.9)–(2.12) we obtain

$$(2.13)\qquad v(t) \ge \frac{1}{2}\psi(t) - B' \qquad (t_0 - \mu \le t < t_0).$$

Taking the g.l.b. of both sides of (2.13) we get

$$\psi(t) \ge -2B' \qquad (t_0 - \mu \le t < t_0),$$

and the boundedness of $v(t)$, for $t_0 - \mu \le t < t_0$, follows. (2.6) is thus established.

Consider the Stefan problem obtained when the condition (1.2) is replaced by

(2.14) $$u_x(0, t) = f(t) \qquad \text{where } f(t) < 0 \text{ and } t > 0.$$

Theorem 2. *Assume that* $f(t)$ $(0 \le t < \infty)$ *is a continuous function and that* $\varphi(x)$ $(0 \le x \le b)$ *is a continuously differentiable function. Then there exists a unique solution* $u(x, t)$, $s(t)$ *of the problem* (1.1), (2.14), (1.3)–(1.5) *for all* $t < \infty$. *Furthermore,* $s(t)$ *is strictly monotone increasing.*

The proof is similar to the proof of Theorem 1 and we therefore omit the details.

From the proofs of Theorems 1, 2 one can see that, in bounded intervals of t, the solution depends continuously on the data f, φ. This fact can be used to extend Theorem 2 to the case where, in (2.14), we only assume that $f \le 0$, by approximating f in bounded t-intervals by negative functions. The free boundary $s(t)$ is now only asserted to be monotone nondecreasing.

3. Asymptotic Behavior of Solutions of Stefan Problems

In the physical applications the units are such that, instead of (1.5),

(3.1) $$\alpha u_x(s(t), t) = -\frac{ds(t)}{dt} \qquad \text{for } t > 0$$

where α is a small positive constant. Theorems 1, 2 remain valid if (1.5) is replaced by (3.1) and α is any positive number, and the proof is the same except for some trivial modifications.

Theorem 3. *Let* $u(x, t)$, $s(t)$ *be a solution of the Stefan problem* (1.1), (2.14), (1.3), (1.4), (3.1). *If*

(3.2) $$\lim_{t\to\infty} t^\delta f(t) = -\gamma \qquad \text{for some } \gamma > 0, \frac{1}{2} < \delta < 1,$$

then

(3.3) $$\lim_{t\to\infty} \frac{s(t)}{t^{1-\delta}} = \frac{\alpha\gamma}{1-\delta}.$$

If (3.2) *holds with* $\delta = 1/2$ *then*

(3.4) $$\lim_{t\to\infty} \frac{s(t) - b}{t^{1/2}} = 2\alpha\gamma(1 + 0(\alpha))$$

where $\alpha^{-1}0(\alpha)$ *is bounded, for* $\alpha \to 0$.

Proof. Integrating (1.1) over the domain $0 < \xi < s(\tau)$, $0 < \tau < t$ and using (2.14), (1.3), (1.4), (3.1), we find

(3.5) $$\frac{s(t)}{\alpha} = \frac{b}{\alpha} - \int_0^t f(\tau)\, d\tau + \int_0^b \varphi(x)\, dx - \int_0^{s(t)} u(x, t)\, dx.$$

We proceed to evaluate the integral

$$I(t) = \int_0^{s(t)} u(x, t)\, dx. \tag{3.6}$$

By the maximum principle, $u \geq 0$. Hence,

$$I(t) \geq 0. \tag{3.7}$$

To find an upper bound on I, consider the problem

$$\begin{aligned} w_{xx} - w_t &= 0 && \text{for } 0 < x < \infty,\ t > 0, \\ w_x(0, t) &= f(t) - \epsilon && \text{for } t > 0 \quad (\epsilon > 0), \\ w(x, 0) &= \Phi(x) && \text{for } 0 < x < \infty, \end{aligned} \tag{3.8}$$

where $\Phi(x) = \varphi(x)$ if $0 \leq x \leq b$, $\Phi(x) = 0$ if $b < x < \infty$. As is easily verified, the function

$$w_\epsilon(x, t) = -\int_0^t [f(\tau) - \epsilon] N(x, t; 0, \tau)\, d\tau + \int_0^b \varphi(\xi) N(x, t; \xi, 0)\, d\xi \tag{3.9}$$

is a solution of (3.8). Here N is the Neumann function (1.23). Since N is a positive function, $w_\epsilon(x, t) \geq 0$. In particular, $w_\epsilon(s(t), t) \geq 0$.

Consider the function $W = w_\epsilon - u$ in the region $0 \leq x \leq s(t)$, $0 \leq t \leq T$ for any $T > 0$. On $x = s(t)$, $W \geq 0$. On $t = 0$, $W = 0$. Finally, on $x = 0$, $W_x = -\epsilon < 0$. Using the maximum principle we conclude that $W \geq 0$ for $0 \leq x \leq s(t)$, $0 \leq t \leq T$. Since this is true for any $T > 0$, $\epsilon > 0$, we get $u(x, t) \leq w(x, t)$ for all $0 \leq x \leq s(t)$, $t > 0$, where $w(x, t)$ is defined by (3.9) with $\epsilon = 0$. Hence,

$$I(t) \leq s(t) \operatorname*{l.u.b.}_{x} \left\{ \int_0^t |f(\tau)| N(x, t; 0, \tau)\, d\tau + \int_0^b \varphi(\xi) N(x, t; \xi, 0)\, d\xi \right\}.$$

Noting next that

$$N(x, t; \xi, \tau) \leq (t - \tau)^{-1/2},$$

we get

$$I(t) \leq s(t) \left\{ \int_0^t (t - \tau)^{1/2} |f(\tau)|\, d\tau + A(t + 1)^{-1/2} \right\} \tag{3.10}$$

for some constant A. Substituting the bound obtained from (3.10), (3.7) into (3.5), proof of the theorem is easily completed.

We shall now consider the asymptotic behavior of $s(t)$ for the two-phase Stefan problem

$$\frac{\partial^2 w_1}{\partial x^2} = a_1^2 \frac{\partial w_1}{\partial t} \qquad \text{for } -\infty < x < s(t),\ t > 0, \tag{3.11}$$

$$\frac{\partial^2 w_2}{\partial x^2} = a_2^2 \frac{\partial w_2}{\partial t} \qquad \text{for } s(t) < x < \infty,\ t > 0, \tag{3.12}$$

$$w_1(x, 0) = \psi_1(x) \geq 0 \qquad \text{for } -\infty < x < 0, \tag{3.13}$$

$$w_2(x, 0) = \psi_2(x) \leq 0 \qquad \text{for } 0 < x < \infty, \tag{3.14}$$

$$w_1(s(t), t) = w_2(s(t), t) = 0 \qquad \text{for } t > 0, \text{ and } s(0) = 0, \tag{3.15}$$

$$\frac{ds(t)}{dt} = -k_1 \frac{\partial w_1(s(t), t)}{\partial x} + k_2 \frac{\partial w_2(s(t), t)}{\partial x} \quad \text{for } t > 0, \tag{3.16}$$

where k_1, k_2, a_1, a_2 are positive constants.

We shall make the assumptions:

$$(\Psi) \qquad \begin{aligned} &\text{as } x \to -\infty, \quad \psi_1(x) \to \gamma_1 \geq 0, \quad d\psi_1(x)/dx \to 0, \\ &\text{as } x \to \infty, \quad \psi_2(x) \to \gamma_2 \leq 0, \quad d\psi_2(x)/dx \to 0, \\ &\int_{-\infty}^{0} |\psi_1(x) - \gamma_1| dx < \infty, \quad \int_{0}^{\infty} |\psi_2(x) - \gamma_2| dx < \infty. \end{aligned}$$

We then require that the solution w_1, w_2, s satisfy:

$$\text{as } x \to \pm\infty, \quad w_i \to \gamma_i \text{ and } \partial w_i/\partial x \to 0 \quad (i = 1, 2), \tag{3.17}$$

uniformly with respect to t in bounded sets.

It is convenient to transform the system (3.11)–(3.17) by setting

$$u_i = w_i - \gamma_i \quad (i = 1, 2). \tag{3.18}$$

The transformed system is:

$$\frac{\partial^2 u_1}{\partial x^2} = a_1^2 \frac{\partial u_1}{\partial t} \quad \text{for } -\infty < x < s(t),\ t > 0, \tag{3.19}$$

$$\frac{\partial^2 u_2}{\partial x^2} = a_2^2 \frac{\partial u_2}{\partial t} \quad \text{for } s(t) < x < \infty,\ t > 0, \tag{3.20}$$

$$u_1(x, 0) = \varphi_1(x) \equiv \psi_1(x) - \gamma_1 \quad \text{for } -\infty < x < 0, \tag{3.21}$$

$$u_2(x, 0) = \varphi_2(x) \equiv \psi_2(x) - \gamma_2 \quad \text{for } 0 < x < \infty, \tag{3.22}$$

$$u_i(s(t), t) = -\gamma_i \quad (i = 1, 2) \quad \text{for } t > 0, \text{ and } s(0) = 0, \tag{3.23}$$

$$\frac{ds(t)}{dt} = -k_1 \frac{\partial u_1(s(t), t)}{\partial x} + k_2 \frac{\partial u_2(s(t), t)}{\partial x} \quad \text{for } t > 0, \tag{3.24}$$

$$\text{as } x \to \pm\infty, \quad u_i \to 0,\ \partial u_i/\partial x \to 0 \quad (i = 1, 2), \tag{3.25}$$

uniformly in t in bounded sets.

The free boundary is not necessarily a monotone function in t. However, by the methods of Secs. 1, 2 one can still prove existence and uniqueness of a solution u_1, u_2, s (under the assumption (Ψ)) provided

$$k_1, \quad k_2, \quad a_1^2 k_1, \quad a_2^2 k_2 \tag{3.26}$$

are sufficiently small. In the physical applications the quantities in (3.26) are indeed sufficiently small to guarantee existence and uniqueness. We shall now derive an asymptotic bound on $s(t)$.

Since, by the maximum principle, $w_1 \geq 0$ and $w_2 \leq 0$ in their respective regions of definition, and since $w_i(s(t), t) = 0$ for $i = 1, 2$, we conclude that $\partial w_i(s(t), t)/\partial x \leq 0$. Hence, $\partial u_i(s(t), t)/\partial x \leq 0$. Introduce the functions

$$(3.27)\quad \lambda_i(t) = -\int_0^t \frac{\partial u_i(s(\tau), \tau)}{\partial x}\, d\tau = \int_0^t \left| \frac{\partial u_i(s(\tau), \tau)}{\partial x} \right| d\tau \qquad (i = 1, 2).$$

The $\lambda_i(t)$ are nonnegative monotone nondecreasing functions. Integrating (3.19) we get

$$(3.28)\qquad \lambda_1(t) = a_1^2 \int_{-\infty}^{0} \varphi_1(x)\, dx - a_1^2 \gamma_1 s(t) - a_1^2 \int_{-\infty}^{s(t)} u_1(x, t)\, dx.$$

We proceed to estimate the integral

$$(3.29)\qquad I_1(t) = a_1^2 \int_{-\infty}^{s(t)} u_1(x, t)\, dx.$$

Introduce the function

$$(3.30)\qquad \sigma(t) = \operatorname*{l.u.b.}_{0<\tau<t} |s(\tau)|$$

and compare u_1 with two functions, z_1 and z_2. z_1 is a solution of (3.19) in the domain $-\infty < \xi < -\sigma(t)$, $0 < \tau < t$, satisfying the conditions

$$(3.31)\qquad \begin{aligned} z_1(\xi, 0) &= \varphi_1(\xi) && \text{for } -\infty < \xi < -\sigma(t), \\ z_1(-\sigma(t), \tau) &= -\gamma_1 && \text{for } 0 < \tau < t, \end{aligned}$$

and z_2 is a solution of (3.19) in the domain $-\infty < \xi < \sigma(t)$, $0 < \tau < t$, satisfying the conditions

$$(3.32)\qquad \begin{aligned} z_2(\xi, 0) &= \varphi_1(\xi) && \text{for } -\infty < \xi < 0, \\ z_2(\xi, 0) &= 0 && \text{for } 0 < \xi < \sigma(t), \\ z_2(\sigma(t), \tau) &= -\gamma_1 && \text{for } 0 < \tau < t. \end{aligned}$$

Introducing the fundamental solution of the equation (3.19),

$$(3.33)\qquad \begin{aligned} K(x, t; \xi, \tau) &= \frac{a_1}{2\pi^{1/2}(t-\tau)^{1/2}} \exp\left\{ -\frac{a_1^2(x-\xi)^2}{4(t-\tau)} \right\} \\ &= a_1 K(a_1^2 x, t; a_1^2 \xi, \tau) \end{aligned}$$

and then Green's function of (3.19) in the half-space $x < 0$,

$$(3.34)\qquad G_1(x, t; \xi, \tau) = K_1(x, t; \xi, \tau) - K_1(-x, t; \xi, \tau),$$

one easily verifies that the function z_1 defined by

$$(3.35)\quad \begin{aligned} a_1^2 z_1(\xi, \tau) = a_1^2 \int_{-\infty}^{-\sigma(t)} & G_1(\xi + \sigma(t), \tau; \xi' + \sigma(t), 0)\varphi_1(\xi')\, d\xi' \\ & + \gamma_1 \int_0^t \frac{\partial}{\partial \xi'} G_1(\xi + \sigma(t), \tau; 0, \tau')\, d\tau' \end{aligned}$$

is a solution of (3.19), (3.31) and, in addition, $z_1(\xi, \tau) \to 0$ as $\xi \to \infty$, uniformly in τ in bounded sets. Employing the maximum principle we conclude that

$$z_1(\xi, \tau) \le u_1(\xi, \tau).$$

Hence,

$$(3.36)\quad -I_1(t) \leq -a_1^2 \int_{-\infty}^{-\sigma(t)} \int_{-\infty}^{-\sigma(t)} G_1(x+\sigma(t), t; \xi+\sigma(t), 0)\varphi_1(\xi)\, d\xi\, dx$$

$$- \gamma_1 \int_{-\infty}^{-\sigma(t)} \int_0^t \frac{\partial}{\partial \xi} G_1(x+\sigma(t), t; 0, \tau)\, d\tau\, dx + 2a_1^2\sigma(t) \underset{0<\tau<t}{\text{l.u.b.}} |u_1|.$$

The first term on the right-hand side of (3.36) is bounded by $a_1^2 \int |\varphi_1(\xi)|\, d\xi$. The third term is bounded by $2a_1^2\sigma(t)B_1$ where $B_1 = \max \{\gamma_1, \text{l.u.b.} |\varphi_1|\}$. Finally, changing the order of integration in the second term on the right-hand side of (3.36) we find that this term is equal to $2\gamma_1 a_1(t/\pi)^{1/2}$. Combining all these facts we get

$$(3.37)\qquad -I_1(t) \leq a_1^2 \int_{-\infty}^{0} |\varphi_1(\xi)|\, d\xi + 2a_1^2 B_1\sigma(t) + \frac{2\gamma_1 a_1}{\pi^{1/2}}\, t^{1/2}.$$

In a similar way we define z_2, analogously to (3.35), in terms of Green's function and then conclude that $u_1(x, t) \geq z_2(x, t)$. Using the definition of z_2 we find, after some calculation, that

$$(3.38)\qquad -I_1 \geq -a_1^2 \int_{-\infty}^{0} |\varphi_1(\xi)|\, d\xi - 2a_1^2 B_1\sigma(t) + \frac{2\gamma_1 a_1}{\pi^{1/2}}\, t^{1/2}.$$

Combining (3.38), (3.37) and substituting the result in (3.28), we obtain

$$(3.39)\quad \lambda_1(t) = \frac{2\gamma_1 a_1}{\pi^{1/2}}\, t^{1/2} - \gamma_1 a_1^2 s(t) + \vartheta_1 a_1^2 \int_{-\infty}^{0} |\varphi_1(x)|\, dx + \vartheta_2 a_1^2 B_1 \sigma(t)$$

where $|\vartheta_1| \leq 2$, $|\vartheta_2| \leq 2$.

Similarly we obtain

$$(3.40)\quad \lambda_2(t) = -\frac{2\gamma_2 a_2}{\pi^{1/2}}\, t^{1/2} - \gamma_2 a_2^2 s(t) + \vartheta_3 a_2^2 \int_0^{\infty} |\varphi_2(x)|\, dx + \vartheta_4 a_2^2 B_2 \sigma(t)$$

where $B_2 = \max \{-\gamma_2, \text{l.u.b.} |\varphi_2|\}$, $|\vartheta_3| \leq 2$, $|\vartheta_4| \leq 2$.

Integrating (3.24) and using the definition of λ_i, in (3.27), and (3.39), (3.40), we find

$$\begin{aligned}(3.41)\qquad s(t) &= k_1\lambda_1(t) - k_2\lambda_2(t) \\ &= \frac{2}{\pi^{1/2}}(k_1 a_1\gamma_1 + k_2 a_2\gamma_2)t^{1/2} - (k_1\gamma_1 a_1^2 - k_2\gamma_2 a_2^2)s(t) \\ &\quad + \vartheta_1 k_1 a_1^2 \int_{-\infty}^{0} |\varphi_1(x)|\, dx - \vartheta_3 k_2 a_2^2 \int_0^{\infty} |\varphi_2(x)|\, dx \\ &\quad + (\vartheta_2 k_2 a_1^2 B_1 - \vartheta_4 k_2 a_2^2 B_2)\sigma(t).\end{aligned}$$

We wish to eliminate $\sigma(t)$ from the right-hand side of (3.41). To do that, we notice that

$$\left|\frac{ds(\tau)}{d\tau}\right| \leq k_1 \left|\frac{\partial}{\partial x} u_1(s(\tau), \tau)\right| + k_2 \left|\frac{\partial}{\partial x} u_2(s(\tau), \tau)\right|\cdot$$

Hence,

(3.42) $$\sigma(t) \leq k_1\lambda_1(t) + k_2\lambda_2(t).$$

Substituting λ_1, λ_2 from (3.39), (3.40) into (3.42), we get

(3.43) $$\sigma(t) \leq \frac{2}{\pi^{1/2}} (k_1\gamma_1 a_1 - k_2\gamma_2 a_2)t^{1/2} - (k_1\gamma_1 a_1^2 + k_2\gamma_2 a_2^2)s(t) + \vartheta_1 k_1 a_1^2 \int_{-\infty}^{0} |\varphi_1(x)|\, dx + \vartheta_3 k_2 a_2^2 \int_0^{\infty} |\varphi_2(x)|\, dx + (\vartheta_2 k_1 a_1^2 B_1 + \vartheta_4 k_2 a_2^2 B_2)\sigma(t).$$

If $k_1a_1^2$, $k_2a_2^2$ are sufficiently small then we get, from (3.43), a bound on $\sigma(t)$ of the form

(3.44) $$\sigma(t) \leq 2(k_1\gamma_1 a_1 - k_2\gamma_2 a_2)t^{1/2} + A(k_1a_1^2 + k_2a_2^2)$$

where A is independent of t. Substituting (3.44) into (3.41) we arrive at the asymptotic formula:

(3.45) $$\frac{s(t)}{t^{1/2}} = \frac{2}{\pi^{1/2}} \{k_1a_1\gamma_1(1 + 0(\beta)) + k_2a_2\gamma_2(1 + 0(\beta))\} + \frac{0(\beta)}{t^{1/2}} \qquad (\beta = k_1a_1^2 + k_2a_2^2),$$

where $\beta^{-1}0(\beta)$ is bounded (as $\beta \to 0$) uniformly with respect to t, $1 \leq t \leq \infty$. Thus, in particular,

(3.46) $$\limsup_{t\to\infty} \left| \frac{s(t)}{t^{1/2}} - \frac{2}{\pi^{1/2}} (k_1a_1\gamma_1 + k_2a_2\gamma_2) \right| = (k_1a_1 + k_2a_2)0(\beta).$$

We sum up:

Theorem 4. *If the functions ψ_1, ψ_2 satisfy the assumption* (Ψ) *then $s(t)$ satisfies the asymptotic formulas* (3.45), (3.46).

Consider the system (3.11)–(3.16) in the special case where

(3.47) $$\psi_1(x) \equiv \gamma_1 \geq 0, \qquad \psi_2(x) \equiv \gamma_2 \leq 0.$$

We try to find an explicit solution of the form

(3.48) $$s(t) = \mu t^{1/2}, \quad w_i(x, t) = f_i(z) \quad (i = 1, 2) \quad \text{where } z = x/t^{1/2}.$$

We get

(3.49) $$f_1(z) = C_1 \int_\mu^z \exp\left[-\frac{a_1^2\zeta^2}{4}\right] d\zeta, \qquad f_2(z) = C_2 \int_\mu^z \exp\left[-\frac{a_2^2\zeta^2}{4}\right] d\zeta,$$

where C_1, C_2, μ satisfy the equations

(3.50) $$\frac{1}{2}\mu = -k_1C_1 \exp\left[-\frac{a_1^2\mu^2}{4}\right] + k_2C_2 \exp\left[-\frac{a_2^2\mu^2}{4}\right],$$

(3.51) $$C_1 \int_\mu^{-\infty} \exp\left[-\frac{a_1^2\zeta^2}{4}\right] d\zeta = \gamma_1, \qquad C_2 \int_\mu^{\infty} \exp\left[-\frac{a_2^2\zeta^2}{4}\right] d\zeta = \gamma_2.$$

For small μ we find the approximate solution

$$\mu = \frac{2}{\pi^{1/2}} (k_1 a_1 \gamma_1 + k_2 a_2 \gamma_2), \tag{3.52}$$

$$C_1 = -\frac{a_1 \gamma_1}{\pi^{1/2}}, \qquad C_2 = \frac{a_2 \gamma_2}{\pi^{1/2}}. \tag{3.53}$$

(3.52) fits in with the asymptotic formula (3.46).

The special case just considered leads to the conjecture that, asymptotically ($t \to \infty$, $\beta \to 0$), $u_1(x, t)$, $u_2(x, t)$ depend only on $z = x/t^{1/2}$ and possibly tend to $f_1(z)$, $f_2(z)$ respectively. This can indeed be proved but the details, which are quite lengthy, will not be given here.

4. Another Method of Solving Stefan Problems

In Secs. 1, 2 we have solved Stefan problems by reducing them to a nonlinear integral equation for $v(t) = u_x(s(t), t)$. This method can be applied also for other types of free boundary problems (see Sec. 5) as well as to more general equations than the heat equations, namely, to parabolic equations

$$a(x, t)u_{xx} + b(x, t)u_x + c(x, t)u - u_t = f(x, t)$$

with sufficiently smooth coefficients. Since, however, the method requires representation of u in terms of fundamental solutions, it appears to be restricted to linear parabolic equations.

Another limitation lies in the fact that if $s(0) = b = 0$ then the integral equation for $v(t)$ may have a nonintegrable singularity at $t = 0$. This may sometimes be overcome by approximating the original problem with problems where $s(0) = b_n \to 0$; see Problems 1–7.

In the present section we give another method for solving Stefan problems. It applies both in case of nonlinear parabolic equations, and in case $s(0) = b = 0$. It is, however, limited to problems where the boundary condition (on the fixed boundary) is given in terms of u_x (but not in terms of u), i.e., the condition is of the form (2.14), but not of the form (1.2).

We shall describe the method in the case of the Stefan problem

$$u_{xx} = u_t \qquad \text{for } 0 < x < s(t),\ t > 0, \tag{4.1}$$

$$u_x(0, t) = f(t) < 0 \qquad \text{for } t > 0, \tag{4.2}$$

$$u(s(t), t) = 0 \qquad \text{for } t \geq 0, \text{ and } s(0) = 0, \tag{4.3}$$

$$u_x(s(t), t) = -\frac{ds(t)}{dt} \qquad \text{for } t > 0. \tag{4.4}$$

Integrating (4.1) and using (4.2)–(4.4) we get

$$s(t) = F(t) - \int_0^{s(t)} u(x, t)\, dx \tag{4.5}$$

where $F(t) = -\int_0^t f(\tau)\, d\tau$.

For any $\lambda > 0$, define a transformation T as follows. Let $\sigma(t)$ be a continuously differentiable monotone nondecreasing function defined in the interval $0 \le t \le \lambda$ and satisfying: $\sigma(0) = 0$, $\sigma(t) > 0$ if $t > 0$. Let $v(x, t)$ be the solution of the system

$$\begin{aligned} v_{xx} &= v_t && \text{for } 0 < x < \sigma(t),\ 0 < t \le \lambda, \\ v_x(0, t) &= f(t) && \text{for } 0 < t \le \lambda, \\ v(\sigma(t), t) &= 0 && \text{for } 0 < t < \lambda. \end{aligned} \tag{4.6}$$

We then set $\rho = T\sigma$ where

$$\rho(t) = F(t) - \int_0^{\sigma(t)} v(x, t)\, dx. \tag{4.7}$$

It is clear that σ is a fixed point of T if and only if the pair v, σ forms a solution of (4.1)–(4.4).

One can show that T is well defined and that $\rho(t)$ is again continuously differentiable and monotone nondecreasing, and $\rho(0) = 0$, $\rho(t) > 0$ for $t > 0$. The proof that T has a unique fixed point can be given with the aid of Theorem 1, Sec. 1, Chap. 3. Since in this book we have not discussed problems of the form (4.6), we shall not further pursue the question of existence; we shall, however, prove the uniqueness of solutions:

Theorem 5. *Assume that* $f(t)$ $(0 \le t < \infty)$ *is a continuous function. Then there exists at most one solution of the Stefan problem* (4.1)–(4.4).

The existence of a solution for (4.1)–(4.4) can be proved also by the methods of Secs. 1, 2 and it is given in Problems 1–7.

Proof. We shall need the following lemma.

Lemma 3. *Let* $\sigma(t)$ $(0 \le t \le \lambda)$ *be a continuous nondecreasing function,* $\sigma(0) = 0$, $\sigma(t) > 0$ *if* $t > 0$, *and let* $v(x, t)$ *be a solution of*

$$\begin{aligned} v_{xx} &= v_t && \text{for } 0 < x < \sigma(t),\ 0 < t < \lambda, \\ v_x(0, t) &= f(t) && \text{for } 0 < t < \lambda, \\ v(\sigma(t), t) &= 0 && \text{for } 0 \le t \le \lambda. \end{aligned} \tag{4.8}$$

Then, for any $B > \underset{0<t<\lambda}{\text{l.u.b.}} |f(t)|$,

$$0 \le v(x, t) \le B(\sigma(t) - x) \qquad \text{for } 0 \le x \le \sigma(t),\ 0 \le t \le \lambda. \tag{4.9}$$

Proof. The inequality $v \ge 0$ follows by using the maximum principle. To prove the second inequality of (4.9), consider, for any $0 < t^* \le \lambda$, the system

$$\begin{aligned} z_{xx} &= z_t && \text{for } 0 < x < \sigma(t^*),\ 0 < t < t^*, \\ z(x, 0) &= 0 && \text{for } 0 < x \le \sigma(t^*), \\ z_x(0, t) &= -B && \text{for } 0 < t \le t^*, \\ z(\sigma(t^*), t) &= 0 && \text{for } 0 < t \le \sigma(t^*). \end{aligned}$$

To prove the existence of z, we take

$$z(x, t) = \int_0^t \psi(\tau)\tilde{G}(x, t; 0, \tau)\, d\tau,$$

where $\tilde{G}$ is Green's function with respect to the half space $-\infty < x < \sigma(t^*)$, and ψ is determined by the condition that $z_x(0, t) = -B$. Using the maximum principle we find that $z \geq 0$. In particular, $z(\sigma(t), t) \geq 0$. Using again the maximum principle we find that $z - v \geq 0$ in the region $0 \leq x \leq \sigma(t)$, $0 \leq t \leq t^*$. It remains to estimate z.

Consider the function $\tilde{z}(x, t) = z(x, t) - (B + \epsilon)(\sigma(t^*) - x)$, for any $\epsilon > 0$. It satisfies the heat equation for $0 < x < \sigma(t^*)$, $0 < t < t^*$, and

$$\begin{aligned} \tilde{z}(x, 0) &\leq 0 && \text{for } 0 \leq x \leq \sigma(t^*), \\ \tilde{z}_x(0, t) &= \epsilon > 0 && \text{for } 0 < t < t^*, \\ \tilde{z}(\sigma(t^*), t) &= 0 && \text{for } 0 < t \leq t^*. \end{aligned}$$

Employing the maximum principle we conclude that $\tilde{z} \leq 0$. Hence, $v(x, t) \leq z(x, t) \leq (B + \epsilon)[\sigma(t^*) - x]$. Taking $t = t^*$, $\epsilon \to 0$ we get

$$v(x, t^*) \leq B(\sigma(t^*) - x).$$

Noting that t^* is an arbitrary point in the interval $(0, \lambda)$, the proof of the second inequality of (4.9) is completed.

We shall also need the following lemma.

Lemma 4. *If u satisfies the heat equation in a domain $0 < x < s(t)$, $0 < t \leq \lambda$ where $s(t)$ is a continuous positive function for $0 \leq t \leq \lambda$, and if $s(0) = 0$, $u(s(t), t) \leq A$ for $0 \leq t \leq \lambda$, $u_x(0, t) \geq 0$ for $0 < t \leq \lambda$, then $u(x, t) \leq A$ for $0 \leq x \leq s(t)$, $0 \leq t \leq \lambda$.*

Proof. For any $\epsilon > 0$, the function $w = u + \epsilon x$ satisfies the heat equation, and $w(s(t), t) \leq A + \epsilon A'$ $(A' = \operatorname{l.u.b.}_{0<t<\lambda} s(t))$ for $0 \leq t \leq \lambda$. Since $w_x(0, t) > 0$, w cannot take its maximum at points $(0, t)$. Hence, by the maximum principle, $w \leq A + \epsilon A'$ for $0 \leq x \leq s(t)$, $0 \leq t \leq \lambda$. Now take $\epsilon \to 0$.

Corollary. *If in Lemma 4, $|u(s(t), t)| \leq A$, $u_x(0, t) = 0$, then $|u(x, t)| \leq A$.*

We now return to the proof of Theorem 5. We shall first prove uniqueness for all $t < \lambda$, where λ is sufficiently small. Suppose u_1, s_1 and u_2, s_2 are two solutions of (4.1)–(4.4) and put

$$\begin{aligned} y(t) &= \min\,(s_1(t), s_2(t)) \\ z(t) &= \max\,(s_1(t), s_2(t)). \end{aligned}$$

By (4.5),

$$(4.10)\quad s_1(t) - s_2(t) = -\int_0^{y(t)} [u_1(x, t) - u_2(x, t)]\, dx + (-1)^i \int_{y(t)}^{z(t)} u_i(x, t)\, dx$$

where u_i is the solution between $y(t)$ and $z(t)$.

By Lemma 3, with $\sigma(\tau) = y(\tau)$,

$$(4.11)\qquad |u_1(y(\tau), \tau) - u_2(y(\tau), \tau)| \le B|s_1(\tau) - s_2(\tau)|.$$

Since $\partial(u_1 - u_2)/\partial x = 0$ on $x = 0$, the corollary to Lemma 4 yields

$$(4.12)\quad |u_1(x, t) - u_2(x, t)| \le B \operatorname*{l.u.b.}_{0 \le \tau \le t} |s_1(\tau) - s_2(\tau)| \qquad (0 \le x \le y(t)).$$

From Lemma 3 we also get,

$$(4.13)\qquad |u_i(x, t)| \le B(z(t) - x)$$

where u_i is the solution appearing on the right-hand side of (4.10). Substituting (4.12), (4.13) into (4.10) and taking the l.u.b. with respect to t, $0 < t < T$, we get

$$(4.14)\qquad S(T) \le BY(T)S(T) + B(S(T))^2$$

where

$$S(T) = \operatorname*{l.u.b.}_{0<t<T} |s_1(t) - s_2(t)|, \qquad Y(T) = \operatorname*{l.u.b.}_{0<t<T} y(t).$$

Since $S(T) \to 0$, $Y(T) \to 0$ as $T \to 0$, the inequality (4.14) cannot hold for T sufficiently small unless $S(T) = 0$. But then, $s_1(t) = s_2(t)$ for all $t < \lambda$ (λ sufficiently small), and therefore also $u_1(x, t) = u_2(x, t)$ if $t < \lambda$.

Having proved uniqueness for $0 \le t \le \lambda$, we can proceed in a similar manner to prove uniqueness step by step on the t-interval. The uniqueness for $\lambda \le t < \infty$ also follows from Theorem 2, Sec. 3, since $s(\lambda) > 0$.

5. Other Free Boundary Problems

Theorem 1, Sec. 1, remains true if (1.5) is replaced by

$$(5.1)\qquad u_x(s(t), t) = -s(t)\frac{ds(t)}{dt} = -\frac{1}{2}\frac{d(s(t))^2}{dt} \qquad \text{for } t > 0.$$

Consider the following 3-dimensional Stefan problem:

(5.2)

$$\begin{aligned}
\Delta c(x, t) &= c_t(x, t) && \text{for } a < x < s(t),\ t > 0 && (a > 0),\\
c(a, t) &= \tilde{f}(t) && \text{where } \tilde{f}(t) \ge 0 \text{ and } t \ge 0,\\
c(x, 0) &= \tilde{\varphi}(x) && \text{where } \tilde{\varphi}(x) \ge 0,\ a < x \le b, \text{ and } \tilde{\varphi}(b) = 0 && (b > a),\\
c(s(t), t) &= 0 && \text{for } t \ge 0 \text{ and } s(0) = b,\\
c_x(s(t), t) &= -\frac{ds(t)}{dt} && \text{for } t > 0,
\end{aligned}$$

where x is the 3-dimensional euclidean distance and Δ is the radial part of the Laplace operator in three variables, i.e., $\Delta = \partial^2/\partial x^2 + 2x^{-1}\,\partial/\partial x$.

Substituting $u(x, t) = xc(x, t)$ we find that $u(x, t)$, $s(t)$ satisfy the system (1.1)–(1.4) with $x = 0$ replaced by $x = a$, $f(t) = a\tilde{f}(t)$, $\varphi(x) = x\tilde{\varphi}(x)$, and the additional equation (5.1). By the remark made at the beginning of this section it follows that there exists a unique solution $c(x, t)$, $s(t)$ of the problem (5.2) for all $t < \infty$, and $s(t)$ is a nondecreasing function of t.

Consider next the following problem. An isolated 3-dimensional drop is surrounded by a totally supersaturated or a totally undersaturated vapor of its own substance. In the first case the drop will grow under the process of condensation, whereas in the second case the drop will decrease due to evaporation. Assume that in either case the drop remains spherical and that no new drops are created in the vapor surrounding the original drop. Assume also that the saturation density g is independent of the radius of the drop at any moment. Denoting the density of the vapor by $c_0(x)$, x being the distance from the center of the drop, and assuming that $\lim_{x\to\infty} c_0(x)$ exists and is $\neq g$, the problem of finding the density of the vapor $c(x, t)$ and the radius of the drop $x = s(t)$, for any $t > 0$, can be reduced (after some transformations) to the following problem:

$$
(5.3)\qquad
\begin{aligned}
u_{xx} &= u_t && \text{for } s(t) < x < \infty,\ t > 0,\\
u(x, 0) &= \varphi(x) && \text{for } b < x < \infty,\\
u(s(t), t) &= s(t) && \text{for } t > 0,\ s(0) = b > 0,\\
\alpha u_x(s(t), t) &= s(t)\,\frac{ds(t)}{dt} + \alpha && \text{for } t > 0.
\end{aligned}
$$

α is a small parameter, and $\alpha < 0$ in the case of condensation whereas $\alpha > 0$ in the case of evaporation. It is easily verified that if $\alpha < 0$ ($\alpha > 0$) then $s(t)$ increases (decreases) with t.

We assume that

$$
\operatorname*{l.u.b.}_{b<x<\infty} |\varphi(x)| < \infty, \quad \varphi'(x) \to 0 \text{ as } x \to \infty, \quad \int_b^\infty |\varphi(x)|\,dx < \infty, \quad \varphi(b) = 0,
$$

and require that

(5.4) u, u_x remain bounded as $x \to \infty$, uniformly in t in bounded sets.

If $|\alpha|$ is sufficiently small—for instance, if $|\alpha|$ l.u.b. $|\varphi| < b/20$ (in the physical applications $|\alpha|$ is in fact much smaller)—then one can establish the following results:

(1) If $\alpha < 0$ then there exists a unique solution of the problem (5.3), (5.4) for all $t < \infty$.

(2) If $\alpha > 0$ then there exists a unique solution of (5.3), (5.4) for all $t < t_0$, and $s(t) \to 0$ as $t \to t_0$ ($t_0 < \infty$).

(3) If $\alpha < 0$,

$$\limsup_{t\to\infty} \left| \frac{s^2(t)}{2|\alpha|t} - 1 \right| = o(1), \tag{5.5}$$

$$\limsup_{t\to\infty} \left| \frac{s(t)\dot{s}(t)}{|\alpha|} - 1 \right| = o(1) \qquad (\dot{s} = ds/dt), \tag{5.6}$$

where $o(1) \to 0$ as $\alpha \to 0$.

(4) Denoting the solution of (5.3), (5.4), for clarity, by $u(x, t; \alpha)$, $s(t; \alpha)$, the functions $s(t; \alpha)$, $ds(t; \alpha)/dt$ are real analytic functions of α in a neighborhood of the (equilibrium) point $\alpha = 0$, when t is restricted to any bounded set $0 \le t \le t^*$. The function $U(x, t; \alpha) \equiv u(x + s(t; \alpha), t; \alpha)$ is also analytic in α in a neighborhood of $\alpha = 0$, when $0 \le x < \infty$, $0 \le t \le t^*$.

All the above results can be derived by a detailed study of the integral equation to which the problem (5.3), (5.4) is reduced by the method of Sec. 1, and by employing comparison arguments based upon the maximum principle. The details are quite lengthy and will not be given here.

We finally mention that with the aid of (5.5), (5.6) one can derive an asymptotic formula for the solution u (compare the remarks following Theorem 4, Sec. 3). Thus one finds that if α is small and t is large then the solution is approximately a function only of $x/t^{1/2}$, α. As an immediate corollary we deduce that for large times, *the density of the supersaturated vapor at any point P is a linear function of the ratio of the radius of the drop to the distance of P from the drop's center.*

PROBLEMS

The purpose of Problems 1–7 is to establish an existence theorem for the Stefan problem (4.1)–(4.4) (note that $s(0) = b = 0$). In view of Theorem 2 it suffices to prove existence for all $t < \sigma$, for some $\sigma > 0$. We may assume that $f(t) \not\equiv 0$ in any interval $0 < t < \epsilon$. Consider, for any $0 < b < 1$, the problem (1.1), (2.14), (1.3), (1.4), (1.5) where $\varphi(x) \equiv 0$; this problem is denoted by π_b and its (unique) solution by u^b, s^b. $v^b(t) = u_x^b(s^b(t), t)$ satisfies the integral equation

$$(\bigstar) \quad v^b(t) = -2\int_0^t f(\tau)N_x(s^b(t)t; 0, \tau)\, d\tau + 2\int_0^t v^b(\tau)N_x(s^b(t), t; s^b(\tau), \tau)\, d\tau.$$

1. Prove that for every $\sigma > 0$, there exists a positive constant independent of b such that $-M \le v^b(t) \le 0$ for $0 \le t \le \sigma$.

   [*Hint:* $\int_0^t x(t-\tau)^{-1}K(x, t; 0, \tau)\, d\tau \le 1$ if $x \ge 0$, and the second integral in ($\bigstar$) is ≥ 0 if $v^b \le 0$.]

2. Prove that the family $\{s^b(t)\}$ is equicontinuous and uniformly bounded.

3. Prove that the family $\{u^b(x, t)\}$ (each function in its respective region $0 \leq x \leq s^b(t)$, $0 \leq t \leq \sigma$) is equicontinuous and uniformly bounded.

4. Taking a sequence $\{b_n\}$ $(b_n \to 0)$ such that $u^{b_n} \to u$, $s^{b_n} \to s$ uniformly, we obtain continuous functions $u(x, t)$, $s(t)$, and $u(s(t), t) = 0$, $s(t)$ nondecreasing. Prove that $s(t) > 0$ if $0 < t \leq \sigma$.

   [*Hint:* Integrate (1.1) for u^b and show that if $s(t) \equiv 0$ then $\int_0^t f(\tau)\, d\tau \equiv 0$.]

5. A well-known theorem asserts that from every subset of a family of bounded functions $\{w^b\}$ $(0 < b < 1)$ one may extract a sequence $\{w^{d_n}\}$ such that, for any (Lebesgue) integrable bounded function ψ,

$$(\bigstar\bigstar) \qquad \int_0^\sigma w^b(t)\psi(t)\, dt \to \int_0^\sigma w(t)\psi(t)\, dt$$

   as $b = d_n \to 0$, where $w(t)$ is some integrable function. Hence,

$$\int_0^{t-\epsilon} v^b(\tau) N_x(s(t), t; s(\tau), \tau)\, d\tau \to \int_0^{t-\epsilon} v(\tau) N_x(s(t), t; s(\tau), \tau)\, d\tau,$$

   for b in some subsequence of $\{b_n\}$ and for any $\epsilon > 0$. Show that also

$$\int_0^{t-\epsilon} v^b(\tau) N_x(s^b(t), t; s^b(\tau), \tau)\, d\tau \to \int_0^{t-\epsilon} v(\tau) N_x(s(t), t; s(\tau), \tau)\, d\tau;$$

   $s(t)$ and $\{b_n\}$ are defined in Problem 4.

6. Show that for any $0 < \delta < \epsilon$,

$$\int_{t-\epsilon}^{t} |v^b(\tau) N_x(s^b(t), t; s^b(\tau), \tau)|\, d\tau \leq M^2\epsilon^{1/2} + A_0 M \int_{\gamma(b)}^{\infty} y^{-1/2} e^{-y} dy,$$

$$\int_{t-\epsilon}^{t-\delta} |v(\tau) N_x(s(\tau), t; s(\tau), \tau)|\, d\tau \leq M^2\epsilon^{1/2} + A_0 M \int_{\gamma}^{\infty} y^{-1/2} e^{-y} dy,$$

   where $\gamma(b) = (s^b(t))^2/4\epsilon$, $\gamma = (s(t))^2/4\epsilon$, and A_0 is a constant; $s(t)$ is defined in Problem 4 and M is the constant appearing in Problem 1.
   [*Hint:* To prove the second inequality, use $(\bigstar\bigstar)$ with $\psi = [\operatorname{sgn}(vN_x)]N_x$ and the first inequality.]

7. Taking $b = b_{n'} \to 0$ where $\{b_{n'}\}$ is a suitable subsequence of the sequence $\{b_n\}$ defined in Problem 4, show that the right-hand side of $(\bigstar)$ tends to the same expression with $b = 0$, $v^0 = v$, and $s^0(t) = s(t) = -\int_0^t v(\tau)\, d\tau$. It then follows that $\{v^{b_{n'}}(t)\}$ converges pointwise to some function $v_0(t)$. Using the Lebesgue convergence theorem verify that $v(t) = v_0(t)$ almost everywhere. Now prove that $u(x, t)$, $s(t)$ (defined in Problem 4) form a solution of (4.1)–(4.4).

CHAPTER 9

FUNDAMENTAL SOLUTIONS FOR PARABOLIC SYSTEMS

Introduction. In this chapter we construct (complex-valued) fundamental solutions for parabolic systems (with complex-valued coefficients) of any order under very weak assumptions on the coefficients. Roughly speaking, we only assume uniform continuity in (x, t), uniform Hölder continuity in x, and boundedness. The fundamental solution is then used, as in Chap. 1, to construct a solution for the Cauchy problem and to prove its uniqueness. In addition we prove in this chapter that the fundamental solution is (roughly) as smooth as the coefficients. Thus, in particular, if the coefficients are infinitely differentiable, the same is true of the fundamental solution.

The reader will note that the parametrix used in this chapter differs from the one used in Chap. 1 in that it is constructed for equations with coefficients depending on t. (In this connection, see the remark at the end of Sec. 6, Chap. 1.) The analysis of the present chapter is somewhat more delicate than that of Chap. 1. We shall assume that the reader is familiar with the contents of Chap. 1, and, therefore, we shall sometimes omit the details of calculations similar to those carried out in that chapter.

1. Definitions

Consider the $M \times M$ system of equations

$$(1.1) \qquad \frac{\partial^{n_i} w_i}{\partial t^{n_i}} = \sum_{j=1}^{M} \sum_{2bk_0+|k| \leq 2bn_j} A_{k_0k}^{ij}(x, t) D_t^{k_0} D_x^k w_j + g_i(x, t) \qquad (i = 1, \ldots, M)$$

where $n_1, \ldots, n_M$ are positive integers, $k = (k_1, \ldots, k_n)$, $|k| = k_1 + \cdots + k_n$, $D_x^k = D_{x_1}^{k_1} \cdots D_{x_n}^{k_n}$, $D_{x_i} = \partial/\partial x_i$, $D_t = \partial/\partial t$. The $g_i(x, t)$ and the coefficients $A_{k_0k}^{ij}(x, t)$ are defined in some set Ω of points (x, t).

The operator

$$(1.2) \qquad L_i^0 w \equiv \sum_{j=1}^{M} \sum_{2bk_0+|k|=2bn_j} A_{k_0k}^{ij}(x, t) D_t^{k_0} D_x^k w_j$$

is called the *principal part* (or the *leading part*) of the operator $L_i w$ which is defined by the double sum on the right-hand side of (1.1). The coefficients of $L_i^0 w$ are called *principal coefficients* (or *leading coefficients*).

Consider the determinant

$$\det\left(\sum_{2bk_0+|k|=2bn_h} A_{k_0k}^{jh}(x,t)\lambda^{k_0}(i\xi)^k - \delta_{jh}\lambda^{n_h}\right) \tag{1.3}$$

where $\xi = (\xi_1, \ldots, \xi_n)$ is a real vector with norm $|\xi| = (\sum_{j=1}^{n} \xi_j^2)^{1/2}$ equal to 1, $\xi^k = \xi_1^{k_1} \cdots \xi_n^{k_n}$, δ_{jh} is the Kronecker symbol, i.e., $\delta_{jh} = 0$ if $j \neq h$ and $\delta_{jh} = 1$ if $j = h$, and $i = \sqrt{-1}$. Denote the roots λ of the polynomial (1.3) by $\lambda_m(\xi; x, t)$ $(m = 1, \ldots, n_1 + \cdots + n_M)$. We say that the system (1.1) is of *parabolic type* (in the sense of Petrowski), or that (1.1) is a *parabolic system* (in the sense of Petrowski), at a point (x^0, t^0) if

$$\max_j \operatorname*{l.u.b.}_{|\xi|=1} \operatorname{Re}\{\lambda_j(\xi; x^0, t^0)\} < 0.$$

If (1.1) is parabolic at each point of Ω then we simply say that (1.1) is parabolic in Ω. $2b$ is called the *parabolic weight* of the system, and $2b \max_j n_j$ is called the *order* of the system.

If there exists a constant $\delta > 0$ such that

$$\max_j \operatorname*{l.u.b.}_{|\xi|=1} \operatorname{Re}\{\lambda_j(\xi; x, t)\} < -\delta \qquad \text{for all } (x, t) \in \Omega, \tag{1.4}$$

then we say that the system (1.1) is *uniformly parabolic* in Ω. (If the principal coefficients are bounded, then the present definition contains the definition of uniform parabolicity for second-order equations as given in Sec. 1, Chap. 1.) δ is called a *module of parabolicity*.

By introducing new dependent functions

$$v_{j1} = w_j, \quad v_{j2} = \frac{\partial w_j}{\partial t}, \quad \ldots, \quad v_{j,n_j-1} = \frac{\partial^{n_j-1} w_j}{\partial t^{n_j-1}} \qquad (j = 1, \ldots, M) \tag{1.5}$$

the system (1.1) is reduced to a system of the form

$$\frac{\partial u_i}{\partial t} = \sum_{j=1}^{N} \sum_{|k|\leq 2p} A_k^{ij}(x,t) D_x^k u_j + f_i(x,t) \qquad (i = 1, \ldots, N). \tag{1.6}$$

The roots of the polynomial

$$\det\left(\sum_{|k|=2p} A_k^{jh}(x,t)(i\xi)^k - \delta_{jh}\lambda\right), \tag{1.7}$$

however, are in general different from the roots of the polynomial (1.3), and (1.6) may not be parabolic even if (1.1) is parabolic (in this connection, see Problems 1, 2).

In this book we consider mainly parabolic systems of the form (1.6). $2p$ is the *order* of the system. We only consider parabolic systems of even order since if $2p$ is an odd number then $\lambda_j(-\xi; x, t) = -\lambda_j(\xi; x, t)$ and consequently the system cannot be of parabolic type.

The *Cauchy problem* for the system (1.1) consists of finding a solution for (1.1) in a strip $R^n \times (0, T]$ satisfying the *initial conditions*

$$\frac{\partial^h w_i(x, 0)}{\partial t^h} = \varphi_{ih}(x) \qquad (h = 0, 1, \ldots, n_i - 1; i = 1, \ldots, M), \tag{1.8}$$

where φ_{ih} are given functions in R^n. For (1.6) the initial conditions are

$$u_i(x, 0) = \varphi_i(x) \qquad (i = 1, \ldots, N). \tag{1.9}$$

In the following sections we construct a fundamental solution for the system (1.6) and then solve the Cauchy problem for (1.6), (1.9). All the considerations can be extended with some modifications to the more general system (1.1), but the details will not be given here.

Throughout the remainder of this section and throughout Secs. 2–4, the set Ω where the parabolic system (1.6) is defined is a cylinder

$$\bar{D} \times [0, T] = \{(x, t); x \in \bar{D}, 0 \leq t \leq T\}$$

where D is an arbitrary domain in R^n.

By a *fundamental solution* (or a *fundamental matrix*) $\Gamma(x, t; \xi, \tau)$ of (1.6) we mean an $N \times N$ matrix of functions defined for $(x, t) \in \Omega$, $(\xi, \tau) \in \Omega$, $t > \tau$ which, as a function of (x, t) $(x \in D, \tau < t \leq T)$, satisfies (1.6) with $f_i \equiv 0$ (i.e., each column is a solution of (1.6) with $f_i \equiv 0$), and is such that

$$\lim_{t \searrow \tau} \int_D \Gamma(x, t; \xi, \tau) f(\xi)\, d\xi = f(x) \qquad \text{for all } x \in D, \tag{1.10}$$

for any continuous function $f(\xi)$ in $\bar{D}$. If D is unbounded, f is further assumed to satisfy the boundedness condition

$$f(x) = 0\{\exp [k|x|^q]\} \qquad \left(q = \frac{2p}{2p - 1}\right) \tag{1.11}$$

for some positive constant k. (The integral in (1.10) is required to exist only if $t - \tau$ is sufficiently small.)

The construction of Γ is based on the parametrix method. In the next two sections we shall construct a parametrix and study some of its properties.

2. The Parametrix

In this section we shall construct a fundamental solution $Z(x, t; \xi, \tau)$ for a parabolic system

$$\frac{\partial u_h}{\partial t} = \sum_{j=1}^{N} \sum_{|k| \leq 2p} A_k^{hj}(t) D_x^k u_j \qquad (h = 1, \ldots, N) \tag{2.1}$$

with coefficients depending only on t. This fundamental solution will be used in Sec. 3 to construct a parametrix for the system (1.6). We assume throughout this section that the coefficients $A_k^{hj}(t)$ of (2.1) are continuous functions for $0 \leq t \leq T$.

We associate with (2.1) the following system of linear ordinary differential equations:

$$\frac{dv_h}{dt} = \sum_{j=1}^{N} \sum_{|k| \leq 2p} A_k^{hj}(t)(i\zeta)^k v_j \qquad (h = 1, \ldots, N). \tag{2.2}$$

Let $V(t; \zeta, \tau) = (V^{hj}(t; \zeta, \tau))$ be the matrix solution of (2.2) satisfying the initial condition

$$V(t; \zeta, \tau)|_{t=\tau} = I \qquad (I = \text{unit matrix}). \tag{2.3}$$

$V(t; \zeta, \tau)$ is called the *Green matrix* of the system (2.2).

Observe that the system (2.2) is obtained, formally, from (2.1) by taking the Fourier transform with respect to x. Note also that the fundamental solution $Z(x, t; \xi, \tau)$ of (2.1) should depend on x, ξ only as a function of $x - \xi$. Hence, the property (1.10) of fundamental solutions becomes

$$\lim_{t \searrow \tau} Z(x, t; \xi, \tau) * f(\xi) = f(x)$$

where "$*$" means convolution. Thus, $\lim_{t \searrow \tau} Z$ operates on f like the Dirac measure. Consequently, the Fourier transform of Z should tend to 1 as $t \searrow \tau$.

It follows that, formally, the Fourier transform of Z should coincide with Green's matrix V. This motivates the following outline for the construction of Z:

Derive appropriate estimates for $V(t; \zeta, \tau)$ and then prove that the functions

$$Z^{jh}(x, t; \xi, \tau) = \frac{1}{(2\pi)^n} \int_{R^n} e^{i\alpha \cdot (x - \xi)} V^{jh}(t; \alpha, \tau)\, d\alpha \tag{2.4}$$

(where $\alpha \cdot x = \alpha_1 x_1 + \cdots + \alpha_n x_n$) are the elements of a fundamental matrix Z.

Following this outline we shall prove the following theorem.

Theorem 1. *Let* (2.1) *be a parabolic system with continuous coefficients for* $0 \leq t \leq T$. *Then there exists a fundamental solution* $Z(x, t; \xi, \tau) \equiv Z(x - \xi; t, \tau)$ *of* (2.1) *in the strip* $0 \leq t \leq T$ *whose elements are given by* (2.4), *and the following inequalities hold for* $0 \leq |m| < \infty$:

$$|D_x^m Z^{jh}(x, t; \xi, \tau)| \leq \frac{C_m}{(t - \tau)^{(n+|m|)/2p}} \exp\left\{-c_m \left(\frac{|x - \xi|^{2p}}{t - \tau}\right)^{1/(2p-1)}\right\} \tag{2.5}$$

where C_m, c_m *are positive constants depending only on* m, *on bounds on the coefficients of* (2.1), *on a modulus of continuity of the principal coefficients, and on a module of parabolicity* δ.

We begin with an elementary lemma.

Lemma 1. *Let $f(z)$ be a continuous real-valued function in the n-dimensional complex space C^n and assume that $f(\lambda z) = \lambda^{2p} f(z)$ for any $\lambda \geq 0$, $z \in C^n$, and that*

$$f(x) \leq -\delta|x|^{2p} \qquad \text{for all } x \in R^n, \tag{2.6}$$

where δ is some positive constant. Then there exists a constant A such that

$$f(x+iy) \leq -\frac{\delta}{2}|x|^{2p} + A|y|^{2p} \qquad \text{for all } x \in R^n,\ y \in R^n. \tag{2.7}$$

Proof. Denote by S the unit hypersphere of R^n. Since $f(x) < -\delta$ on S, for every $x \in S$ there exists a neighborhood of x in C^n such that $f < -\delta/2$ in this neighborhood. Covering S with a finite number of these neighborhoods, we conclude that for some $\epsilon_0 > 0$,

$$f(x+iy) < -\frac{\delta}{2} \qquad \text{for } |x+iy| = 1,\ |y| < \epsilon_0.$$

Hence, if $x \in R^n$, $y \in R^n$ and $|y| < \epsilon_0|x+iy|$, then

$$f(x+iy) = f\left(\frac{x+iy}{|x+iy|}\right)|x+iy|^{2p} < -\frac{\delta}{2}|x+iy|^{2p}. \tag{2.8}$$

The function $f(x+ie) + \frac{1}{2}\delta|x+ie|^{2p}$ as a function of x is bounded from above, independently of e, where $e \in R^n$, $|e| = 1$. Indeed, if $|x|\epsilon_0 > 1$ then it becomes negative (by (2.8)), whereas if $|x|\epsilon_0 \leq 1$ then it is bounded by a constant independent of e because it is a continuous function of x, e. Denoting by A its least upper bound which is independent of e, we then have

$$f(x+iy) + \frac{\delta}{2}|x+iy|^{2p} = |y|^{2p}\left[f(x'+ie) + \frac{\delta}{2}|x'+ie|^{2p}\right]$$
$$\leq A|y|^{2p} \qquad (y \neq 0)$$

where $x' = x/|y|$, and (2.7) is thereby proved.

The next fact we need is concerned with estimating the matrix $\exp[tP(s)]$ in terms of the eigenvalues $\lambda_j(s)$ of the $P(s)$, where $P(s)$ is an $N \times N$ matrix whose elements are polynomials in $s = (s_1, \ldots, s_n)$, $s \in C^n$. Denoting by $|B|$ the sum of the absolute values of the elements of a matrix B we have:

$$|\exp[tP(s)]| \leq B_0(1 + t^{1/2p} + t^{1/2p}|s|)^{2p(N-1)} \exp\{t \max_j \operatorname{Re}[\lambda_j(s)]\} \tag{2.9}$$

where B_0 is a constant depending only on bounds on the coefficients of the polynomials of the elements of $P(s)$. For proof, see [44; 168–171].

Set

$$P(t,\zeta) = P_0(t,\zeta) + P_1(t,\zeta)$$

where

$$P_0(t,\zeta) = \left(\sum_{|k|=2p} A_k^{hj}(t)(i\zeta)^k\right),$$

$$P_1(t,\zeta) = \left(\sum_{0\leq|k|<2p} A_k^{hj}(t)(i\zeta)^k\right).$$

Applying Lemma 1 to the real part of each eigenvalue $\lambda_{0j}(\tau, \zeta)$ of the matrix $P_0(\tau, \zeta)$ we obtain a bound on $\max_j \text{Re}\,[\lambda_{0j}(\tau, \zeta)]$. Substituting this bound into (2.9) (with $tP(s)$ replaced by $(t - \tau)P_0(\tau, \zeta)$) and using the inequality $t^{N-1}e^{-\epsilon t} \leq$ const. for $\epsilon = \delta/4$ and $0 \leq t < \infty$, we obtain

$$(2.10) \quad |\exp (t - \tau)P_0(\tau, \zeta)| \leq B_1 \exp [(t - \tau)Q(\xi, \eta)] \qquad (0 \leq \tau \leq t \leq T, \zeta = \xi + i\eta)$$

where

$$(2.11) \qquad Q(\xi, \eta) = -\frac{\delta}{4}|\xi|^{2p} + A|\eta|^{2p}$$

and where B_1, A are constants independent of t, τ, ζ.

Write the system (2.2), (2.3) in the form

$$(2.12) \qquad \frac{dV}{dt} = P_0(t^*, \zeta)V + g(t, \zeta), \qquad V|_{t=\tau} = I,$$

where

$$(2.13) \qquad g(t, \zeta) = \{[P_0(t, \zeta) - P_0(t^*, \zeta)] + P_1(t, \zeta)\} V(t; \zeta, \tau).$$

The solution of (2.12) can then be written in the form

$$(2.14) \quad V(t; \zeta, \tau) = \exp [(t - \tau)P_0(t^*, \zeta)] + \int_\tau^t \exp [(t - \sigma)P_0(t^*, \zeta)]g(\sigma, \zeta)\, d\sigma.$$

Using (2.10) we then obtain

$$(2.15) \quad |V(t; \zeta, \tau)| \leq B_1 \exp [(t - \tau)Q(\xi, \eta)] + B_1 \int_\tau^t \exp [(t - \sigma)Q(\xi, \eta)]|g(\sigma, \zeta)|\, d\sigma.$$

Choosing $t^* = \tau$ and taking $\tau \leq t \leq \tau + \epsilon$, we obtain from (2.13),

$$(2.16) \qquad |g(t, \zeta)| \leq [\omega(\epsilon)|\zeta|^{2p} + B_2|\zeta|^{2p-1}]|V(t; \zeta, \tau)|$$

where $\omega(\epsilon) \searrow 0$ if $\epsilon \searrow 0$ and B_2 is a constant independent of t, τ, ζ, ϵ. Using

$$B_2|\zeta|^{2p-1} \leq \omega(\epsilon)|\zeta|^{2p} \qquad \text{if } |\zeta| \geq B_2/\omega(\epsilon),$$

we can simplify the right-hand side of (2.16) and then, upon substitution into (2.15), get

$$(2.17) \quad |V(t; \zeta, \tau)| \leq B_1 \exp [(t - \tau)Q(\xi, \eta)] + 2B_1\omega(\epsilon)|\zeta|^{2p} \int_\tau^t \exp [(t - \sigma)Q(\xi, \eta)]|V(\sigma; \zeta, \tau)|\, d\sigma.$$

We now need the following lemma (for proof see Problem 3).

Lemma 2. *Let φ, ψ, χ be real-valued continuous functions for $\tau \leq t \leq T$ and let $\chi(t) \geq 0$ for $\tau \leq t \leq T$. If*

$$\varphi(t) \leq \psi(t) + \int_\tau^t \chi(\sigma)\varphi(\sigma)\, d\sigma \qquad \text{for } \tau \leq t \leq T$$

then

$$\varphi(t) \leq \psi(t) + \int_\tau^t \chi(\sigma)\psi(\sigma) \exp\left[\int_\sigma^t \chi(\rho)\, d\rho\right] d\sigma \qquad \text{for } \tau \leq t \leq T.$$

Taking $\varphi(t) = |V(t; \zeta, \tau)| \exp[-(t - \tau)Q(\xi, \eta)]$, $\psi(t) = B_1$, $\chi(t) = 2B_1\omega(\epsilon)|\zeta|^{2p}$ and noting that

$$(2.18) \quad \int_\sigma^t \chi(\rho)\, d\rho = 2B_1\omega(\epsilon)(t - \sigma)|\zeta|^{2p} \leq (t - \sigma)\left(\frac{\delta}{8}|\xi|^{2p} + |\eta|^{2p}\right) \qquad (\zeta = \xi + i\eta)$$

if ϵ is sufficiently small (independently of t, σ, ζ), we get from (2.17),

$$|V(t; \zeta, \tau)| \exp[-(t - \tau)Q(\xi, \eta)] \leq B_1 + 2B_1^2\omega(\epsilon)|\zeta|^{2p} \int_\tau^t \exp\left[(t - \sigma)\left(\frac{\delta}{8}|\xi|^{2p} + |\eta|^{2p}\right)\right] d\sigma.$$

Hence,

$$|V(t; \zeta, \tau)| \leq B_3 \exp[(t - \tau)Q(\xi, \eta)] \cdot \left\{1 + (t - \tau)|\zeta|^{2p} \exp\left[(t - \tau)\left(\frac{\delta}{8}|\xi|^{2p} + |\eta|^{2p}\right)\right]\right\}.$$

It follows that

$$(2.19) \qquad |V(t; \xi + i\eta, \tau)| \leq B_4 \exp\{(t - \tau)[-\delta_0|\xi|^{2p} + A_0|\eta|^{2p}]\},$$

where B_4, A_0, δ_0 are positive constants independent of t, τ, ξ, η. (2.19) was proved under the assumption that $|\zeta| \geq B_2/\omega(\epsilon)$ where ϵ is now fixed so that (2.18) holds. If $|\zeta| < B_2/\omega(\epsilon)$ then (2.19) clearly holds for an appropriate constant B_4.

Having estimated $V(t; \zeta, \tau)$ in the interval $\tau \leq t \leq \tau + \epsilon$, we can now estimate it in the interval $\tau + \epsilon \leq t \leq \tau + 2\epsilon$ by using the identity (see Problem 4)

$$(2.20) \qquad V(t; \zeta, \tau) = V(t; \zeta, \tau + \epsilon)V(\tau + \epsilon; \zeta, \tau).$$

Thus,

$$|V(t; \zeta, \tau)| \leq |V(t; \zeta, \tau + \epsilon)|\, |V(\tau + \epsilon; \zeta, \tau)|$$

and each factor on the right-hand side can be estimated by using the inequality (2.19) with appropriate t, τ. Proceeding similarly step by step we thus derive the inequality (2.19) for all $\tau \leq t \leq T$.

It may be mentioned that if we begin with the equality

$$V(t; \zeta, \tau) = I + \int_\tau^t P(\sigma, \zeta)V(\sigma; \zeta, \tau)\, d\sigma,$$

take absolute values and thus get

$$|V(t; \zeta, \tau)| \leq N + \int_\tau^t |P(\sigma, \zeta)|\, |V(\sigma; \zeta, \tau)|\, d\sigma,$$

and then use the inequality $|P(\sigma, \zeta)| \leq \text{const.}\ (1 + |\zeta|^{2p})$ and Lemma 2, we arrive at the inequality

$$|V(t; \zeta, \tau)| \leq \text{const.} \exp [\text{const.}\ (t - \tau)|\zeta|^{2p}] \tag{2.21}$$

where the constants are positive. (2.21) is a weaker inequality than (2.19).

We shall now use (2.19) to estimate the functions Z^{jh} defined in (2.4). Since the coefficients of (2.2) are entire functions of ζ, $\zeta \in C^n$, by a standard theorem for ordinary differential equations it follows that $V^{jh}(t; \zeta, \tau)$ are also entire functions of ζ. Using Cauchy's theorem and (2.19) we find that the function

$$Z^{jh}(x, t; \xi, \tau) = \frac{1}{(2\pi)^n} \int_{R^n} e^{i(\alpha + i\beta)\cdot(x-\xi)} V^{jh}(t; \alpha + i\beta, \tau)\, d\alpha \qquad (t > \tau)$$

is independent of β and, hence, coincides with the function defined in (2.4).

Using (2.19) again, we get,

$$|Z^{jh}(x, t; \xi, \tau)| \leq B_5 e^{-\beta\cdot(x-\xi)} \exp [A_0(t - \tau)|\beta|^{2p}] \cdot \int_{R^n} \exp [-\delta_0(t - \tau)|\alpha|^{2p}]\, d\alpha. \tag{2.22}$$

We shall need the fact (see Problem 5) that for any $\kappa > 0$ there exists a $\mu > 0$ such that

$$\mu\kappa = \frac{\mu^{2p}}{2p} + \frac{\kappa^q}{q} \qquad \left(q = \frac{2p}{2p - 1}\right). \tag{2.23}$$

Taking $\beta_k = |\beta_k| \operatorname{sgn} (x_k - \xi_k)$, $|\beta_1| = \cdots = |\beta_n|$,

$$\kappa = \frac{|x - \xi|}{n[2pA_0(t - \tau)]^{1/2p}}, \tag{2.24}$$

and then choosing $|\beta|$ so that (2.23) holds for $\mu = [2pA_0(t - \tau)]^{1/2p}|\beta|$, we obtain

$$-\beta\cdot(x - \xi) + (t - \tau)A_0|\beta|^{2p} \leq -|\beta| \frac{|x - \xi|}{n} + (t - \tau)A_0|\beta|^{2p} = -A_1 \left(\frac{|x - \xi|^{2p}}{t - \tau}\right)^{1/(2p-1)},$$

for some constant $A_1 > 0$. Substituting this into (2.22) we get

$$|Z^{jh}(x, t; \xi, \tau)| \leq B_5 \exp \left\{-A_1 \left(\frac{|x - \xi|^{2p}}{t - \tau}\right)^{1/(2p-1)}\right\} \cdot \int_{R^n} \exp [-\delta_0(t - \tau)|\alpha|^{2p}]\, d\alpha. \tag{2.25}$$

Taking polar coordinates and then substituting $\rho = |\alpha|(t - \tau)^{1/2p}$ we find that the integral on the right-hand side of (2.25) is bounded by a constant times

$$(t - \tau)^{-n/2p} \int_0^\infty \rho^{n-1} \exp [-\delta_0 \rho^{2p}]\, d\rho \leq \text{const.}\ (t - \tau)^{-n/2p}.$$

Substituting this into (2.25) we obtain (2.5) for $m = 0$.

To prove (2.5) for $|m| > 0$, we begin with the equality

$$D_x^m Z^{jh}(x, t; \xi, \tau) = \frac{1}{(2\pi)^n} \int_{R^n} (i(\alpha + i\beta))^m e^{i(\alpha+i\beta)\cdot(x-\xi)} V^{jh}(t; \alpha + i\beta, \tau)\, d\alpha.$$

It follows that

$$\begin{aligned}(2.26)\quad |D_x^m Z^{jh}(x, t; \xi, \tau)| &\leq B_6 |\beta|^{|m|} e^{-\beta\cdot(x-\xi)} \exp\,[A_0(t-\tau)|\beta|^{2p}] \\ &\quad \cdot \int_{R^n} \exp\,[-\delta_0(t-\tau)|\alpha|^{2p}]\, d\alpha \\ &\quad + B_6 e^{-\beta\cdot(x-\xi)} \exp\,[A_0(t-\tau)|\beta|^{2p}] \int_{R^n} |\alpha|^{|m|} \\ &\quad \cdot \exp\,[-\delta_0(t-\tau)|\alpha|^{2p}]\, d\alpha \\ &\equiv I + J.\end{aligned}$$

To estimate J we proceed as in the case $m = 0$ and note that

$$\begin{aligned}\int_{R^n} |\alpha|^{|m|} \exp\,[-\delta_0(t-\tau)|\alpha|^{2p}]\, d\alpha &\leq \text{const.}\ (t-\tau)^{-(n+|m|)/2p} \int_0^\infty \rho^{n+|m|-1} \\ &\quad \cdot \exp\,[-\delta_0 \rho^{2p}]\, d\rho \\ &\leq \text{const.}\ (t-\tau)^{-(n+|m|)/2p}.\end{aligned}$$

Hence J is bounded by the right-hand side of (2.5).

As for I, we also proceed as before, but note (see Problem 5) that, with κ given, the μ for which (2.23) holds is $\mu = \kappa^{q-1}$. Since $|\beta|$ is chosen so that $\mu = [2pA_0(t-\tau)]^{1/2p}|\beta|$, and since $q - 1 = 1/(2p-1)$, we get

$$(2.27)\quad |\beta| = A_2 \left(\frac{|x-\xi|}{(t-\tau)^{1/2p}}\right)^{1/(2p-1)} (t-\tau)^{-1/2p} \qquad (A_2 \text{ positive constant}).$$

Replacing the factor $|\beta|^{|m|}$ in I by the $|m|$th power of the right-hand side of (2.27) and estimating the remaining factors of I as in the case of $m = 0$ we find, upon using the inequality $t^{1/2p}e^{-\epsilon t} \leq$ const. for any fixed $\epsilon > 0$ and $0 \leq t < \infty$, that I is also bounded by the right-hand side of (2.5). In view of (2.26), the proof of (2.5) for any m is thereby completed. The assertion concerning the constants C_m, c_m follows from the previous proof.

Since the $V^{jh}(x, \xi + i\eta, \tau)$ decrease exponentially in ξ as $|\xi| \to \infty$, the standard rules for the Fourier transform are valid and thus, by taking the inverse Fourier transform of (2.2) and using (2.4), we conclude that $Z(x, t; \xi, \tau)$, as a function of (x, t), satisfies (2.1).

To complete the proof of Theorem 1, it remains to show that if $f(x)$ is any continuous function in R^n satisfying (1.11) then

$$(2.28)\qquad \lim_{t \searrow \tau} \int_{R^n} Z(x-\xi; t, \tau) f(\xi)\, d\xi = f(x).$$

Since

$$\int_{R^n} Z(x-\xi; t, \tau)\, d\xi = V(t; 0, \tau) \to 1 \qquad \text{as } t \searrow \tau,$$

writing $f(\xi) = f(x) + [f(\xi) - f(x)]$ we conclude that it suffices to prove that

$$\int_{R^n} Z(x - \xi; t, \tau)[f(\xi) - f(x)]\, d\xi \to 0 \qquad \text{as } t \searrow \tau. \tag{2.29}$$

We break the last integral into two parts: I_1 with $|\xi - x| < \delta$ and I_2 with $|\xi - x| > \delta$, and take δ such that $|f(\xi) - f(x)| < \epsilon$ if $|\xi - x| < \delta$. Here ϵ is any fixed positive number, and δ is a fixed positive number depending on ϵ. If we use (2.5) to estimate $|Z|$, and then substitute $|x - \xi| = \rho(t - \tau)^{1/2p}$, then we obtain

$$\int_{R^n} |Z(x - \xi; t, \tau)|\, d\xi \le C \tag{2.30}$$

where C is independent of t, τ. Similarly, we find that

$$\int_{|\xi - x| > \delta} |Z(x - \xi; t, \tau)| \exp\,[2k|x - \xi|^q]\, d\xi \to 0 \qquad \text{if } t \searrow \tau. \tag{2.31}$$

Using (2.30) it follows that

$$|I_1| \le \epsilon \int_{|\xi - x| < \delta} |Z(x - \xi; t, \tau)|\, d\xi \le \epsilon \int_{R^n} |Z(x - \xi; t, \tau)|\, d\xi \le C\epsilon. \tag{2.32}$$

Using (2.31) and the inequality

$$|f(\xi)| \le \text{const.} \exp\,[2k|x - \xi|^q]$$

where the constant depends on x, we find that $I_2 \to 0$ if $t \searrow \tau$ (for each fixed x). Hence, $|I_2| < \epsilon$ if $t - \tau$ is sufficiently small. Combining this with (2.32) we get $|I_1 + I_2| \le (C + 1)\epsilon$ if $t - \tau$ is sufficiently small. Since ϵ is arbitrary, (2.29) follows.

3. The Parametrix for Equations with Parameters

Consider a parabolic system

$$\frac{\partial u_i}{\partial t} = \sum_{j=1}^{N} \sum_{|k| \le 2p} A_k^{ij}(y, t) D_x^k u_j \equiv P_i(y, t, D_x)u \qquad (i = 1, \ldots, N) \tag{3.1}$$

with coefficients $A_k^{ij}(y, t)$ which depend on t and on a parameter y; y varies in R^n. Set

$$\begin{aligned} P_{0i}(y, t, D_x)u &= \sum_{j=1}^{N} \sum_{|k| = 2p} A_k^{ij}(y, t) D_x^k u_j, \\ P_{1i}(y, t, D_x)u &= \sum_{j=1}^{N} \sum_{|k| < 2p} A_k^{ij}(y, t) D_x^k u_j, \end{aligned} \tag{3.2}$$

so that $P_i = P_{0i} + P_{1i}$. P_{0i} is the principal part of P_i. We shall need the following assumptions:

(A) The coefficients of (3.1) are bounded functions in $\Omega = \{(y, t);\ y \in R^n,\ 0 \le t \le T\}$, and are continuous in t; the coefficients of P_{0i} are continuous in t, uniformly with respect to (y, t) in Ω.

(B) The system (3.1) is uniformly parabolic with respect to (y, t) in Ω.

Denoting the fundamental solution constructed in Sec. 2 for the system (3.1) (with y fixed) by $Z(x - \xi; t; y, \tau)$ we have:

(3.3)
$$|D_x^m Z(x - \xi, t; y, \tau)| \leq \frac{C_m}{(t - \tau)^{(n+|m|)/2p}} \exp\left\{-c_m \left(\frac{|x - \xi|^{2p}}{t - \tau}\right)^{1/(2p-1)}\right\}$$

where C_m, c_m are positive constants independent of y.

Lemma 3. *Consider two $N \times N$ parabolic systems of order $2p$ of the form*

$$\frac{\partial u_i}{\partial t} = P_i(y, t, D_x)u \qquad (i = 1, \ldots, N), \tag{3.4}$$

$$\frac{\partial u_i}{\partial t} = Q_i(y, t, D_x)u \qquad (i = 1, \ldots, N), \tag{3.5}$$

satisfying (A), (B) *and denote by $Z(x - \xi, t; y, \tau)$, $\hat{Z}(x - \xi, t; y, \tau)$ the fundamental solutions of* (3.4), (3.5) *respectively (constructed in Sec.* 2). *Assume that the coefficients $A_k^{ij}(y, t)$ and $B_k^{ij}(y, t)$ of* (3.4) *and* (3.5) *respectively have r derivatives with respect to y, which are bounded continuous functions of (y, t) in Ω, and let*

$$|D_y^s[A_k^{ij}(y, t) - B_k^{ij}(y, t)]| \leq \epsilon \qquad \text{in } \Omega \qquad (0 \leq |s| \leq r). \tag{3.6}$$

Then Z and $\hat{Z}$ has r continuous derivatives with respect to y, and each such derivative has derivatives of any order with respect to x. Furthermore,

$$|D_x^m D_y^s[Z(x - \xi, t; y, \tau) - \hat{Z}(x - \xi, t; y, \tau)]| \leq \frac{C_m \epsilon}{(t - \tau)^{(n+|m|)/2p}} \tag{3.7}$$
$$\cdot \exp\left\{-c_m \left(\frac{|x - \xi|^{2p}}{t - \tau}\right)^{1/(2p-1)}\right\} \qquad (0 \leq |m| < \infty, 0 \leq |s| \leq r)$$

where C_m, c_m depend only on bounds on $D_y^s A_k^{ij}(y, t)$, $D_y^s B_k^{ij}(y, t)$ $(0 \leq |s| \leq r)$, on moduli of continuity (in t) for the principal coefficients of (3.4), (3.5), *and on modules of parabolicity of* (3.4), (3.5) *(all these quantities are taken to be independent of y).*

Proof. Let $V = V(y, t; \zeta, \tau)$, $\hat{V} = \hat{V}(y, t; \zeta, \tau)$ be the Fourier transforms of Z, $\hat{Z}$ respectively. V satisfies the system

$$\begin{aligned} &\frac{\partial V^i}{\partial t} = P_i(y, t, \zeta)V \qquad (i = 1, \ldots, N), \\ &V|_{t=\tau} = I. \end{aligned} \tag{3.8}$$

By a standard theorem for ordinary differential equations it follows that $D_y V$ exists. Differentiating (3.8) with respect to y we get, for $D_y V$, the system

$$\frac{\partial(D_y V^i)}{\partial t} = P_i(y, t, \zeta)(D_y V) + [D_y P_i(y, t, \zeta)]V, \tag{3.9}$$
$$D_y V|_{t=\tau} = 0.$$

Using the estimate (2.19) which now holds uniformly with respect to y, and employing the considerations of Sec. 2 for the system (3.9), we derive for $D_y V$ the bound (compare (2.19))

$$|D_y V(y, t; \xi + i\eta, \tau)| \leq B \exp\{(t - \tau)[-\delta_0|\xi|^{2p} + A_0|\eta|^{2p}]\} \tag{3.10}$$

for some positive constants B, δ_0, A_0. Differentiating (3.9) once more with respect to y and using the inequalities already derived for $|V|$, $|D_y V|$, we get, by the considerations of Sec. 2, a bound on $D_y^2 V$ of the same form as (3.10). We can proceed in this manner to derive bounds on $D_y^s V$ for all $|s| \leq r$. Taking the inverse Fourier transform we find that $D_y^s Z(x - \xi, t; y, \tau)$ exist and are bounded by the right-hand side of (2.5) with $m = 0$. Bounds on $D_x^m D_y^s Z$ are next obtained by the same arguments as in Sec. 2. We conclude that all the derivatives $D_x^m D_y^s Z$ $(0 \leq |m| < \infty, 0 \leq |s| \leq r)$ exist and are continuous functions, and

$$|D_x^m D_y^s Z(x - \xi, t; y, \tau)| \leq \frac{B_m}{(t - \tau)^{(n+|m|)/2p}} \exp\left\{-b_m\left(\frac{|x - \xi|^{2p}}{t - \tau}\right)^{1/(2p-1)}\right\} \tag{3.11}$$

for $0 \leq |m| < \infty$, $0 \leq |s| \leq r$, where B_m, b_m are some positive constants depending only on upper bounds on $D_y^s A_k^{ij}(y, t)$ $(0 \leq |s| \leq r)$, on moduli of continuity (in t) of the principal coefficients of P_i, and on a module of parabolicity of (3.4); all these quantities are taken to be independent of y.

The same considerations are valid also for $\hat{Z}(x - \xi, t; y, \tau)$.

Consider now the matrix $W = V - \hat{V}$. Its elements W^{hj} satisfy the system

$$\frac{dW^{hj}}{dt} = \sum_{m=1}^{N} \sum_{|k| \leq 2p} A_k^{hm}(y, t)(i\zeta)^k W^{mj}$$
$$+ \sum_{m=1}^{N} \sum_{|k| \leq 2p} [A_k^{hm}(y, t) - B_k^{hm}(y, t)](i\zeta)^k \hat{V}^{mj}, \tag{3.12}$$
$$W^{hj}|_{t=\tau} = 0.$$

Writing (3.12) in the matrix form

$$\frac{dW}{dt} = P(y, t, \zeta)W + g(y, t, \zeta), \tag{3.12$'$}$$
$$W|_{t=\tau} = 0,$$

and using the assumption (3.6) and the inequalities for $D_y^s \hat{V}$ which are the same as those derived above for $D_y^s V$, we get

$$|D_y^s g(y, t, \zeta)| \leq B'\epsilon(1 + |\zeta|^{2p}) \exp\{(t - \tau)[-\delta_0'|\xi|^{2p} + A_0'|\eta|^{2p}]\} \tag{3.13}$$

where B', δ_0', A_0' are positive constants independent of ζ, ϵ. Using (3.13) for $s = 0$ we obtain, by applying to W considerations similar to those used in estimating the solution of (2.2), (2.3),

$$(3.14) \qquad |W(y, t; \zeta, \tau)| \leq B''\epsilon \exp \{(t - \tau)[-\delta'|\xi|^{2p} + A'|\eta|^{2p}]\}$$

for some positive constants B'', δ' A' independent of ζ, ϵ.

Differentiating (3.12′) once with respect to y and using (3.13) for $0 \leq |s| \leq 1$ and (3.14) we find that $D_y W$ is bounded by the right-hand side of (3.14) (with different constants).

By further differentiation of (3.12′) with respect to y we obtain bounds on all the y-derivatives of W of orders $\leq r$; the bounds are the same as in (3.14), but with different constants. Using these bounds and taking the inverse Fourier transform of $D_y^s W$, we get the inequalities (3.7) for $m = 0$. To derive the inequalities (3.7) for any m, we use the same argument as in Sec. 2 (in connection with deriving (2.5) in case $|m| > 0$).

Taking in Lemma 3

$$Q_i(y, t, D_x) = P_i(y + h, t, D_x)$$

we obtain the following result.

Lemma 4. *Assume that the system* (3.1) *satisfies* (A), (B) *and that* $D_y^s A_k^{ij}(y, t)$ $(0 \leq |s| \leq r)$ *are continuous bounded functions in* Ω *satisfying a Hölder condition (exponent* α*) in* y*, uniformly in* Ω*. Then, for all* $0 \leq |m| < \infty$, $0 \leq |s| \leq r$,

(3.15)

$$|D_x^m D_y^s Z(x - \xi, t; y, \tau)| \leq \frac{C_m}{(t - \tau)^{(n+|m|)/2p}} \exp \left\{-c_m \left(\frac{|x - \xi|^{2p}}{t - \tau}\right)^{1/(2p-1)}\right\},$$

$$(3.16) \quad |D_x^m D_y^s [Z(x - \xi, t; y + h, \tau) - Z(x - \xi, t; y, \tau)]|$$
$$\leq \frac{C_m |h|^\alpha}{(t - \tau)^{(n+|m|)/2p}} \exp \left\{-c_m \left(\frac{|x - \xi|^{2p}}{t - \tau}\right)^{1/(2p-1)}\right\}$$

where C_m, c_m *are positive constants (independent of* y, x, ξ, t, τ, h).

Remark. From the proof of (3.11) it follows that (3.15) (or (3.11)) remains true even if the assumption on the Hölder continuity of $D_y^s A_k^{ij}(y, t)$ is omitted.

Lemma 5. *Assume that the system* (3.1) *satisfies the assumptions of Lemma* 4 *with* $r = 0$. *Let* $f(x, t)$ *be a continuous function in* Ω, *satisfying*

$$(3.17) \qquad |f(x, t)| \leq \text{const.} \exp [a|x|^q] \qquad \left(q = \frac{2p}{2p - 1}\right)$$

for some $a < cT^{-1/(2p-1)}$ *where* $c = \min_{|m| \leq 2p} \{c_m\}$, *the* c_m *are the constants appearing in* (3.15), (3.16). *Assume further that* $f(x, t)$ *is Hölder continuous* *in x, uniformly with respect to (x, t) in bounded sets of Ω. Then the integral*

$$\Phi(x, t) = \int_0^t d\tau \int_{R^n} Z(x - \xi, t; \xi, \tau) f(\xi, \tau)\, d\xi \tag{3.18}$$

is convergent for $0 \le t \le T$, $D_x^k \Phi(x, t)$ $(0 \le |k| \le 2p)$ and $D_t\Phi(x, t)$ exist and are continuous functions for $0 < t \le T$, and

$$D_x^k\Phi(x, t) = \int_0^t d\tau \int_{R^n} D_x^k Z(x - \xi, t; \xi, \tau) f(\xi, \tau)\, d\xi, \tag{3.19}$$

$$D_t\Phi(x, t) = f(x, t) + \int_0^t d\tau \int_{R^n} D_t Z(x - \xi, t; \xi, \tau) f(\xi, \tau)\, d\xi. \tag{3.20}$$

Proof. The proof is based on the same method as in Chap. 1, Sec. 3; the breaking of integrals into several parts is similar. The fact that the estimates for Z and its derivatives (taken from Lemma 4) are now different does not present any difficulty. The details are therefore omitted.

Note that the integrals of (3.18)–(3.20) with respect to τ are improper at $\tau = t$.

4. Construction of Fundamental Solutions. The Cauchy Problem

We shall consider the system (1.6) and make the following assumptions:

(A_1) The coefficients of (1.6) are continuous bounded functions in $\Omega = R^n \times [0, T]$ and, furthermore, the principal coefficients are continuous in t uniformly with respect to (x, t) in Ω.

(A_2) The coefficients of (1.6) are Hölder continuous (exponent α) in x, uniformly with respect to (x, t) in bounded subsets of Ω, and, furthermore, the principal coefficients are Hölder continuous (exponent α) in x uniformly with respect to (x, t) in Ω.

Theorem 2. *Assume that* (1.6) *is uniformly parabolic in $\Omega = R^n \times [0, T]$ and that (A_1), (A_2) hold. Then there exists a fundamental solution $\Gamma(x, t; \xi, \tau)$ of* (1.6) *satisfying the inequalities:*

$$|D_x^m \Gamma(x, t; \xi, \tau)| \le \frac{C}{(t - \tau)^{(n+|m|)/2p}} \exp\left\{-c\left(\frac{|x - \xi|^{2p}}{t - \tau}\right)^{1/(2p-1)}\right\} \tag{4.1}$$

for $0 \le |m| < 2p$, where C, c are positive constants.

Proof. Write the system (1.6), for $f_i \equiv 0$, in the form

$$\frac{\partial u_i}{\partial t} = P_i(x, t, D_x)u = P_{0i}(x, t, D_x)u + P_{1i}(x, t, D_x)u \tag{4.2}$$

where P_{0i} is the principal part of P_i. Let $Z(x - \xi, t; y, \tau)$ be the fundamental solution of the system

$$\frac{\partial u_i}{\partial t} = P_{0i}(y, t, D_x)u \tag{4.3}$$

constructed in Sec. 2. Following the parametrix method we try to construct Γ in the form

(4.4)

$$\Gamma(x, t; \xi, \tau) = Z(x - \xi, t; \xi, \tau) + \int_\tau^t d\sigma \int_{R^n} Z(x - y, t; y, \sigma)\Phi(y, \sigma; \xi, \tau)\, dy$$

where Φ is an $N \times N$ matrix.

Let K be the $N \times N$ matrix whose ith column K^i is given by

$$\begin{aligned}(4.5)\quad K^i(x, t; \xi, \tau) &= [P_i(x, t, D_x) - \delta_i D_t]Z(x - \xi, t; \xi, \tau) \\ &= [P_{0i}(x, t, D_x) - P_{0i}(\xi, t, D_x)]Z(x - \xi, t; \xi, \tau) \\ &\quad + P_{1i}(x, t, D_x)Z(x - \xi, t; \xi, \tau),\end{aligned}$$

where $\delta_i D_t Z = D_t Z^i$ by definition. Using (3.3) and (A_2) we get

$$(4.6)\quad |K(x, t; \xi, \tau)| \leq \frac{A_1}{(t - \tau)^{(n+2p-\alpha)/2p}} \exp\left\{-a_1 \left(\frac{|x - \xi|^{2p}}{t - \tau}\right)^{1/(2p-1)}\right\},$$

where A_i, a_i are used to denote appropriate positive constants.

If Φ is Hölder continuous so that Lemma 5 can be applied to the integral on the right-hand side of (4.4), then Γ satisfies the system (4.2) (as a function of (x, t)) if and only if

$$(4.7)\quad \Phi(x, t; \xi, \tau) = K(x, t; \xi, \tau) + \int_\tau^t d\sigma \int_{R^n} K(x, t; y, \sigma)\Phi(y, \sigma; \xi, \tau)\, dy.$$

The series

$$(4.8)\qquad \Phi(x, t; \xi, \tau) = \sum_{m=1}^{\infty} K_m(x, t; \xi, \tau)$$

where $K_1 = K$,

$$(4.9)\qquad K_m(x, t; \xi, \tau) = \int_\tau^t d\sigma \int_{R^n} K_1(x, t; y, \sigma)K_{m-1}(y, \sigma; \xi, \tau)\, dy$$

is a formal solution of (4.7). To prove that the series is convergent and is a solution of (4.7), we shall need the following lemma.

Lemma 6. *Let*

$$I_a = \int_{-\infty}^{\infty} \frac{1}{[(t - \sigma)(\sigma - \tau)]^{1/2p}} \exp\{-af(x, \xi, y; t, \tau, \sigma)\}\, dy$$

where $\tau < \sigma < t$, $-\infty < x < \infty$, $-\infty < y < \infty$,

$$f(x, \xi, y; t, \tau, \sigma) = \left(\frac{|x - y|^{2p}}{t - \sigma}\right)^{1/(2p-1)} + \left(\frac{|y - \xi|^{2p}}{\sigma - \tau}\right)^{1/(2p-1)}$$

and a *is a positive number. For any* $0 < \epsilon < 1$ *there exists a constant* M *depending only on* ϵ, a, p, *such that*

$$(4.10)\qquad I_a \leq \frac{M}{(t - \tau)^{1/2p}} \exp\left\{-a(1 - \epsilon)\left(\frac{|x - \xi|^{2p}}{t - \tau}\right)^{1/(2p-1)}\right\}.$$

Proof. The function f, as a function of σ $(\tau < \sigma < t)$, obtains a minimum at the point σ where $y = [x(\sigma - \tau) + \xi(t - \sigma)]/(t - \tau)$, and the minimum is

$$[|x - \xi|^{2p}/(t - \tau)]^{1/(2p-1)}$$

Hence,

$$I_a \leq I_{\epsilon a} \exp\left\{-a(1 - \epsilon)\left(\frac{|x - \xi|^{2p}}{t - \tau}\right)^{1/(2p-1)}\right\} \tag{4.11}$$

where $I_{\epsilon a}$ is defined as I_a with a replaced by ϵa. If $\tau \leq \sigma \leq \tau + (t - \tau)/2$ then we use the inequality

$$f(x, \xi, y; t, \tau, \sigma) \geq \left(\frac{|y - \xi|^{2p}}{\sigma - \tau}\right)^{1/(2p-1)}$$

and substitute, in $I_{\epsilon a}$, $|y - \xi| = \rho(\sigma - \tau)^{1/2p}$. Since $t - \sigma \geq (t - \tau)/2$, we get

$$I_{\epsilon a} \leq \text{const.}/(t - \tau)^{1/2p}.$$

Substituting this into (4.11), the inequality (4.10) follows. The proof of (4.10) in case $[\tau + (t - \tau)/2] \leq \sigma \leq t$ is similar.

We shall now extend Lemma 6 to the case of n variables. Introducing the norm

$$||x|| = \left(\sum_{i=1}^{n} |x_i|^q\right)^{1/q} \qquad \text{where } q = \frac{2p}{2p - 1}, \tag{4.12}$$

we define, for $\tau < \sigma < t$,

$$f_n(x, \xi, y; t, \tau, \sigma) = \left(\frac{||x - y||^{2p}}{t - \sigma}\right)^{1/(2p-1)} + \left(\frac{||y - \xi||^{2p}}{\sigma - \tau}\right)^{1/(2p-1)}$$

and we then have the following extension of Lemma 6.

Lemma 7. *Let*

$$I_a = \int_{R^n} \frac{1}{[(t - \sigma)(\sigma - \tau)]^{n/2p}} \exp\{-af_n(x, \xi, y; t, \tau, \sigma)\}\, dy$$

where $\tau < \sigma < t$, $x \in R^n$, $\xi \in R^n$, *and* a *is any positive number. For any* $0 < \epsilon < 1$ *there exists a constant* M *depending only on* ϵ, a, p, n *such that*

$$I_a \leq \frac{M}{(t - \tau)^{n/2p}} \exp\left\{-a(1 - \epsilon)\left(\frac{||x - \xi||^{2p}}{t - \tau}\right)^{1/(2p-1)}\right\}.$$

Proof. Noting that for $b = (1 - \epsilon)a$

$$\exp\{-bf_n(x, \xi, y; t, \tau, \sigma)\} = \prod_{j=1}^{n} \exp\{-bf_1(x_j, \xi_j, y_j; t, \tau, \sigma)\},$$

it follows that (4.11) remains valid if $|x - \xi|$ in the braces is replaced by $||x - \xi||$. The estimation of $I_{\epsilon a}$ is similar to that in the case $n = 1$.

Observing that (4.6) implies

$$(4.13)\quad |K(x, t; \xi, \tau)| \le \frac{A_1}{(t-\tau)^{(n+2p-\alpha)/2p}} \exp\left\{-a_2\left(\frac{\|x-\xi\|^{2p}}{t-\tau}\right)^{1/(2p-1)}\right\},$$

and using Lemma 7, we get

$$|K_2(x, t; \xi, \tau)| \le A_2 \int_\tau^t \frac{1}{[(t-\sigma)(\sigma-\tau)]^{(2p-\alpha)/2p}} I_{a_2}\, d\sigma$$

$$(4.14)\qquad \le \frac{A_3}{(t-\tau)^{(n+2p-2\alpha)/2p}} \exp\left\{-a_3\left(\frac{\|x-\xi\|^{2p}}{t-\tau}\right)^{1/(2p-1)}\right\}.$$

Proceeding step by step in this manner we arrive at some positive integer m_0 such that

$$|K_{m_0}(x, t; \xi, \tau)| \le A_4 \exp\left\{-a_4\left(\frac{\|x-\xi\|^{2p}}{t-\tau}\right)^{1/(2p-1)}\right\}.$$

We can now prove by induction on m that

$$(4.15)\quad |K_{m+m_0}(x, t; \xi, \tau)| \le A_4[A_5(t-\tau)^{\alpha/2p}]^m \frac{1}{\Gamma(1+m\alpha/2p)} \cdot \exp\left\{-a_5\left(\frac{\|x-\xi\|^{2p}}{t-\tau}\right)^{1/(2p-1)}\right\}$$

where a_5 is any positive constant smaller than $\min\{a_2, a_4\}$. Indeed, from (4.13), (4.15) and the inequality (see the proofs of Lemmas 6, 7)

$$\exp\{-a_5 f_n(x, \xi, y; t, \tau, \sigma)\} \le \exp\left\{-a_5\left(\frac{\|x-\xi\|^{2p}}{t-\tau}\right)^{1/(2p-1)}\right\},$$

it follows, if we take $a_5 < a_2$, that

$$|K_{m+m_0+1}(x, t; \xi, \tau)| \le A_6 A_4 A_5^m \cdot \exp\left\{-a_5\left(\frac{\|x-\xi\|^{2p}}{t-\tau}\right)^{1/(2p-1)}\right\} J/\Gamma\left(1+\frac{m\alpha}{2p}\right)$$

where

$$J = \int_\tau^t \left\{\int_{R^n} \frac{1}{(\sigma-\tau)^{n/2p}} \exp\left\{-(a_2-a_5)\left(\frac{\|y-\xi\|^{2p}}{\sigma-\tau}\right)^{1/(2p-1)}\right\} dy\right\} \cdot \frac{(t-\sigma)^{m\alpha/2p}}{(\sigma-\tau)^{(2p-\alpha)/2p}}\, d\sigma.$$

The inner integral of J is bounded by some constant A_7. Since

$$\int_\tau^t (t-\sigma)^{m\alpha/2p}(\sigma-\tau)^{(\alpha-2p)/2p}\, d\sigma = (t-\tau)^{(m+1)\alpha/2p}\,\Gamma\left(1+\frac{m\alpha}{2p}\right)\Gamma\left(\frac{\alpha}{2p}\right)/\Gamma\left(1+\frac{(m+1)\alpha}{2p}\right),$$

taking $A_5 \ge A_6 A_7 \Gamma(\alpha/2p)$, the inequality (4.15), with m replaced by $m+1$, follows. The proof of (4.15) for all $m \ge 0$ is thus completed.

From (4.15) it follows that the series in (4.8) is convergent and that it satisfies the equation (4.7). Furthermore, from the estimates for the K_m (also for $m \leq m_0$) we find that

$$(4.16)\quad |\Phi(x, t; \xi, \tau)| \leq \frac{A_8}{(t-\tau)^{(n+2p-\alpha)/2p}} \exp\left\{-a_6\left(\frac{|x-\xi|^{2p}}{t-\tau}\right)^{1/(2p-1)}\right\}.$$

We next need to establish the Hölder continuity of $\Phi(x, t; \xi, \tau)$ in x. We claim that for every $0 < \beta < \alpha$,

$$(4.17)\quad |\Phi(x, t; \xi, \tau) - \Phi(y, t; \xi, \tau)| \leq \frac{\text{const. } |x-y|^{\beta}}{(t-\tau)^{(n+2p-\alpha+\beta)/2p}}$$

$$\cdot\left\{\exp\left[-a_7\left(\frac{|x-\xi|^{2p}}{t-\tau}\right)^{1/(2p-1)}\right] + \exp\left[-a_7\left(\frac{|y-\xi|^{2p}}{t-\tau}\right)^{1/(2p-1)}\right]\right\}.$$

The proof of (4.17) is similar to the proof of (1.4.17), and the details are therefore omitted.

Using (4.16), (4.17) and writing $\Gamma(x, t; \xi, \tau)$ in a form analogous to (1.4.26), we find, upon applying Lemma 5, that $\Gamma(x, t; \xi, \tau)$ is a solution of the system (4.2).

To prove the second property of fundamental solutions, namely (1.10), one first establishes the analogue of (1.4.30). In addition one has to show that $\int_{R^n} Z(x - \xi, t; \xi, \tau)\, d\xi \to 1$ as $t \searrow \tau$. To prove the last property we recall the relation $\int_{R^n} Z(x - \xi, t; 0, \tau)\, d\xi = 1$. Thus it suffices to show that, as $t \searrow \tau$, $\int_{R^n} [Z(x - \xi, t; \xi, \tau) - Z(x - \xi, t; 0, \tau)]\, d\xi \to 0$. The last relation follows by breaking the integral into two parts corresponding to $|\xi| < \delta$ and $|\xi| \geq \delta$, and making use of Lemma 4.

It remains to prove the inequalities (4.1). By (4.4) and Lemma 5,

$$(4.18)\quad D_x^m\Gamma(x, t; \xi, \tau) = D_x^m Z(x-\xi, t; \xi, \tau)$$

$$+ \int_\tau^t d\sigma \int_{R^n} D_x^m Z(x-y, t; y, \sigma)\Phi(y, \sigma; \xi, \tau)\, dy.$$

We now evaluate the right-hand side by making use of (3.3), (4.16), and Lemma 7.

If in Theorem 2 we also assume that all the coefficients A_k^{ij} of (1.6) are Hölder continuous (exponent α) in x, uniformly in Ω, then (4.1) holds also for $|m| = 2p$, and

$$(4.19)\quad |D_x^m\Gamma(x, t; \xi, \tau) - D_x^m\Gamma(y, t; \xi, \tau)| \leq \frac{\text{const. } |x-y|^{\beta}}{(t-\tau)^{(n+|m|+\beta)/2p}}$$

$$\cdot\left\{\exp\left[-a_8\left(\frac{|x-\xi|^{2p}}{t-\tau}\right)^{1/(2p-1)}\right] + \exp\left[-a_8\left(\frac{|y-\xi|^{2p}}{t-\tau}\right)^{1/(2p-1)}\right]\right\},$$

where β is any positive number ≤ 1 if $|m| < 2p$ and $< \alpha$ if $|m| = 2p$.

Since both (4.19) and (4.1) with $|m| = 2p$ will not be needed in the future, their proofs are omitted.

From the form of Γ it follows that if f is a continuous function satisfying

(3.17) with $a < c'T^{-1/(2p-1)}$ for some positive constant c' depending on the P_i (but not on T), then $\int \Gamma f$ can be written as a sum (compare Sec. 5, Chap. 1) $\int Zf + \int Z\hat{f}$ where $\hat{f} = \int \Phi f$ is also a continuous function satisfying (3.17) (with a different a). If $f(x, t)$ is Hölder continuous in x, uniformly in bounded subsets of Ω, then the same is true of $\hat{f}(x, t)$ (as follows by using (4.17)). Hence:

Lemma 8. *Lemma* 5 *remains true if* $Z(x - \xi, t; \xi, \tau)$ *is replaced by* $\Gamma(x, t; \xi, \tau)$, *provided the assumptions of Theorem* 2 *are satisfied and provided, in* (3.17), $a < c'T^{-1/(2p-1)}$ *for some constant* c' *(independent of* T*).*

We shall now solve the Cauchy problem for the system (1.6), (1.9). By a *solution* of (1.6), (1.9) in a strip $0 \le t \le t_0$ we mean a function u which satisfies (1.6) for $0 < t \le t_0$ ($D_t u$ and $D_x^m u$ $(0 \le |m| \le 2p)$ are assumed to exist and be continuous for $0 < t \le t_0$), and which is continuous for $0 \le t \le t_0$ and satisfies (1.6) for $x \in R^n$. Writing (1.6), (1.9) in the matrix form

$$\frac{\partial u}{\partial t} = P(x, t, D_x)u + f(x, t), \tag{4.20}$$

$$u(x, 0) = \varphi(x), \tag{4.21}$$

we shall prove the following theorem.

Theorem 3. *Let* (4.20) *be a uniformly parabolic system in the strip* $\Omega = R^n \times [0, T]$ *and let* (A_1), (A_2) *be satisfied. Let* $f(x, t)$ *be a continuous function in* Ω, *Hölder continuous in* x *uniformly in bounded subsets of* Ω, *and let* $\varphi(x)$ *be a continuous function in* R^n. *Finally, assume that, for* $q = 2p/(2p - 1)$,

$$|f(x, t)| \le A \exp [a|x|^q] \quad \text{in } \Omega, \tag{4.22}$$

$$|\varphi(x)| \le A \exp [a|x|^q] \quad \text{in } R^n. \tag{4.23}$$

Then there exists a solution of the Cauchy problem (4.20), (4.21) *in the strip* $0 \le t \le t_0$, *where* $t_0 = \min \{T, (\bar{c}/a)^{2p-1}\}$ *and where* $\bar{c}$ *is a constant depending on* P *(but not on* T*), and*

$$|u(x, t)| \le \text{const.} \exp [a'|x|^q] \quad \text{for } x \in R^n, 0 \le t \le t_0 \tag{4.24}$$

for some constant a'.

Proof. In view of Lemma 8 and Theorem 2, the function

$$-\int_0^t \int_{R^n} \Gamma(x, t; \xi, \tau) f(\xi, \tau) \, d\xi \, d\tau \tag{4.25}$$

is a solution of (4.20), (4.21) with $\varphi \equiv 0$, in some strip $0 \le t \le t_1$. Next the proof of Theorem 11, Sec. 6, Chap. 1, can obviously be extended to the present case of parabolic systems. Hence the function

$$\int_{R^n} \Gamma(x, t; \xi, 0)\varphi(\xi)\, d\xi \tag{4.26}$$

is a solution of (4.20), (4.21) with $f \equiv 0$, in some strip $0 \le t \le t_2$.

Combining (4.25), (4.26) it follows that

$$u(x, t) = \int_{R^n} \Gamma(x, t; \xi, 0)\varphi(\xi)\, d\xi - \int_0^t \int_{R^n} \Gamma(x, t; \xi, \tau) f(\xi, \tau)\, d\xi\, d\tau \tag{4.27}$$

is a solution of the Cauchy problem (4.20), (4.21) in some strip $0 \le t \le t_0$. Since both t_1 and t_2 are of the form min $\{T, (c^*/a)^{(2p-1)}\}$ where c^* is a constant depending only on P (but not on T), the same is true of t_0. It thus remains to prove (4.24). This follows from the following lemma.

Lemma 9. *For any positive numbers q, A, B with $B < A$, there exist positive numbers C, C' such that, for all $x \in R^n$,*

$$\int_{R^n} \exp\, [-A|x - \xi|^q] \exp\, [B|\xi|^q]\, d\xi \le C' \exp\, [C|x|^q]. \tag{4.28}$$

Proof. Writing

$$\exp\, [-A|x - \xi|^q] = \exp\, [-\epsilon|x - \xi|^q] \exp\, [-(A - \epsilon)|x - \xi|^q]$$

for some $\epsilon > 0$ such that $A - \epsilon > B$, we see that it suffices to prove the inequality

$$B|\xi|^q \le (A - \epsilon)|x - \xi|^q + C|x|^q \qquad \text{for all } x \in R^n,\ \xi \in R^n. \tag{4.29}$$

If $\xi = 0$ then (4.29) is certainly satisfied. Assume then that $\xi \ne 0$. Dividing both sides by $|\xi|$ and setting $y = x/|\xi|$, $e = \xi/|\xi|$, (4.29) reduces to

$$B \le (A - \epsilon)|y - e|^q + C|y|^q. \tag{4.30}$$

Since $A - \epsilon > B$, (4.30) holds if $|y| < \eta$ for some sufficiently small constant $\eta > 0$. If $|y| \ge \eta$ then (4.30) holds if $C\eta^q = B$.

Using (4.1) we get:

Corollary. *The solution $u(x, t)$ of Theorem 3 satisfies:*

$$|D_x^k u(x, t)| \le \frac{\text{const.}}{t^{|k|}} \exp\, [a'|x|^q] \qquad \text{for } 0 \le |k| < 2p. \tag{4.31}$$

We conclude this section with a brief discussion of fundamental solutions in cylinders $\Omega = \overline{D} \times [0, T]$ where D is an arbitrary domain in R^n. The construction of a fundamental solution in this case is obtained from the construction in the case of $D = R^n$ by obvious modifications. Thus, the parametrix is the same, but in the definition of Γ and Φ the integration over R^n (in (4.4), (4.7)) is replaced by integration over D. We thus arrive at the following result.

Theorem 4. *Assume that the system* (1.6) *is uniformly parabolic in $\Omega = \overline{D} \times [0, T]$ and that* (A_1), (A_2) *hold in $\Omega = \overline{D} \times [0, T]$. Then there exists a fundamental solution $\Gamma(x, t; \xi, \tau)$ of* (1.6) *and the inequalities* (4.1) *hold.*

5. The Adjoint System

Given a system

$$\frac{\partial u_i}{\partial t} = \sum_{j=1}^{N} \sum_{|k| \le 2p} A_k^{ij}(x, t) D_x^k u_j \qquad (i = 1, \ldots, N) \tag{5.1}$$

in $\Omega = \overline{D} \times [0, T]$ where D is any domain in R^n, we define the *adjoint system* to be

$$\frac{\partial v_i}{\partial t} = - \sum_{j=1}^{N} \sum_{|k| \le 2p} (-1)^{|k|} D_x^k [A_k^{ji}(x, t) v_j] \qquad (i = 1, \ldots, N). \tag{5.2}$$

If we write the systems (5.1), (5.2) in the form

$$\frac{\partial u_i}{\partial t} = P_i(x, t, D_x) u_i, \qquad \frac{\partial v_i}{\partial t} = Q_i(x, t, D_x) v$$

then

$$\sum_{i=1}^{N} \int_\Omega [v_i P_i(x, t, D_x) u + u_i Q_i(x, t, D_x) v] \, dx \, dt = 0 \tag{5.3}$$

holds for all infinitely differentiable functions u_i, v_i having compact supports contained in the interior of Ω. Indeed, substituting Q_i from (5.2) and integrating by parts we find that (5.3) holds. The property (5.3) determines uniquely the linear differential operators Q_i, i.e., there cannot be two distinct systems of operators Q_i and Q_i^* $(i = 1, \ldots, N)$ for which (5.3) holds. Indeed, in the contrary case it would follow that $(Q_i - Q_i^*)v \equiv 0$ for all v as in (5.3). Hence, by Theorem 14, Sec. 8, Chap. 1, $Q_i \equiv Q_i^*$, a contradiction.

Write the system (5.2) in the form

$$\frac{\partial v_i}{\partial t} = - \sum_{j=1}^{N} \sum_{|k| \le 2p} A_k^{*ij}(x, t) D_x^k v_j \qquad (i = 1, \ldots, N) \tag{5.4}$$

and assume that the coefficients of (5.4) satisfy the assumptions (A_1), (A_2) of Sec. 4, with R^n replaced by $\overline{D}$, and that (5.1) is uniformly parabolic in $\Omega = \overline{D} \times [0, T]$.

A *fundamental solution* (or a *fundamental matrix*) of (5.4) is an $N \times N$ matrix $\Gamma^*(x, t; \xi, \tau)$ defined for $(x, t) \in \Omega$, $(\xi, \tau) \in \Omega$, $t < \tau$, satisfying (5.4) as a function of (x, t) $(x \in D, 0 \le t < \tau)$, and satisfying the relation

$$\lim_{t \nearrow \tau} \int_D \Gamma^*(x, t; \xi, \tau) f(\xi) \, d\xi = f(x) \qquad \text{for all } x \in D, \tag{5.5}$$

for any continuous function $f(x)$ in $\overline{D}$ for which (1.11) holds with some $k > 0$.

Γ^* can be constructed in the form

$$\begin{aligned} \Gamma^*(x, t; \xi, \tau) &= Z^*(x - \xi, t; \xi, \tau) \\ &\quad + \int_t^\tau d\sigma \int_D Z^*(x - \xi, t; y, \sigma) \Phi^*(y, \sigma; \xi, \tau) \, dy \end{aligned} \tag{5.6}$$

where Z^* is the fundamental solution for the adjoint of the system for which Z was a fundamental solution (see Sec. 4). The properties of Γ^* are similar to those of Γ.

We shall need the following assumption:

(A_3) The derivatives $D_x^h A_k^{ij}(x, t)$ $(0 \leq |h| \leq |k|)$ are continuous bounded functions in $\Omega = R^n \times [0, T]$, and Hölder continuous (exponent α) in x, uniformly with respect to (x, t) in bounded subsets of Ω.

Theorem 5. *Let the system* (5.1) *be uniformly parabolic in* $\Omega = R^n \times [0, T]$ *and let the assumptions* (A_1), (A_2), (A_3) *hold. Then*

$$\Gamma^*(\xi, \tau; x, t) = (\Gamma(x, t; \xi, \tau))^{\sim} \tag{5.7}$$

where $A^{\sim}$ *denotes the transpose of* A.

Proof. If u, v are two smooth scalar functions and $D_x = \partial/\partial x_i$, then

$$\begin{aligned} vD_x^m u &= D_x(vD_x^{m-1}u) - (D_x v)(D_x^{m-1}u) \\ &= D_x(vD_x^{m-1}u) - D_x[(D_x^{m-2}u)] + (D_x^2 v)(D_x^{m-2}u) \\ &= \cdots \\ &= D_x\{vD_x^{m-1}u - (D_x v)(D_x^{m-2}u) + \cdots + (-1)^{m-1}(D_x^{m-1}v)(D_x u)\} \\ &\quad + (-1)^m(D_x^m v)u. \end{aligned}$$

Hence,

$$vD_x^m u - (-1)^m u D_x^m v = D_x B(u, v) \tag{5.8}$$

where $B(u, v)$ is a bilinear function in u, v and their derivatives up to an order $\leq m - 1$.

If D_x^m is any partial derivative of order m, then we obtain in a similar manner,

$$vD_x^m u - (-1)^{|m|} u D_x^m v = \sum_{j=1}^{n} \frac{\partial}{\partial x_j} B_j(u, v). \tag{5.9}$$

Denote by $w \cdot v$ the scalar product $\sum w_i v_i$ of any two vectors w, v. Setting $A_k = (A_k^{ij})$,

$$Pu = \sum_{|k| \leq 2p} A_k D_x^k u, \qquad P^*v = \sum_{|k| \leq 2p} (-1)^{|k|} D_x^k(A_k^{\sim} v) \tag{5.10}$$

it is clear that (5.1), (5.2) can be written in the form

$$\frac{\partial u}{\partial t} = Pu, \qquad \frac{\partial v}{\partial t} = -P^*v. \tag{5.11}$$

Using (5.9) we obtain

$$v \cdot A_k D_x^k u - (-1)^{|k|} u \cdot D_x^k(A_k^{\sim} v) = \sum \frac{\partial}{\partial x_j} B_{jk}[u, v] \tag{5.12}$$

where $B_{jk}[u, v]$ are bilinear expressions in the components of u, v and their derivatives up to an order $\leq |k| - 1$. Summing over k we find that

$$(5.13)\quad v\cdot\left(Pu-\frac{\partial u}{\partial t}\right)-u\cdot\left(P^*v+\frac{\partial v}{\partial t}\right)=\Sigma\frac{\partial}{\partial x_j}B_j[u,v]-\frac{\partial}{\partial t}(u\cdot v),$$

where $B_j = \Sigma B_{jk}$. The identity (5.13) is called *Green's identity*. If u, v are matrices U, V then (5.13) is replaced by

(5.14)

$$V\left(PU-\frac{\partial U}{\partial t}\right)-\left(P^*V^\sim+\frac{\partial V^\sim}{\partial t}\right)^\sim U=\sum_{j=1}^{n}\frac{\partial}{\partial x_j}\hat{B}_j(U,V)-\frac{\partial(VU)}{\partial t},$$

where the elements of $\hat{B}_j$ are some bilinear expressions in the elements of U, V and their derivatives up to an order $\leq 2p-1$.

Taking, in (5.14), $U = \Gamma(x, t; x^0, t^0)$, $V = (\Gamma^*(x, t; \bar{x}, \bar{t}))^\sim$ and integrating over the set $t^0 + \epsilon < t < \bar{t} - \epsilon$, $|x| < R$ for any $\epsilon > 0$, $R > 0$, and then taking $R \to \infty$ and $\epsilon \to 0$, the assertion (5.7) follows (compare Sec. 8, Chap. 1).

Using Green's formula with a solution u of the Cauchy problem

$$\frac{\partial u}{\partial t}=Pu,\qquad u(x,0)=0,$$

and with $v = h\Gamma$ (h a vector-function), we can employ the method of proof of Theorem 16, Sec. 9, Chap. 1, and thus obtain the following uniqueness theorem for the Cauchy problem.

Theorem 6. *Let the system* (5.1) (*or* (1.6)) *be uniformly parabolic in* $\Omega = R^n \times [0, T]$ *and let the assumptions* (A_1), (A_2), (A_3) *hold. Then there exists at most one solution to the Cauchy problem* (1.6), (1.9) *satisfying, for some* $k > 0$,

$$(5.15)\quad \int_0^T\int_{R^n}|u(x,t)|\exp\left[-k|x|^q\right]dx\,dt<\infty\qquad\left(q=\frac{2p}{2p-1}\right).$$

6. Differentiability of Fundamental Solutions

We shall need the following assumption:

(C) The derivatives $D_x^h A_k^{ij}(x, t)$ ($0 \leq |h| \leq r$; r a positive integer) exist and are bounded continuous functions of (x, t) in $\Omega = R^n \times [0, T]$. Furthermore, the principal coefficients of (1.6) are continuous in t, uniformly with respect to (x, t) in Ω.

Theorem 7. *Let* (1.6) *be a uniformly parabolic system in* $\Omega = R^n \times [0, T]$ *and let* (C) *hold. Let* $\Gamma(x, t; \xi, \tau)$ *be the fundamental solution of* (1.6) *constructed in Sec.* 4. *Then* $D_x^{m+a}D_\xi^b\Gamma(x, t; \xi, \tau)$ *exist and are continuous functions for all* $0 \leq |a| + |b| \leq r$, $0 \leq |m| < 2p$, *and*

(6.1) $$|D_x^{m+a}D_\xi^b\Gamma(x, t; \xi, \tau)| \leq \frac{\text{const.}}{(t-\tau)^{(|m|+|a|+|b|+n)/2p}} \exp\left\{-d\left(\frac{|x-\xi|^{2p}}{t-\tau}\right)^{1/(2p-1)}\right\},$$

(6.2) $$|D_x^m D_\xi^b\Gamma(\xi + x, t; \xi, \tau)| \leq \frac{\text{const.}}{(t-\tau)^{(n+|m|)/2p}} \exp\left\{-d\left(\frac{|x|^{2p}}{t-\tau}\right)^{1/(2p-1)}\right\},$$

for some constant $d > 0$.

Proof. Consider the function K defined in (4.5). By the assumption (C) and the inequalities of (3.15),

(6.3) $$|D_x^a D_\xi^b K(x, t; \xi, \tau)| \leq \frac{\text{const.}}{(t-\tau)^{(n+2p-1+|a|+|b|)/2p}} \exp\left\{-b_1\left(\frac{|x-\xi|^{2p}}{t-\tau}\right)^{1/(2p-1)}\right\},$$

(6.4) $$|D_\xi^b K(z + \xi, t; \xi, \tau)| \leq \frac{\text{const.}}{(t-\tau)^{(n+2p-1)/2p}} \exp\left\{-b_1\left(\frac{|z|^{2p}}{t-\tau}\right)^{1/(2p-1)}\right\},$$

where b_i are used to denote positive constants.

Consider next K_2 defined by (4.9). Write

(6.5) $$\begin{aligned} K_2(x, t; \xi, \tau) &= \int_\tau^{\tau+(t-\tau)/2} d\sigma \int_{R^n} K(x, t; y, \sigma)K(y, \sigma; \xi, \tau)\, dy \\ &\quad + \int_{\tau+(t-\tau)/2}^t d\sigma \int_{R^n} K(x, t; y, \sigma)K(y, \sigma; \xi, \tau)\, dy \\ &\equiv K_{21}(x, t; \xi, \tau) + K_{22}(x, t; \xi, \tau). \end{aligned}$$

For K_{21}, $t - \sigma \geq (t - \tau)/2 > 0$. Hence

$$D_x^a K_{21}(x, t; \xi, \tau) = \int_\tau^{\tau+(t-\tau)/2} d\sigma \int_{R^n} [D_x^a K(x, t; y, \sigma)]K(y, \sigma; \xi, \tau)\, dy.$$

Substituting $y = \xi + z$ we get

(6.6) $$D_x^a K_{21}(x, t; \xi, \tau) = \int_\tau^{\tau+(t-\tau)/2} d\sigma \int_{R^n} [D_x^a K(x, t; \xi + z, \sigma)]K(\xi + z, \sigma; \xi, \tau)\, dz.$$

Applying D_ξ^b to both sides of (6.6) and assuming that D_ξ^b commutes with the integrals on the right-hand side, we obtain, after making use of (6.3), (6.4),

$$\begin{aligned} |D_x^a D_\xi^b K_{21}(x, t; \xi, \tau)| &\leq \text{const.} \int_\tau^{\tau+(t-\tau)/2} d\sigma \int_{R^n} (t-\sigma)^{-(n+2p-1+|a|+|b|)/2p} \\ &\quad \cdot (\sigma-\tau)^{-(n+2p-1)/2p} \exp\left\{-b_1\left(\frac{|x-\xi-z|^{2p}}{t-\sigma}\right)^{1/(2p-1)}\right\} \\ &\quad \cdot \exp\left\{-b_1\left(\frac{|z|^{2p}}{\sigma-\tau}\right)^{1/(2p-1)}\right\} dz. \end{aligned}$$

Using Lemma 7, Sec. 4, we find that

$$(6.7)\quad |D_x^a D_\xi^b K_{21}(x, t; \xi, \tau)| \leq \frac{\text{const.}}{(t-\tau)^{(n+2p-2+|a|+|b|)/2p}} \exp\left\{-b_2\left(\frac{|x-\xi|^{2p}}{t-\tau}\right)^{1/(2p-1)}\right\}.$$

The proof that $D_x^a D_\xi^b K_{21}$ exists and that D_ξ^b commutes with the integrals on the right-hand side of (6.6) can be given by the method of proof of Theorem 3, Sec. 3, Chap. 1.

Since K_{22} can be treated similarly to K_{21}, we obtain

$$(6.8)\quad |D_x^a D_\xi^b K_2(x, t; \xi, \tau)| \leq \frac{\text{const.}}{(t-\tau)^{(n+2p-2+|a|+|b|)/2p}} \exp\left\{-b_2\left(\frac{|x-\xi|^{2p}}{t-\tau}\right)^{1/(2p-1)}\right\},$$

If we substitute in the integral of $K_2(\xi + z, t; \xi, \tau)$ (see (4.9)) $y = \xi + z'$, we get

$$K_2(\xi + z, t; \xi, \tau) = \int_\tau^t d\sigma \int_{R^n} K(\xi + z, t; \xi + z', \sigma) K(\xi + z', \sigma; \xi, \tau)\, dz'.$$

Applying D_ξ^b to both sides and using (6.4), we get

$$(6.9)\quad |D_\xi^b K_2(\xi + z, t; \xi, \tau)| \leq \frac{\text{const.}}{(t-\tau)^{(n+2p-2)/2p}} \cdot \exp\left\{-b_2\left(\frac{|z|^{2p}}{t-\tau}\right)^{1/(2p-1)}\right\}.$$

The inequalities (6.8), (6.9) are analogous to (6.3), (6.4).

We can now proceed to estimate the K_m inductively and prove that for all $1 \leq m < \infty$,

$$|D_x^a D_\xi^b K_m(x, t; \xi, \tau)| \leq \frac{B_m}{(t-\tau)^{(n+2p-m+|a|+|b|)/2p}} \cdot \exp\left\{-b_3\left(\frac{|x-\xi|^{2p}}{t-\tau}\right)^{1/(2p-1)}\right\},$$

$$|D_\xi^b K_m(\xi + z, t; \xi, \tau)| \leq \frac{B_m}{(t-\tau)^{(n+2p-m)/2p}} \exp\left\{-b_3\left(\frac{|z|^{2p}}{t-\tau}\right)^{1/(2p-1)}\right\},$$

where $B_m = B^m/\Gamma\left(1 + \dfrac{m\alpha}{2p}\right)$, B a constant. From (4.8) it then follows that

$$(6.10)\quad |D_x^a D_\xi^b \Phi(x, t; \xi, \tau)| \leq \frac{\text{const.}}{(t-\tau)^{(n+2p-1+|a|+|b|)/2p}} \cdot \exp\left\{-b_3\left(\frac{|x-\xi|^{2p}}{t-\tau}\right)^{1/(2p-1)}\right\},$$

$$(6.11)\quad |D_\xi^b \Phi(\xi + z, t; \xi, \tau)| \leq \frac{\text{const.}}{(t-\tau)^{(n+2p-1)/2p}} \cdot \exp\left\{-b_3\left(\frac{|z|^{2p}}{t-\tau}\right)^{1/(2p-1)}\right\}.$$

Using (3.15) and (6.10), (6.11) we can treat the integral on the right-hand side (4.4) in the same manner that we have treated $K_2(x, t; \xi, \tau)$ above. Denoting this integral by $I(x, t; \xi, \tau)$ we thus find that $D_x^{m+a} D_\xi^b I$ are continuous functions (for $0 \leq |a| + |b| \leq r$, $0 \leq |m| < 2p$) and the inequalities (6.1), (6.2) hold for Γ replaced by I. Using (3.15) and the definition of Γ in (4.4), the proof of Theorem 7 follows.

Since

$$\frac{\partial \Gamma(x, t; \xi, \tau)}{\partial t} = P(x, t, D_x)\Gamma(x, t; \xi, \tau), \tag{6.12}$$

we can conclude from Theorem 7 that if the coefficients of (1.6) are sufficiently smooth in (x, t), then $\Gamma(x, t; \xi, \tau)$ is also sufficiently smooth in (x, t, ξ). From Theorem 5 we then can also deduce the smoothness of $\Gamma(x, t, \xi, \tau)$ with respect to τ. In particular we have the following theorem.

Theorem 8. *If the coefficients of a uniformly parabolic system* (1.6) *in* $\Omega = R^n \times [0, T]$ *are infinitely differentiable in* Ω, *and if all the derivatives are bounded functions, then* $\Gamma(x, t; \xi, \tau)$ *is infinitely differentiable in* $(x, t; \xi, \tau)$, *and*

$$|D_x^a D_\xi^b D_t^c D_\tau^d \Gamma(x, t; \xi, \tau)| \leq C'(t - \tau)^{-(n+|a|+|b|+2pc+2pd)/2p} \cdot \exp\left\{-C\left(\frac{|x - \xi|^{2p}}{t - \tau}\right)^{1/(2p-1)}\right\}, \tag{6.13}$$

for all nonegative integers a, b, c, d where C, C′ are positive constants depending on a, b, c, d.

If a parabolic system with infinitely differentiable coefficients is defined in a cylinder $D_0 \times [0, T]$ where D_0 is a domain in R^n, and if D is a bounded domain in R^n with $\bar{D} \subset D_0$, then we can modify the definition of the system outside some neighborhood of $\bar{D} \times [0, T]$ in the strip $0 \leq t \leq T$ so that the new system satisfies the assumptions of Theorem 8. The proof is similar to that of Lemma 1, Sec. 7, Chap. 3. Applying Theorem 8 to the modified system we obtain the following theorem.

Theorem 9. *If* (1.6) *is a parabolic system with infinitely differentiable coefficients in a cylinder* $D_0 \times [0, T]$ *and if D is a bounded domain in* R^n *with* $\bar{D} \subset D_0$, *then there exists a fundamental solution* $\Gamma(x, t; \xi, \tau)$ *of* (1.6) *in* $\bar{D} \times [0, T]$ *which is infinitely differentiable in* $(x, t; \xi, \tau)$ *and* (6.2), (6.13) *hold.*

Let $u(x, t)$ be a solution of a parabolic system

$$\frac{\partial u}{\partial t} = P(x, t, D_x)u + f(x, t) \tag{6.14}$$

in a domain G of the space (x, t) and suppose that $\Omega = \bar{D} \times [0, T]$ is a closed bounded subdomain of G, and that the boundary ∂D of D is sufficiently smooth.

Integrating Green's identity (5.13) with v being any column of a fundamental matrix $\Gamma^*(x, t; \xi, \tau)$, we get

$$(6.15) \qquad u(\xi, \tau) = - \int_0^\tau \int_D (\Gamma^*(x, t; \xi, \tau))^\sim f(x, t)\, dx\, dt + I(\xi, \tau)$$

where $I(\xi, \tau)$ is a sum of boundary integrals taken over $t = 0$ and over $\partial D \times [0, \tau]$. If the coefficients of P are infinitely differentiable then, by Theorems 5, 9, we may assume that $(\Gamma^*(x, t; \xi, \tau))^\sim = \Gamma(\xi, \tau; x, t)$ and that Γ is an infinitely differentiable function satisfying (6.2), (6.13).

The integral $I(\xi, \tau)$ is then infinitely differentiable for $\xi \in D, 0 < \tau \leq T$. As for the first integral on the right-hand side of (6.15), we write it in the form

$$\int_0^{\tau/2} \int_D \Gamma(\xi, \tau; x, t) f(x, t)\, dx\, dt + \int_{\tau/2}^{\tau} \int_D \Gamma(\xi, t; x, t) f(x, t)\, dx\, dt$$
$$\equiv J_1(\xi, \tau) + J_2(\xi, \tau).$$

$J_1(\xi, \tau)$ is clearly infinitely differentiable. As for J_2, notice first that the differentiability of $J_2(\xi, \tau)$ for ξ bounded away from ∂D is unaffected if we modify the definition of $f(x, t)$ for x in a small neighborhood of ∂D. If $f(x, t)$ is an infinitely differentiable function in the original domain G, we can modify f in J_2 so that it remains infinitely differentiable and it vanishes identically for $x \notin D$. We now substitute, in $J_2(\xi, \tau)$, $x = \xi + z$ and obtain

$$J_2(\xi, \tau) = \int_{\tau/2}^{\tau} \int_{R^n} \Gamma(\xi, \tau; \xi + z, t) f(\xi + z, t)\, dz\, dt.$$

Using (6.2) with $m = 0$ it follows that $D_\xi^b J_2(\xi, \tau)$ exists (for any $|b| > 0$) and is a continuous function.

We have thus proved that $D_\xi^b u(\xi, \tau)$ exists and is a continuous function, for any $|b| > 0$. From the differential system (6.14) we now deduce that $D_\tau D_\xi^b u$, $D_\tau^2 D_\xi^b u$, etc., exist and are continuous functions. We have thus proved the following theorem.

Theorem 10. *If* (6.14) *is a parabolic system with infinitely differentiable coefficients in a domain G, then every solution of* (6.14) *in G is an infinitely differentiable function in G.*

From the proof of Theorem 10 we can deduce:

Corollary. *If* (6.14) *is a parabolic system in a domain G and if the coefficients of P and of its adjoint P^* have r continuous x-derivatives in G, $r \geq 1$, and if f has r continuous x-derivatives in G, then any solution of* (6.14) *in G has $r + 2p - 1$ continuous x-derivatives.*

Indeed, in treating J_2, first differentiate it $|m|$ times with respect to ξ, for any $0 \leq |m| \leq 2p - 1$, and then use the substitution $x = \xi + z$.

If f and the coefficients of P, P^* are differentiable to a certain order with respect to (x, t), then we can deduce the differentiability of $u(x, t)$

to an appropriate order (with respect to (x, t)) by using the equation (6.14) and the corollary.

In Chap. 3 we have proved differentiability theorems for solutions of second-order parabolic equations; see, in particular, Theorem 11, Sec. 5. The proof of that theorem was based upon the interior a priori estimates of Theorem 5, Sec. 2, Chap. 3. Now, Theorem 5 and its proof extend also to parabolic systems of any order, with the distance function $d(P, Q)$ defined by

$$(6.16)\quad d(P, Q) = [|x - x^0|^2 + |t - t^0|^{1/p}]^{1/2} \qquad (P = (x, t),\ Q = (x^0, t^0)),$$

provided we employ, instead of the fundamental solution for the heat equation, the fundamental solutions for parabolic systems with constant coefficients which were constructed in Sec. 2. Using the extension of Theorem 5, Sec. 2, Chap. 3, to parabolic systems, the proof of Theorem 11, Sec. 5, Chap. 3, can then be extended to such systems with obvious modifications, and we arrive at the following result.

Theorem 11. *Let* (1.6) *be a parabolic system in a domain* G *and assume that the functions*

$$(6.17)\quad D_t^m D_x^h f(x, t), \qquad D_t^m D_x^h A_k^{ij}(x, t) \qquad (0 \le 2pm + |h| \le r,\ 0 \le m \le s)$$

are Hölder continuous (exponent α*) in* G*, in the metric* (6.16)*, for some* $r \ge 1$*,* $s \ge 0$*. If* u *is a solution of* (1.6) *in* G *and if* $D_x^h u$ *are Hölder continuous (exponent* α*) functions for* $0 \le |h| \le 2p$*, then the functions*

$$(6.18)\quad D_t^m D_x^h u \qquad (0 \le 2pm + |h| \le r + 2p,\ 0 \le m \le s + 1)$$

exist and are Hölder continuous (exponent α*).*

Theorem 11 can also be extended to nonlinear parabolic systems, by modifying the proof of Theorem 13, Sec. 5, Chap. 3.

7. Elliptic Equations

Consider a system of differential equations

$$(7.1)\qquad \sum_{j=1}^{N} \sum_{|k| \le m} A_k^{ij}(x) D_x^k u_j = f_i(x) \qquad (i = 1, \ldots, N),$$

and form the matrix

$$(7.2)\qquad P(x, \xi) = \left(\sum_{|k|=m} A_k^{ij}(x) \xi^k \right).$$

If for any real vector $\xi \neq 0$, $\det P(x, \xi) \neq 0$, then we say that the system (7.1) is of *elliptic type*, or, that (7.1) is an *elliptic system*. m is the *order* of the system. If the coefficients of the system are real, then m must be an even number. Indeed, if m is odd, then from $\det P(x, -\xi) = -\det P(x, \xi)$ follows the existence of real vectors $\xi^0 \neq 0$ such that $\det P(x, \xi^0) = 0$.

Consider now, for simplicity, one elliptic equation

$$Lu \equiv \sum_{|k| \leq m} A_k(x) D_x^k u = f(x) \tag{7.3}$$

in a domain D. Assume that the coefficients $A_k(x)$ belong to $C^{|k|}(D)$ so that the adjoint L^* of L exists. We then have *Green's identity* (see (5.13))

$$vLu - uL^*v = \sum_{i=1}^{n} \frac{\partial}{\partial x_i} B_i[u, v] \tag{7.4}$$

where $B_i[u, v]$ are bilinear expressions in u, v and their derivatives up to an order $\leq m - 1$.

Let R be a bounded domain with $\overline{R} \subset D$, and with a sufficiently smooth boundary ∂R. Integrating (7.4) over R we obtain

$$\int_R (vLu - uL^*v)\, dx = \int_{\partial R} B[u, v]\, dS_x, \tag{7.5}$$

where dS_x is the surface element on ∂R and $B[u, v] = \sum \nu_i B_i[u, v]$, ν_i being the ith cosine direction of the outward normal to ∂R at x $(x \in \partial R)$.

A *fundamental solution* in D is a function $K(x, z)$ defined for $x \in D$, $z \in D$, $x \neq z$ and satisfying the following property: For every function $v(x)$ in $C^m(D)$ and for every domain R with sufficiently smooth boundary ∂R, such that $z \in R$, $\overline{R} \subset D$,

$$v(z) = \int_R K(x, z) L^* v(x)\, dx + \int_{\partial R} B[K(x, z), v(x)]\, dS_x. \tag{7.6}$$

We say that $K(x, z)$ has a *pole* at $x = z$.

In view of (7.5), the condition (7.6) is equivalent to the condition:

$$v(z) = \lim_{\epsilon \to 0} \left\{ \int_{R_\epsilon} v(x) LK(x, z)\, dx + \int_{|x-z|=\epsilon} B[K(x, z), v(x)] dS_x^\epsilon \right\}, \tag{7.7}$$

where R_ϵ is the complement in R of a ball with center z and radius ϵ, and dS_x^ϵ is the surface element on the boundary of this ball. It follows that $LK(x, z) = 0$ for all $x \in D$, $x \neq z$. The condition (7.7) also implies a certain behavior of $K(x, z)$ near the pole $x = z$.

The fundamental solutions defined in Chap. 5, Sec. 6, for second-order elliptic equations are also fundamental solutions in the sense of the present definition.

In the construction of fundamental solutions for elliptic equations there appear two difficulties which are not encountered in the parabolic case. The first difficulty occurs in the construction of fundamental solutions for equations with constant coefficients. The method of first constructing the Fourier transform of K and then taking the inverse Fourier transform cannot be readily applied because the fundamental solution is, in general, not integrable at ∞.

The second difficulty occurs in carrying out the parametrix method and thus obtaining an integral equation of Fredholm type (compare Sec. 6, Chap. 5). This integral equation may not always have a solution. It can, however, be shown that if the domain D has a sufficiently small diameter then the integral equation has a solution. Another case when the integral equation has a solution is when ∂D and the coefficients of L are sufficiently smooth, and when the following property holds:

Every solution u of $Lu = 0$ in D, which vanishes on the boundary ∂D of D with its first $m - 1$ derivatives, is identically zero, i.e., the only solution of the Cauchy problem

$$(7.8) \qquad Lu = 0 \quad \text{in } D, \qquad \frac{\partial^j u}{\partial \nu^j} = 0 \quad \text{on } \partial D \qquad (0 \le j \le m - 1)$$

(where ν is the normal to ∂D) is the zero solution.

Thus, in both cases mentioned above, a fundamental solution exists in D.

We shall state some results concerning the structure of fundamental solutions with constant coefficients. Since $K(x, z)$ is only a function of $x - z$ if the coefficients of L are constants, we write $K(x, z) = K(x - z)$. Setting $r = |x|$ we have:

$$(7.9) \qquad K(x) = \begin{cases} A\left(\dfrac{x}{r}, r\right) r^{m-n} & \text{for } n \text{ odd}, \\[2ex] B\left(\dfrac{x}{r}, r\right) r^{m-n} + C\left(\dfrac{x}{r}, r\right) r^{m-n} \log r & \text{for } n \text{ even}, \end{cases}$$

where $A(\xi, r)$, $B(\xi, r)$, $C(\xi, r)$ are analytic functions in (ξ, r) in a neighborhood of $|\xi| = 1$, $r = 0$, and $C\left(\dfrac{x}{r}, r\right) r^{m-n}$ is a function $C(x)$ analytic at $x = 0$.

If the elliptic operator (with constant coefficients) is also homogeneous (i.e., if it coincides with its principal part), then

$$(7.10) \quad K(x) = \begin{cases} A\left(\dfrac{x}{r}\right) r^{m-n} & \text{for } n \text{ odd, or for } m < n, \\[2ex] B\left(\dfrac{x}{r}\right) r^{m-n} + C(x) \log r & \text{for } n \text{ even and } m \ge n, \end{cases}$$

where $A(\xi)$, $B(\xi)$ are analytic functions in a neighborhood of $|\xi| = 1$ and $C(x)$ is a polynomial of degree $m - n$.

We finally remark that the question of differentiability of solutions of elliptic equations can be treated analogously to the case of parabolic equations, either by using fundamental solutions or by using interior estimates of Schauder's type. By the latter method it can be shown that if the coefficients of the equation belong to $C^{p+\alpha}$ for some integer p and

$0 < \alpha < 1$, and if $f \in C^{p+\alpha}$, then the solutions of $Lu = f$ belong to $C^{m+p+\alpha}$. These results hold true also for elliptic systems.

PROBLEMS

1. Prove that if we take in (1.3) and in (1.7) $0 \leq 2bk_0 + |k| \leq 2bn_h$ and $0 \leq |k| \leq 2p$ respectively, then the roots of the two determinants coincide. ((1.6) is obtained from (1.1) by (1.5)).
[*Hint:* λ is a root for the first (second) determinant, if and only if there exists a nontrivial solution $w_j = C_j \exp [\lambda t + i\xi \cdot x]$ ($v_{jh} = C_{jh} \exp [\lambda t + i\xi \cdot x]$) of the system (1.1) ((1.6)). Show that if v_{jh} is such a solution of (1.6) then $w_j = C_{j1} \exp [\lambda t + i\xi \cdot x]$ is a solution of (1.1).]

2. Let $P(\xi)$ be the matrix obtained from the matrix in (1.7) when $\lambda = 0$ and when $|k| = 2p$ is replaced by $0 \leq |k| \leq 2p$. Prove that (1.6) is parabolic if and only if
$$\max_j \operatorname{Re} \{\lambda_j(\xi)\} \leq -\delta|\xi|^{2p} + C$$
for some positive constants δ, C and for arbitrary real vector ξ, where $\lambda_j(\xi)$ are the roots of $P(\xi)$.
[*Hint:* Let $P_0(\xi)$ be the matrix in (1.7) when $\lambda = 0$. Take $\xi_0 = \xi/|\xi|$ and write
$$\det (P(\xi) - \lambda I) = |\xi|^{2Np} \det (P_0(\xi_0) + \epsilon(\xi) - \lambda^* I)$$
where $\lambda^* = \lambda/|\xi|^{2p}$, $\epsilon(\xi) \to 0$ if $|\xi| \to \infty$.]

3. Prove Lemma 2, Sec. 2.
[*Hint:* $\rho(t) = \int_\tau^t \chi(\sigma)\varphi(\sigma)\, d\sigma$ satisfies $\rho' - \chi\rho \leq \chi\psi$.]

4. Prove (2.20).
[*Hint:* Both sides satisfy (2.2), (2.3).]

5. Prove that for every $p > 1$, $q = p/(p-1)$, $\alpha > 0$, $\beta > 0$,
$$\alpha\beta \leq \frac{\alpha^p}{p} + \frac{\beta^q}{q},$$
and that equality holds if and only if $\beta = \alpha^{p-1}$.
[*Hint:* The right-hand side represents a sum of two areas: one bounded by $y = x^{p-1}$, $x = \alpha$, $y = 0$ and the other bounded by $y = x^{p-1}$, $x = 0$, $y = \beta$.]

6. Consider a parabolic system of order $2p$ with constant coefficients
$$\frac{\partial u}{\partial t} = P(D_x)u$$
which is homogeneous (i.e., P coincides with its principal part). Show that the fundamental solution $Z(x, t; \xi, \tau) = Z(x - \xi; t, \tau)$ constructed in Sec. 2 satisfies (2.5) for all $-\infty < \tau < t < \infty$, $x \in R^n$, $\xi \in R^n$. (C_m, c_m are independent of x, ξ, t, τ.)

7. Prove the following (Liouville type) theorem: If u is a solution in $-\infty < t < 0$ of a parabolic system as in Problem 6, and if

$$|u(x, t)| \leq \text{const.}\ (1 + |t|)^{\beta}(1 + |x|)^{\gamma} \qquad (-\infty < t < 0, x \in R^n)$$

for some $\beta \geq 0, \gamma \geq 0$, then $u(x, t)$ is a polynomial of degree $\leq [\gamma]$ in $x_1, \ldots, x_n$ and of degree $\leq \min\{[\beta], [\gamma/2p]\}$ in t.
[*Hint:* Represent u analogously to (4.3.7) where $f \equiv 0$ and G is the fundamental solution of Problem 6. Show (as in the estimation of $D_x^i J$ following (4.3.7)) that

$$|D_t^i D_x^j u(P)| \leq K d^{-2pi-j} \underset{N}{\text{l.u.b.}}\ |u|,$$

and take $d \to \infty$.]

8. Prove that $K(x) = r^{2p-n}(A_{pn} \log r + B_{pn})$ is a fundamental solution of the p-harmonic equation $\Delta^p u = 0$ (where $\Delta = \sum_{j=1}^{n} \partial^2/\partial x_j^2$). Here $r = |x|$, A_{pn}, B_{pn} are constants, and $A_{pn} = 0$ if $2p < n$ or if n is odd whereas $B_{pn} = 0$ if $2p \geq n$ and n is even.

CHAPTER 10

BOUNDARY VALUE PROBLEMS FOR ELLIPTIC AND PARABOLIC EQUATIONS OF ANY ORDER

Introduction. In this chapter we consider the Dirichlet problem for "strongly elliptic" operators of any order $2m$, i.e., the problem of finding a solution of the strongly elliptic equation

$$Lu = f \qquad \text{in a domain } D \text{ of } R^n, \tag{0.1}$$

satisfying the boundary conditions

$$\frac{\partial^j u}{\partial \nu^j} = \varphi_j \qquad \text{on the boundary } \partial D \qquad (j = 0, 1, \ldots, m-1), \tag{0.2}$$

where ν is the normal to the boundary ∂D. D is taken to be bounded. The main result is that either the homogeneous problem ($f \equiv 0$, $\varphi_j \equiv 0$) has a unique solution (which is $u \equiv 0$) and then the problem (0.1), (0.2) also has a unique solution for any f, φ_j, or the homogeneous problem has non-zero solutions and then the problem (0.1), (0.2) has a solution only for f and φ_j satisfying some finite set of "orthogonality" relations.

This and other results for strongly elliptic equations are then used to solve the first initial-boundary value problem in cylinders for parabolic equations, i.e., the problem of finding a solution u of the parabolic equation

$$Lu - \frac{\partial u}{\partial t} = f \qquad \text{in } D \times (0, T], \tag{0.3}$$

satisfying the initial condition

$$u(x, 0) = \psi(x) \qquad \text{for } x \in D, \tag{0.4}$$

and the boundary conditions

$$\frac{\partial^j u}{\partial \nu^j} = \varphi_j \qquad \text{for } x \in \partial D,\ 0 < t \leq T \qquad (j = 0, 1, \ldots, m-1). \tag{0.5}$$

f, φ_j, and the coefficients of L (in (0.3), (0.5)) are functions of (x, t).

The method used to solve (0.1), (0.2) is entirely different from the methods used in previous chapters to solve boundary value problems. It can be outlined as follows: (a) A "generalized" Dirichlet problem is formulated and then solved with the aid of a representation theorem for functionals in a Hilbert space; the reduction of the problem to the representation theorem is done with the aid of a certain inequality. (b) The "generalized" solution is shown to be sufficiently smooth in D (provided f, the coefficients of L, and the data are sufficiently smooth), so that it satisfies (0.1) in the usual sense. (c) The "generalized" solution is shown to be sufficiently smooth near the boundary, so that it also satisfies (0.2) in the usual sense.

The concepts of weak derivatives, strong derivatives, and mollifiers are most useful in carrying out the details of the previous outline. They are introduced in Sec. 1. Some general differential inequalities are given in Sec. 2. In Secs. 3, 4, 5 the proofs of (a), (b), (c), respectively, are given. In Sec. 6 we establish an existence theorem for equations of the form $dx/dt + A(t)x = f(t)$ in a Hilbert space, and this theorem is used in Sec. 7, together with the results for elliptic equations, to solve the problem (0.3)–(0.5). Some properties of solutions of higher-order parabolic equations are mentioned in Sec. 8.

1. Weak and Strong Derivatives. Mollifiers

Throughout this chapter D is always a bounded domain in R^n and ∂D is its boundary. $L^2(D)$, or simply L^2, is the Hilbert consisting of all complex-valued measurable functions $u(x)$ in D with the norm

$$|u|_0^D = \left\{\int_D |u(x)|^2\,dx\right\}^{1/2}$$

and with the scalar product

$$(u, v)_0^D = \int_D u(x)\overline{v(x)}\,dx.$$

We denote by $C^m(D)$ $(0 \le m \le \infty)$ the set of all functions with m continuous derivatives in D, by $C^m(\overline{D})$ the set of all functions whose first m derivatives are uniformly continuous in D (and therefore can be considered to be defined by continuity also on ∂D), and by $C_c^m(D)$ those functions of $C^m(D)$ which have compact supports in D. We define $C_c^m(\overline{D}) = C_c^m(D)$.

If $u \in C^j(D)$ then, for any $\varphi \in C_c^\infty(D)$, $|\alpha| \le j$,

$$\int_D u(x)D^\alpha\varphi(x)\,dx = (-1)^{|\alpha|}\int_D v(x)\varphi(x)\,dx \tag{1.1}$$

for $v(x) = D^\alpha u(x)$, where $D^\alpha = D_1^{\alpha_1}\cdots D_n^{\alpha_n}$, $D_i = \partial/\partial x_i$, $\alpha = (\alpha_1, \ldots, \alpha_n)$, $|\alpha| = \alpha_1 + \cdots + \alpha_n$. Formula (1.1) motivates the following definition.

Definition. Let u, v belong to $L^2(A)$ for any compact subdomain A of D. If (1.1) holds for any $\varphi \in C_c^\infty(D)$ then we say that v is the αth *weak derivative* of u and write $D^\alpha u = v$ (w.d.).

If $D^\alpha u = v$ (w.d.) and $D^\alpha u = w$ (w.d.) then $\int (v - w)\varphi \, dx = 0$ for any $\varphi \in C_c^\infty(D)$; hence $v = w$ (i.e., $v(x) = w(x)$ almost everywhere). We conclude that any αth weak derivative is uniquely defined. In particular, $D^\alpha 0 = 0$ (w.d.).

Let $\hat{C}^j(D)$ $(0 \leq j < \infty)$ be the subset of $C^j(D)$ consisting of all functions $u(x)$ with finite norm

$$|u|_j^D = \left\{ \sum_{|\alpha| \leq j} \int_D |D^\alpha u(x)|^2 \, dx \right\}^{1/2}. \tag{1.2}$$

$\hat{C}^j(D)$ is a scalar-product space (or pre-Hilbert space) with the scalar product

$$(u, v)_j^D = \sum_{|\alpha| \leq j} \int_D D^\alpha u(x) \cdot D^\alpha \overline{v(x)} \, dx. \tag{1.3}$$

When there is no confusion, the superindex "D" will be omitted.

We denote by $H^j(D)$, or simply H^j, the completion of $\hat{C}^j(D)$ (with respect to the norm (1.2)). $\{u_m\}$ is a Cauchy sequence in $\hat{C}^j(D)$ if and only if for any α, $0 \leq |\alpha| \leq j$,

$$\int_D |D^\alpha u_m(x) - D^\alpha u_k(x)|^2 \, dx \to 0 \qquad \text{as } m, k \to \infty. \tag{1.4}$$

Since $L^2(D)$ is a complete space, there exist functions $u^\alpha \in L^2(D)$ such that

$$\int_D |D^\alpha u_m(x) - u^\alpha(x)|^2 \, dx \to 0 \qquad \text{as } m \to \infty. \tag{1.5}$$

Since two Cauchy sequences in $\hat{C}^j(D)$ determine the same elements u^α of $L^2(D)$, for $0 \leq |\alpha| \leq j$, if and only if they are equivalent Cauchy sequences, we can identify the elements of H^j with the vectors $\{u^\alpha; 0 \leq |\alpha| \leq j\}$ obtained in the above manner.

Lemma 1. *If* $\{u^\alpha; 0 \leq |\alpha| \leq j\}$ *belongs to* H^j *then* $u^\alpha = D^\alpha u^0$ *(w.d.).*

Proof. Let $\{u_m\}$ be a Cauchy sequence in $\hat{C}^j(D)$ such that (1.5) holds for $0 \leq |\alpha| \leq j$. Using Schwarz' inequality one easily shows that for any $\varphi \in C_c^\infty(D)$,

$$\int_D D^\alpha u_m(x) \cdot \varphi(x) \, dx \to \int_D u^\alpha(x)\varphi(x) \, dx, \tag{1.6}$$

$$\int_D u_m(x) D^\alpha \varphi(x) \, dx \to \int_D u^0(x) D^\alpha \varphi(x) \, dx. \tag{1.7}$$

Now, by partial integration,

$$\int_D u_m(x) D^\alpha \varphi(x) \, dx = (-1)^{|\alpha|} \int_D D^\alpha u_m(x) \cdot \varphi(x) \, dx.$$

Taking $m \to \infty$ in the last equality and using (1.6), (1.7) we get the relation (1.1) with $u = u^0$, $v = u^\alpha$. Hence $D^\alpha u^0 = u^\alpha$ (w.d.).

Corollary 1. *The vector* $\{u^\alpha; 0 \leq |\alpha| \leq j\}$ *is uniquely determined by its component* u^0.

Indeed, we have to show that if $u^0 = 0$ then $u^\alpha = 0$. But this follows from Lemma 1 and from the relation $D^\alpha 0 = 0$ (w.d.).

From the corollary it follows that the elements of $H^j(D)$ can be identified with the elements of some linear subspace of $L^2(D)$, i.e.,

Corollary 2. *An element $u \in L^2(D)$ belongs to $H^j(D)$ if and only if there exists a Cauchy sequence $\{u_m\}$ in $\hat{C}^j(D)$ such that $u_m \to u$ in $L^2(D)$.*

This identification of $H^j(D)$ with a linear subspace of $L^2(D)$ will be assumed from now on.

Definition. If $u \in H^j(A)$ for any compact subdomain A of D then we say that u has *strong derivatives* up to order j in D, or that u has j strong derivatives in D.

Thus, for every compact subdomain A of D there exists a sequence $\{u_m\}$ in $\hat{C}^j(A)$ such that $|u_m - u|_0^A \to 0$ and such that $|D^\alpha u_m - u^\alpha|_0^A \to 0$ for some $u^\alpha \in L^2(A)$ $(0 < |\alpha| \leq j)$. From Corollary 1 it follows that the u^α are independent of A. We call u^α the αth *strong derivative* of u in D and write $D^\alpha u = u^\alpha$ (s.d.). By Lemma 1, u^α is then also the αth weak derivative of u. Thus, strong derivatives are also weak derivatives.

If $u \in C^j(D)$ then the (usual) derivative $D^\alpha u$ $(0 \leq |\alpha| \leq j)$ is also the αth strong derivative of u in D.

Definition. The completion of the space $C_c^\infty(D)$ under the norm (1.2) is denoted by $\mathring{H}^j(D)$, or simply $\mathring{H}^j$.

We identify $\mathring{H}^j$ with a linear subspace of H^j. Note that $H^0 = \mathring{H}^0 = L^2$. For $j = 0$ we shall occasionally use the notation

$$|u| = |u|_0, \qquad (u, v) = (u, v)_0.$$

We now make a few remarks concerning convergence in Hilbert spaces. A sequence $\{u_m\}$, in a Hilbert space H with scalar product (,), is said to be *weakly convergent* (to u) if the sequence $\{(u_m, f)\}$ is convergent (to (u, f)) for any $f \in H$. A weakly convergent sequence is bounded. From any bounded sequence $\{u_m\}$ in H one can extract a weakly convergent subsequence. If $\{u_m\}$ is weakly convergent to u, then there exists a subsequence $\{u_{m'}\}$ whose arithmetic means converge to u in the norm of H (see Problem 1).

Lemma 2. *Let $u \in L^2(D)$ and suppose that there exists a sequence $\{u_m\}$ in $C^j(D)$ such that*

(a) *$\{u_m\}$ is weakly convergent to u in $L^2(D)$, and*
(b) *$|u_m|_j^D \leq$ const. independent of m.*

Then $u \in H^j(D)$ and, for any $0 \leq |\alpha| \leq j$, the strong derivative $D^\alpha u$ is the weak limit (in $L^2(D)$) of $\{D^\alpha u_m\}$.

Proof. Because of (b), we can choose a subsequence $\{u_{m'}\}$ which is weakly convergent in $H^j(D)$. Choose a subsequence $\{u_{m''}\}$ of $\{u_{m'}\}$ whose arithmetic means $v_{m''}$ converge in $H^j(D)$. Since, by (a), these arithmetic means $v_{m''}$ converge also weakly to u in $L^2(D)$, $|v_{m''} - u|_0^D \to 0$. From Corollary 2 to Lemma 1 it follows that $u \in H^j(D)$.

We next have, for any $\varphi \in C_c^\infty(D)$,

$$\int_D D^\alpha u_m \cdot \varphi \, dx = (-1)^{|\alpha|} \int_D u_m D^\alpha \varphi \, dx \to (-1)^{|\alpha|} \int_D u D^\alpha \varphi \, dx = \int_D u^\alpha \varphi \, dx$$

where $u^\alpha = D^\alpha u$ (s.d.). Since $C_c^\infty(D)$ is dense in $L^2(D)$, we easily get (using (b))

$$\int_D D^\alpha u_m \cdot f \, dx \to \int_D u^\alpha f \, dx \qquad \text{for any } f \in L^2(D),$$

i.e., $D^\alpha u_m \to u^\alpha$ weakly in $L^2(D)$.

We shall now introduce a smoothing operator, or mollifier, which will be a very useful tool in the study of strong derivatives.

Definition. Let $\rho(x)$ be a C^∞ function in R^n satisfying the following properties (for the construction of ρ, see Problem 1, Chap. 1):

(1) $\rho(x) = 0 \qquad \text{if } |x| \geq 1,$

(2) $\rho(x) \geq 0 \qquad \text{if } |x| \leq 1,$

(3) $\int_{R^n} \rho(x) \, dx = 1.$

For any $\epsilon > 0$, the operator

$$(J_\epsilon u)(x) = \frac{1}{\epsilon^n} \int_{|y-x|<\epsilon} \rho\left(\frac{x-y}{\epsilon}\right) u(y) \, dy \tag{1.8}$$

is called a *mollifier* of u. The function u is defined on D, whereas the mollifier is considered only on a compact subdomain A of D. ϵ is taken to be $< \epsilon_0$, where ϵ_0 is the distance from A to the boundary ∂D of D. The integral in (1.8) is then well defined since if $x \in A$, $|y - x| < \epsilon$ then $y \in D$. Furthermore,

$$(J_\epsilon u)(x) = \frac{1}{\epsilon^n} \int_D \rho\left(\frac{x-y}{\epsilon}\right) u(y) \, dy. \tag{1.9}$$

If $u \in L^2(D)$ then $J_\epsilon u$ belongs to $C^\infty(A)$. If u belongs to $C^m(D)$ and $|\alpha| \leq m$, then

$$D^\alpha(J_\epsilon u) = J_\epsilon(D^\alpha u), \tag{1.10}$$

as is easily verified.

Lemma 3. *If* $u \in L^2(D)$ *then*

$$|J_\epsilon u|_0^A \leq |u|_0^D, \tag{1.11}$$

$$\lim_{\epsilon \to 0} |J_\epsilon u - u|_0^A = 0. \tag{1.12}$$

Proof. The proof of (1.11) is straightforward:

$$(|J_\epsilon u|_0^A)^2 = \int_A \left| \frac{1}{\epsilon^n} \int_D \rho\left(\frac{x-y}{\epsilon}\right) u(y)\, dy \right|^2 dx$$

$$\leq \int_A \left\{ \left[\frac{1}{\epsilon^n} \int_D \rho\left(\frac{x-y}{\epsilon}\right) dy\right] \left[\frac{1}{\epsilon^n} \int_D \rho\left(\frac{x-y}{\epsilon}\right) |u(y)|^2\, dy\right] \right\} dx$$

$$= \int_D |u(y)|^2 \left[\frac{1}{\epsilon^n} \int_A \rho\left(\frac{x-y}{\epsilon}\right) dx\right] dy \leq \int_D |u(y)|^2\, dy = (|u|_0^D)^2.$$

To prove (1.12) consider first the case when u is continuous in D. Writing

$$(J_\epsilon u)(x) - u(x) = \frac{1}{\epsilon^n} \int_D \rho\left(\frac{x-y}{\epsilon}\right) [u(y) - u(x)]\, dy$$

and setting $\delta(\epsilon) = \text{l.u.b. } |u(x) - u(y)|$ where x varies in A, y varies in D, and $|y - x| \leq \epsilon$, we have

$$|(J_\epsilon u)(x) - u(x)| \leq \delta(\epsilon) \frac{1}{\epsilon^n} \int_D \rho\left(\frac{x-y}{\epsilon}\right) dy = \delta(\epsilon).$$

Hence,

$$\int_A |J_\epsilon u - u|^2\, dx \to 0 \qquad \text{as } \epsilon \to 0.$$

To prove (1.12) in case u is only assumed to belong to $L^2(D)$, let v be a continuous function in D such that $|u - v|_0^D < \delta$ where δ is any given positive number. Writing

$$J_\epsilon u - u = J_\epsilon(u - v) + (J_\epsilon v - v) + (v - u)$$

and noting, by (1.11), that $|J_\epsilon(u - v)|_0^A \leq \delta$, we get

$$|J_\epsilon u - u|_0^A \leq 2\delta + |J_\epsilon v - v|_0^A.$$

Since (1.12) was already proved for continuous functions, we find that $|J_\epsilon u - u|_0^A \leq 3\delta$ if ϵ is sufficiently small, and the proof of (1.12) is thereby completed.

Lemma 4. *If u is j times strongly differentiable, and if its jth derivatives are k times strongly differentiable, then u has $j + k$ strong derivatives.*

Proof. Take subdomains A, B of D with $\overline{A} \subset B \subset \overline{B} \subset D$ and let ϵ_0 be the distance from $\overline{A}$ to the boundary of B. Since u belongs to $L^2(B)$, if $\epsilon < \epsilon_0$ then $J_\epsilon u$ is defined on A and, by Lemma 3, $|J_\epsilon u - u|_0^A \to 0$ as $\epsilon \to 0$. In view of Lemma 2 it remains to prove that

$$|J_\epsilon u|_{j+k}^A \leq \text{const. independent of } \epsilon. \tag{1.13}$$

If $|\gamma| > j$ then $\gamma = \alpha + \beta$ where $|\beta| = j$. For any $x \in A$,

$$(1.14)\quad (D^\gamma J_\epsilon u)(x) = \frac{1}{\epsilon^n}\int_B D_x^{\alpha+\beta}\,\rho\left(\frac{x-y}{\epsilon}\right)\cdot u(y)\,dy$$
$$= \frac{1}{\epsilon^n}\int_B D_x^\alpha\rho\left(\frac{x-y}{\epsilon}\right)\cdot D^\beta u(y)\,dy$$

where in the last equality we have used the definition of the strong derivative $D^\beta u$ and the fact that $D_x^\alpha\rho\left(\frac{x-y}{\epsilon}\right)$ is in $C_c^\infty(B)$ for any fixed $x \in A$. In the same way we find that the right-hand side of (1.14) is equal to

$$\frac{1}{\epsilon^n}\int_B \rho\left(\frac{x-y}{\epsilon}\right) D^\alpha(D^\beta u(y))\,dy = [J_\epsilon(D^\alpha(D^\beta u))](x).$$

We have thus proved that

$$(1.15)\qquad D^\gamma(J_\epsilon u) = J_\epsilon(D^\alpha(D^\beta u)).$$

From (1.15) and (1.11) we get

$$|D^\gamma(J_\epsilon u)|_0^A \le |D^\alpha(D^\beta u)|_0^B < \infty.$$

If $|\gamma| \le j$ then by a part of the previous reasoning we get $D^\gamma(J_\epsilon u) = J_\epsilon(D^\gamma u)$, and using (1.11) we find that

$$|D^\gamma(J_\epsilon u)|_0^A \le |D^\gamma u|_0^B < \infty.$$

The proof of (1.13) is thereby completed.

Remark 1. Since, by Lemma 2, $\{D^\gamma J_\epsilon\}$ converges weakly in $L^2(A)$ to the strong derivative $D^\gamma u$, and since, by (1.12), $J_\epsilon(D^\alpha(D^\beta u)) \to D^\alpha(D^\beta u)$ in the norm of $L^2(A)$, it follows from (1.15) that

$$(1.16)\qquad D^{\alpha+\beta}u = D^\alpha(D^\beta u)\qquad \text{(s.d.)}.$$

Remark 2. The argument used to justify the second equality in (1.14) shows that (1.10) holds also if u is only assumed to have an αth weak derivative.

The relation

$$(1.17)\qquad \int_D Du\cdot\varphi\,dx = -\int_D u\cdot D\varphi\,dx$$

which holds for any $\varphi \in C_c^1(D)$, $u \in C^1(D)$ holds also if u is only assumed to have first strong derivatives in D. This follows by completion, i.e., by taking a sequence $\{u_m\}$ of functions in $C^1(A)$ which converges to u in the norm of $H^1(A)$, writing down the relation (1.17) for every u_m and then taking $m \to \infty$. Here A is any domain containing the support of φ, with $\bar{A} \subset D$.

We have thus established the standard rule (1.17) of integration by parts also for functions u which have only strong derivatives. The above reasoning can be used to extend also other standard rules of calculus to functions with strong derivatives.

Partition of Unity. Let $D_1, \ldots, D_N$ be open subsets of D such that $D = \bigcup_{i=1}^{N} D_i$ (the D_i are said to form an *open covering* of D). Then there exist functions φ_i such that:

(1) $\varphi_i \in C_c^\infty(R^n)$ and (support $\varphi_i) \cap D \subset D_i$,

(2) $\varphi_i \geq 0$,

(3) $\Sigma\, \varphi_i \equiv 1$ in D.

The φ_i are said to form a *partition of unity* subordinate to the open covering $\{D_i\}$. For proof see, for instance, [44; 45–47].

Lemma 5. *Let $D_1, \ldots, D_N$ be subdomains of D and let $D = \bigcup_{i=1}^{N} D_i$. If a function u belongs to $H^j(D_i)$ for every i then u belongs to $H^j(D)$.*

Proof. Let $\{\varphi_i\}$ be a partition of unity subordinate to the covering $\{D_i\}$. Since $u \in H^j(D_i)$, there exists a sequence $\{u_{i,m}\}$ in $H^j(D_i)$ such that

$$(1.18) \quad |u_{i,m} - u_{i,k}|_j^{D_i} \to 0, \qquad |u_{i,m} - u|_0^{D_i} \to 0 \qquad \text{as } m, k \to \infty.$$

Setting $u_m = \sum_{i=1}^{N} \varphi_i u_{i,m}$ we have

$$D^\alpha u_m - D^\alpha u_k = \sum_{i=1}^{N} D^\alpha[\varphi_i(u_{i,m} - u_{i,k})].$$

Using Leibniz' rule to expand the right-hand side and then applying (1.18), we get

$$|D^\alpha u_m - D^\alpha u_k|_0^D \leq \text{const.} \sum_{i=1}^{N} \sum_{|\beta| \leq |\alpha|} |D^\beta u_{i,m} - D^\beta u_{i,k}|_0^{D_i} \to 0$$

if $m, k \to \infty$, where $0 \leq |\alpha| \leq j$. Since also

$$|u_m - u|_0^D = \left| \sum_{i=1}^{N} \varphi_i(u_{i,m} - u) \right|_0^D \leq \sum_{i=1}^{N} \left| u_{i,m} - u \right|_0^{D_i} \to 0$$

as $m \to \infty$, it follows that $u \in H^j(D)$.

Lemma 5 shows that the concept of strong derivatives is a local concept (even though it is not a pointwise concept). Thus:

Corollary. *If for each point $x \in D$ there exists a neighborhood in which u has j strong derivatives, then u has j strong derivatives in D.*

Lemma 6. *Let D be a bounded domain whose boundary consists of two disjoint sets, Γ_1 and Γ_2, where Γ_2 is an open domain on a hyperplane $x_n = c$. If u has j continuous strong derivatives in D and if all its strong derivatives of orders $\leq j$ belong to $L^2(D)$, then u belongs to $H^j(A)$ for any subdomain A of D whose closure lies in $D + \Gamma_2$.*

Proof. Let B be a subdomain of D whose closure lies in $D + \Gamma_2$, such that $A \subset B$, and let the boundary of B consist of two disjoint sets, ∂B_1

and ∂B_2, where ∂B_2 is an open set on $x_n = c$. We may assume that the boundary of A decomposes similarly into two disjoint sets ∂A_1 and ∂A_2, ∂A_2 being an open set on $x_n = c$. We take the domain B such that $\overline{\partial A_2} \subset \partial B_2$ and $\overline{\partial A_1} \cap \overline{\partial B_1} = \emptyset$. Finally, we may assume that B lies in the half-space $x_n > c$.

We now define mollifiers somewhat differently than in (1.8), (1.9), namely,

$$(J'_\epsilon u)(x) = \frac{1}{\epsilon^n} \int_D \rho\left(\frac{x_\epsilon - y}{\epsilon}\right) u(y)\, dy \tag{1.19}$$

where $x_\epsilon = (x_1, \ldots, x_{n-1}, x_n + 2\epsilon)$ if $x = (x_1, \ldots, x_{n-1}, x_n)$ and $0 < 3\epsilon < \epsilon_0$, ϵ_0 being the distance from ∂A_1 to ∂B_1.

By obvious modifications of the proof of Lemma 3 we find that if $u \in L^2(B)$ then

$$|J'_\epsilon u|_0^A \le |u|_0^B, \tag{1.20}$$

$$|J'_\epsilon u - u|_0^A \to 0 \qquad \text{if } \epsilon \to 0. \tag{1.21}$$

Since $\rho((x_\epsilon - y)/\epsilon)$ belongs to $C_c^\infty(B)$ for any fixed $x \in A$, we get

$$\begin{aligned}(D^\alpha J'_\epsilon u)(x) &= \frac{1}{\epsilon^n} \int_D D_x^\alpha \rho\left(\frac{x_\epsilon - y}{\epsilon}\right) \cdot u(y)\, dy \\ &= \frac{1}{\epsilon^n} \int_D \rho\left(\frac{x_\epsilon - y}{\epsilon}\right) D^\alpha u(y)\, dy \\ &= (J'_\epsilon(D^\alpha u))(x)\end{aligned} \tag{1.22}$$

for any $0 \le |\alpha| \le j$. Since the strong derivatives $D^\alpha u$ belong to $L^2(D)$, employing the property (1.21) we conclude from (1.22) that

$$|D^\alpha(J'_\epsilon u) - D^\alpha u|_0^A \to 0 \qquad \text{as } \epsilon \to 0;$$

hence $u \in H^j(A)$.

We recall (see Sec. 8, Chap. 3) that ∂D is said to belong to class C^j if we can cover ∂D by a finite number of open sets N_k such that each set $N_k \cap \partial D$ can be represented in the form (3.8.6) with h in class C^j.

Theorem 1. *Let ∂D be of class C^j $(j \ge 1)$. If u has j strong derivatives in D and if all its strong derivatives $D^\alpha u$ $(0 \le |\alpha| \le j)$ belong to $L^2(D)$, then u belongs to $H^j(D)$.*

Note that the converse is trivial, i.e., if $u \in H^j(D)$ then all the first j strong derivatives of u exist and belong to $L^2(D)$.

Proof. In view of Lemma 5 it suffices to show that every point $x^0 \in \partial D$ has a neighborhood N_0 such that u belongs to $H^j(N_0 \cap D)$. We may assume that for some neighborhood N of x^0, $N \cap \partial D$ can be represented in the form

$$x_n = h(x_1, \ldots, x_{n-1})$$

where $h \in C^j$, and that $N \cap D$ lies on the side $x_n > h(x_1, \ldots, x_{n-1})$. Perform a transformation

$$(1.23)\qquad \begin{aligned} y_i &= x_i \qquad \text{for } i = 1, \ldots, n-1, \\ y_n &= x_n - h(x_1, \ldots, x_{n-1}) \quad \text{or} \quad x_n = y_n + h(y_1, \ldots, y_{n-1}). \end{aligned}$$

If $V(y) = v(x)$ where $v \in C^1$, then

$$(1.24)\qquad \frac{\partial V}{\partial y_i} = \sum_{k=1}^{n} \frac{\partial v}{\partial x_k}\frac{\partial x_k}{\partial y_i}, \qquad \frac{\partial v}{\partial x_i} = \sum_{k=1}^{n} \frac{\partial V}{\partial y_k}\frac{\partial y_k}{\partial x_i}.$$

It follows that the L^2 norm of $\Sigma\, |\partial v(x)/\partial x_i|$ is bounded from above and from below by positive constants times the L^2 norm of $\Sigma\, |\partial V(y)/\partial y_i|$. Consequently, $\{\partial v_m/\partial x_i\}$ are Cauchy sequences in L^2, for $i = 1, \ldots, n$, if and only if the same is true of $\{\partial V_m/\partial y_i\}$ for $i = 1, \ldots, n$. It follows that $v(x)$ has first strong derivatives (belongs to $H^1(N \cap D)$) if and only if $V(y)$ has first strong derivatives (belongs to $H^1(M^*)$, where M^* is the image of $N \cap D$ in the y-space). Furthermore, the relations (1.24) hold also for strong derivatives (by completion).

The above remarks hold also for all the strong derivatives of orders $\leq j$. Hence, if we show that $U(y) = u(x)$ belongs to $H^j(M^{**})$ where M^{**} is related to M^* in the same manner that the domain A was related to the domain B in the proof of Lemma 6, then it follows that $u(x)$ belongs to $H^j(N_0 \cap D)$ for some neighborhood N_0 of x^0. The fact that $U \in H^j(M^{**})$ follows from Lemma 6.

2. Differential Inequalities

Lemma 7. *For any $\epsilon > 0$ and for any integer $j > 0$ there exists a constant $C = C(\epsilon, j)$ such that the inequality*

$$(2.1)\qquad |u|_{j-1}^2 \leq \epsilon |u|_j^2 + C|u|_0^2$$

holds for any $u \in \mathring{H}^j(D)$.

Note that C is independent of D.

Proof. Consider first the case where $u \in C_c^\infty(D)$. Let $\tilde{u}$ be the Fourier transform of u, i.e.,

$$\tilde{u}(\xi) = \int_{R^n} e^{-ix\cdot\xi}\, u(x)\, dx.$$

Then

$$(2.2)\qquad \widetilde{D^\alpha u}(\xi) = (i\xi)^\alpha \tilde{u}(\xi).$$

By Plancherel's theorem, if $v \in L^2(R^n)$ then $\tilde{v} \in L^2(R^n)$ and

$$(2.3)\qquad \int_{R^n} |\tilde{v}(\xi)|^2\, d\xi = (2\pi)^n \int_{R^n} |v(x)|^2\, dx.$$

Hence, using (2.2) we get

$$(2.4) \qquad (2\pi)^n \int_{R^n} |D^\alpha u(x)|^2\, dx = \int_{R^n} |\xi^\alpha|^2 |\tilde{u}(\xi)|^2\, d\xi.$$

It follows that

$$(2.5) \qquad \begin{aligned} |u|_{j-1}^2 &= (2\pi)^{-n} \sum_{|\alpha| \le j-1} \int_{R^n} |\xi^\alpha|^2 |\tilde{u}(\xi)|^2\, d\xi, \\ |u|_j^2 &= (2\pi)^{-n} \sum_{|\alpha| \le j} \int_{R^n} |\xi^\alpha|^2 |\tilde{u}(\xi)|^2\, d\xi. \end{aligned}$$

Since for any $\epsilon > 0$,

$$\sum_{|\alpha| \le j-1} |\xi^\alpha|^2 \le \epsilon \sum_{|\alpha| \le j} |\xi^\alpha|^2 + C \qquad (\text{for all } \xi \in R^n)$$

for some constant C depending only on ϵ, j, we obtain from (2.5)

$$|u|_{j-1}^2 \le \epsilon |u|_j^2 + (2\pi)^{-n}\, C \int_{R^n} |\tilde{u}(\xi)|^2\, d\xi.$$

Since, by (2.3), the last integral is equal to $(2\pi)^n |u|_0^2$, (2.1) follows.

Having proved (2.1) for functions u in $C_c^\infty(D)$, we can now prove it for any $u \in \mathring{H}^j(D)$ by completion, i.e., we take a sequence $\{u_m\}$ in $C_c^\infty(D)$ which converges to u in $\mathring{H}^j(D)$, apply (2.1) to each u_m and then take $m \to \infty$.

In the next lemma we extend the inequality (2.1) to functions in $H^j(D)$.

Lemma 8. *Assume that ∂D is of class C^2. For any ϵ and for any integer $j > 0$ there exists a constant $C = C(\epsilon, j, D)$ such that the inequality* (1.1) *holds for any $u \in H^j(D)$.*

Proof. It suffices to prove (1.1) for functions u in $\hat{C}^j(D)$, since the proof for any $u \in H^j(D)$ then follows by completion. (1.1) is a consequence of the following inequality:

$$(2.6) \qquad \sum_{|\alpha| = i} \int_D |D^\alpha u|^2\, dx \le \epsilon \sum_{|\beta| = j} \int_D |D^\beta u|^2\, dx + \frac{C(j, D)}{\epsilon^{i/(j-i)}} \int_D |u|^2\, dx \qquad (i < j),$$

where $0 < \epsilon \le \epsilon_0$ and ϵ_0, $C(j, D)$ depend only on j, D.

We first prove (2.6) for $i = 1$, $j = 2$, $n = 1$, and $0 < \epsilon < 4|D|^2$, $|D|$ being the length of the interval D. Divide D into subintervals of length $< (\epsilon/2)^{1/2}$ and $> (\epsilon/4)^{1/2}$ (the inequality $\epsilon < 4|D|^2$ is being used here). For each such subinterval $a < x < b$ set $\alpha = (b - a)/4$ and let x_1, x_2 be variable points in the intervals $a < x < a + \alpha$, $a + 3\alpha < x < b$ respectively. By the mean value theorem,

$$(2.7) \qquad \frac{u(x_2) - u(x_1)}{x_2 - x_1} = Du(x_{12}) \qquad (x_1 < x_{12} < x_2).$$

Writing, for any $x \in (a, b)$,

$$Du(x) = Du(x_{12}) + \int_{x_{12}}^x D^2 u(\xi)\, d\xi$$

and using (2.7), we get

$$|Du(x)| \le \frac{|u(x_1)| + |u(x_2)|}{2\alpha} + \int_a^b |D^2u(\xi)|\, d\xi.$$

Integrating with respect to x_1, x_2 in their respective intervals $(a, a + \alpha)$, $(a + 3\alpha, b)$, we find

$$\alpha^2|Du(x)| \le \frac{1}{2}\int_a^b |u(\xi)|\, d\xi + \alpha^2 \int_a^b |D^2u(\xi)|\, d\xi.$$

Taking squares and using Schwarz' inequality, we get

$$|Du(x)|^2 \le \frac{2}{\alpha^3}\int_a^b |u(\xi)|^2\, d\xi + 8\alpha \int_a^b |D^2(\xi)|^2\, d\xi.$$

Integrating with respect to x in the interval (a, b), we obtain

$$\int_a^b |Du|^2\, dx \le 2(b-a)^2 \int_a^b |D^2u|^2\, dx + \frac{2^7}{(b-a)^2}\int_a^b |u|^2\, dx$$

$$\le \epsilon \int_a^b |D^2u|^2\, dx + \frac{2^9}{\epsilon}\int_a^b |u|^2\, dx.$$

Summing over all the subintervals we conclude that

$$\int_D |Du|^2\, dx \le \epsilon \int_D |D^2u|^2\, dx + \frac{2^9}{\epsilon}\int_D |u|^2\, dx. \tag{2.8}$$

(2.6) will now be proved for $n > 1$, $i = 1$, $j = 2$. Consider first the case where D is a cube with edges parallel to the coordinate axes, and set $\partial u/\partial x_1 = Du$. Considering u as a function of x_1 with $x_2, \ldots, x_n$ as parameters, we can apply (2.8) and then, after integrating with respect to the parameters $x_2, \ldots, x_n$, find that $\int_D |\partial u/\partial x_1|^2\, dx$ is bounded by the right-hand side of (2.6). Since a similar bound holds for any $\partial u/\partial x_k$, (2.6) follows for $i = 1, j = 2$. If D is not a cube, then we cover D by a finite number of domains Γ_λ, Δ_μ where Γ_λ are cubes with edges parallel to the coordinate axes, and each Δ_μ can be mapped by a one-to-one transformation $y = y(x)$ onto a cube with sides parallel to the coordinate axes in the y-space; the mapping $y = y(x)$ and its inverse having two continuous derivatives. (2.6) for $i = 1, j = 2$ holds for each Γ_λ (with $C(j, D)$ depending only on j) and for each Δ_μ (with $C(j, D)$ depending on j, D). Summing over λ, μ the inequality (2.6) for $i = 1, j = 2$ (and appropriate $C(j, D)$) follows.

We shall now prove (2.6) for all i, j by induction on j. For $j = 1$, (2.6) was already proved for $i = 1$ and it is trivial if $i = 0$. Assuming (2.6) to hold for all $j \le k$ we shall prove it for $j = k + 1$. The notation $|D^m u|^2 = \sum_{|\alpha| = m} |D^\alpha u|^2$ will be used.

If $i = k$ then applying (2.6) with $i = 1, j = 2$ to the $(k-1)$th derivatives of u we get, for any sufficiently small $\epsilon > 0$,

$$\int_D |D^k u|^2 \, dx \leq \frac{\epsilon}{2} \int_D |D^{k+1}u|^2 \, dx + \frac{C_1}{\epsilon} \int_D |D^{k-1}u|^2 \, dx$$

where C_m are used to denote positive constants depending only on k, D. Using the inductive assumption with $j = k$, $i = k - 1$, we have

$$\int_D |D^{k-1}u|^2 \, dx \leq \delta \int_D |D^k u|^2 \, dx + \frac{C_2}{\delta^{k-1}} \int_D |u|^2 \, dx.$$

Substituting this in the previous inequality and taking $\delta = \epsilon/2C_1$ we get (2.6) with $i = k$, $j = k + 1$.

If $i < k$ then, by the inductive assumption,

$$\int_D |D^i u|^2 \, dx \leq \delta \int_D |D^k u|^2 \, dx + \frac{C_3}{\delta^{i/(k-i)}} \int_D |u|^2 \, dx.$$

By (2.6) with $i = k$, $j = k + 1$,

$$\int_D |D^k u|^2 \, dx \leq \mu \int_D |D^{k+1}u|^2 \, dx + \frac{C_4}{\mu^k} \int_D |u|^2 \, dx.$$

Substituting the last inequality into the previous one and taking $\delta = \epsilon^{(k-i)/(k+1-i)}$, $\mu = \epsilon^{1/(k+1-i)}$, we obtain (2.6) for $j = k + 1$.

In the following two theorems uniform bounds are obtained on a function u in terms of the norm $|u|_i$ for any $i \geq [n/2] + 1$.

Theorem 2 (Sobolev's Lemma). *For any bounded domain D of R^n and for any function u in $\hat{C}^j(D)$, where $j = [n/2] + 1$,*

$$|u(y)|^2 \leq C \sum_{|\alpha| \leq j} R^{2|\alpha| - n} \int_D |D^\alpha u(x)|^2 \, dx \tag{2.9}$$

for all $y \in D$, where R is the distance from y to the boundary of D and C is a constant depending only on n.

Proof. Let $g(t)$ be a C^∞ function of a real variable t, such that $g(t) = 1$ if $t \leq \frac{1}{2}$ and $g(t) = 0$ if $t \geq 1$. The function $h(t) = g(t/R)$ satisfies

$$\left| \frac{d^k h(t)}{dt^k} \right| \leq \frac{A_k}{R^k} \tag{2.10}$$

where A_k are constants depending only on g and k. Setting $r = |x - y|$ and observing that $h(0) = 1$, $h(R) = 0$, we get

$$u(y) = -\int_0^R \frac{\partial}{\partial r} [h(r)u(x)] \, dr.$$

Integrating over the $(n - 1)$-dimensional unit sphere Ω with center y, we obtain

$$\Omega_n u(y) = -\int_\Omega \int_0^R \frac{\partial}{\partial r} (hu) \, dr \, d\Omega \qquad \left(\Omega_n = \int_\Omega d\Omega \right).$$

Integrating by parts $j - 1$ times with respect to r, we find

$$\Omega_n u(y) = \frac{(-1)^j}{(j-1)!} \int_\Omega \int_0^R r^{j-1} \frac{\partial^j}{\partial r^j}(hu)\, dr\, d\Omega.$$

Writing $r^{j-1} = r^{j-n}r^{n-1}$ and using Schwarz' inequality, it follows that

$$|u(y)|^2 \le C_1 \left\{ \int \left| \frac{\partial^j}{\partial r^j}(hu) \right|^2 dV \right\} \left\{ \int r^{2(j-n)}\, dV \right\} \tag{2.11}$$

where $dV = r^{n-1}\, dr\, d\Omega$ is the volume element of D, and C_i are used to denote positive constants depending only on n. The integrals in (2.11) are taken over the ball $r < R$. Since $2j > n$, the last integral on the right-hand side of (2.11) is bounded by $C_2 R^{2j-n}$. To evaluate the first integral, we use Leibniz' rule and (2.10). We then get

$$|u(y)|^2 \le C_3 \sum_{k=0}^{j} R^{2k-n} \int_D \left| \frac{\partial^k u}{\partial r^k} \right|^2 dV. \tag{2.12}$$

Noting that

$$\left| \frac{\partial^k u(x)}{\partial r^k} \right| \le C_4 \sum_{|\alpha|=k} |D^\alpha u(x)| \qquad (0 \le k \le j),$$

the inequality (2.9) follows from (2.12).

The inequality (2.9) is not useful for estimating $u(y)$ near the boundary of D since R becomes arbitrarily small. In order to obtain a good bound near the boundary, we have to impose some condition on D.

Definition. We say that D satisfies the *cone condition* if every point y in $\bar{D}$ is the vertex of a finite cone $\Gamma_y(\rho)$, i.e., the intersection of a cone with a sphere of radius ρ about the vertex y, such that $\Gamma_y(\rho)$ lies in $\bar{D}$ and its volume is $\ge \gamma\rho^n$, where ρ, γ are positive numbers independent of y.

Theorem 2′. *Assume that D satisfies the cone condition. Then for any $u \in \hat{C}^j(D)$, where $j = [n/2] + 1$,*

$$|u(y)|^2 \le C' \sum_{|\alpha| \le j} \int_D |D^\alpha u(x)|^2\, dx \tag{2.13}$$

for all $y \in D$, where C' is a constant depending only on n, ρ, γ.

The proof is obtained by slightly modifying the previous proof, replacing R by ρ and integrating only over the solid angle of the cone $\Gamma_y(\rho)$ (instead of Ω).

The reader may verify that the cone condition is satisfied if D is a convex set; also if ∂D is of class C^1.

Suppose D satisfies the cone condition and $u \in H^j(D)$, $j = [n/2] + 1$. Let $\{u_m\}$ be a sequence of functions in $\hat{C}^j(D)$ which converges to u in the norm of $H^j(D)$. Applying (2.13) to $u_m - u_k$ we see that $\{u_m\}$ is a uniformly convergent sequence in D. Its pointwise limit $u^*(x)$ is therefore a continuous function and it satisfies (2.13). Since $u(x) = u^*(x)$ almost everywhere, we arrive at the following result.

Corollary 1. *Let D satisfy the cone condition. If $u \in H^j(D)$, $j = [n/2] + 1$, then u can be identified with a uniformly continuous function $u(x)$ in D, and* (2.13) *holds.*

In a similar way one can prove:

Corollary 2. *Let D satisfy the cone condition. If $u \in H^j(D)$ where $j \geq [n/2] + 1$ then u can be identified with a function $u(x)$ which has $j - [n/2] - 1$ uniformly continuous derivatives in D.*

Theorem 3 (Rellich's Lemma). *Let $\{u_m\}$ be a sequence of functions in $\mathring{H}^1(D)$ (D bounded) such that $|u_m|_1^D \leq$ const. $< \infty$. There exists a subsequence $\{u_{m'}\}$ such that*

$$|u_{m'} - u_{k'}|_0^D \to 0 \qquad \text{as } m' \to \infty,\ k' \to \infty,$$

i.e., bounded sets of $\mathring{H}^1(D)$ are precompact sets of $L^2(D)$.

Proof. It suffices to prove the theorem in case the u_m are real-valued functions. We first establish two lemmas.

Lemma 9 (Poincaré's Inequality). *Let Q be a cube $0 \leq x_i \leq \sigma$ ($i = 1, \ldots, n$) and let u be a real-valued function belonging to $C^1(Q)$. Then*

$$|u|_0^2 \leq \frac{1}{\sigma^n}\left(\int_Q u(x)\,dx\right)^2 + \frac{n}{2}\,\sigma^2|u|_1^2. \tag{2.14}$$

Proof. For any $x \in Q$, $y \in Q$,

$$u(x_1, x_2, \ldots, x_n) - u(y_1, y_2, \ldots, y_n)$$
$$= \int_{y_1}^{x_1} \frac{\partial}{\partial \xi_1} u(\xi_1, x_2, \ldots, x_n)\,d\xi_1 + \int_{y_2}^{x_2} \frac{\partial}{\partial \xi_2} u(y_1, \xi_2, x_3, \ldots, x_n)\,d\xi_2$$
$$+ \cdots + \int_{y_n}^{x_n} \frac{\partial}{\partial \xi_n} u(y_1, \ldots, y_{n-1}, \xi_n)\,d\xi_n.$$

Taking squares and using Schwarz' inequality, we get

$$u^2(x) + u^2(y) - 2u(x)u(y)$$
$$\leq n\sigma \int_0^\sigma \left(\frac{\partial}{\partial \xi_1} u(\xi_1, x_2, \ldots, x_n)\right)^2 d\xi_1 + n\sigma \int_0^\sigma \left(\frac{\partial}{\partial \xi_2} u(y_1, \xi_2, x_3, \ldots, x_n)\right)^2 d\xi_2$$
$$+ \cdots + n\sigma \int_0^\sigma \left(\frac{\partial}{\partial \xi_n} u(y_1, \ldots, y_{n-1}, \xi_n)\right)^2 d\xi_n.$$

Integrating with respect to $x_1, \ldots, x_n, y_1, \ldots, y_n$, we get

$$2\sigma^n \int_Q u^2(x)\,dx - 2\left(\int_Q u(x)\,dx\right)^2 \leq n\sigma^{n+2} \sum_{i=1}^n \int_Q \left(\frac{\partial u(x)}{\partial x_i}\right)^2 dx,$$

from which (2.14) follows.

Lemma 10 (Friedrich's Inequality). *For any $\epsilon > 0$ there exists an integer $M > 0$ and real-valued functions $w_1, \ldots, w_M$ in $L^2(D)$ (D bounded) with $|w_j|_0 = 1$, such that for any real-valued function u in $\mathring{H}^1(D)$*

$$(2.15) \qquad |u|_0^2 \leq \epsilon |u|_1^2 + \sum_{j=1}^{M} (u, w_j)^2.$$

Proof. It suffices to prove the lemma for $u \in C_c^1(D)$, since its validity for all $u \in \mathring{H}^1(D)$ then follows by completion. Extend u outside D by setting $u = 0$, and let Q be a cube containing $\overline{D}$, whose edges are parallel to the coordinate axes. Let σ_0 be the length of each edge. Divide Q into cubes by introducing hyperplanes $x_i = y_{mi}$, so that $y_{m+1,i} - y_{mi} = \sigma$ for all m, i; σ is taken such that σ_0/σ is an integer. Denote by $Q_1, Q_2, \ldots, Q_M$ $(M = (\sigma_0/\sigma)^2)$ the cubes thus obtained.

By Lemma 9,

$$(|u|_0^{Q_j})^2 \leq \frac{1}{\sigma^n}\left(\int_{Q_j} u\, dx\right)^2 + \frac{n}{2}\sigma^2(|u|_1^{Q_j})^2.$$

Summing over j, and introducing the functions

$$w_j = \begin{cases} \sigma^{-n/2} & \text{in } Q_j, \\ 0 & \text{outside } Q_j, \end{cases}$$

we get (2.15) provided σ is such that $n\sigma^2/2 \leq \epsilon$.

We can now complete the proof of Theorem 3. Let $h = 1, 2, \ldots$, and take for any $\epsilon = 1/h$ a finite sequence w_{jh} $(j = 1, \ldots, M(h))$ as in Lemma 10. Let $\{u_{m1}\}$ be a subsequence of $\{u_m\}$ such that $\{(u_{m1}, w_{j1})_0\}$ are convergent sequences for $j = 1, 2, \ldots, M(1)$. Similarly we define $\{u_{m,h+1}\}$, inductively, to be a subsequence of $\{u_{mh}\}$ such that $\{(u_{m,h+1}, w_{j,h+1})_0\}$ are convergent sequences for $j = 1, 2, \ldots, M(h + 1)$. Let $\{u_{m'}\}$ be the diagonal sequence $\{u_{mm}\}$. Then $(u_{m'}, w_{jh})_0 \to 0$ as $m' \to \infty$, for any h, j.

Given $\epsilon = 1/h$ choose m_0' such that

$$\sum_{j=1}^{M(h)} (u_{m'} - u_{k'}, w_{jh})^2 < \epsilon \qquad \text{if } m' \geq m_0',\ k' \geq m_0'.$$

Then, by Lemma 10,

$$|u_{m'} - u_{k'}|_0^2 \leq \epsilon + \epsilon(|u_{m'}|_1 + |u_{k'}|_1)^2 \leq (1 + 4K^2)\epsilon$$

where $K = \text{l.u.b. } |u_m|_1$. Since ϵ can be made arbitrarily small, the proof is completed.

The following two lemmas are concerned with estimating norms of difference quotients.

Lemma 11. *Let B be a subdomain of a bounded domain D, with $\overline{B} \subset D$, and let h_0 be the distance from B to the boundary of D. Let $j \geq 1$ and let $u \in H^j(D)$. Then the difference quotient, say with respect to x_1,*

$$u^h(x) = \frac{1}{h}(u(x_1 + h, x_2, \ldots, x_n) - u(x_1, x_2, \ldots, x_n))$$

is well defined on B if $|h| < h_0$, and

(2.16) $$|u^h|_{j-1}^B \le |u|_j^D,$$

(2.17) $$\lim_{h\to 0} \left| u^h - \frac{\partial u}{\partial x_i} \right|_0^B = 0.$$

Lemma 11′. *Let D, j be as in Lemma 11, and let $u \in \mathring{H}^j(D)$. Extend u outside D by setting $u = 0$. Then*

(2.18) $$|u^h|_{j-1}^D \le |u|_j^D,$$

(2.19) $$\lim_{h\to 0} \left| u^h - \frac{\partial u}{\partial x_1} \right|_0^D = 0,$$

where u^h is defined as in Lemma 11.

Proof of Lemma 11. If $u \in \hat{C}^j(D)$, $0 \le |\alpha| < j$,

$$\int_B |D^\alpha u^h(x)|^2\, dx = \int_B \left| \frac{1}{h} \int_{x_1}^{x_1+h} D^\alpha D_{\xi_1} u(\xi_1, x_2, \ldots, x_n)\, d\xi_1 \right|^2 dx.$$

Using Schwarz' inequality and then substituting $\xi_1 = \xi_1' + x_1$, we get

$$\int_B |D^\alpha u^h(x)|^2\, dx \le \int_B \frac{1}{h} \int_0^h |D^\alpha D_{\xi_1'} u(\xi_1' + x_1, x_2, \ldots, x_n)|^2\, d\xi_1'\, dx$$

$$= \frac{1}{h} \int_0^h \left\{ \int_B |D^\alpha D_{x_1} u(\xi_1' + x_1, x_2, \ldots, x_n)|^2\, dx \right\} d\xi_1'.$$

Substituting $x_1 + \xi_1' \to x_1$ in the inner integral and noting that B is then mapped into a subset of D (since $|h| < h_0$) we conclude that the inner integral is bounded by

$$\int_D |D^\alpha D_{x_1} u(x)|^2\, dx.$$

Hence,

$$\int_B |D^\alpha u^h(x)|^2\, dx \le \int_D |D^\alpha D_{x_1} u(x)|^2\, dx.$$

Summing over α, (2.16) follows. Having proved (2.16) for $u \in \hat{C}^j(D)$, its validity for $u \in H^j(D)$ follows by completion. Indeed, taking a sequence $\{u_m\}$ in $\hat{C}^j(D)$ with $|u_m - u|_j^D \to 0$ and noting that, if $0 \le |\alpha| < j$, $|h| < h_0$,

$$\int_B |D^\alpha u_m(x_1 + h, x_2, \ldots, x_n) - D^\alpha u(x_1 + h, x_2, \ldots, x_n)|^2\, dx$$

$$\le \int_D |D^\alpha u_m(x) - D^\alpha u(x)|\, dx$$

as follows by substituting $x_1 + h \to x_1$ in the left-hand side, we find that for fixed h, $|h| < h_0$,

$$|u_m^h - u^h|_{j-1}^B \to 0 \qquad \text{as } m \to \infty.$$

Since (2.16) holds for all the u_m, it therefore holds also for u.

To prove (2.17), note first that it holds for functions u in $\hat{C}^j(D)$ since then $u^h(x) \to \partial u(x)/\partial x_1$ uniformly in compact subsets of D. If $u \in H^j(D)$, we can find for any $\epsilon > 0$ a function $v \in \hat{C}^j(D)$ such that $|v - u|_j^D \leq \epsilon$. By (2.16) we also have $|u^h - v^h|_0^B \leq \epsilon$. Hence,

$$\left| u^h - \frac{\partial u}{\partial x_1} \right|_0^B \leq |u^h - v^h|_0^B + \left| v^h - \frac{\partial v}{\partial x_1} \right|_0^B + \left| \frac{\partial v}{\partial x_1} - \frac{\partial u}{\partial x_1} \right|_0^B$$

$$\leq 2\epsilon + \left| v^h - \frac{\partial v}{\partial x_1} \right|_0^B \leq 3\epsilon$$

if $|h|$ is sufficiently small, and the proof is completed.

Proof of Lemma 11′. u belongs to $H^j(G)$ for any bounded domain G containing $\overline{D}$. Indeed, if $\{u_m\}$ is a sequence of functions in $C_c^\infty(D)$ which converges to u in the $H^j(D)$ norm, and if we extend the u_m outside D by setting $u_m = 0$, then $u_m \in C_c^\infty(G)$ and

$$|u_m - u|_j^G \to 0 \qquad \text{as } m \to \infty.$$

Noting that $|u|_j^G = |u|_j^D$, Lemma 11′ now follows from Lemma 11 applied to the pair D, G (instead of B, D).

We conclude this section with a basic result on the boundary behavior of functions in $\mathring{H}^1$.

Lemma 12. *Let D be a bounded domain with ∂D of class C^1. If u is a continuous function in $\overline{D}$ and if $u \in \mathring{H}^1(D)$, then $u = 0$ on ∂D.*

Proof. Let $x^0 \in \partial D$ and consider first a special case where there exists a neighborhood N of x^0 such that $N \cap \partial D$ lies on the hyperplane $x_n = 0$ and $N \cap D$ lies in the half-space $x_n > 0$. Let S_k be an n-dimensional right spherical cylinder with base B lying on $N \cap \partial D$ and with axis parallel to the x_n-axis, and let x^0 be the center of B and k the height of S_k. If w belongs to $\hat{C}^j(S_k)$ and is continuous on $\overline{S}_k$ and vanishes on B, then

$$|w(x)|^2 \leq \left(\int_0^{x_n} \left| \frac{\partial w}{\partial x_n} \right| dx_n \right)^2 \leq k \int_0^{x_n} \left| \frac{\partial w}{\partial x_n} \right|^2 dx_n.$$

Integrating over S_k we get

$$\int_{S_k} |w|^2 \, dx \leq k^2 \int_{S_k} \left| \frac{\partial w}{\partial x_n} \right|^2 dx. \tag{2.20}$$

Consider now the general case. $N \cap \partial D$ is not necessarily planar. We may assume that $N \cap \partial D$ can be represented in the form $x_n = h(x_1, \ldots, x_{n-1})$ where $h \in C^1$. The transformation (1.23) maps $N \cap \partial D$ onto an open set lying on $y_n = 0$ and we may assume that $N \cap D$ is mapped into a domain lying in $y_n > 0$. Setting $W(y) = w(x)$ we have, by (2.20),

$$\int_{S_k^*} |W(y)|^2 \, dy \leq k^2 \int_{S_k^*} \left| \frac{\partial W(y)}{\partial y_n} \right|^2 dy \tag{2.21}$$

where S_k^* is a right spherical cylinder defined analogously to S_k in the y-space. Its base B^* has the image y^0 of x^0 for its center, and is such that

S_k^* lies in the image of $D \cap N$ under the transformation (1.23), for all k sufficiently small.

Returning to the x-coordinates we obtain from (2.21),

$$\int_{\tilde{S}_k} |w(x)|^2 \, dx \leq Ck^2 \int_{\tilde{S}_k} \sum_{i=1}^{n} \left| \frac{\partial w(x)}{\partial x_i} \right|^2 dx \tag{2.22}$$

where C is a constant and $\tilde{S}_k$, S_k^* correspond to each other under the mapping (1.23). C depends on bounds on the first derivatives of the function h, but not on k. (2.22) was proved under the assumption that $w \in \hat{C}^1(\tilde{S}_k)$, w is continuous on $\bar{\tilde{S}}_k$, and $w = 0$ on $N \cap \partial D$. Since $u \in \mathring{H}^1(D)$, it can be approximated in the norm of $\mathring{H}^1(D)$ by a sequence of functions in $C_c^\infty(D)$. Since (2.22) holds for each of these functions, it also holds for u. Dividing both sides of (2.22) (with $w = u$) by k and taking $k \to 0$, we get

$$\frac{1}{k} \int_{\tilde{S}_k} |u(x)|^2 \, dx \to 0 \qquad \text{as } k \to 0. \tag{2.23}$$

By going back to the y-coordinates we find that

$$\begin{aligned} \frac{1}{k} \int_{\tilde{S}_k} |u(x)|^2 \, dx &\geq \frac{C'}{k} \int_{S_k^*} |U(y)|^2 \, dy \\ &\to C' \int_{B^*} |U(y_1, \ldots, y_{n-1}, 0)|^2 \, dy_1 \cdots dy_{n-1} \end{aligned} \tag{2.24}$$

as $k \to 0$, where C' is a positive constant. Since also

$$\int_{B^*} |U(y_1, \ldots, y_{n-1}, 0)|^2 \, dy_1 \cdots dy_{n-1} \geq C'' \int_B |u(x)|^2 \, dB_x$$

where B corresponds to B^* under the transformation (1.23), dB_x is the surface element of B, and C'' is some positive constant, we obtain from (2.23), (2.24),

$$\int_B |u(x)|^2 \, dB_x = 0,$$

i.e., $u(x) = 0$ on B. In particular, $u(x^0) = 0$. Since x^0 was an arbitrary point on ∂D, $u = 0$ on ∂D.

3. Existence Theory for the Dirichlet Problem for Elliptic Equations

In this section we consider equations of order $2m$ having a *divergence form*, i.e.,

$$Lu \equiv \sum_{0 \leq |\rho|, |\sigma| \leq m} (-1)^{|\rho|} D^\rho (a^{\rho\sigma}(x) D^\sigma u) = f(x). \tag{3.1}$$

It may be noted that every equation of order $2m$ can be written in a divergence form if the coefficients of the derivatives D^α, for $|\alpha| > m$, belong to $C^{|\alpha|-m}$. Conversely, (3.1) can be written in the form $\sum A_\alpha D^\alpha u = f$ if $a^{\rho\sigma} \in C^{|\rho|}$.

We say that the equation (3.1) is *elliptic* (or of *elliptic type*) at a point x if

$$\sum_{|\rho|=|\sigma|=m} \xi^\rho a^{\rho\sigma}(x)\xi^\sigma \neq 0 \tag{3.2}$$

for any real vector $\xi \neq 0$. If the equation is elliptic at each point of a domain D, then it is said to be elliptic in D. L is an *elliptic operator* if $Lu = f$ is an elliptic equation.

If the equation (3.1) can be written in the form $\Sigma A_\alpha D^\alpha u = f$, then the present definition of ellipticity coincides with the one given in Sec. 7, Chap. 9.

In what follows we shall need a condition stronger than (3.2), namely,

$$\operatorname{Re}\{\sum_{|\rho|=|\sigma|=m} \xi^\rho a^{\rho\sigma}(x)\xi^\sigma\} \geq c_0|\xi|^{2m} \qquad (c_0 > 0) \tag{3.3}$$

for any real vector ξ and $x \in D$. If (3.3) is satisfied then we say that (3.1) is a *strongly elliptic equation* in D, and L is a *strongly elliptic operator* in D. c_0 in (3.3) is called a *module* of strong ellipticity.

Note that ellipticity means only that

$$|\sum_{|\rho|=|\sigma|=m} \xi^\rho a^{\rho\sigma}(x)\xi^\sigma| \geq c_1|\xi|^{2m} \qquad (c_1 = c_1(x) > 0), \tag{3.4}$$

and *uniform ellipticity* means that c_1 is independent of $x \in D$ and that the left-hand side of (3.4) is bounded by $|\xi|^{2m}/c_1$. c_1 is called a *module* of ellipticity. If the principal coefficients of L are bounded real-valued functions, then L is strongly elliptic in D if and only if either L or $-L$ is uniformly elliptic in D.

The following assumptions will be made throughout this section.

(A_1) L is strongly elliptic in a bounded domain D, with a module of strong ellipticity $c_0 > 0$.

(A_2) The coefficients of L are bounded (in D) by a constant c_1.

(A_3) The principal coefficients of L have a modulus of continuity $c_2(t)$ in D (i.e., $|a^{\rho\sigma}(x) - a^{\rho\sigma}(y)| \leq c_2(|x - y|)$ for all $|\rho| = |\sigma| = m$, $x \in D, y \in D$; $c_2(t) \searrow 0$ if $t \searrow 0$).

We also assume, temporarily:

(A_4) The coefficients $a^{\rho\sigma}$ belong to $C^{|\rho|}$ and to $C^{|\sigma|}$.

Definition. A function u is said to be a *classical solution* (or, simply, a solution) of the equation

$$Lu = f \qquad \text{in } D \tag{3.5}$$

if $u \in C^{2m}(D)$ and if (3.5) is satisfied at each point of D. A function u is said to be a *weak solution* of (3.5) if $u \in L^2(A)$ for any compact subdomain A of D and if

$$(L^*\varphi, u) = (\varphi, f) \tag{3.6}$$

holds for any $\varphi \in C_c^\infty(D)$. Here L^* is the adjoint of L. Thus, since we are considering L of the form (3.1), L^* is given by

$$L^*v = \sum_{0 \le |\rho|,|\sigma| \le m} (-1)^{|\sigma|} D^\sigma(\overline{a^{\rho\sigma}} D^\rho v). \tag{3.7}$$

If u is a classical solution of (3.5) then it is clearly also a weak solution of (3.5). Conversely, if u is a weak solution and if, in addition, $u \in C^{2m}(D)$, then u is a classical solution since integration by parts on the left-hand side of (3.6) yields $(\varphi, Lu) = (\varphi, f)$ for any $\varphi \in C_c^\infty(D)$, i.e., (3.5) holds pointwise (assuming that $f(x)$ is a continuous function).

The *(classical) Dirichlet problem* consists of finding a classical solution u of (3.5) satisfying the boundary conditions

$$\frac{\partial^j u}{\partial \nu^j} = \varphi_j \qquad \text{on } \partial D \qquad (j = 0, 1, \ldots, m-1) \tag{3.8}$$

where ν is the normal to ∂D. u is assumed to belong to $C^{m-1}(\overline{D})$ and ∂D is assumed to be at least of class C^1 so that the derivatives $\partial^j u/\partial \nu^j$ are well defined. The φ_j are often referred to as the *Dirichlet data.*

Let u_0 be any function in $C^{m-1}(\overline{D})$ satisfying (3.8). Then the boundary conditions (3.8) are equivalent to the condition that the function $w = u - u_0$ vanishes on ∂D together with its first $m - 1$ derivatives.

Now, by Lemma 12, Sec. 2, a continuous function w which belongs to $\mathring{H}^1(D)$ must vanish on ∂D (∂D of class C^1). It can also be shown that, conversely, a continuous function in $H^1(D)$ which vanishes on ∂D belongs to $\mathring{H}^1(D)$. Both statements extend also to classes $\mathring{H}^m(D)$, i.e., if ∂D is of class C^m and if $w \in C^{m-1}(\overline{D})$, then $w \in \mathring{H}^m(D)$ if and only if $w \in H^m(D)$ and it vanishes on ∂D together with its first $m - 1$ derivatives.

In view of the observations made in the last two paragraphs, one is motivated to consider the condition that $u - u_0$ belongs to $\mathring{H}^m(D)$ as a weak form of the conditions (3.8). We now define:

The Generalized Dirichlet Problem (first formulation). Given a function $u_0 \in H^m(D)$ and a function $f \in H^0(D)$, find a weak solution of (3.5) such that $u - u_0$ belongs to $\mathring{H}^m(D)$.

In this section we shall solve the generalized Dirichlet problem. In Secs. 4, 5 it will be shown that the solution is also a solution of the classical Dirichlet problem provided f, φ_j, ∂D, and the coefficients of L are sufficiently smooth.

We shall give a useful reformulation of the generalized Dirichlet problem. First we observe that if u is a classical solution of (3.5) then, by partial integration,

$$B[\varphi, u] \equiv \sum_{0 \le |\rho|,|\sigma| \le m} (D^\rho \varphi, a^{\rho\sigma} D^\sigma u) = (\varphi, f) \tag{3.9}$$

for all $\varphi \in C_c^\infty(D)$, or

$$B[\varphi, u - u_0] = (\varphi, f) - B[\varphi, u_0]. \tag{3.10}$$

$B[u, v]$ is linear in u, antilinear in v and, by Schwarz' inequality,

$$|B[u, v]| \leq \text{const. } |u|_m |v|_m \tag{3.11}$$

where the constant depends only on bounds on the coefficients of L. Note that $B[u, v]$ is defined (and (3.11) holds) for any u, v in H^m.

From the previous definition of the generalized Dirichlet problem it follows that the solution u is necessarily in H^m. We claim that u satisfies

$$(L^*\varphi, u) = B[\varphi, u] \qquad \text{for all } \varphi \in C_c^\infty(D). \tag{3.12}$$

Indeed, for $u \in C^m(D)$ this follows by partial integration, whereas for $u \in H^m(D)$ it then follows by approximating it, in the norm of $H^m(D)$, by functions in $C^m(D)$.

Since the solution u of the generalized Dirichlet problem satisfies (3.6), it also satisfies (in view of (3.12)) the relation (3.10). Conversely, (3.10) implies (3.6). We have thus proved:

Lemma 13. *u is a solution of the generalized Dirichlet problem if and only if $u - u_0 \in \mathring{H}^m(D)$ and (3.10) holds for all $\varphi \in C_c^\infty(D)$.*

On the basis of Lemma 13 we can reformulate the generalized Dirichlet problem as the problem of finding a function u such that $u - u_0 \in \mathring{H}^m(D)$ and (3.10) holds for all $\varphi \in C_c^\infty(D)$.

In proving Lemma 13 we have used the assumption (A_4). From the form of $B[\varphi, v]$, however, we see that the new formulation of the generalized Dirichlet problem is meaningful even if the assumption (A_4) is omitted. We shall solve in this section the generalized Dirichlet problem in its new formulation, without assuming (A_4). For the sake of future reference we state:

The Generalized Dirichlet Problem (second formulation). Given a function $u_0 \in H^m(D)$ and a function $f \in H^0(D)$, find a function u such that $u - u_0 \in \mathring{H}^m(D)$ and such that (3.10) holds for all $\varphi \in C_c^\infty(D)$.

The following theorem will play a decisive role in solving the generalized Dirichlet problem.

Theorem 4 (Gårding's Inequality). *If L satisfies the assumptions (A_1)–(A_3) then there exist constants $c > 0$ and k_0 such that*

$$\text{Re } B[\varphi, \varphi] \geq c|\varphi|_m^2 - k_0|\varphi|_0^2 \tag{3.13}$$

for all $\varphi \in \mathring{H}^m(D)$. The constants c, k_0 depend only on c_0, c_1, c_2, and D.

Proof. It suffices to prove (3.13) for $\varphi \in C_c^\infty(D)$. Indeed, for any $u \in \mathring{H}^m(D)$ we then take a sequence $\{\varphi_j\}$ in $C_c^\infty(D)$ such that $|u - \varphi_j|_m \to 0$, apply (3.13) to each φ_j and note that $B[\varphi_j, \varphi_j] \to B[u, u]$ as $j \to \infty$, so that (3.13) holds also for u.

We introduce the notation

$$\widehat{|\varphi|}_j = \left\{ \sum_{|\alpha|=j} \int_D |D^\alpha \varphi(x)|^2 \, dx \right\}^{1/2}, \tag{3.14}$$

and consider first the case that L is homogeneous (i.e., L coincides with its principal part) and its coefficients are constants.

Using the rules (2.2), (2.3) we have

$$\begin{aligned} B[\varphi, \varphi] &= \sum_{|\rho|=|\sigma|=m} (D^\rho \varphi, a^{\rho\sigma} D^\sigma \varphi) = \sum_{|\rho|=|\sigma|=m} ((i\xi)^\rho \tilde{\varphi}, a^{\rho\sigma}(i\xi)^\sigma \tilde{\varphi}) \\ &= \int_{R^n} |\tilde{\varphi}(\xi)|^2 \left[\sum_{|\rho|=|\sigma|=m} \xi^\rho a^{\rho\sigma} \xi^\sigma \right] d\xi. \end{aligned} \tag{3.15}$$

Using (3.3) and the relation $|\xi|^{2m} = \sum_{|\rho|=m} |\xi^{2\rho}|$, we get

$$\operatorname{Re} B[\varphi, \varphi] \geq c_0 \int_{R^n} |\tilde{\varphi}(\xi)|^2 \left[\sum_{|\rho|=m} \xi^\rho \xi^\rho \right] d\xi.$$

Since the integral on the right-hand side coincides with the right-hand side of (3.15) when $a^{\rho\sigma} = \delta^{\rho\sigma}$ ($\delta^{\rho\sigma} = 0$ if $\rho \neq \sigma$, $\delta^{\rho\sigma} = 1$ if $\rho = \sigma$), we get, upon using (3.15) with $a^{\rho\sigma} = \delta^{\rho\sigma}$,

$$\operatorname{Re} B[\varphi, \varphi] \geq c_0 \sum_{|\rho|=m} (D^\rho \varphi, D^\rho \varphi),$$

i.e.,

$$\operatorname{Re} B[\varphi, \varphi] \geq c_0 \widehat{|\varphi|}_m^2. \tag{3.16}$$

Consider next the case where L is still homogeneous but its coefficients are not assumed to be constants. We shall estimate $\operatorname{Re} B[\varphi, \varphi]$ in case the support of φ lies in an open neighborhood A of a point $x^0 \in D$ ($A \subset D$) and the diameter $|A|$ of A is sufficiently small.

Write

$$\begin{aligned} B[\varphi, \varphi] &= \sum_{|\rho|=|\sigma|=m} (D^\rho \varphi, a^{\rho\sigma}(x^0) D^\sigma \varphi) \\ &\qquad + \sum_{|\rho|=|\sigma|=m} (D^\rho \varphi, [a^{\rho\sigma}(x) - a^{\rho\sigma}(x^0)] D^\sigma \varphi) \\ &\equiv I + J. \end{aligned}$$

By (3.16),

$$\operatorname{Re} I \geq c_0 \widehat{|\varphi|}_m^2.$$

If $|A|$ is sufficiently small (depending on c_0 and on the function c_2 which occurs in the condition (A_3)), then

$$|J| < \frac{1}{2} c_0 \widehat{|\varphi|}_m^2.$$

We conclude that

$$\operatorname{Re} B[\varphi, \varphi] \geq \frac{1}{2} c_0 \widehat{|\varphi|}_m^2 \qquad \text{for all } \varphi \in C_c^\infty(A). \tag{3.17}$$

Consider now the general case and cover D by a finite number of domains A_j each having a sufficiently small diameter so that (3.17) holds for $A = A_j$. Let $\{\alpha_j\}$ be a partition of unity subordinate to the covering $\{A_j\}$, and set $\beta_j = \sqrt{\alpha_j}$. Write

$$(3.18)\quad B[\varphi, \varphi] = \sum_{|\rho|=|\sigma|=m} \int_D D^\rho\varphi\cdot\overline{a^{\rho\sigma}D^\sigma\bar{\varphi}}\,dx + \sum_{\substack{|\rho|,|\sigma|\le m\\ |\rho|+|\sigma|<2m}} \int_D D^\rho\varphi\cdot\overline{a^{\rho\sigma}D^\sigma\bar{\varphi}}\,dx$$
$$\equiv H + K.$$

Clearly,

$$(3.19)\qquad |K| \le \text{const. } |\varphi|_m|\varphi|_{m-1}.$$

Next,

$$(3.20)\qquad H = \sum_j \sum_{|\rho|=|\sigma|=m} \int_D [\beta_j D^\rho\varphi][\beta_j\overline{a^{\rho\sigma}}D^\sigma\bar{\varphi}]\,dx$$
$$= \sum_j \sum_{|\rho|=|\sigma|=m} \int_D D^\rho(\beta_j\varphi)\cdot\overline{a^{\rho\sigma}}D^\sigma(\beta_j\bar{\varphi})\,dx$$
$$- \sum_j \sum \int_D c_{\rho_1\rho_2\sigma_1\sigma_2} D^{\rho_1}\beta_j\cdot D^{\rho_2}\varphi\cdot\overline{a^{\rho\sigma}}D^{\sigma_1}\beta_j\cdot D^{\sigma_2}\bar{\varphi}\,dx$$
$$\equiv H_1 - H_2.$$

In H_2 the c's are bounded functions and $\rho_1 + \sigma_1 > 0$ so that $\rho_2 + \sigma_2 < 2m$. Hence,

$$(3.21)\qquad |H_2| \le \text{const. } |\varphi|_m|\varphi|_{m-1}$$

where the constant depends only on bounds on the $a^{\rho\sigma}$.

H_1 is a sum of terms

$$\sum_{|\rho|=|\sigma|=m} \int_D D^\rho\psi_j\cdot\overline{a^{\rho\sigma}D^\sigma\bar{\psi}_j}\,dx \qquad \text{with } \psi_j = \varphi\beta_j \in C_c^\infty(A_j).$$

By the result of (3.17) with $A = A_j$ we get

$$\text{Re } H_1 \ge \frac{1}{2}c_0 \sum_j \widehat{|\psi_j|}_m^2 = \frac{1}{2}c_0 \sum_j \sum_{|\rho|=m} \int_D D^\rho(\beta_j\varphi)\cdot D^\rho(\beta_j\bar{\varphi})\,dx.$$

The double sum on the right-hand side coincides with the double sum of H_1 (on the right-hand side of (3.20)) when $a^{\rho\sigma} = \delta^{\rho\sigma}$. Hence

$$(3.22)\qquad \text{Re } H_1 \ge \frac{1}{2}c_0\left\{\sum_j \sum_{|\rho|=m} \int_D [\beta_j D^\rho\varphi][\beta_j D^\rho\bar{\varphi}]\,dx + H_2'\right\}$$

where H_2' coincides with H_2 when $a^{\rho\sigma} = \delta^{\rho\sigma}$ and, therefore, by (3.21),

$$(3.23)\qquad |H_2'| \le \text{const. } |\varphi|_m|\varphi|_{m-1}.$$

Combining (3.18)–(3.23), we obtain

$$(3.24)\qquad \text{Re } B[\varphi, \varphi] \ge \frac{1}{2}c_0\widehat{|\varphi|}_m^2 - C_1|\varphi|_m|\varphi|_{m-1}$$

where C_1 is a constant depending only on c_0, c_1, c_2.

Using the elementary inequality

$$C_1|\varphi|_m|\varphi|_{m-1} \leq \delta|\varphi|_m^2 + \frac{C_1}{4\delta}|\varphi|_{m-1}^2 \qquad (\delta > 0)$$

and applying Lemma 7, Sec. 7, we get

$$(3.25) \qquad C_1|\varphi|_m|\varphi|_{m-1} \leq \delta|\varphi|_m^2 + \frac{C_1}{4\delta}(\epsilon|\varphi|_m^2 + C_2|\varphi|_0^2) \qquad (\epsilon > 0)$$

where C_2 depends only on ϵ. By Lemma 7 we also have

$$(3.26) \qquad \frac{1}{2}c_0\widehat{|\varphi|}_m^2 \geq \frac{1}{3}c_0|\varphi|_m^2 - \text{const. } |\varphi|_0^2.$$

Taking $\delta = c_0/9$ and then $\epsilon = (4\delta/C_1)(c_0/9)$ and substituting (3.25), (3.26) into (3.24), the inequality (3.13) (with $c = c_0/6$) follows.

We shall need the following representation theorem for functionals in a Hilbert space.

Theorem 5. *Let $B[x, y]$ be a bilinear form (i.e., linear in x and anti-linear in y) in a Hilbert space H with norm $|\ \ |$ and scalar product $(\ ,\ )$ and assume that $B[x, y]$ is bounded, i.e.,*

$$(3.27) \qquad |B[x, y]| \leq \text{const. } |x|\,|y| \qquad \text{for all } x, y \text{ in } H.$$

Suppose further that

$$(3.28) \qquad |B[x, x]| \geq c|x|^2 \qquad \text{for all } x \in H,$$

for some positive constant c. Then every bounded linear functional $F(x)$ in H can be represented in the form

$$(3.29) \qquad F(x) = B[x, v] = \overline{B[w, x]}$$

for some elements v, w in H which are uniquely determined by F.

Proof. For fixed v, $B[x, v]$ is a bounded linear functional in x. Hence there exists a unique y such that

$$B[x, v] = (x, y).$$

Set $y = Av$. Then A is a linear operator on H. Since

$$c|v|^2 \leq |B[v, v]| \leq |(v, y)| \leq |v|\,|y|,$$

i.e., $|v| \leq |y|/c$, A has a bounded inverse. It follows that the range of A, $R(A)$, is a closed linear subspace of H. We claim that $R(A) = H$. Indeed, if $R(A) \neq H$ then there exists an element $z \neq 0$ which is orthogonal to $R(A)$, i.e., $(z, Av) = 0$ for all $v \in H$. This implies that $B[z, v] = (z, Av) = 0$. Taking $v = z$ we get $B[z, z] = 0$; hence, by (3.28), $z = 0$, which contradicts the choice of z.

Consider now the functional $F(x)$. It can be represented in the form

(x, a) for some $a \in H$. Since $R(A) = H$, there exists an element v such that $Av = a$, i.e.,

$$B[x, v] = (x, a) = F(x).$$

v is uniquely determined since if also $B[x, v'] = F(x)$ for some $v' \in H$ and for all x, then $B[x, v - v'] = 0$. Taking $x = v - v'$ and using (3.28) we get $v = v'$.

The representation $F(x) = \overline{B[w, x]}$ follows by applying the previous result to the bilinear form $\overline{B[y, x]}$.

We are now ready to establish an existence theorem for the generalized Dirichlet problem in case $k_0 = 0$ in (3.13).

Theorem 6. *Assume that L satisfies* (A_1)–(A_3) *and that the bilinear form $B[\varphi, u]$ defined in* (3.9) *satisfies, for some constant $c > 0$,*

$$\text{Re}\, B[\varphi, \varphi] \geq c|\varphi|_m^2 \qquad \text{for all } \varphi \in C_c^\infty(D). \tag{3.30}$$

Then there exists a unique solution of the generalized Dirichlet problem (second formulation).

Proof. By completion, (3.30) holds for all $\varphi \in \mathring{H}^m(D)$. Recalling also (3.11) we see that $B[u, v]$ satisfies the assumptions of Theorem 5 with $H = \mathring{H}^m$. Consider the functional

$$F(\psi) = (\psi, f) - B[\psi, u_0]$$

defined for $\psi \in \mathring{H}^m$. Since it is clearly a bounded linear functional on $\mathring{H}^m$, we can apply Theorem 5 and thus conclude that there exists a unique $v \in \mathring{H}^m$ such that $F(\psi) = B[\psi, v]$. The function $u = u_0 + v$ is then the unique solution of the generalized Dirichlet problem, in its second formulation.

From Theorems 4, 6 we obtain:

Theorem 7. *Assume that L satisfies* (A_1)–(A_3). *Then there exists a constant k_0 (depending only on c_0, c_1, c_2, D) such that for any $k \geq k_0$ the generalized Dirichlet problem (second formulation) for $L + k$ has a unique solution.*

Before stating the main existence theorem of this section, we introduce some concepts and state the Fredholm alternative in Hilbert spaces.

An operator T mapping a Hilbert space H into itself is said to be *completely continuous* if it maps each bounded sequence $\{x_n\}$ into a sequence which has a convergent subsequence. The *adjoint* T^* of a continuous linear operator T defined on H (with values in H) is defined by $(T^*x, y) = (x, Ty)$ where (x, z) is the scalar product of x, z in H. If T is completely continuous, then T^* is also completely continuous.

Consider the equations

$$x - \lambda Tx = f, \tag{3.31}$$

$$x - \lambda Tx = 0, \tag{3.32}$$

where λ is a complex parameter, and the *adjoint equations*

$$y - \bar{\lambda}T^*y = g, \tag{3.33}$$

$$y - \bar{\lambda}T^*y = 0. \tag{3.34}$$

The elements f, g are given. λ is said to be an *eigenvalue* of (3.32) (or of T) if there exists an $x^0 \neq 0$ such that $x^0 - \lambda Tx^0 = 0$. x^0 is called an *eigenvector* corresponding to λ. Denote by $X(\lambda)$ the linear space spanned by the eigenvectors corresponding to λ, and let $N(\lambda)$ denote the dimension of $X(\lambda)$. $X^*(\bar{\lambda})$, $N^*(\bar{\lambda})$ are similarly defined with respect to (3.34).

Fredholm Alternative. Let T be a continuous linear operator defined on H (with values in H) and assume that T is also completely continuous. Then,

(α) λ is an eigenvalue of (3.32) if and only if $\bar{\lambda}$ is an eigenvalue of (3.34), and then the spaces $X(\lambda)$, $X^*(\bar{\lambda})$ are finite dimensional and $N(\lambda) = N^*(\bar{\lambda})$.

(β) If λ is not an eigenvalue of (3.32) then for any f, g in H there exist unique solutions x, y of (3.31), (3.33) respectively.

(γ) If λ is an eigenvalue of (3.32) then there exists a solution of (3.31) if and only if f is orthogonal to $X^*(\bar{\lambda})$.

It may be mentioned that the Fredholm alternative (stated in Sec. 5, Chap. 5 (statements (a), (b)) for the case of integral equations may be considered as a special case of the present Fredholm alternative if we take $H = L^2(G)$.

We can now state the main existence theorem of this section.

Theorem 8. *Assume that L satisfies* (A_1)–(A_3) *and that* $u_0 = 0$. *Then the Fredholm alternative holds for the generalized Dirichlet problem (second formulation). More precisely, either for every* $f \in H^0(D)$ *there exists a unique solution of*

$$B[\varphi, u] = (\varphi, f) \qquad \text{for all } \varphi \in C_c^\infty(D),\ u \in \mathring{H}^m(D), \tag{3.35}$$

or there is a finite number of linearly independent solutions v_j ($j = 1, \ldots, h$) *of* $B[v, \varphi] = 0$ *for all* $\varphi \in C_c^\infty(D)$, $v \in \mathring{H}^m(D)$, *and then there exists a solution of* (3.35) *if and only if* $(f, v_j) = 0$ *for* $j = 1, \ldots, h$. *In the second case the solution, if existing, is not unique.*

Proof. Take a fixed $k \geq k_0$, $k \neq 0$, where k_0 is the constant appearing in Theorems 4, 7, and set $L_k = L + k$. The bilinear form $B_k[u, v] = B[u, v] + k(u, v)$ is associated with L_k in the same manner that $B[u, v]$ is associated with L. By Theorem 7, for any $g \in \mathring{H}(D)$ there exists a unique solution w of

$$B_k[\varphi, w] = (\varphi, g) \qquad \text{for all } \varphi \in C_c^\infty(D),\ w \in \mathring{H}^m(D).$$

Set $w = L_k^{-1}g$. Then, since (3.35) is equivalent to

$$B_k[\varphi, u] = (\varphi, ku + f) \qquad \text{for all } \varphi \in C_c^\infty(D),\ u \in \mathring{H}^m(D),$$

u satisfies (3.35) if and only if $u = L_k^{-1}(ku + f)$, i.e., if and only if

$$u - Tu = f_1 \qquad (T = kL_k^{-1}, f_1 = L_k^{-1}f). \tag{3.36}$$

We claim that T is continuous and completely continuous in H^0. It suffices to prove this for L_k^{-1}. Now, if g varies in H^0 then $w = L_k^{-1}g$ varies in $\mathring{H}^m$ and it satisfies

$$B_k[\varphi, w] = (\varphi, g) \qquad \text{for all } \varphi \in C_c^\infty(D).$$

By completion, this relation holds also for all $\varphi \in \mathring{H}^m$. It therefore holds, in particular, for $\varphi = w$. Hence,

$$c|w|_m^2 \leq |B_k[w, w]| \leq |(w, g)| \leq |w|_0|g|_0 \leq |w|_m|g|_0,$$

i.e., $c|w|_m \leq |g|_0$. It follows that if g varies in bounded sets of H^0 then $w = L_k^{-1}g$ varies in bounded sets of $\mathring{H}^m$. L_k^{-1} is therefore a continuous transformation in H^0. Furthermore, since by Rellich's lemma (Theorem 3, Sec. 2), bounded sequences in $\mathring{H}^m$ $(m \geq 1)$ have subsequences which are Cauchy sequences in H^0, L_k^{-1} is also completely continuous.

The Fredholm alternative can thus be applied. It follows that:

(a) either for every f_1 there exists a unique solution of (3.36), or
(b) there are nontrivial solutions of $u - Tu = 0$, and the equation (3.36) has a solution if and only if $(f_1, v_j) = 0$ for all the solutions v_j of $v - T^*v = 0$.

In case (a) holds there exists a solution of the generalized Dirichlet problem for every f and it is unique (since $f = 0$ implies $f_1 = 0$ and, consequently, $u = 0$).

To consider the case (b), note first that Theorems 4, 7 are true also for L^* with possibly a different constant k_0. Taking k also larger than this constant k_0, we claim that

$$T^* = k(L^* + k)^{-1}. \tag{3.37}$$

Indeed, by definition of T^*,

$$(T^*v, g) = (v, Tg) \qquad \text{for all } v, g \text{ in } H^0. \tag{3.38}$$

Set $Tg = h$, $T^*v = w$. Then

$$B_k[\varphi, h] = k(\varphi, g)$$

for all $\varphi \in C_c^\infty(D)$ and, by completion, also for $\varphi = w$ which belongs to $\mathring{H}^m(D)$. Thus,

$$B_k[w, h] = k(w, g). \tag{3.39}$$

Set $S = k(L^* + k)^{-1}$, $Sv = w_1$. Then w_1 satisfies

$$B_k[w_1, \psi] = k(v, \psi) \qquad \text{for all } \psi \in C_c^\infty(D),\ w_1 \in \mathring{H}^m(D).$$

By completion we conclude that the last relation holds also for $\psi = h$. Hence,

$$B_k[w_1, h] = k(v, h) = k(v, Tg) = k(T^*v, g) = k(w, g). \tag{3.40}$$

Combining (3.39), (3.40) we obtain

$$B_k[w - w_1, h] = 0.$$

Since h can be any element of $\mathring{H}^m(D)$, $B_k[w - w_1, w - w_1] = 0$, i.e., $w = w_1$. Thus, $Sv = T^*v$ and (3.37) follows.

From (3.37) we conclude that $T^*v - v = 0$ if and only if

$$B[v, \varphi] = 0 \qquad \text{for all } \varphi \in C_c^\infty(D),\ v \in \mathring{H}^m(D). \tag{3.41}$$

Thus, to complete the proof of the theorem it remains to show that the conditions $(f_1, v_j) = 0$ are equivalent to the conditions $(f, v_j) = 0$. This follows from the equalities

$$(f_1, v_j) = (L_k^{-1}f, v_j) = (f, (L_k^{-1})^*v_j) = \frac{1}{k}(f, T^*v_j) = \frac{1}{k}(f, v_j).$$

4. Differentiability of Weak Solutions in the Interior

In this section we shall prove that weak solutions of strongly elliptic equations $Lu = f$ in D are smooth in (the interior of) D to any given order if f and the coefficients of L are sufficiently smooth. The assumptions (A_1)–(A_3) made in Sec. 3 are tacitly assumed throughout this section. We first prove a few lemmas.

Lemma 14. *Assume that $a^{\rho\sigma} \in C^1(D)$ and let u have m strong derivatives in D. If*

$$|B[\varphi, u]| \leq \text{const. } |\varphi|_{m-1} \tag{4.1}$$

for all $\varphi \in C_c^\infty(D)$ then u has $m + 1$ strong derivatives in D.

Proof. Let A, B, C be subdomains of D such that $\bar{A} \subset B$, $\bar{B} \subset C$, $\bar{C} \subset D$ and let $\zeta(x)$ be a function which belongs to $C_c^\infty(B)$ and satisfies: $\zeta(x) = 1$ if $x \in A$, $0 \leq \zeta \leq 1$ in B. Set $v = \zeta u$, $x^h = (x_1 + h, x_2, \ldots, x_n)$ if $x = (x_1, x_2, \ldots, x_n)$,

$$g^h(x) = \frac{g(x^h) - g(x)}{h}.$$

Then

$$D^\rho v^h = [D^\rho(\zeta u)]^h = (\zeta D^\rho u)^h + \sum_{|\lambda| < |\rho|} (b_\lambda D^\lambda u)^h$$

where the coefficients b_λ vanish outside B. By Lemma 11, Sec. 2, if $|h|$ is sufficiently small,

$$|D^\rho v^h - (\zeta D^\rho u)^h|_0^D \leq K_1 \sum_{|\lambda| < m} |D^\lambda u|_1^C \leq K_2 |u|_m^C \leq K_3 \tag{4.2}$$

where K_i are used to denote constants independent of h and of φ below.

Using (4.2) we get

$$(4.3)\qquad B[\varphi, v^h] = \Big| \sum_{|\rho|,|\sigma| \le m} (D^\rho\varphi, a^{\rho\sigma}D^\sigma v^h) \Big|$$
$$\le \Big| \sum_{|\rho|,|\sigma| \le m} (D^\rho\varphi, a^{\rho\sigma}(\zeta D^\sigma u)^h) \Big| + K_4|\varphi|_m.$$

Using the identity

$$(4.4)\qquad f(x)g^h(x) = (f(x)g(x))^h - f^h(x)g(x^h),$$

we obtain from (4.3)

$$(4.5)\qquad |B[\varphi, v^h]| \le \Big| \sum_{|\rho|,|\sigma| \le m} (D^\rho\varphi, (a^{\rho\sigma}\zeta D^\sigma u)^h) \Big|$$
$$+ \sum_{|\rho|,|\sigma| \le m} |(D^\rho\varphi(x), (a^{\rho\sigma}(x))^h\zeta(x^h)D^\rho u(x^h))| + K_4|\varphi|_m.$$

Since $(a^{\rho\sigma}(x))^h$ is uniformly bounded for $x \in C$, $|h|$ sufficiently small, and since

$$(4.6)\qquad \int_{B_1} |g(x^h)|^2\, dx \le \int_{B_2} |g(x)|^2\, dx$$

for any two bounded domains B_1, B_2 with $\overline{B}_1 \subset B_2$, provided $|h|$ is sufficiently small, we obtain from (4.5)

$$(4.7)\qquad |B[\varphi, v^h]| \le \Big| \sum_{|\rho|,|\sigma| \le m} (\zeta D^\rho\varphi^{-h}, a^{\rho\sigma}D^\sigma u) \Big| + K_5|\varphi|_m;$$

here the identity

$$(4.8)\qquad (f, g^h) = (f^{-h}, g)$$

has been used.

Writing

$$\zeta D^\rho\varphi^{-h} = D^\rho(\zeta\varphi^{-h}) + \sum_{|\lambda| < |\rho|} d_\lambda D^\lambda\varphi^{-h}$$

and using the inequality $|D^\lambda\varphi^{-h}|_0^D \le |D^\lambda\varphi|_1^D$ which follows by Lemma 11′, Sec. 2, we get from (4.7)

$$(4.9)\qquad |B[\varphi, v^h]| \le |B[\zeta\varphi^{-h}, u]| + K_6|\varphi|_m.$$

If we apply (4.1) with φ replaced by $\zeta\varphi^{-h}$ and use the inequality

$$|\zeta\varphi^{-h}|_{m-1} \le K_7|\varphi|_m$$

which follows by using Leibniz' rule and Lemma 11′, then we find from (4.9) that

$$(4.10)\qquad |B[\varphi, v^h]| \le K_8|\varphi|_m.$$

(4.10) was proved for any $\varphi \in C_c^\infty(D)$. By completion it then holds also for any $\varphi \in \mathring{H}^m(D)$. The mollifiers $J_{1/j}(\zeta u)$ (defined as in (1.8)) vanish outside C if j is sufficiently large and, as easily verified (using Lemma 3, Sec. 1, the rule (1.10) which holds also for strong derivatives, and the fact that $\zeta u \in H^m(C)$), $J_{1/j}(\zeta u) \to \zeta u$ in the norm of $H^m(C)$. Hence, $\zeta u \in \mathring{H}^m(C)$.

But then $(\zeta u)^h$ belongs to $\mathring{H}^m(D)$ if $|h|$ is sufficiently small indeed it is the limit, in the norm of $\mathring{H}^m(D)$, of the functions $[J_{1/j}(\zeta u)]^h$.

We may therefore take $\varphi = v^h$ in (4.10) and, using Gårding's inequality, thus obtain

$$c|v^h|_m^2 \leq |B[v^h, v^h]| + k_0|v^h|_0^2 \leq K_8|v^h|_m + k_0|v^h|_0^2.$$

Since $m \geq 1$, $k_0|v^h|_0^2 \leq k_0|v|_1^2 \leq K_9$ by Lemma 11′, and we find that

$$(4.11) \qquad |v^h|_m \leq K_{10}.$$

By Lemma 11′, $v^h \to \partial v/\partial x_1$ in $L^2(D)$ and, in view of (4.11), Lemma 2, Sec. 1, can be used to conclude that $\partial v/\partial x_1$ belongs to $H^m(D)$. Similarly one proves that $\partial v/\partial x_i$ belongs to $H^m(D)$ for $i = 2, \ldots, n$. Employing Lemma 4, Sec. 1, it follows that $v \in H^{m+1}(D)$. Since $u = v$ on A, u then belongs to $H^{m+1}(A)$. Finally, since A can be any subdomain of D, u has $m + 1$ strong derivatives in D.

Lemma 15. *Assume that $a^{\rho\sigma} \in C^{\max(1,|\rho|-m+j)}$ for some $1 \leq j \leq m$. If u has m strong derivatives in D, and if*

$$(4.12) \qquad |B[\varphi, u]| \leq \text{const. } |\varphi|_{m-j}$$

for all $\varphi \in C_c^\infty(D)$, then u has $m + j$ strong derivatives in D.

Proof. The case $j = 1$ coincides with Lemma 14. Proceeding by induction we assume the lemma to hold for $j - 1$ (where $2 \leq j \leq m$) and for any bounded domain D, and shall prove it for j and for any bounded domain D. Since (4.12) implies that

$$|B[\varphi, u]| \leq \text{const. } |\varphi|_{m-j-1},$$

by the inductive assumption it follows that u has $m + j - 1$ strong derivatives.

Replacing φ by $D\varphi$ in (4.12) we get

$$(4.13) \quad |B[D\varphi, u]| \leq \text{const. } |D\varphi|_{m-j} \leq \text{const. } |\varphi|_{m-j+1}.$$

Next,

$$(4.14) \quad B[D\varphi, u] = -\sum_{|\rho|,|\sigma|\leq m} (D^\rho\varphi, D[a^{\rho\sigma}D^\sigma u]) = -B[\varphi, Du] + I$$

where

$$I = \sum_{|\rho|,|\sigma|\leq m} \int_D D^\rho\varphi \cdot D\overline{a^{\rho\sigma}} \cdot D^\sigma\overline{u}\, dx.$$

Integrating by parts $\max(0, |\rho| - m + j - 1)$ times and using the fact that $|u|_{m+j-1}^A < \infty$ for any compact subdomain A of D we get

$$\left| \int_D D^\rho\varphi \cdot [D\overline{a^{\rho\sigma}} \cdot D^\sigma\overline{u}]\, dx \right| \leq \text{const. } |\varphi|_{m-j+1}$$

where the constant is independent of φ provided $\varphi \in C_c^\infty(B)$ where B is any fixed subdomain of D with $\overline{B} \subset D$. The same bound is then obtained for I. Using this bound in (4.14) and combining the result with (4.13), we arrive at the inequality

$$|B[\varphi, Du]| \leq \text{const. } |\varphi|_{m-j+1} \qquad \text{for any } \varphi \in C_c^\infty(B).$$

The inductive assumption can now be applied to Du in B, since Du has strong derivatives up to order $m + j - 2 \geq m$. It follows that Du has $m + j - 1$ strong derivatives in B. Using Lemma 4, Sec. 1, we conclude that u has $m + j$ strong derivatives in B, and since B is an arbitrary subdomain of D with $\overline{B} \subset D$, u has $m + j$ strong derivatives in D.

Lemma 16. *Assume that $a^{\rho\sigma} \in C^{\max(1,|\rho|+p)}$ where p is an integer ≥ 0, and that f has p strong derivatives in D. If u has m strong derivatives in D and if u is a solution of* (3.9), *then u has $2m + p$ strong derivatives in D.*

Proof. Since

$$B[\varphi, u] = (\varphi, f) \qquad \text{for all } \varphi \in C_c^\infty(D), \tag{4.15}$$

we can apply Lemma 15 with $j = m$ and thus conclude that u has $2m$ strong derivatives.

Next, substituting $D\varphi$ for φ in the relation (4.15) and integrating by parts we find that

$$B[\varphi, Du] + (\varphi, L'u) = (\varphi, Df)$$

where

$$L'u = \sum_{|\rho|,|\sigma| \leq m} (-1)^{|\rho|} D^\rho[(Da^{\rho\sigma})D^\sigma u].$$

Hence,

$$|B[\varphi, Du]| \leq \text{const. } |\varphi|_0$$

for all $\varphi \in C_c^\infty(A)$ where A is any fixed subdomain of D with $\overline{A} \subset D$ and where the constant is independent of φ. Applying Lemma 15 with $j = m$ we find that Du has $2m$ strong derivatives in A and, hence, by Lemma 4, Sec. 1, u has $2m + 1$ strong derivatives in A. Since A is an arbitrary subdomain of D with $\overline{A} \subset D$, u has $2m + 1$ strong derivatives in A.

The next step consists of using (4.15) with φ replaced by $D^2\varphi$. Integration by parts yields

$$B[\varphi, D^2u] + (\varphi, L''u) = (\varphi, D^2f)$$

where L'' is a linear operator of order $\leq 2m + 1$ with bounded coefficients. By the same argument as before we deduce that D^2u is m times strongly differentiable in A and, hence, by Lemma 4, Sec. 1, u has $m + 2$ strong derivatives. Proceeding similarly step by step, the proof of the lemma is completed.

From Lemma 16 we immediately obtain the following theorem:

Theorem 9. *Let L satisfy the assumptions* (A_1)–(A_3) *and let $a^{\rho\sigma} \in C^{\max(1,|\rho|+p)}$ for some integer $p \geq 0$. If f has p strong derivatives in D and if*

$f \in H^0(D)$, then all the solutions of the generalized Dirichlet problem (second formulation), in particular, those established in Theorems 7, 8, have $2m + p$ strong derivatives in D. If $p \geq [n/2] + 1$ then the solutions are classical solutions of $Lu = f$ in D, and they belong to $C^{2m+p-[n/2]-1}(D)$.

The last part follows from Corollary 2 of Theorem 2′, Sec. 2.

We shall now establish the interior differentiability of any weak solution of $Lu = f$.

Theorem 10. *Let L satisfy (A_1)–(A_3) and let $a^{\rho\sigma} \in C^{m+1}$. If for any compact subdomain A of D, $u \in H^0(A)$ and*

$$|(L^*\varphi, u)| \leq \text{const.}\ |\varphi|_{2m-j} \qquad \text{for all } \varphi \in C_c^\infty(A) \tag{4.16}$$

where $0 \leq j \leq 2m$, then u has j strong derivatives in D.

Theorem 11. *Let L satisfy (A_1)–(A_3) and let $a^{\rho\sigma} \in C^{\max(m+1,|\rho|+p)}$ for some integer $p \geq 0$. If f has p strong derivatives in D and u is a weak solution in D of $Lu = f$, then u has $2m + p$ strong derivatives in D. If $p \geq [n/2] + 1$ then u is a classical solution of $Lu = f$ in D, and it belongs to*

$$C^{2m+p-[n/2]-1}(D).$$

For $p = 0$ Theorem 11 follows from Theorem 10. For $p > 0$ it then follows by combining the result for $p = 0$ with Lemma 16 and Corollary 2 to Theorem 2′. It thus remains to prove Theorem 10.

Proof of Theorem 10. The proof is by induction on j. For $j = 0$ the assertion is trivial. We now assume that the assertion holds for $j - 1$ $(0 \leq j - 1 < 2m)$ and prove it for j. Since (4.16) implies

$$|(L^*\varphi, u)| \leq \text{const.}\ |\varphi|_{2m-j+1} \qquad \text{for all } \varphi \in C_c^\infty(A),$$

it follows by the inductive assumption that u has $j - 1$ strong derivatives in D. If $j - 1 \geq m$ then the assertion for j follows from Lemma 15 applied with j replaced by $j - m$. It thus remains to consider the case where $j - 1 < m$.

Let Δ be the Laplace operator and consider the operator

$$L_q = (-1)^q\Delta^q + 1. \tag{4.17}$$

From the first part of the proof of Theorem 4, Sec. 3 and from the inequality

$$\sum_{|\alpha| \leq m} |\xi^\alpha|^2 \leq \text{const.} \left\{ \sum_{|\beta| = m} |\xi^\beta|^2 + 1 \right\},$$

one obtains the inequality

$$(\varphi, L_q\varphi) \geq c'|\varphi|_q^2 \qquad \text{for all } \varphi \in C_c^\infty(R^n) \tag{4.18}$$

where c' is a positive constant. Theorem 6, Sec. 3, can therefore be applied. It follows that for any subdomain B of D with $\overline{B} \subset D$ there exists a solution h of the generalized Dirichlet problem (the two formulations coincide)

$$L_q h = u \qquad \text{in } B, \qquad h \in \mathring{H}^q(B).$$

We take $q = m - j + 1$. Since u has $j - 1$ strong derivatives in D, Theorem 9 implies that h has $2q + j - 1 = m + q$ strong derivatives in B. Substituting $u = L_q h$ into (4.16) we get

$$(4.19) \qquad |(L^*\varphi, L_q h)| \le \text{const.}\ |\varphi|_{2m-j}$$

for any $\varphi \in C_c^\infty(B)$.

Now, LL_q is an elliptic operator of order $2m + 2q$. Writing it in the form

$$LL_q v \equiv \sum_{|\rho|,|\sigma| \le m} \sum_{|\tau|=q} (-1)^{|\rho|+q} D^\rho[a^{\rho\sigma} D^\sigma(D^\tau D^\tau v)] + Lv$$

we find, by partial integration, that

$$(4.20) \quad (L^*\varphi, L_q\psi) = (\varphi, LL_q\psi) = \sum_{|\alpha|,|\beta| \le m+q} (D^\alpha \varphi, b^{\alpha\beta} D^\beta \psi) + B[\varphi, \psi] \equiv \hat{B}[\varphi, \psi]$$

for any $\varphi \in C_c^\infty(B)$, $\psi \in C^{2m+2q}(B)$. $\hat{B}[\varphi, \psi]$ is a bilinear form associated with the elliptic operator LL_q, and the $b^{\alpha\beta}$ are sums of terms $\pm D^k a^{\rho\sigma}$ where $|\alpha| = |\rho| + q - |k|$, $|\beta| = |\sigma| + q$, and $0 \le |k| \le q$. It follows that $b^{\alpha\beta} \in C^1$.

Integrating by parts, in $(L^*\varphi, L_q\psi)$, in a slightly different manner than in (4.20) (not transferring the derivatives D^ρ occurring in $L^*\varphi$ to the right-hand side of the scalar product), we find that

$$(L^*\varphi, L_q\psi) = \hat{B}[\varphi, \psi]$$

if ψ is only in $C^{m+q}(B)$. By completion it follows that

$$(L^*\varphi, L_q h) = \hat{B}[\varphi, h]$$

for any $\varphi \in C_c^\infty(B)$, since h has $m + q$ strong derivatives in B. Recalling (4.19) we obtain

$$|\hat{B}[\varphi, h]| \le \text{const.}\ |\varphi|_{2m-j} = \text{const.}\ |\varphi|_{m+q-1}$$

for all $\varphi \in C_c^\infty(B)$. Applying Lemma 14 with m replaced by $m + q$ we conclude that h has $m + q + 1$ strong derivatives in B. Hence $u = L_q h$ has $(m + q + 1) - 2q = m + 1 - q = j$ strong derivatives in B and, since B is an arbitrary subdomain of D with $\bar{B} \subset D$, u has j strong derivatives in D.

From Theorem 11 one easily obtains:

Corollary. *Let L be a strongly elliptic operator in D of order $2m$, written in the form $L = \Sigma\, a_\alpha D^\alpha$, and let $a_\alpha \in C^q(D)$ where $q = \max(2m + 1, m + p)$ for some integer $p \ge 0$. If $f \in C^p(D)$ then any weak solution of $Lu = f$ in D belongs to $C^{2m+p-[n/2]-1}(D)$. If $a_\alpha \in C^\infty(D)$, $f \in C^\infty(D)$ then $u \in C^\infty(D)$.*

The question of differentiability of classical solutions of elliptic equations (and not just strongly elliptic equations) was briefly mentioned at the end of Sec. 7, Chap. 9.

5. Differentiability Near the Boundary

In this section we shall prove that if ∂D, f, and the coefficients of L are sufficiently smooth then the same is true of the solution of the generalized Dirichlet problem (in either formulation)

$$Lu = f \quad \text{in } D, \qquad u \in \mathring{H}^m(D) \tag{5.1}$$

in $\overline{D}$. More precisely, we shall prove the following theorem.

Theorem 12. *Let L satisfy* (A_1)–(A_3) *and suppose that*

$$a^{\rho\sigma} \in C^{\max(1,|\rho|+m+p-2)}(\overline{D}),\ \partial D \in C^{3m+p-2},\ f \in H^{p-1}(D)$$

for some integer $p \geq 1$. Then any solution u of the generalized Dirichlet problem (5.1) *belongs to $H^{m+p}(D)$.*

Using Corollary 2 to Theorem 2′, Sec. 2, we get:

Corollary 1. *If the assumptions of Theorem* 12 *hold with $p \geq [n/2]$ then $u \in C^{m-1}(\overline{D})$ and, therefore (by Lemma* 12, *Sec.* 2) *u satisfies the boundary conditions*

$$\frac{\partial^j u}{\partial \nu^j} = 0 \qquad \text{on } \partial D \qquad (j = 0, 1, \ldots, m-1)$$

in the usual sense. If $p \geq [n/2] + 2$ then (by Theorem 9) *u is also a classical solution of $Lu = f$ in D, and $u \in C^{m+p-[n/2]-1}(\overline{D}) \cap C^{2m+p-[n/2]-2}(D)$.*

Thus, if the assumptions of Theorem 12 hold with $p \geq [n/2] + 2$ then the solutions of the generalized Dirichlet problem are also solutions of the classical Dirichlet problem.

Corollary 2. *Let L be a strongly elliptic operator with coefficients in $C^\infty(\overline{D})$ and assume further that $f \in C^\infty(\overline{D})$, that $\partial D \in C^\infty$, and that $\varphi_0, \ldots, \varphi_{m-1}$ are C^∞ functions on ∂D. Then any solution of classical Dirichlet problem* (3.5), (3.8) *belongs to $C^\infty(\overline{D})$.*

If $\varphi_0 \equiv \cdots = \varphi_{m-1} \equiv 0$ then the corollary follows from Theorem 12 and Corollary 2 to Theorem 2′, Sec. 2. Therefore, if we can construct a function $\Phi(x)$ in $C^\infty(\overline{D})$ which satisfies the conditions (3.8), then, by applying the previous special case to $u - \Phi$, the proof is completed. To construct Φ, denote by $\sigma(x)$ the distance of any point $x \in D$ to ∂D. If $\sigma(x)$ is sufficiently small then x lies on a unique inward normal to ∂D, at some point x^*, and $\sigma(x) = |x - x^*|$ (compare Sec. 8, Chap. 3). If ν in (3.8) designates the inward normal, then we take

$$\Phi(x) = \zeta(x) \sum_{\lambda=0}^{m-1} \frac{(\sigma(x))^\lambda}{\lambda!} \varphi_\lambda(x^*),$$

where ζ is a C^∞ function which vanishes outside a sufficiently small open neighborhood N of ∂D which equals 1 in some neighborhood N_0 of ∂D with $\overline{N}_0 \subset N$.

Proof of Theorem 12. From Theorem 9 we already know that $u \in H^{2m+p-1}(A)$ for any compact subdomain A of D. If we show that for any $x^0 \in \partial D$ there exists a neighborhood N^* such that $u \in H^{m+p}(N^* \cap D)$ then, using Lemma 5, Sec. 1, the proof of the theorem is completed. Since u has $m + p$ strong derivatives in D, in view of Theorem 1, Sec. 1, it only remains to show that all these derivatives belong to $H^0(N^* \cap D)$.

It suffices to consider the special case where x^0 has a neighborhood N such that $N \cap \partial D$ lies on the hyperplane $x_n = 0$ and $N \cap D$ lies in the half-space $x_n > 0$. Indeed, the general case can be reduced to the special case by performing a local transformation as in the proof of Theorem 1, Sec. 1.

Let $\partial N_2 = \overline{N \cap D} \cap \{x_n = 0\}$ and let ∂N_1 be a complementary part of the boundary of $N \cap D$. Let A be a subdomain of $N \cap D$ whose boundary consists of a closed subdomain ∂A_2 contained in the interior of ∂N_2 and a set ∂A_1 contained in $N \cap D$ so that ∂A_1 is bounded away from ∂N_1. The following notation will be used in the present proof:

$$x = (x_1, \ldots, x_{n-1}), \quad y = x_n, \quad D_x^\alpha = D_1^{\alpha_1} \cdots D_{n-1}^{\alpha_{n-1}}, \quad D^\alpha = D_1^{\alpha_1} \cdots D_n^{\alpha_n}.$$

We claim that

$$|D_x^\alpha u|_m^A < \infty \qquad \text{for all } |\alpha| \le m + p - 1, \tag{5.2}$$

i.e.,

$$D_x^\alpha D^\beta u \in H^0(A) \qquad \text{for all } |\alpha| \le m + p - 1,\ |\beta| \le m. \tag{5.3}$$

To verify (5.2) for $|\alpha| = 1$ we proceed similarly to the proof of Lemma 14, Sec. 4, and take ζ to vanish in a neighborhood of ∂N_1 and to be equal to 1 on A. We take the difference quotient with respect to any x_i $(1 \le i \le n - 1)$. Since u is the limit in the norm $|\ \ |_m^{N \cap D}$ of infinitely differentiable functions vanishing near the boundary ∂N_2, the same is true of ζu and of $(\zeta u)^h$. Hence, after deriving an inequality of the form (4.10) we may substitute $\varphi = v^h$. Using Lemmas 11′, 2 of Sec. 2 we then find that $D_x u$ belongs to $H^m(A)$. Furthermore, $D_x u$ is the limit in the norm $|\ \ |_m^A$ of the arithmetic means of some sequence $\{v^{h_i}\}$ and, hence, also of a sequence of infinitely differentiable functions vanishing near the boundary ∂N_2.

We can next proceed analogously to Lemma 15, by considering $B[D_x\varphi, u]$ and, after partial integration, applying the previous result to

$D_x u$. Proceeding similarly step by step, as in the proofs of Lemmas 15, 16, the proof of (5.2) (or (5.3)) is completed.

To prove Theorem 12 we have to show that

$$(5.4) \qquad D^\gamma u \text{ belong to } H^0(A) \qquad \text{for all } |\gamma| \leq m + p.$$

In view of (5.3), it remains to consider only those derivatives $D^\gamma u$ which do not occur in (5.3). For the sake of clarity consider first the case $m = 1$, i.e.,

$$(5.5) \quad Lu \equiv \sum_{|\alpha| \leq 2} a_\alpha D_x^\alpha + \sum_{|\alpha| \leq 1} a_{1\alpha} D_y D_x^\alpha + a_2 D_y^2 u = f \text{ (almost everywhere).}$$

Suppose first that $p = 1$. $D_y^2 u$ is the only derivative in (5.4) which does not occur in (5.3). Since $a_2 \neq 0$, $D_y^2 u$ can be estimated from (5.5) and, thus, (5.4) is valid.

If $p = 2$, apply D_x to both sides of (5.5). Then

$$(5.6) \qquad L(D_x u) = -L'u + D_x f$$

where L' is a linear operator of order $\leq 2m = 2$. $D_y^2 D_x u$ is the only derivative in (5.6) which is not yet known to belong to $H^0(A)$. Since its coefficient $a_2 \neq 0$, it also belongs to $H^0(A)$. Applying next D_y to both sides of (5.5) we find that $D_y^3 u \in H^0(A)$ and the proof of (5.4) for $p = 2$ is completed.

For arbitrary p, we first apply D_x^{p-1} to both sides of (5.4) and conclude that $D_x^{p-1} D_y^2 u \in H^0(A)$, then apply $D_x^{p-2} D_y$, etc.

Now let $m > 1$ and consider first the case $p = 1$. The only derivative which occurs in (5.4) but not in (5.3) is $D_y^{m+1} u$ and, in contrast to the case $m = 1$, it cannot be evaluated immediately from the equation $Lu = f$. All the derivatives occurring in Lu but not in (5.3) are of the form $D_y^{m+1} D^\beta u$ where $0 \leq |\beta| \leq m - 1$. Their sum can therefore be written in the form

$$(5.7) \qquad D_y^m g \qquad \text{where } g = \sum_{|\beta| \leq m} b_\beta D^\beta u$$

and the coefficient δ of $D_y^m u$ in g is $\neq 0$ in A. Writing the equation $Lu = f$ in the form

$$(5.8) \qquad L_0 u + D_y^m g = f,$$

we conclude that $D_y^m g \in H^0(A)$. g also belongs to $H^0(A)$. Assuming, as we may, that A is a rectangle and using Lemma 8, Sec. 2, for the function g (as a function of the variable y) in closed rectangles A^* contained in A and then integrating with respect to x and taking $A^* \to A$, we find that $D_y^i g \in H^0(A)$ for any $i = 1, 2, \ldots, m - 1$. Except for $\delta D_y^{m+1} u$ all the terms in $D_y g$ are already known to belong to $H^0(A)$ (by (5.3)). Hence $D_y^{m+1} u \in H^0(A)$.

Consider next the case $p = 2$. Applying D_x to both sides of (5.8) we get

$$(5.9) \qquad D_x L_0 u + D_y^m D_x g = D_x f.$$

All the derivatives of u in D_xL_0u already occur in (5.3). Hence $D_y^m(D_xg) \in H^0(A)$. Since also $D_xg \in H^0(A)$, it follows (by applying Lemma 8, Sec. 2, as in the case $p = 1$) that $D_yD_xg \in H^0(A)$. Except for $\delta D_y^{m+1}D_xu$ all the terms in D_yD_xg are already known to belong to $H^0(A)$ (by (5.3)). Hence $D_y^{m+1}D_xu \in H^0(A)$.

Next we use the fact that $D_y^2g \in H^0(A)$. Recalling (5.3) and the fact that $D_y^{m+1}D_xu \in H^0(A)$, we find that $D_y^{m+2}u \in H^0(A)$. This completes the proof for $p = 2$.

For general $p \leq m$, we first apply D_x^{p-1} to both sides of (5.8) and, using the fact that $D_yD_x^{p-1}g \in H^0(A)$, conclude that $D_x^{p-1}D_y^{m+1}u \in H^0(A)$. We next apply D_x^{p-2} to both sides of (5.8) and, using the fact that $D_y^2D_x^{p-2}g \in H^0(A)$, conclude that $D_x^{p-2}D_y^{m+2} \in H^0(A)$, etc. Finally, using the fact that $D_y^pg \in H^0(A)$ it follows that $D_y^{m+p}u \in H^0(A)$.

It remains to consider the case $p > m$. If $p = m + 1$, applying D_x to both sides of $Lu = f$ we get

$$L(D_xu) = L'u + D_xf \equiv \tilde{f}$$

where L' is a linear operator of order $\leq 2m$, so that $\tilde{f} \in H^0(A)$. Applying the result for $p = m$ we conclude that $D_xD^\alpha u \in H^0(A)$ for all $|\alpha| \leq 2m$. Applying D_y to both sides of $Lu = f$ we see that all the terms, with the exception of $\delta D_y^{2m+1}u$, are already known to belong to $H^0(A)$. Hence, $D_y^{2m+1}u \in H^0(A)$.

If $p = m + 2$, we first apply D_x^2 to $Lu = f$ and find that $D_x^2D^\alpha u \in H^0(A)$ for all $|\alpha| \leq 2m$. Next we apply D_xD_y and find that $D_xD_yD^\alpha u \in H^0(A)$ for all $|\alpha| \leq 2m$. Finally, applying D_y^2 it also follows that $D_y^{m+2}u \in H^0(A)$ and the proof is completed.

In general, if $p = m + k$, we first apply D_x^k to $Lu = f$, then $D_x^{k-1}D_y$, etc.

In the remaining part of this section we shall give some additional results on elliptic equations which will be needed later on in treating parabolic equations. We shall need the following definition:

Definition. An elliptic operator $L(x, D_x)$ of order $2m$ is said to satisfy the *root condition* in a domain B if for every $x \in B$ and for each pair of linearly independent real vectors ξ, η, the polynomial in λ, $L_0(x, \xi + \lambda\eta)$, has precisely m roots with positive imaginary part. Here L_0 is the principal part of L.

The root condition is satisfied if $n \geq 3$, or if the coefficients of L_0 are real.

We now state an important result concerning a priori estimates.

Theorem 13. *Let $L = \Sigma\, a_\alpha D^\alpha$ be an elliptic operator with module of ellipticity $\geq \delta$ in a bounded domain D, and assume that the $a_\alpha \in C^p(\overline{D})$ and are bounded together with their first p derivatives by a constant H, where p is* *an integer* ≥ 0, *that the principal coefficients have a modulus of continuity* ω, *and that* L *satisfies the root condition in* $\overline{D}$. *Let* $f \in H^p(D)$ *and assume that* ∂D *can be covered by a finite number of neighborhoods* N_j *such that each* $N_j \cap \partial D$ *can be represented by a function whose first* $2m + p$ *derivatives are continuous and bounded by a constant* K.

If $u \in C^{2m}(D) \cap C^{m-1}(\overline{D})$ *and* $Lu = f$ *in* D, $\partial^j u/\partial \nu^j = 0$ *on* ∂D *for* $j = 0, 1, \ldots, m - 1$, *and if* $|u|_{2m} < \infty$, *then*

$$|u|_{2m+p} \leq M|f|_p + M'|u|_0 \tag{5.10}$$

where M, M' *are constants depending only on* H, ω, δ, K, p. *If, further, the Dirichlet problem has at most one solution, then* $M' = 0$, *i.e.*,

$$|u|_{2m+p} \leq M|f|_p. \tag{5.11}$$

For a detailed proof the reader is referred to [1; 704–706].

With the aid of Theorem 13 and Corollary 1 to Theorem 12 we shall establish the following:

Theorem 14. *Let* p *be an integer* ≥ 1 *and let* L *be a strongly elliptic operator in* $\overline{D}$ *satisfying the root condition. Assume that* ∂D *is sufficiently smooth, that the coefficients of* L *are sufficiently smooth in* $\overline{D}$, *and that there is at most one solution of the Dirichlet problem. Then for any* $f \in H^p(D)$ *there exists a unique solution* u *of the generalized Dirichlet problem* (5.1), *and* (5.11) *holds.*

Note that in Theorem 13 u belongs to $C^{2m}(D) \cap C^{m-1}(\overline{D})$, which need not be the case for the solution u of the present theorem.

Proof. For any compact subdomain A of D, f can be approximated in the norm of $H^p(A)$ by mollifiers $J_\epsilon f$. In small neighborhoods of the boundary we can perform transformations which map ∂D onto a portion lying on a hyperplane and then approximate f by mollifiers of the form $J'_\epsilon f$ introduced in the proof of Lemma 6, Sec. 1. Using partition of unity as in the proof of Lemma 5, Sec. 1, we can then construct a sequence of $C^\nu(\overline{D})$ functions f_j such that $|f_j - f|_p^D \to 0$ as $j \to \infty$, provided ∂D is of class C^ν.

Consider the Dirichlet problem

$$Lu_j = f_j \qquad \text{in } D,$$

$$\frac{\partial^k u_j}{\partial \nu^k} = 0 \qquad \text{on } \partial D, \qquad \text{for } k = 0, 1, \ldots, m - 1.$$

By the assumption of uniqueness it follows, using Theorem 8, Sec. 3, and Corollary 1 to Theorem 12, that u_j exists and is sufficiently smooth in $\overline{D}$. Theorem 13 can therefore be applied. We get,

$$|u_i - u_j|_{2m+p} \leq M|f_i - f_j|_p.$$

Thus, the u_j form a Cauchy sequence in $H^{p+2m}(D)$. Since the limit is the unique solution of the generalized Dirichlet problem (5.1), and since (5.11) holds for each pair u_j, f_j, (5.11) holds also for u.

Remark. By approximating ∂D and L by sufficiently smooth boundaries and operators respectively, one can improve Theorem 14 with regard to the differentiability assumptions on ∂D and L. This, however, will not be needed in the future.

We finally consider the case where the coefficients of L depend on a real parameter t which varies in a bounded interval $a \leq t \leq b$. Thus,

$$L(t)u \equiv \sum_{|\alpha| \leq 2m} a_\alpha(x, t) D^\alpha u. \tag{5.12}$$

We assume that for each t the Dirichlet problem (5.1) has at most one solution. The solution is denoted by $L^{-1}(t)f$.

Theorem 15. *Let k, p be any integers ≥ 0 and let $L(t)$ be strongly elliptic for all t ($a \leq t \leq b$) and satisfying the root condition. Assume that ∂D is sufficiently smooth and that the coefficients $a_\alpha(x, t)$ are sufficiently smooth for $x \in \bar{D}$, $a \leq t \leq b$. Then the mapping $t \to L^{-1}(t)f$ is k times continuously differentiable from the interval (a, b) into $H^{2m+p}(D)$.*

Proof. Consider the function

$$u^h(x, t) = \frac{u(x, t + h) - u(x, t)}{h}$$

where $u(x, t) = (L^{-1}(t)f)(x)$. Using the rule (4.4) we find that

$$Lu^h = F_h \qquad \text{in } D, \qquad u^h \in \mathring{H}^m(D)$$

where $F_h(x, t) = -\sum a_\alpha^h(x, t) D^\alpha u(x, t + h)$. Using Theorem 14 it follows that $|u^h|_{2m+p}$ is bounded independently of h, provided $t + h$ remains in a closed subset of (a, b). Hence,

$$|u(\cdot, t + h) - u(\cdot, t)|_{2m+p} \leq \text{const.}\ |h| \tag{5.13}$$

where $u(\cdot, t)$ is the function whose value at each point x is $u(x, t)$. (5.13) shows that the mapping $t \to L^{-1}(t)f$ is a continuous mapping from (a, b) into $H^{2m+p}(D)$. The case $k = 0$ is thus proved.

To prove the theorem for $k = 1$, we differentiate the equation $L(t)u = f$ with respect to t and find that $v = \partial u/\partial t$, if existing, must satisfy

$$L(t)v = \hat{F} \qquad \text{in } D, \qquad v \in \mathring{H}^m(D), \tag{5.14}$$

where $\hat{F}(x, t) = -\sum [\partial a_\alpha(x, t)/\partial t] D^\alpha u(x, t)$. Observing that $\hat{F} \in H^p(D)$, it follows that a unique solution v of (5.14) exists. Furthermore, since

$$|F_h - \hat{F}|_p \to 0 \qquad \text{as } h \to 0,$$

as is easily verified using (5.13), applying Theorem 14 we obtain

$$|u^h(\cdot, t) - v(\cdot, t)|_{2m+p} \to 0 \qquad \text{as } h \to 0.$$

Thus, the mapping $t \to L^{-1}(t)f$, from (a, b) into $H^{2m+p}(D)$, is differentiable in t. Next, $v(\cdot, t)$ varies continuously in t, as a function with values in $H^{2m+p}(D)$. This can be proved by estimating $v^h(\cdot, t)$ using Theorem 14 and (5.13). The proof of the theorem for $k = 1$ is thereby completed.

Proceeding similarly step by step, the proof of the theorem for any k can thus be obtained.

From the previous proof one obtains:

Corollary. *The inequalities*

$$|D_t^j L^{-1}(t)f|_{2m+p} \leq \text{const. } |f|_p$$

hold for $a < t < b$, $j = 0, 1, \ldots, k$.

6. Abstract Existence Theorems

The main result of this section is Theorem 17.

Let H, V be two Hilbert spaces. Denote by $|f|$ the norm of $f \in H$ and by $||u||$ the norm of $u \in V$. The scalar product of elements f, g in H is denoted by (f, g) and the scalar product of elements u, v in H is denoted by $((u, v))$. Assume that V is a linear subspace of H and that

$$||u|| \geq k|u| \qquad \text{for any } u \in V \qquad (k \text{ constant}).$$

Finally, assume that V is dense in H.

Example: $H = L^2(D)$, $V = \mathring{H}^m(D)$.

Let $a(u, v)$ be a continuous bilinear form on V, i.e., the mapping $(u, v) \to a(u, v)$ from $V \times V$ into the field of complex numbers is continuous and is linear in u and antilinear in v. As is well known, a bilinear form $a(u, v)$ is continuous if and only if it is bounded, i.e., if and only if

$$|a(u, v)| \leq c_1 ||u|| \, ||v|| \qquad (c_1 \text{ constant}). \tag{6.1}$$

For fixed v, $a(u, v)$ is a continuous linear functional of $u \in V$ and therefore there exists a unique element in V, say $A_0 v$, such that

$$a(u, v) = ((u, A_0 v)) \tag{6.2}$$

for all $u \in V$. A_0 is a linear operator on V (into V) and by (6.1) it follows that A_0 is also continuous.

Similarly,

$$a(u, v) = ((A_1 u, v)) \tag{6.3}$$

where A_1 is a continuous linear operator on V.

Denote by N the set of elements $u \in V$ for which the antilinear mapping

$$v \to a(u, v) \tag{6.4}$$

is continuous on V when V is provided with the topology (i.e., the norm) of H. Since V is dense in H, (6.4) can be extended into a continuous antilinear functional on H and, consequently, there exists a unique element in H, say Au, such that

$$a(u, v) = (Au, v). \tag{6.5}$$

A is a linear operator, unbounded in general, with domain $d(A) = N$.

Lemma 17. *If for all $v \in V$*

$$|a(v, v)| \geq \alpha||v||^2 \qquad (\alpha \text{ positive constant}) \tag{6.6}$$

then for every $f \in H$ there exists a unique element $u \in d(A)$ such that $Au = f$.

Proof. $F(v) \equiv \overline{(f, v)}$ is a continuous linear functional on H and hence also on V. Now apply Theorem 5, Sec. 3.

In the future we shall need a more delicate variant of Lemma 17, which we proceed to derive.

Let F be a Hilbert space and denote by $(u, v)_F$ the scalar product of elements u, v in F and by $||u||_F$ the norm of $u \in F$. Let Φ be a linear subspace of F and denote by $(((\varphi, \psi)))$ a scalar product of elements φ, ψ in Φ and by $|||\varphi|||$ the norm of $\varphi \in \Phi$, i.e., $|||\varphi||| = \{(((\varphi, \varphi)))\}^{1/2}$. We assume that for all $\varphi \in \Phi$

$$||\varphi||_F \leq c_1|||\varphi||| \qquad (c_1 \text{ constant}) \tag{6.7}$$

but we do not assume that Φ is a complete space (i.e., a Hilbert space) or that Φ is dense in F.

Let $E(u, \varphi)$ be a bilinear form on $F \times \Phi$ and assume:

(6.8) for every $\varphi \in \Phi$ the mapping $u \to E(u, \varphi)$ is continuous on F,

(6.9) $|E(\varphi, \varphi)| \geq \alpha|||\varphi|||^2$ for all $\varphi \in \Phi$ (α positive constant).

Theorem 16. *Let* (6.8), (6.9) *hold. If $\varphi \to L(\varphi)$ is a continuous antilinear functional on Φ then there exists an element u in F such that*

$$L(\varphi) = E(u, \varphi) \qquad \text{for all } \varphi \in \Phi. \tag{6.10}$$

Proof. By (6.8), for every $\varphi \in \Phi$,

$$E(u, \varphi) = (u, K\varphi)_F \tag{6.11}$$

and K is a linear operator from Φ into F. K is a one-to-one mapping since $K\varphi = 0$ implies

$$0 = |(\varphi, K\varphi)_F| = |E(\varphi, \varphi)| \geq \alpha|||\varphi|||^2,$$

i.e., $\varphi = 0$. Set $\mathfrak{A} = K\Phi$. The inverse R_0 of K is then defined on $\mathfrak{A}$ and its range is Φ. Setting $K\varphi = a$, $\varphi = R_0a$ we have

$$\alpha|||R_0a|||^2 \leq |E(\varphi, \varphi)| = |(\varphi, K\varphi)_F| \leq ||\varphi||_F||K\varphi||_F \leq c_1|||\varphi|||\ ||K\varphi||_F,$$

i.e., $|||R_0a||| \leq (c_1/\alpha)||a||_F$. Thus R_0 is a continuous linear operator from $\mathfrak{A}$ onto Φ. We can therefore extend R_0 into a continuous linear operator

$\overline{R}_0$ from the closure $\bar{\mathfrak{a}}$ (taken with respect to the topology of F) into $\hat{\Phi}$, the completion of Φ (with respect to the norm $|||\varphi|||$).

The given functional $L(\varphi)$ can also be extended into a continuous antilinear functional on $\hat{\Phi}$, and therefore $L(\varphi) = (((\xi_L, \varphi)))$ for some $\xi_L \in \hat{\Phi}$. In view of (6.11), (6.10) is then equivalent to $(((\xi_L, \varphi))) = (u, K\varphi)_F$ for all $\varphi \in \Phi$, i.e.,

$$(u, a)_F = (((\xi_L, R_0 a))) \qquad \text{for all } a \in \mathfrak{a}. \tag{6.12}$$

It remains to find u satisfying (6.12).

Let P be the projection operator corresponding to $\bar{\mathfrak{a}}$ (as a closed subspace of F). Then $R = \overline{R}_0 P$ is a continuous linear operator from F into $\hat{\Phi}$. Its adjoint R^* (defined by $(((\hat{\varphi}, Rf))) = (R^*\hat{\varphi}, f)_F$ for all $\hat{\varphi} \in \hat{\Phi}$, $f \in F$) is a continuous linear operator from $\hat{\Phi}$ into F, and (6.12) becomes

$$(u, a)_F = (((\xi_L, Ra))) = (R^*\xi_L, a)_F \qquad \text{for all } a \in \mathfrak{a}. \tag{6.13}$$

Thus $u = R^*\xi_L$ satisfies (6.13) and hence it also satisfies (6.10).

We shall now derive some differential inequalities which will be needed in proving the main result of this section, namely, Theorem 17.

Let X be a Hilbert space and let $|v|_X$ be the norm of $v \in X$. A function $f(t)$ from an interval $a < t < b$ into X is said to belong to $L^2(a, b; X)$ if

$$\int_a^b |f(t)|_X^2 \, dt < \infty.$$

We set $L^2(X) = L^2(-\infty, \infty; X)$,

$$||f||_{L^2(X)} = \left\{ \int_{-\infty}^{\infty} |f(t)|_X^2 \, dt \right\}^{1/2}, \tag{6.14}$$

and denote by $D^k(X)$, for any integer $k \geq 0$, the set of functions $f(t)$ all of whose derivatives $f'(t), \ldots, f^{(k)}(t)$ are continuous and belong to $L^2(X)$. $L^2(X)$ is a Hilbert space with the norm (6.14) and $D^k(X)$ is a scalar-product space with the norm

$$||f||_{D^k(X)} = \left\{ \sum_{p=0}^{k} \int_{-\infty}^{\infty} |f^{(p)}(t)|_X^2 \, dt \right\}^{1/2}. \tag{6.15}$$

Note that the derivative $f'(t)$ is defined as the limit, in X, of

$$[f(t + h) - f(t)]/h$$

when $h \to 0$, and that the completeness of $L^2(X)$ can be proved by expressing the functions $f(t)$ in terms of a fixed orthonormal basis of X and then using the completeness of the space $L^2(-\infty, \infty)$ (of complex-valued functions).

The set of all functions $f(t)$ in $D^k(X)$ satisfying $f(t) = 0$ for all $-\infty < t < 0$ is denoted by $D_+^k(X)$. The set of all infinitely differentiable functions $f(t)$ from $-\infty < t < \infty$ into X which vanish outside bounded intervals is denoted by $D(X)$. Finally, we set $Df(t) = f'(t)$.

Lemma 18. *Let $\varphi \in D(X)$. For any $\gamma > 0$*

$$\|e^{-\gamma t}\varphi(t)\|_{L^2(X)} \leq \frac{1}{\gamma}\|e^{-\gamma t}D\varphi(t)\|_{L^2(X)}. \tag{6.16}$$

Proof. By partial integration (which can be justified, for instance, by expanding the functions in terms of a fixed orthonormal basis of X)

$$2\,\text{Re}\int_{-\infty}^{\infty} e^{-2\gamma t}(D\varphi(t), \varphi(t))_X\,dt = 2\gamma\int_{-\infty}^{\infty} e^{-2\gamma t}|\varphi(t)|_X^2\,dt.$$

Hence,

$$\gamma\|e^{-\gamma t}\varphi\|_{L^2(X)}^2 \leq \|e^{-\gamma t}D\varphi\|_{L^2(X)}\|e^{-\gamma t}\varphi\|_{L^2(X)}$$

and (6.16) follows.

Corollary. *If $\varphi \in D(X)$, then for any integer $k > 0$*

$$\|e^{-\gamma t}\varphi(t)\|_{L^2(X)} \leq \frac{1}{\gamma^k}\|e^{-\gamma t}D^k\varphi(t)\|_{L^2(X)}. \tag{6.17}$$

Lemma 19. *Let $A(t)$ $(-\infty < t < \infty)$ be a family of linear continuous operators from X into itself and assume that $t \to (A(t)f, g)_X$ is a k times continuously differentiable function for each pair f, g in X and that, for all $-\infty < t < \infty$,*

$$|D^j(A(t)f, g)_X| \leq \text{const.}\,|f|_X|g|_X \qquad (0 \leq j \leq k). \tag{6.18}$$

Assume, finally, that for all $-\infty < t < \infty$, $f \in X$,

$$\text{Re}\,(A(t)f, f) \geq \alpha|f|_X^2 \qquad (\alpha \text{ positive constant}). \tag{6.19}$$

Then there exists a constant $\gamma_0 > 0$ such that for all $\gamma \geq \gamma_0$ and for all $\varphi \in D(X)$,

$$\text{Re}\int_{-\infty}^{\infty}(e^{-\gamma t}D^k(A(t)\varphi(t)), e^{-\gamma t}D^k\varphi(t))_X\,dt \geq \frac{\alpha}{2}\|e^{-\gamma t}D^k\varphi\|_{L^2(X)}. \tag{6.20}$$

It should be clarified that the notation

$$(D^k(A(t)\varphi(t)), g)_X = D^k(A(t)\varphi(t), g)_X \qquad (g \in X) \tag{6.21}$$

is being used in (6.20), and that if $\varphi(t)$ is k times continuously differentiable then the right-hand side of (6.21) in fact exists and Leibniz' rule holds. Thus,

$$(D^k(A(t)\varphi(t)), g)_X = \sum_{j=0}^{k}\binom{k}{j}(A^{(j)}(t)\varphi^{(k-j)}(t), g)_X \tag{6.22}$$

where (in accordance with (6.21))

$$(A^{(j)}(t)f, g)_X = (D^jA(t)f, g)_X = D^j(A(t)f, g)_X. \tag{6.23}$$

Proof. Using the rule (6.22) we find that the left-hand side of (6.20) is equal to

$$\text{Re}\int_{-\infty}^{\infty}(e^{-2\gamma t}A(t)D^k\varphi(t), D^k\varphi(t))_X\,dt + S(\varphi) \tag{6.24}$$

where

$$S(\varphi) = \sum_{j=1}^{k} \binom{k}{j} \operatorname{Re} \int_{-\infty}^{\infty} (e^{-2\gamma t} A^{(j)}(t) D^{k-j}\varphi(t), D^k\varphi(t))_X \, dt.$$

In view of (6.19), the first term in (6.24) is not less than

$$\alpha \| e^{-\gamma t} D^k \varphi \|^2_{L^2(X)}.$$

Hence it suffices to show that

$$|S(\varphi)| \leq \frac{\alpha}{2} \| e^{-\gamma t} D^k \varphi \|^2_{L^2(X)}. \tag{6.25}$$

Now, from (6.18), (6.23) it follows that $A^{(j)}(t)$ $(0 \leq j \leq k)$ are bounded operators, uniformly with respect to t. Hence, for some constant b_1,

$$|S(\varphi)| \leq b_1 \sum_{j=1}^{k} \| e^{-\gamma t} D^{k-j} \varphi \|_{L^2(X)} \| e^{-\gamma t} D^k \varphi \|_{L^2(X)}.$$

Using (6.17) we get

$$|S(\varphi)| \leq b_1 \left(\sum_{j=1}^{k} \frac{1}{\gamma^j} \right) \| e^{-\gamma t} D^k \varphi \|^2_{L^2(X)}.$$

Taking γ_0 such that $(b_1 k/\gamma_0) \leq \alpha/2$, (6.25) follows if $\gamma \geq \gamma_0$.

Remark 1. Suppose now that φ belongs to the space $D_+(X)$ of all infinitely differentiable functions $\varphi(t)$ defined for $0 \leq t < \infty$ and vanishing for all t sufficiently large. It is convenient to assume that φ is extended as zero to $-\infty < t < 0$. If $\varphi(0) = 0$ then (6.16) holds (with the same proof). Similarly, if $\varphi \in D_+(X)$ and

$$\varphi(0) = D\varphi(0) = \cdots = D^{k-1}\varphi(0) = 0, \tag{6.26}$$

then (6.17) holds and therefore also (6.20) is valid (with the same proof).

Remark 2. If φ is infinitely differentiable for $0 \leq t < \infty$, satisfies (6.26), vanishes for $t < 0$ but does not have a compact support, and if

$$\sum_{m=0}^{k} \int_0^{\infty} |e^{-\gamma t} D^m \varphi(t)|^2_X \, dt < \infty,$$

then by applying (6.20) to $\zeta_T \varphi$ where ζ_T is an infinitely differentiable function satisfying: $\zeta_T = 1$ for $t \leq T$, $\zeta_T = 0$ for $t \geq T + 1$, $|D^m \zeta_T(t)| \leq$ const. for $T \leq t \leq T + 1$, $0 \leq m \leq k$, and then taking $T \to \infty$, the inequality (6.20) follows.

In proving Lemma 19 and the two remarks following the proof, we used the fact that φ is k times continuously differentiable; the fact that φ is differentiable to higher orders was not used at all. Hence:

Lemma 20. *If $A(t)$ is as in Lemma 19 then the inequality (6.20) holds for any φ such that $e^{-\gamma t}\varphi \in D^k_+(X)$.*

We now introduce the concept of weak derivatives.

Definition. Let $g \in L^2(a, b; X)$ and consider the antilinear mapping

$$\psi(t) \to -\int_a^b (g(t), \psi'(t))_X \, dt$$

for all continuously differentiable functions ψ with $\psi(a) = \psi(b) = 0$ (if $b = \infty$ we require ψ to vanish for all t sufficiently large; similar requirement is made in case $a = -\infty$). If the mapping is continuous then there exists a unique function $v(t)$ in $L^2(a, b; X)$ such that

$$-\int_a^b (g(t), \psi'(t))_X \, dt = \int_a^b (v(t), \psi(t))_X \, dt.$$

We then say that $v(t)$ is the *weak derivative* of $g(t)$ and write $g'(t) = v(t)$ (w.d.). Similarly, $g^{(k)}(t) = D^k g(t) = w(t)$ (w.d.) if

$$(-1)^k \int_a^b (g(t), \psi^{(k)}(t))_X \, dt = \int_a^b (w(t), g(t))_X \, dt$$

for all k times continuously differentiable functions $\psi(t)$ whose first $k - 1$ derivatives vanish at $x = a$, $x = b$. Note that $D^k g = 0$ (w.d.) if $g = 0$.

We have previously defined spaces $D^k(X)$. These are not complete spaces. Their completion will be denoted by $H^k(-\infty, \infty; X)$. More generally, we introduce the norm

$$||f||_{H^k(a,b;X)} = \left\{ \sum_{p=0}^{k} \int_a^b |f^{(p)}(t)|_X^2 \, dt \right\}^{1/2} \tag{6.27}$$

and denote the completion (with respect to this norm) of the k times continuously differentiable functions with finite norm by $H^k(a, b; X)$. These spaces are analogous to the spaces $H^k(D)$ introduced in Sec. 1. $H^k(a, b; X)$ can be identified with a subspace of $L^2(a, b; X)$. If $f \in H^k(\alpha, \beta; X)$ for any $a < \alpha < \beta < b$ then we say that $f(t)$ has k *strong derivatives*. If $\{u_m(t)\}$ is a Cauchy sequence in the norm of $H^k(\alpha, \beta; X)$ (the $u_m(t)$ are k times continuously differentiable) which converges to $f(t)$ in $L^2(\alpha, \beta; X)$, then we call the limit, in $L^2(\alpha, \beta; X)$, of $\{D^j u_m(t)\}$ the jth *strong derivative* of $f(t)$ and write $D^j f(t)$ (s.d.), or $f^{(j)}(t)$ (s.d.). Strong derivatives are weak derivatives. The reader may verify (compare Theorem 1, Sec. 1) that if $f(t)$ has k strong derivatives in a finite interval (a, b), and if all these derivatives belong to $L^2(a, b; X)$, then $f \in H^k(a, b; X)$.

The relation

$$\int_a^b (u'(t), v(t))_X \zeta(t) \, dt = -\int_a^b (u(t), v'(t))_X \zeta(t) \, dt - \int_a^b (u(t), v(t))_X \zeta'(t) \, dt \tag{6.28}$$

holds for any continuously differentiable real-valued function $\zeta(t)$ with $\zeta(a) = \zeta(b) = 0$ and for any continuously differentiable $v(t)$, where $u'(t)$ is the weak derivative of u. By completion it also holds for any

$$v \in H^1(a, b; X).$$

If, in particular, $u \in H^1(a, b; X)$ then by taking $v = u$ we get

$$(6.29) \quad (u'(t), u(t))_X + (u(t), u'(t))_X = \frac{d}{dt}(u(t), u(t))_X \quad \text{almost everywhere,}$$

where the derivative on the right-hand side is a weak derivative (of a real-valued function).

For later references we write down the relation

$$(6.30) \quad \frac{d}{dt}(u(t), v)_X = (u'(t), v)_X \quad \text{almost everywhere} \quad \text{(w.d.)}$$

which is a special case of (6.28).

We proceed to the main result of this section. We are given two Hilbert spaces H and V as at the beginning of the present section, and a family $a(t; u, v)$ $(-\infty < t \leq T$, for some $0 < T < \infty)$ of continuous bilinear forms on V, and we assume:

(B_1) For every u, v in V the function $t \to a(t; u, v)$ is k times $(k \geq 1)$ continuously differentiable in t $(-\infty < t \leq T)$, and

$$(6.31) \quad |D_t^j a(t; u, v)| \leq M\|u\|\,\|v\| \qquad (u \in V, v \in V, 0 \leq j \leq k).$$

(B_2) There exists a constant λ such that

$$(6.32) \quad \operatorname{Re} a(t; u, v) + \lambda|v|^2 \geq \alpha\|v\|^2 \qquad (v \in V,\ \alpha \text{ positive constant}).$$

Theorem 17. *Let* (B_1), (B_2) *be satisfied and let* $g(t)$ *be a function satisfying:*

$$(6.33) \qquad g, g', \ldots, g^{(k)} \text{ (s.d.) belong to } L^2(-\infty, T; V),$$
$$g(t) = 0 \qquad \text{if } t < 0.$$

Then there exists a unique function $u(t)$ *satisfying:*

$$(6.34) \qquad u, u', \ldots, u^{(k)} \text{ (s.d.) belong to } L^2(-\infty, T; V),$$
$$u(t) = 0 \qquad \text{if } t < 0,$$

and for any $v \in V$

$$(6.35) \qquad a(t; u(t), v) + \frac{d}{dt}(u(t), v) = (g(t), v) \quad \text{(w.d.)}.$$

Proof. We may assume that $\lambda = 0$ since otherwise we first perform a transformation $u = e^{\lambda t}w$ which transforms (6.35) into an equivalent problem with $a(t; u, v)$ replaced by $a(t; u, v) + \lambda(u, v)$.

To prove uniqueness we suppose $g = 0$ and prove that $u = 0$. Using (6.30) we get from (6.35)

$$a(t; u(t), v) + (u'(t), v) = 0.$$

Choosing $v = u(t)$, then taking the real parts of both sides and noting, by (6.29), that

$$2 \text{ Re } (u'(t), u(t)) = \frac{d}{dt} (||u(t)||^2),$$

we get, upon using (6.32) with $\lambda = 0$,

$$\frac{d}{dt} ||u(t)||^2 \leq 0 \quad \text{(w.d.)}.$$

Since $u(t) = 0$ for $t < 0$, it can be verified (see Problem 3) that $u(t) = 0$ almost everywhere, i.e., $u = 0$.

We now prove existence. We first extend (see Problem 4) $a(t; u, v)$ to $t \geq T$ so that the assumptions (B_1), (B_2) remain valid for $-\infty < t < \infty$ (with possibly different constants α, λ, M). We may assume that $\lambda = 0$. Now, for each t we have, by (6.3),

$$a(t; u, v) = ((A(t)u, v)) \tag{6.36}$$

and, in view of (B_1), (B_2) (with $\lambda = 0$), $A(t)$ satisfies the assumptions of Lemmas 19, 20 with $X = V$.

Denote by $H_+^k(V)$ the completion of $D_+^k(V)$. In view of Lemma 20, (6.20) holds also if $e^{-\gamma t}\varphi \in H_+^k(V)$, i.e.,

$$\text{Re} \int_0^\infty ((e^{-\gamma t}D^k(A(t)\varphi(t)), e^{-\gamma t}D^k\varphi(t)))\, dt \tag{6.37}$$

$$\geq \frac{\alpha}{2} \int_0^\infty ||e^{-\gamma t}D^k\varphi(t)||^2\, dt, \qquad \text{if } e^{-\gamma t}\varphi \in H_+^k(V).$$

Let F be the space of functions u with $e^{-\gamma t}u \in H_+^k(V)$, γ fixed and $\geq \gamma_0$. We provide F with the norm

$$||u||_F = \left\{ \int_0^\infty ||e^{-\gamma t}D^k u(t)||^2\, dt \right\}^{1/2}$$

where $D^k u$ is a strong derivative. We denote by Φ the space of all $\varphi \in F$ with

$$e^{-\gamma t}D^{k+1}\varphi \in L^2(0, \infty; H)$$

and introduce in Φ the norm

$$|||\varphi||| = \{||\varphi||_F^2 + |D^k\varphi(0)|^2\}^{1/2}.$$

(By expanding $D^k\varphi(t)$, $D^{k+1}\varphi(t)$ in terms of a fixed orthonormal basis of H and using Corollary 2 to Theorem 2′, Sec. 2 for $n = 1$, we find that $D^k\varphi(t)$ is a continuous function of t, so that $D^k\varphi(0)$ is uniquely defined by φ.)

For $u \in F$, $\varphi \in \Phi$ consider the bilinear form

$$E(u, \varphi) = \int_0^\infty ((e^{-\gamma t}D^k(A(t)u(t)), e^{-\gamma t}D^k\varphi(t)))\, dt$$
$$- \int_0^\infty (D^k u(t), D[e^{-2\gamma t}D^k\varphi(t)])\, dt$$

and the antilinear functional

$$L(\varphi) = \int_0^\infty ((e^{-\gamma t}D^k g(t), e^{-\gamma t}D^k\varphi(t)))\, dt$$

where $g(t)$ is also extended to $t \geq T$, in such a manner that its first k strong derivatives exist and $e^{-\gamma t}D^k g \in L^2(0, \infty; V)$.

$u \to E(u, \varphi)$ is a continuous linear functional on F. In view of (6.37) one easily gets

$$\operatorname{Re} E(\varphi, \varphi) \geq \alpha_1 |||\varphi|||^2 \qquad (\alpha_1 > 0).$$

Theorem 16 can therefore be applied. It follows that there exist $u \in F$ such that

$$E(u, \varphi) = L(\varphi) \qquad \text{for all } \varphi \in \Phi. \tag{6.38}$$

Let $Z(t)$ be a C^∞ complex-valued function with a compact support and let $z(t)$ be its restriction to $0 \leq t < \infty$. Define

$$Y_k(t) = \begin{cases} 0 & \text{if } t < 0, \\ t^{k-1}/(k-1)! & \text{if } t \geq 0, \end{cases}$$

and set $\psi = Y_k * z$ where "$*$" means convolution. Then

$$e^{-\gamma t}D^j\psi \in L^2(0, \infty) \qquad \text{for } 0 \leq j \leq k,$$

$$\psi(0) = \cdots = \psi^{(k-1)}(0) = 0, \qquad D^k\psi = z.$$

Taking $\varphi(t) = \psi(t)v$, $v \in V$ in (6.38), we get

$$\int_0^\infty e^{-2\gamma t}((D^k(A(t)u(t)), v))\overline{z(t)}\, dt - \int_0^\infty (D^k u(t), v) D[e^{-2\gamma t}\overline{z(t)}]\, dt = \int_0^\infty ((D^k g(t), v))e^{-2\gamma t}\overline{z(t)}\, dt. \tag{6.39}$$

Let $\tilde{u}(t) = u(t)$ if $t \geq 0$, $\tilde{u}(t) = 0$ if $t < 0$. Define $\tilde{g}$ in a similar manner. Then $D^k\tilde{u} = (D^k u)^\sim$ (s.d.).

Using the notation

$$\langle f, g\rangle = \int_{-\infty}^\infty f(t)g(t)\, dt$$

we can now write (6.39) in the form

$$\langle D^k((A(t)\tilde{u}, v)), e^{-2\gamma t}\overline{Z}\rangle - \langle D^k(\tilde{u}, v), D(e^{-2\gamma t}\overline{Z})\rangle = \langle D^k((\tilde{g}, v)), e^{-2\gamma t}\overline{Z}\rangle.$$

Consequently,

$$D^k[((A(t)\tilde{u}, v)) + D(\tilde{u}, v) - ((\tilde{g}, v))] = 0.$$

It follows (see Problem 5) that

$$((A(t)\tilde{u}, v)) + D(\tilde{u}, v) = ((\tilde{g}, v)) \tag{6.40}$$

for all $v \in V$ and for almost all t. Restricting (6.40) to $0 \leq t \leq T$ and using (6.36) we find that u satisfies (6.35). That u satisfies (6.34) is obvious from its definition.

Remark. The uniqueness assertion of Theorem 17 remains true also if $k = 0$; the proof, however, is more involved. This will not be used in the future.

7. The First Initial-boundary Value Problem for Parabolic Equations

In this section we shall apply Theorem 17 with

$$V = \overset{\circ}{H}{}^m(D), \qquad H = L^2(D) \tag{7.1}$$

where D is a bounded domain in R^n. Let $a(t; u, v)$ be a bilinear form in $u, v \in V$ for each value of t in a bounded interval $0 \le t \le T$, and assume:

(E_1) $t \to a(t; u, v)$ is k times continuously differentiable in t $(0 \le t \le T)$ and

$$|D_t^j a(t; u, v)| \le M \|u\| \, \|v\| \qquad (0 \le j \le k) \tag{7.2}$$

for all u, v in V.

(E_2) $$\text{Re } a(t; u, v) \ge \alpha \|v\|^2 \qquad \text{for all } v \in V,$$

where α is a positive constant.

By Lemma 17, Sec. 6, for any $f \in H$ there exists a unique element u in V satisfying

$$a(t; u, v) = (f, v) \qquad \text{for all } v \in V,$$

and

$$A(t)u = f$$

where $A(t)$ is defined by (6.5) with $a(u, v)$ replaced by $a(t; u, v)$. Set $u = A^{-1}(t)f$ and assume:

(E_3) For any $f \in H^p(D)$ $(p = 0, 1, \ldots, r)$, $A^{-1}(t)f$ belongs to $H^{2m+p}(D)$ and the function $t \to A^{-1}(t)f$ is k times continuously differentiable from $[0, T]$ into $H^{2m+p}(D)$; furthermore,

$$|D^j A^{-1}(t)f|_{2m+p} \le \text{const. } |f|_p$$

for all $0 \le t \le T$, $j = 0, 1, \ldots, k$.

The notation $H^k(a, b; X)$ was already introduced in Sec. 6. We now introduce $H_0^k(0, T; X)$ as the subspace of $H^k(-\infty, T; X)$ consisting of all functions $v(t)$ with $v(t) = 0$ for $-\infty < t < 0$. From the assumption (E_3) it follows that $f(t) \to A^{-1}(t)f(t)$ is a continuous linear mapping from $H^k(0, T; H^p(D))$ into $H^k(0, T; H^{2m+p}(D))$ and from $H_0^k(0, T; H^p(D))$ into $H_0^k(0, T; H^{2m+p}(D))$, for $p = 0, 1, \ldots, r$.

Theorem 18. *Let* (E_1)–(E_3) *and* (7.1) *hold and let* $f \in H_0^k(0, T; H^r(D))$, *for some* $k \ge 1$, $r \ge 0$. *Then there exists a unique function* u *satisfying:*

$$u \in H_0^k(0, T; V),$$

$$u(t) \in d(A(t)) \quad \text{almost everywhere,} \quad \text{and} \quad A(t)u(t) \in L^2(0, T; H),$$

$$A(t)u(t) + u'(t) = f(t) \quad \text{almost everywhere} \quad \text{(w.d.)}.$$

The solution has the following properties:

(i) *if* $0 \leq r \leq m$ *then* $u \in H^{k-1}(0, T; H^{2m+r}(D))$;

(ii) *if* $m < r \leq 3m$ *then* $u \in H^{k-1}(0, T; H^{3m}(D))$, *and if* $k \geq 2$ *then also* $u \in H^{k-2}(0, T; H^{2m+r}(D))$, *and similarly for* r *in the intervals* $(2m, 4m)$, $(3m, 5m)$, *etc.*

Proof. The existence and uniqueness follow from Theorem 17. We next write

$$u(t) = A^{-1}(t)(f(t) - u'(t)). \tag{7.3}$$

Since $f \in H_0^k(0, T; H^r(D))$ and $u' \in H_0^{k-1}(0, T; H^m(D))$, if $0 \leq r \leq m$ then $f - u'$ belongs to $H^{k-1}(0, T; H^r(D))$. Using (7.3) and the remark preceding Theorem 18, (i) follows.

Next, if $m \leq r \leq 3m$ then $f - u'$ belongs to $H^{k-1}(0, T; H^m(D))$ and, consequently, $u \in H^{k-1}(0, T; H^{3m}(D))$. If further $k \geq 2$, then

$$u' \in H^{k-2}(0, T; H^{3m}(D))$$

and $f - u'$ then belongs to $H^{k-2}(0, T; H^r(D))$. Hence,

$$u \in H^{k-2}(0, T; H^{2m+r}(D))$$

and the proof of (ii) is completed. It is now clear how to proceed step by step for r in intervals $(2m, 4m)$, $(3m, 5m)$, etc.

It can be shown (see Problem 7) that if $u \in H^j(0, T; H^j(D))$ then $u \in H^j(\Omega)$ where $\Omega = D \times [0, T]$. Using Corollary 2 to Theorem 2′, Sec. 2, we deduce:

Corollary. *For any positive integer* q *there exist positive integers* k, r *such that if the assumptions* (E_1)–(E_3) *hold with these* k, r *then the solution* u *of Theorem* 18 *belongs to* $C^q(\bar{\Omega})$ *where* $\Omega = D \times (0, T)$.

Consider now a parabolic equation with complex coefficients

$$\frac{\partial u}{\partial t} = P(x, t, D_x)u + f \qquad \text{in } D \times (0, T]. \tag{7.4}$$

The condition of uniform parabolicity (in the sense of Petrowski) coincides with the condition that $-P$ is a strongly elliptic operator.

Together with (7.4) we consider the initial condition

$$u(x, 0) = 0 \qquad \text{for } x \in D, \tag{7.5}$$

and the boundary conditions

$$\frac{\partial^j u(x, t)}{\partial \nu^j} = 0 \qquad \text{for } x \in \partial D, 0 < t \leq T \qquad (j = 0, 1, \ldots, m - 1), \tag{7.6}$$

where $2m$ is the order of P, and ν is the normal to ∂D. The problem of solving (7.4)–(7.6) is called the *first initial-boundary value problem* (with zero initial and boundary data).

If all the first $2m$ x-derivatives of u and the first t-derivative of u are continuously differentiable in $D \times (0, T]$ and (7.4) is satisfied, then we say that u is a *classical solution* of (7.4). If, in addition, all the first $m - 1$ x-derivatives of u are uniformly continuous in $D \times [\epsilon, T]$ for any $\epsilon > 0$, u is continuous in $\bar{D} \times [0, T]$ and (7.5), (7.6) hold (the derivatives in (7.6) are defined by continuity), then we say that u is a *classical solution of the first initial-boundary value problem* (7.4)–(7.6). The concept of weak solution can be given analogously to the elliptic case.

We associate a bilinear form $a(t; u, v)$ with P in the same way that we have associated the form $B[\varphi, u]$ with L in Sec. 3 (see (3.9)). In view of Gårding's inequality, (6.32) is satisfied. In solving the first initial-boundary value problem we may assume that $\lambda = 0$ since otherwise we first perform a transformation $u = e^{\lambda t}w$. Thus, we may assume that (E_2) holds.

For any positive integers k, r, if the coefficients of P are sufficiently smooth in $\bar{D} \times [0, T]$ and if ∂D is sufficiently smooth, then (E_1) is satisfied and by Theorem 15, Sec. 5, and its corollary, also (E_3) holds. Applying Theorem 18 and its corollary we see that the unique solution $u(x, t)$ established there is a "weak" solution of the equation (7.4), it belongs to $\mathring{H}^m(D)$ for each t, it belongs to $H_0^k(0, T; \mathring{H}^m(D))$, and it is continuously differentiable in $\bar{D} \times [0, T]$ up to some order N, where $N \to \infty$ if $k \to \infty$, $r \to \infty$. Hence u is a classical solution of (7.4). In view of Lemma 12, Sec. 2, u satisfies (7.6) in the usual sense. Finally, from the definition of $H_0^k(0, T; V)$ and from the continuity of $u(x, t)$ in t it follows that (7.5) is also satisfied.

Noting that the assumption that $f \in H_0^k(0, T; H^r(D))$ is satisfied if f is sufficiently smooth in $\bar{D} \times [0, T]$, and

$$\frac{\partial^j f(x, 0)}{\partial t^j} = 0 \qquad \text{for } x \in D,\ j = 0, 1, \ldots, N_0 \tag{7.7}$$

for some N_0 sufficiently large, we can sum up our results as follows:

Theorem 19. *Let* (7.4) *be a uniformly parabolic equation in* $\bar{D} \times [0, T]$ *and assume that the coefficients of* P *are sufficiently smooth in* $\bar{D} \times [0, T]$, *that the root condition is satisfied and that* ∂D *is sufficiently smooth. Then for any sufficiently smooth function* f *in* $\bar{D} \times [0, T]$ *satisfying* (7.7) *there exists a unique classical solution* $u(x, t)$ *of the system* (7.4)–(7.6).

u belongs to $C^p(\bar{D} \times [0, T])$, *for any given integer* $p > 0$, *provided* ∂D, f, *and the coefficients of* P *are sufficiently smooth* (*depending on* p) *and provided* (7.7) *holds with* N_0 *sufficiently large* (*depending on* p).

We shall now motivate the condition (7.7). Let u be a smooth solution (in $\bar{D} \times [0, T]$) of (7.4)–(7.6). Then (7.4) is satisfied also at the points of the set $B = \{(x, 0); x \in \partial D\}$ and, using (7.5), (7.6), we get $f(x, 0) = 0$ for $x \in \partial D$. Now, in the methods used above, f was considered as an element $f(t)$ of a Hilbert space. Therefore, a reasonable way to impose the necessary condition that $f = 0$ on B seems to be by requiring that $f(0) = 0$. This, however, implies that $f(x, 0) = 0$ for all $x \in D$. From (7.4) it then follows that $\partial u(x, 0)/\partial t = 0$ for $x \in D$. We next differentiate (7.4) once with respect to t and proceed as before. We get $\partial f(x, 0)/\partial t = 0$ on D. It is now clear how (7.7) is obtained as a necessary condition for the success of the previous methods.

If we want the solutions of (7.4)–(7.6) to be smooth in $\bar{D} \times [0, T]$, then we must require that $D_t^i D_x^j f(x, 0) = 0$ for $x \in \partial D$ and $i, j = 0, 1, \ldots, N_0$ (for an appropriate N_0). Introducing the somewhat stronger condition

$$\text{(7.8)} \qquad \frac{\partial^j f(x, 0)}{\partial t^j} = 0 \qquad \text{for } x \text{ in some neighborhood of } \partial D,\ 0 \le j \le N_0,$$

we shall prove that Theorem 19 remains true if (7.7) is replaced by (7.8). We begin by decomposing f into a sum $f_1 + f_2$, where f_1 satisfies (7.7) and f_2 vanishes near $\partial D \times [0, T]$. Let $K(x, t; \zeta, \tau)$ be a fundamental solution of (7.4). Then

$$u_1(x, t) = -\int_0^t \int_D K(x, t; \xi, \tau)\, f_2(\xi, \tau)\, d\xi\, d\tau$$

satisfies:

$$\frac{\partial u_1}{\partial t} = P(x, t, D_x)u_1 + f_2, \qquad u_1(x, 0) = 0,$$

and it tends to zero, with some of its derivatives, as $t \to 0$ and x restricted to be near ∂D. Let w be a function having the same Dirichlet data (on $\partial D \times [0, T]$) as u_1, such that $[\partial w/\partial t - Pw]$ satisfies (7.7) (with a different N_0). By Theorem 19, there exists a solution u_2 of

$$\frac{\partial u_2}{\partial t} = P(x, t, D_x)u_2 + \left[\frac{\partial w}{\partial t} - P(x, t, D_x)w\right] + f_1$$

satisfying (7.5), (7.6). It is clear that $u = u_1 - w + u_2$ is a solution of (7.4)–(7.6). It is a smooth function in $\bar{D} \times [0, T]$ since each of its components is smooth in $\bar{D} \times [0, T]$. We have thus proved:

Corollary to Theorem 19. *Theorem* 19 *remains true if the assumption* (7.7) *is replaced by the weaker assumption* (7.8).

Remark. If the conditions (7.5), (7.6) are replaced by nonhomogeneous conditions, then by introducing a function Φ satisfying these conditions and considering $u - \Phi$, we can obtain the extension of Theorem 19 to the present case.

8. Further Results on Higher-order Equations

Asymptotic Behavior of Solutions. Let u satisfy a parabolic equation

$$\frac{\partial u}{\partial t} = P(x, t, D_x)u + f(x, t) \qquad \text{in } \Omega = D \times (0, \infty) \tag{8.1}$$

and boundary conditions

$$\frac{\partial^j u}{\partial \nu^j} = \varphi_j(x, t) \qquad \text{on } \partial D \times (0, \infty) \qquad (j = 0, 1, \ldots, m - 1), \tag{8.2}$$

where $2m$ is the order of P. Assume for simplicity that P has real coefficients. Analogously to results of Chap. 6 it can be shown that if, as $t \to \infty$,

$$\varphi_j(x, t) \to \varphi_j(x), \qquad f(x, t) \to f(x), \qquad P(x, t, \xi) \to P(x, \xi) \tag{8.3}$$

in some appropriate sense, then also

$$\int_D |u(x, t) - v(x)|^2 \, dx \to 0 \qquad \text{as } t \to \infty,$$

where $v(x)$ is the solution of the Dirichlet problem

$$P(x, D_x)v = f(x) \qquad \text{in } D, \tag{8.4}$$

$$\frac{\partial^j v}{\partial \nu^j} = \varphi_j(x) \qquad \text{on } \partial D \qquad (j = 0, 1, \ldots, m - 1). \tag{8.5}$$

In addition to various differentiability assumptions, it is also assumed that L is *positive*, i.e., for some positive constant γ,

$$- \int_D \varphi(x) P(x, t, D_x) \varphi(x) \, dx \geq \gamma \int_D (\varphi(x))^2 \, dx \tag{8.6}$$

for any real-valued function φ in $C^{2m}(\overline{D})$ which vanishes on ∂D together with its first $m - 1$ normal derivatives.

Under further assumptions on the nature of the convergence in (8.3), it can be shown that, as $t \to \infty$,

$$u(x, t) \to v(x) \qquad \text{uniformly in } \Omega.$$

These results can be extended also to noncylindrical domains, as in Chap. 6. For detailed statements and proofs, see [43].

Uniqueness for Backward Parabolic Equations. Analogously to results of Sec. 7, Chap. 6, the following holds:

If u is a solution of a parabolic equation

$$\frac{\partial u}{\partial t} = P(x, t, D_x)u \qquad \text{in } D \times [0, T],$$

satisfying the boundary conditions

$$\frac{\partial^j u}{\partial \nu^j} = 0 \qquad \text{on } \partial D \times [0, T] \qquad (j = 0, 1, \ldots, m - 1)$$

where $2m$ is the order of P, and if $u(x, T) \equiv 0$ for $x \in D$, then $u(x, t) \equiv 0$ in $D \times [0, T]$. Here P is assumed to be a sum of a self-adjoint operator and an operator of order $\leq m$. For details, see [78], [1]. For extension of results of Sec. 8, Chap. 6 to higher order parabolic equations, see [1].

Analyticity of Solutions. If $f(x, t)$ and the coefficients of $P(x, t, D_x)$ are analytic functions of (x, t) in some domain Ω, then the solutions of the parabolic equation (8.1) (which are already known to be infinitely differentiable functions in Ω) are analytic functions of x.

A more precise result can be stated. First we introduce some notation.

Let $\{M_q\}$ be a monotone increasing sequence of positive numbers. An infinitely differentiable function in a domain Ω is said to belong to the class $C\{M_q; 2m; \Omega\}$ if for any compact subdomain A of Ω there exist constants B_0, B_1, B_2 such that for all $(x, t) \in A$

$$|D_t^p D_x^q f(x, t)| \leq B_0 B_1^p B_2^{|q|} M_{|q|+2mp} \qquad (0 \leq p, |q| < \infty).$$

Similarly we say that $f(x) \in C\{M_q; D\}$ if

$$|D_x^q f(x)| \leq B_0 B^{|q|} M_{|q|} \qquad (0 \leq |q| < \infty)$$

for every compact subdomain of D. The reader may verify that $f(x) \in C(q!;D)$ if and only if $f(x)$ is analytic in D.

We now assume that

$$\binom{p}{i} M_i M_{p-i} \leq \text{const. } M_p \qquad \text{for all } 0 \leq i \leq p,\ 0 \leq p < \infty. \tag{8.7}$$

We then have the following result:

If f and the coefficients of P belong to $C\{M_q; 2m; \Omega\}$ then the same is true of any solution of (8.1).

The special case $M_q = q!$ gives an improvement of the theorem (stated above) for the analytic case.

The proof of the italicized statement is given in [27] in case the principal coefficients are independent of x. The proof in the general case, however, is quite similar provided the cubes with respect to the euclidean norms $(|x|^2 + |t|^2)^{1/2}$ used in [27; § 3] are replaced by cubes with respect to the norm $(|x|^2 + |t|^{1/m})^{1/2}$.

The italicized statement holds (with the same proof) also for parabolic systems. Its analogue for elliptic equations (with $C\{M_q; 2m; \Omega\}$ replaced by $C\{M_q; D\}$) is also valid. In particular, *if $f(x)$ and the coefficients of an elliptic operator $P(x, D_x)$ are analytic, then the same is true of the solutions of $P(x, D_x)u = f(x)$.* This result holds also up to the boundary provided the boundary and the Dirichlet data are analytic; see [88].

PROBLEMS

1. Prove that if $\{u_m\}$ is weakly convergent to u in a Hilbert space H, then there exists a subsequence $\{v_m\}$ whose arithmetic means converge to u in the norm of H.
[*Hint:* Suppose $u = 0$. Set $m_1 = 1$. Let m_2 ($> m_1$) be such that $|(u_{m_1}, u_m)| < \frac{1}{2}$ if $m \geq m_2$. Let m_3 ($> m_2$) be such that $|(u_{m_1}, u_m)| < \frac{1}{3}$, $|(u_{m_2}, u_m)| < \frac{1}{3}$ if $m \geq m_3$, etc. Take $v_n = u_{m_n}$.]

2. Introducing the norm
$$|u|_{j,p} = \left\{ \sum_{|\alpha| \leq j} \int_D |D^\alpha u(x)|^p \, dx \right\}^{1/p}$$
for any positive number $p \geq 1$, extend Lemma 8, Sec. 2, by proving that, for any $\epsilon > 0$, $p \geq 1$,
$$|u|_{j-1,p}^2 \leq \epsilon |u|_{j,p}^2 + C|u|_{0,p}^2$$
where C depends only on ϵ, j, p, D.
[*Hint:* Modify the proof of Lemma 8.]

3. Let $g(t) \in L^2(a, b)$, $g'(t) \in L^2(a, b)$ (w.d.) and assume that $g(t) \geq 0$, $g'(t) \leq 0$ almost everywhere in (a, b), and $g(t) \equiv 0$ in some interval (a, c) with $a < c < b$. Prove that $g(t) = 0$ almost everywhere in (a, b).
[*Hint:* $h(t) = \int_a^t g'(\tau)\, d\tau$ satisfies $g' = h'$ (w.d.). Deduce that $g(t) = h(t) +$ const. almost everywhere, and note that the constant is zero.]

4. Prove that the bilinear form $a(t; u, v)$ satisfying (B_1), (B_2) (Sec. 6) can be extended to $T < t < \infty$ so that the extended form satisfies (B_1), (B_2) in $-\infty < t < \infty$ (with possibly different constants α, λ, M).
[*Hint:* Extend $a(t; u, v)$ by
$$a(T; u, v) + \zeta(t) \sum_{j=1}^{k} \frac{(t-T)^j}{j!} D_t^j a(T; u, v),$$
for appropriate $\zeta(t)$.]

5. Prove: if $f \in L^2(D)$ and all the weak derivatives of f up to order p exist in D, and if all the weak derivatives of order p are zero, then f is a polynomial of degree $\leq p$.
[*Hint:* The mollifiers $J_\epsilon f$ are polynomials of degree $\leq p$.]

6. Let P be a parabolic operator with real coefficients and let (8.6) hold. Let u be a solution in $C^{2m}(\bar{\Omega})$ of (8.1) and of (8.2) with $\varphi_j \equiv 0$, and let
$$\int_D (f(x, t))^2 \, dx \to 0 \qquad \text{as } t \to \infty,$$
u and f being real. Prove that
$$\int_D (u(x, t))^2 \, dx \to 0 \qquad \text{as } t \to \infty.$$

[*Hint:* Set $\psi(t) = \int (u(x, t))^2\,dx$, $\epsilon(t) = \int (f(x, t))^2\,dx$. Using (8.6) derive:

$$\psi'(t) + 2\gamma\psi(t) \leq 2 \int f(x, t)u(x, t)\,dx;$$

then $\psi'(t) + \gamma\psi(t) \leq \dfrac{1}{\gamma}\,\epsilon^2(t)$. Show that $\psi(t) \to 0$.]

7. Prove that if $u \in H^j(a, b; H^j(D))$ then $u \in H^j(\Omega)$ where $\Omega = D \times (a, b)$; more precisely, there exists a function $U(x, t)$ in $H^j(\Omega)$ such that for each t it is equal to $u(t)$, as an element of $H^j(D)$.

[*Hint:* Write $u = u(t) = u(x, t)$ in the form $u(t) = \sum_{m=1}^{\infty} u_m(t)\,e_m$ in $H^j(D)$, where $\{e_m\}$ is an orthonormal basis in $H^j(D)$. $u_m(t) = (u(t), e_m)_j^D$ has j strong derivatives in $L^2(a, b)$, and

$$D_t^i u(t) = \sum_{m=1}^{\infty} [D_t^i u_m(t)]\,e_m \qquad \text{in } H^j(D), \text{ for } i \leq j.$$

Set $W = H^j(a, b; H^j(D))$ and denote the norm of this space by $|\ |_W$. Then

$$\infty > |u|_W^2 = \sum_{i=0}^{j} \int_a^b [|D_t^i u(t)|_0^D]^2\,dt = \sum_{m=1}^{\infty} \sum_{i=0}^{j} \int_a^b |D_t^i u_m(t)|^2\,dt.$$

For any $M < N$, $U_{MN}(x, t) = \sum_{m=M}^{N} u_m(t)\,e_m(x)$ belongs to $H^j(\Omega)$ and

$$|U_{MN}(x, t)|_j^\Omega \leq |U_{MN}(x, t)|_W \to 0 \qquad \text{if } M \to \infty.$$

Hence there exists a function $U(x, t)$ in $H^j(\Omega)$ such that $|U_{1N}(x, t) - U(x, t)|_0^\Omega \to 0$ if $N \to \infty$. By Fatou's lemma, for almost all t

$$\liminf_{N\to\infty} \int_D |U_{1N}(x, t) - U(x, t)|^2\,dx = 0.$$

Noting that for every t,

$$\int_D |U_{1N}(x, t) - u(x, t)|^2\,dx \to 0 \qquad \text{if } N \to \infty,$$

deduce that for almost all t, $U(x, t) = u(x, t)$ almost everywhere.]

APPENDIX

NONLINEAR EQUATIONS

Numbers in brackets will refer to the bibliography for the appendix.

We shall give here a brief account of recent developments in the theory of nonlinear second-order elliptic and parabolic equations in any number of variables. Some a priori estimates on the Hölder exponents and the Hölder coefficients of the solutions play a fundamental role in these developments. We first describe the estimates. Consider an elliptic equation of the form

$$\sum_{i,j=1}^{n} \frac{\partial}{\partial x_i}\left(a_{ij}(x)\frac{\partial u}{\partial x_j}\right) = 0 \tag{1}$$

in a bounded domain Ω of R^n, and let $a_{ij}(x)$ be measurable functions in Ω and

$$\lambda^{-1}\sum_{i=1}^{n} \xi_i^2 \le \sum_{i,j=1}^{n} a_{ij}(x)\xi_i\xi_j \le \lambda \sum_{i=1}^{n} \xi_i^2 \tag{2}$$

for all $x \in \Omega$, ξ real, where $\lambda > 1$ is some constant. By a solution u of (1) we mean a strongly differentiable function $u(x)$ satisfying

$$\sum \int a_{ij} \frac{\partial u}{\partial x_j}\frac{\partial \varphi}{\partial x_i} = 0$$

for any $\varphi \in C_c^\infty(\Omega)$.

Theorem 1 (de Giorgi). *Let $u(x)$ be a solution of* (1) *such that*

$$\int_\Omega u^2(x)\,dx \le M^2.$$

Then there exist positive constants A and α $(0 < \alpha < 1)$ depending only on λ, n such that

$$|u(x) - u(y)| \le \frac{AM}{\delta^{\alpha+n/2}}|x - y|^\alpha$$

for any $0 < \delta < 1$ such that the balls of radius δ about x and y lie in Ω.

Theorem 1 is due to de Giorgi [3]. The important feature is the fact that A, α depend only on λ, n.

Consider a parabolic equation of the form

$$\frac{\partial u}{\partial t} - \sum_{i,j=1}^{n} \frac{\partial}{\partial x_i}\left(a_{ij}(x,t)\frac{\partial u}{\partial x_j}\right) = 0 \tag{3}$$

in a strip $0 < t \leq T$, and assume that

$$\lambda^{-1} \sum_{i=1}^{n} \xi_i^2 \leq \sum_{i,j=1}^{n} a_{ij}(x, t)\xi_i\xi_j \leq \lambda \sum_{i=1}^{n} \xi_i^2 \tag{4}$$

for all $x \in R^n$, $0 \leq t \leq T$, ξ real, where $\lambda > 1$ is a constant. It is further assumed that the a_{ij} have, say, three continuous bounded derivatives in the whole strip $0 \leq t \leq T$.

Theorem 2 (Nash). *Let $u(x, t)$ be a (classical) solution of (3) in $0 \leq t \leq T$ and let $|u| \leq M$. There exist positive constants A and α $(0 < \alpha < 1)$ such that for all $0 < \tau < t \leq T$, $x \in R^n$, $y \in R^n$,*

$$|u(x, t) - u(y, \tau)| \leq AM \left\{ \left(\frac{|x - y|}{\sqrt{\tau}} \right)^{\alpha} + \left(\frac{t - \tau}{\tau} \right)^{\alpha/(2\alpha+2)} \right\}.$$

Theorem 2′ (Nash). *Let $u(x)$ be a (classical) solution of (1) in a bounded domain Ω and assume that the coefficients a_{ij} are sufficiently smooth (say, $a_{ij} \in C^3(\bar{\Omega})$) and that (2) holds. There exist positive constants A and α $(0 < \alpha < 1)$ depending only on λ, n such that if $|u| \leq M$ in Ω, then*

$$|u(x) - u(y)| \leq AM \left(\frac{|x - y|}{d} \right)^{\alpha}$$

for all x, y in Ω, where $d = \min(d_x, d_y)$ and d_x is the distance from x to the boundary of Ω.

Theorems 2, 2′ are due to Nash [14]. The methods used by de Giorgi and Nash are entirely different from each other. Nash derived Theorem 2 by using fundamental solutions, and then derived Theorem 2′ by using Theorem 2. The method of de Giorgi, which is simpler, is based upon inequalities, obtained directly from the differential equation for u, by choosing appropriate test functions. Moser [10] has found a simpler proof of de Giorgi's theorem.

By further extending the method of de Giorgi, Stampacchia [18], [19] and Morrey [9] have extended Theorem 1 to general elliptic equations of the second order

$$\sum_{i,j=1}^{n} \frac{\partial}{\partial x_i} \left(a_{ij}(x) \frac{\partial u}{\partial x_j} \right) + \sum_{i=1}^{n} b_i(x) \frac{\partial u}{\partial x_i} + c(x)u = f(x) \tag{5}$$

with coefficients b_i, c in some L^p classes. Morrey derived the corresponding boundary estimates and then solved the Dirichlet problem under very weak assumptions on the coefficients. The Neumann problem was considered by Stampacchia [19].

Ladyzhenskaja and Uraltseva [6], [7] extended the results of de Giorgi to parabolic equations

$$\sum_{i,j=1}^{n} \frac{\partial}{\partial x_i} \left(a_{ij}(x, t) \frac{\partial u}{\partial x_j} \right) + \sum_{i=1}^{n} b_i(x, t) \frac{\partial u}{\partial x_i} + c(x, t)u - \frac{\partial u}{\partial t} = f(x, t) \tag{6}$$

in a bounded cylinder $\{(x, t); x \in \Omega, 0 < t < T\}$ and also to some systems of parabolic equations. The method of Moser was extended to elliptic equations of the form (5) and also to parabolic equations of the form (6) (in bounded cylinders) by Kruzhkov [4*a*] (see also Moser [11]). Oleinik and Kruzhkov [15] extended Theorems 2, 2′ to parabolic equations (in bounded cylinders) and to elliptic equations of the general forms (6) and (5) respectively.

The classical Harnack inequality states that given a domain $\Omega \subset R^n$, for any compact subdomain Ω_0 there corresponds a positive constant B depending only on Ω_0 such that the inequality

$$u(x) \leq Bu(y) \tag{7}$$

holds for all x, y in Ω_0 and for all harmonic functions u which are positive in Ω. (7) is equivalent to

$$\max_{\Omega_0} u \leq B \min_{\Omega_0} u. \tag{7'}$$

Moser [12] extended Harmack's inequality to positive solutions of (1) with the constant B depending only on Ω_0 and λ. He also obtained in [13] the analogous results for the parabolic equation (3). The theorem of de Giorgi follows fairly easily from the Harnack inequality.

The estimates of de Giorgi and Nash and their extensions to general second-order equations, as well as the Harnack inequalities of Moser, have important applications for both linear and nonlinear equations. We first mention some of the applications to linear equations.

Existence theorems and properties of the solutions of linear equations with discontinuous coefficients were established by Morrey [9], Stampacchia [18], [19], Ladyzhenskaja and Uraltseva [6], [7], and Littman, Stampacchia, and Weinberger [8]. In the course of proving Theorem 1 (see [3], [10]) one obtains an inequality from which the following theorem of Liouville type follows immediately.

Theorem 3. *Let $u(x)$ be a solution in R^n of* (1) *and let* (2) *hold for all $x \in R^n$. If u is bounded then $u \equiv$* const.

The following theorem can be derived with the aid of Harnack's inequality (see [12]).

Theorem 4. *Let $u(x)$ be a bounded solution of* (1) *in the domain $|x| > 1$ and let* (2) *hold for all x with $|x| > 1$. Then $\lim_{|x| \to \infty} u(x)$ exists.*

The Harnack inequality proved in [13] states:

Theorem 5 (Moser). *Let D be a domain in R^n and let D_0 be a convex subdomain of D which has a distance $\geq d$ from the boundary of D, $d > 0$. If u is a positive solution of* (3) *for $x \in D$, $0 < t \leq T$ and if* (4) *holds, then*

$$\log \frac{u(x', t')}{u(x'', t'')} \leq A \left(\frac{|x'' - x'|^2}{t'' - t'} + \frac{t'' - t'}{d^2} + 1 \right) \tag{8}$$

for $0 < d^2 \leq t' < t'' \leq T$, $x' \in D_0$, $x'' \in D_0$ *where* A *is a positive constant depending only on* λ, n.

The coefficients a_{ij} in Theorem 5 are only assumed to be measurable functions, and the solution u is taken in the sense that

$$\int \left[\Sigma\, a_{ij} \frac{\partial u}{\partial x_j} \frac{\partial \varphi}{\partial x_i} + \frac{\partial u}{\partial t} \varphi \right] dx\, dt = 0$$

for all $\varphi \in C^\infty$ with compact support in $D \times (0, T)$.

We shall give here an extension of Theorem 5 and an application to the Cauchy problem for positive solutions. Consider the parabolic equation

$$L_0 u \equiv \sum_{i,j=1}^{n} \frac{\partial}{\partial x_i} \left(a_{ij}(x, t) \frac{\partial u}{\partial x_j} \right) + \sum_{i=1}^{n} b_i(x, t) \frac{\partial u}{\partial x_i} - \frac{\partial u}{\partial t} = 0 \tag{9}$$

and assume that a_{ij}, b_i are measurable functions for $x \in D$, $0 \leq t \leq T$, that (4) holds, and that the b_i are bounded, i.e.,

$$|b_i(x, t)| \leq B.$$

If $u(x, t)$ satisfies (9) for $x \in D$, $0 < t \leq T$, then it also satisfies the equation

$$\sum_{i,j=1}^{n} \frac{\partial}{\partial x_i} \left(a_{ij}(x, t) \frac{\partial u}{\partial x_j} \right) + \sum_{i=1}^{n} \frac{\partial}{\partial y} \left(b_i(x, t) y \frac{\partial u}{\partial x_i} \right) + \sum_{i=1}^{n} \frac{\partial}{\partial x_i} \left(b_i(x, t) y \frac{\partial u}{\partial y} \right) + \frac{\partial^2 u}{\partial y^2} - \frac{\partial u}{\partial t} = 0 \tag{10}$$

in the cylinder $G_d = \{(x, y, t); x \in D,\ 0 \leq y \leq d,\ 0 < t \leq T\}$, for any $d > 0$. (This observation was made in [15].) The equation (10) has the form (3) and it satisfies the condition of uniform parabolicity (i.e., the analogue of (4)) in G_d if d is sufficiently small, i.e., if

$$d \leq \frac{1}{2B} \min \left(\frac{1}{\lambda}, \frac{1}{n} \right). \tag{11}$$

Applying Theorem 5 to the solution $u(x, t)$ of (10) we get:

Theorem 5′. *Let* D *be a domain in* R^n *and let* D_0 *be a convex subdomain of* D *which has a distance* $\geq d$ *from the boundary of* D, *where* d *is any positive number satisfying* (11). *If* u *is a positive solution of* (9) *for* $x \in D$, $0 < t \leq T$, *then for any* $0 < d^2 \leq t' < t'' \leq T$, $x' \in D_0$, $x'' \in D_0$, *the inequality* (8) *holds;* A *is a constant depending only on* λ, B, n.

Corollary 1. *If the assumptions* $(A_1)'$, $(A_3)'$ *of Chap. 1, Secs. 6, 8, are satisfied with* $D = R^n$, $T_0 = 0$, $T_1 = T$, *then the inequality* (2.4.14) *holds.*

Consequently, Theorem 13 of Sec. 4, Chap. 2, and its corollary 2 remain true even if the assumption (2.4.7) is omitted (see the last paragraph of Sec. 4, Chap. 2).

Proof. It suffices to prove (2.4.14) for $L_0 + c_0$ where L_0 is defined by (9) and $c_0 = \Sigma\, \partial b_i/\partial x_i$. Indeed, the proof for $L_0 + c$, with general c, then follows by the proof of the italicized statement preceding Theorem 12, Sec. 4, Chap. 2.

Now, by Theorem 15, Sec. 8, Chap. 1, as a function of (ξ, τ) $v(\xi, \tau) \equiv \Gamma(0, t; \xi, \tau)$ satisfies the equation which is the adjoint of $(L_0 + c_0)u = 0$, i.e.,

$$\sum_{i,j=1}^{n} \frac{\partial}{\partial \xi_i}\left(a_{ij}(\xi, \tau)\, \frac{\partial v}{\partial \xi_j}\right) - \sum_{i=1}^{n} b_i(\xi, \tau)\, \frac{\partial v}{\partial \xi_i} + \frac{\partial v}{\partial \tau} = 0.$$

Substituting $-\tau$ for τ we see that Theorem 5′ can be applied with $D = R^n$, D_0 arbitrary, and d sufficiently small (depending on l.u.b. $|b_i|$). Taking $x' = 0$, $x'' = \xi$, we obtain

$$v(\xi, \tau) \geq v(0, t - d^2) \exp\left\{-A\left[\frac{|\xi|^2}{t - d^2 - \tau} + \frac{t - d^2 - \tau}{d^2} + 1\right]\right\}$$

for $0 \leq \tau < t - d^2$. Hence, if $0 \leq \tau \leq t - 2d^2$, $2d^2 < t < T$, $\xi \in R^n$,

$$v(\xi, \tau) \geq A' \exp\left[-A\, \frac{|\xi|^2}{d^2}\right]$$

where A' is a positive constant. This completes the proof of (2.4.14).

Corollary 2. *Let L (given by* (2.1.1)*) be uniformly parabolic in* $\Omega = R^n \times [0, T]$, let a_{ij}, $\partial a_{ij}/\partial x_k$, b_i *be uniformly bounded continuous functions in* Ω, *and let* $c \equiv 0$ *in* Ω. *If* $Lu = 0$ *in* $\Omega_0 = R^n \times (0, T]$ *and* $u(x, 0) \equiv 0$ *on* R^n, *and if u satisfies* (2.4.2), *then* $u(x, t) \equiv 0$ *in.*

Proof. It was shown by Aronson [0] that Theorem 5 remains true even if u is taken as a solution in some weaker sense. Thus, in particular, if $u(x, 0) \equiv 0$ and if u is extended by zero to $t < 0$, then (8) holds for $-\infty < t' < t'' \leq T$. Theorem 5′ can be extended in the same way. Now, by Theorem 9, Sec. 4, Chap. 2, $u(x, t) \geq 0$ in Ω. Writing (2.1.1) (with $c \equiv 0$) in the form (9) and using the extended form of Theorem 5′, it follows that, in any domain $R^n \times (0, T_0]$ $(T_0 < T)$, $u(x, t) \leq B_0 \exp [\beta_0 |x|^2]$ where B_0, β_0 are constants depending on T_0. By Theorem 10, Sec. 4, Chap. 2, $u \equiv 0$ in Ω.

Aronson [0] has proved that if u is a weak non-negative solution of (3) and if $u(x, 0) \equiv 0$ on R^n, then $u \equiv 0$ in Ω. He assumes only that the a_{ij} are measurable and satisfy (4).

We now mention the main applications of the previous a priori estimates (and the methods of deriving them) to nonlinear equations. Ladyzhenskaja and Uraltseva [5] established existence theorems for quasi-linear elliptic equations of the form

$$\sum_{i,j=1}^{n} \frac{\partial}{\partial x_i}\left(a_{ij}(x, u, u_x)\frac{\partial u}{\partial x_j}\right) = f(x, u, u_x), \tag{12}$$

where $u_x = (\partial u/\partial x_1, \ldots, \partial u/\partial x_n)$. They have established in [6], [7] analogous results for quasi-linear parabolic equations of either the form

$$\sum_{i,j=1}^{n} \frac{\partial}{\partial x_i}\left(a_{ij}(x, t, u, u_x)\frac{\partial u}{\partial x_j}\right) - \frac{\partial u}{\partial t} = f(x, t, u, u_x) \tag{13}$$

or the form

$$\sum_{i,j=1}^{n} a_{ij}(x, t, u, u_x)\frac{\partial^2 u}{\partial x_i\, \partial x_j} - \frac{\partial u}{\partial t} = f(x, t, u, u_x), \tag{14}$$

and for some systems of quasi-linear parabolic equations. Oleinik and Kruzhkov [15] have derived analogous existence theorems for parabolic equations of the form

$$\sum_{i,j=1}^{n} \frac{\partial}{\partial x_i}\left(a_{ij}(x, t, u)\frac{\partial u}{\partial x_j}\right) - a(x, t, u)\frac{\partial u}{\partial t} = f(x, t, u, u_x). \tag{15}$$

In [7] the crucial step is the derivation of an a priori bound on u_x, where u is a solution of (14). By extending the method of de Giorgi, Ladyzhenskaja and Uraltseva obtained such a bound in terms of $l.u.b.\ |u_x|$ on the lateral boundary. The latter quantity however can be estimated by a device of S. Bernstein which involves the maximum principle. The considerations for (13) in [6] and for (15) in [15] are of a similar nature as those of [7], although they are quite different technically.

The results of [7] show, in particular, that Theorem 9 of Sec. 4, Chap. 7 remains true for cylindrical domains if the condition (7.4.17) is replaced by the conditions

$$\begin{gathered} |f(x, t, u, w)| + \sum\left|\frac{\partial}{\partial x_i} f(x, t, u, w)\right|(1 + |w|) \\ + \sum\left|\frac{\partial}{\partial w_j} f(x, t, u, w)\right|(1 + |w|) \le A(|u|)(1 + |w|)^2, \\ \left|\frac{\partial}{\partial u} f(x, t, u, w)\right| \le A(|u|)(1 + |w|)^{2-\epsilon} \qquad (\text{for some } \epsilon > 0), \end{gathered} \tag{16}$$

and if L, S, ψ are sufficiently smooth. The theorem remains true (for cylindrical domains and L, S, ψ sufficiently smooth) also in case (7.4.17) is replaced by

$$|f(x, t, u, w)| \le A(|u|)(1 + |w|^{2-\epsilon}) \qquad (\text{for some } \epsilon > 0). \tag{17}$$

This in fact is a special case of an existence theorem for semilinear parabolic equations of any order proved by Friedman [2].

Other applications of the a priori estimates were derived by Serrin [16],

[17] in his study of the local behavior of solutions of quasi-linear elliptic equations. He also obtained a Harnack inequality for quasi-linear equations.

Solutions of quasi-linear parabolic equations may not exist for all $t < \infty$; see, for instance, Problem 8, Chap. 2; also [2]. Filippov [1] and Kruzhkov [4] proved the existence of solutions for all $t < \infty$ for certain classes of quasilinear parabolic equations in case $n = 1$. From [1] it follows that Theorem 9, Sec. 4, Chap. 3, does not remain true if (7.4.17) is replaced by $|f(x, t, u, w)| \leq A(|u|)(1 + |w|)^{2+\epsilon}$ for some $\epsilon > 0$.

APPENDIX BIBLIOGRAPHY

[0] D. G. Aronson, "Uniqueness of positive solutions of second order parabolic equations," *Ann. Polon. Math.*, **16** (1965).

[1] A. F. Filippov, "On conditions for existence of solutions of quasi-linear parabolic equations," *Doklady Akad. Nauk SSSR* (N.S.), **141** (1961), 568–570.

[2] A. Friedman, "Remarks on nonlinear parabolic equations," *Nonlinear Partial Differential Equations and Applications,* Proc. Sympos. Amer. Math. Soc., 1965.

[3] E. de Giorgi, "Sulla differenziabilita e l'analicita delle estremali degli integrali multipli regolari," *Mem. Accad. Sci. Torino. Cl. Sci. Mat. Nat.*, Ser. 3, **3** (1957), 25–43.

[4] S. N. Kruzhkov, "On the Cauchy problem in the large for some nonlinear differential equations of the second order," *Doklady Akad. Nauk SSSR* (N.S.), **132** (1960), 36–39.

[4*a*] ———, "A priori estimates for generalized solutions of second order elliptic and parabolic equations," *Doklady Akad. Nauk SSSR* (N.S.), **150** (1963), 748–751.

[5] O. A. Ladyzhenskaja and N. N. Uraltseva, "Quasi-linear elliptic equations and variational problems with many independent variables," *Uspehi Math. Nauk SSSR,* **16,** no. 1 (1961), 19–60.

[6] ———, "Boundary problems for linear and quasi-linear parabolic equations, I, II," *Izvest. Akad. Nauk SSSR,* **26** (1962), 5–52, 753–780.

[7] ———, "Boundary problems for linear and quasi-linear equations and systems of parabolic type, III," *Izvest. Akad. Nauk SSSR,* **27** (1963), 161–240.

[8] W. Littman, G. Stampacchia, and H. Weinberger, "Regular points for elliptic equations with discontinuous coefficients," *Ann. Scuola Norm. Sup. Pisa,* **17** (1963), 47–79.

[9] C. B. Morrey, "Second order elliptic equations in several variables and Hölder continuity," *Math. Zeit.*, **72** (1959), 146–164.

[10] J. Moser, "A new proof of de Giorgi's theorem concerning the regularity problem for elliptic differential equations," *Comm. Pure Appl. Math.*, **13** (1960), 457–468.

[11] ———, "On the regularity problem for elliptic and parabolic differential equations," *Symp. Partial Differential Equations and Continuum Mechanics*, Univ. Wisconsin Press, 1961, 159–169.

[12] ———, "On Harnack's theorem for elliptic equations," *Comm. Pure Appl. Math.*, **14** (1961), 577–591.

[13] ———, "A Harnack's inequality for parabolic differential equations," *Comm. Pure Appl. Math.*, **17** (1964), 101–134.

[14] J. Nash, "Continuity of solutions of parabolic and elliptic equations," *Amer. J. Math.*, **80** (1958), 931–953.

[15] O. A. Oleinik and S. N. Kruzhkov, "Quasi-linear parabolic equations of the second order with many independent variables," *Uspehi Math. Nauk SSSR*, **16,** no. 5 (1961), 115–155.

[16] J. Serrin, "A Harnack inequality for nonlinear equations," *Bull. Amer. Math. Soc.*, **69** (1963), 481–486.

[17] ———, "Local behavior of solutions of quasi-linear equations," to appear.

[18] G. Stampacchia, "Contributi alla regolarizzazione dell soluzioni dei problemi al contorno per equazioni del secondo ordine ellitiche," *Ann. Scuola Norm. Sup. Pisa*, **12** (1958), 223–245.

[19] ———, "Problemi al contorno ellitichi con dati discontinui dotati di soluzioni Holderiane," *Annali Mat. Pure Appl.*, **51** (1960), 1–38.

BIBLIOGRAPHICAL REMARKS

Chapters 1, 9. The parametrix method is due to Levi [75]. Fundamental solutions for second-order parabolic equations were constructed by Gevrey [48] for $a_{ij} = \delta_{ij}$, Rothe [106] for coefficients independent of t, Dressel [20], [21] for sufficiently smooth coefficients, and Pogorzelski [96] and Aronson [3] for Hölder continuous coefficients. The construction in Chap. 1 is based essentially on [96]. Fundamental solutions for parabolic systems of any order were constructed by Petrowski [95], Ladyzhenskaja [70], Eidelman [23]–[26], Slobodetski [112], Aronson [4], [5] and Pogorzelski [102], [103]. In [95], [70], [23] the coefficients depend only on t, in [24] the coefficients are sufficiently smooth, and in [25], [26], [112], [4], [5], [102], [103] the coefficients are only Hölder continuous (essentially as in Chap. 9).

Existence for the Cauchy problem by the method of semigroups was given by Mizohata [83]. Uniqueness for the Cauchy problem was first established by Tychonov [116] for the heat equation. Petrowski [95] proved uniqueness, for one parabolic equation of any order with coefficients depending only on t, for bounded solutions. Ladyzhenskaya [70] extended this result to solutions bounded by $0[\exp h|x|^q]$, $q = 2p/(2p - 1)$. Uniqueness theorems for general parabolic systems were proved in [23]–[26], [112], [4], [5] and by Friedman [34], under various pointwise and integral growth conditions similar to that of [70].

Differentiability of solutions of parabolic systems was proved by Eidelman [24] (using fundamental solutions), Mizohata [84], and Friedman [29] (using a priori estimates).

Fundamental solutions for general elliptic systems were constructed by Lopatinski and John; see [58], [60] and the references given there. Problem 7, Chap. 9, is based on [24], [29].

Chapter 2. The strong maximum principle for elliptic equations is due to Hopf [51], and for parabolic equations (Secs. 1, 2) it is due to Nirenberg [89]. Theorems 9, 10 are due to Krzyzanski [65], [66]. Theorem 13 is due to Friedman [34]. Theorem 14 was proved by Viborni [117] and Friedman [31]. Theorem 16 is due to Westphal [120]. Theorem 21 was proved by Hopf [53] and Oleinik [92]. Problems 2–4 are based on [54] and Problems 5, 6 are based on [67].

Chapters 3, 4. The Schauder estimates for second-order elliptic equations were derived by him in [108], [109]. The proof was simplified by Miranda [82]. Using a method of Hopf [52], Douglis and Nirenberg [19] further simplified the proof and also extended the interior estimates to very general elliptic systems. Agmon,

Douglis and Nirenberg [2] derived boundary estimates for elliptic systems satisfying general boundary conditions; see also Browder [11], [12].

For parabolic equations of the second order, Ciliberto [13] derived estimates of the Schauder type in the case of one space dimension. Barrar [6], [7] derived the estimates in the general case, and Friedman [29], [30] derived them by a simplified method. In [29] the interior estimates were derived also for general parabolic systems.

The method of continuity as presented in Secs. 3, 4 is based on Friedman [30] and is analogous to a method for elliptic equations due to Nirenberg [91]; see also Courant and Hilbert [17]. The results of Sec. 5, Chap. 3, are based on Friedman [29]; the method used is due to Hopf [52] and was used by Douglis and Nirenberg [19] for elliptic systems.

For references on Green's function, see [33].

The material of Chap. 4 is based on Friedman [29], [30]. Problems 1–13 follow Friedman [33].

For the treatment of the Laplace equation by the method analogous to that of Problems 1–13, see Kellogg [63] (this is the *method of balayage* due to Poincaré). For a modified method due to Perron, see Courant and Hilbert [17]. The method of integral equations in solving both the Dirichlet and the Neumann problems is given in Kellogg [63] for the Laplace equation, and in Miranda [82] for general second-order elliptic equations. For details on barriers, see [63].

Chapter 5. The results of Secs. 2–4 and Problems 2–4 are based on Pogorzelski [98]–[101]. For results analogous to those of Secs. 2–4 for elliptic equations, see Miranda [82] and Pogorzelski [97], [104]. Green's and Neumann's functions were constructed by Itô [55]–[57] using the parametrix method.

The third boundary value problem for elliptic equation was solved by Giraud [49]. This problem leads to integral equations which are not of Fredholm type. Pagni [94] solved the third initial-boundary value problem for the heat equation. Similar results for general second-order parabolic equations were sketched by Lipko [79]. More general problems were treated by Lions [77] but the solution is usually taken in some generalized sense.

Chapter 6. The results of Secs. 1–6 are due to Friedman [35], [36], [43]. The results of Sec. 7 are due to Lees and Protter [73]. The results of Sec. 8 are due to Protter [105]; overlapping results were established by Cohen and Lees [14] and, in particular, Problem 7 is based on [14]. Agmon and Nirenberg [1] studied solutions of ordinary differential equations and inequalities in a Banach space and, as an application, they derived results on the asymptotic behavior of solutions of equations of various types. Mizohata [85] proved a uniqueness theorem for the Cauchy problem for backward parabolic equations.

Chapter 7. Schauder's fixed point theorem was proved in [107]. The fixed point theorem of Leray and Schauder was proved in [74]. The results of Secs. 2–4 are based on Friedman [32], [42], and the results of Sec. 5 are based on Friedman [37]. Kaplan [61] derived asymptotic bounds on solutions of quasi-linear differential inequalities. Cordes [15], [16] derived a $(1 + \delta)$-estimate for solutions of elliptic equations which is stronger than the analogous estimate for parabolic equations. His result can be used to solve some quasi-linear elliptic equations.

Chapter 8. The developments of Secs. 1–3, 5 are due to Friedman [38]–[41]. The method given in Sec. 4 is due to Douglas [18]. It was further developed by Kyner [68], [69] who proved an existence theorem also for nonlinear equations. For other methods, see Kolodner [64], Sestini [110], [111], and Oleinik [93]. Oleinik considers many-phase problems for nonlinear equations, but the solutions satisfy the boundary condition only in some generalized sense.

Chapter 10. The concepts of weak and strong solutions and of mollifiers were introduced by Sobolev [113] and Friedrichs [45]. Lemma 8 was proved by Ehrling [22]. The present proof is due to Nirenberg [90]. Theorem 2 was proved by Sobolev in [113]. The Hilbert space approach to the Dirichlet problem was originated by Weyl [121]. It was developed by Vishik [118], [119] and Gårding [47]. Theorem 4 was proved in [47] and Theorem 5 was introduced by Lax and Milgram [72]. The interior differentiability of weak solutions (of elliptic systems) was proved by different methods by Friedrichs [46], John [59], Browder [8], Lax [71], and Nirenberg [90]. The differentiability near the boundary was proved by Nirenberg [90] and Browder [9]. Previously Morrey [86] had proved the differentiability up to the boundary for second-order elliptic systems.

In [2], [12] Theorem 13 was established not only for the L^2 norm but also for any L^p norm and, in [2], also under general boundary conditions. Overlapping results were obtained by other authors; see [12] for detailed references.

Sections 1–5 are based on Nirenberg [90], and Secs. 6, 7 are based on Lions [77]. In [76] Lions solved the first initial-boundary value problem also in noncylindrical domains. Lemma 19 is due to Trèves [115; 102–104]. The existence and differentiability of weak solutions (for parabolic equations) were studied also by Browder [10]. Lax and Milgram [72] solved the first initial-boundary value problem for equations with coefficients depending only on x by the method of semigroups. The case of general variable coefficients was treated by Sobolevski [114] who also treated quasi-linear equations.

The analyticity, up to the boundary, of solutions of analytic nonlinear elliptic equations was established by Morrey [87] and Friedman [28].

BIBLIOGRAPHY

[1] S. Agmon and L. Nirenberg, "Properties of solutions of ordinary differential equations in Banach spaces," *Comm. Pure Appl. Math.*, **16** (1963), 121–239.

[2] S. Agmon, A. Douglis, and L. Nirenberg, "Estimates near the boundary for solutions of elliptic partial differential equations satisfying general boundary conditions I," *Comm. Pure Appl. Math.*, **12** (1959), 623–727.

[3] D. G. Aronson, "The fundamental solution of a linear parabolic equation containing a small parameter," *Ill. J. Math.*, **3** (1959), 580–619.

[4] ———, "On the initial value problem for parabolic systems of differential equations," *Bull. Amer. Math. Soc.*, **65** (1959), 310–318.

[5] ———, "Uniqueness of solutions of the initial value problem for parabolic systems of differential equations," *J. Math. and Mech.*, **11** (1962), 403–420.

[6] R. B. Barrar, *Some estimates for solutions of linear parabolic equations*, University of Michigan Thesis, 1952.

[7] ———, "Some estimates for solutions of parabolic equations," *J. Math. Analys. and Appl.*, **3** (1961), 373–397.

[8] F. E. Browder, "Strongly elliptic systems of differential equations," Contributions to the Theory of Partial Differential Equations, *Ann. Math. Studies*, no. 33 (Princeton University Press, 1954), 15–51.

[9] ———, "On the regularity properties of solutions of elliptic differential equations," *Comm. Pure Appl. Math.*, **9** (1956), 351–361.

[10] ———, "Parabolic systems of differential equations with time-dependent coefficients," *Proc. Nat. Acad. Sci.*, **42** (1956), 914–917.

[11] ———, "A priori estimates for solutions of elliptic boundary value problems I, II," *Koninkl. Ned. Akad. Wetenschap.*, **22** (1960), 145–159, 160–167.

[12] ———, "On the spectral theory of elliptic differential operators I," *Math. Ann.*, **142** (1961), 20–130.

[13] C. Ciliberto, "Formule de maggiorazione e teoremi di essistenza per le soluzioni delle equazioni paraboliche in due variabili," *Ricerche di Mat.*, **3** (1954), 40–75.

[14] P. J. Cohen and M. Lees, "Asymptotic decay of solutions of differential inequalities," *Pacific J. Math.*, **11** (1961), 1235–1249.

[15] O. H. Cordes, "Über die erste Randwertaufgabe bei quasilinearen Differentialgleichungen zweiter Ordnung in mehr als zwei Variablen," *Math. Ann.*, **131** (1956), 278–312.

[16] ———, "Vereinfachter Beweis der Existenz einer Apriori-Hölderkonstanten," *Math. Ann.*, **138** (1959), 155–178.

[17] R. Courant and D. Hilbert, *Methods of Mathematical Physics*, vol. 2: *Partial Differential Equations* (Interscience Publishers, 1962).

[18] J. Douglas, "A uniqueness theorem for the solution of a Stefan problem," *Proc. Amer. Math. Soc.*, **8** (1957), 402–408.

[19] A. Douglis and L. Nirenberg, "Interior estimates for elliptic systems of partial differential equations," *Comm. Pure Appl. Math.*, **8** (1955), 503–538.

[20] F. G. Dressel, "The fundamental solution of the parabolic equation," *Duke Math. J.*, **7** (1940), 186–203.

[21] ———, "The fundamental solution of the parabolic equation II," *Duke Math. J.*, **13** (1946), 61–70.

[22] G. Ehrling, "On a type of eigenvalue problems for certain elliptic differential operators," *Math. Scand.*, **2** (1954), 267–285.

[23] S. D. Eidelman, "Bounds for solutions of parabolic systems and some applications," *Math. Sbornik*, **33** (1953), 359–382.

[24] ———, "On the fundamental solution of parabolic systems," *Math. Sbornik*, **38** (1956), 51–92.

[25] ———, "The fundamental matrix of general parabolic systems," *Doklady Akad. Nauk SSSR* (N.S.), **120** (1958), 980–983.

[26] ———, "On the fundamental solution of parabolic systems," *Math. Sbornik*, **95** (1961), 73–136.

[27] A. Friedman, "Classes of solutions of linear systems of partial differential equations of parabolic type," *Duke Math. J.*, **24** (1957), 433–442.

[28] ———, "On the regularity of solutions of nonlinear elliptic and parabolic systems of partial differential equations," *J. Math. and Mech.*, **7** (1958), 43–60.

[29] ———, "Interior estimates for parabolic systems of partial differential equations," *J. Math. and Mech.*, **7** (1958), 393–418.

[30] ———, "Boundary estimates for second order parabolic equations and their applications," *J. Math. and Mech.*, **7** (1958), 771–792.

[31] ———, "Remarks on the maximum principle for parabolic equations and its applications," *Pacific J. Math.*, **8** (1958), 201–211.

[32] ———, "On quasi-linear parabolic equations of the second order," *J. Math. and Mech.*, **7** (1958), 793–810.

[33] ———, "Parabolic equations of the second order," *Trans. Amer. Math. Soc.*, **93** (1959), 509–530.

[34] ———, "On the uniqueness of the Cauchy problem for parabolic equations," *Amer. J. Math.*, **81** (1959), 503–511.

[35] ———, "Convergence of solutions of parabolic equations to a steady state," *J. Math. and Mech.*, **8** (1959), 57–76.

[36] ———, "Asymptotic behavior of solutions of parabolic equations," *J. Math. and Mech.*, **8** (1959), 372–392.

[37] ———, "Generalized heat transfer between solids and gases under nonlinear boundary conditions," *J. Math. and Mech.*, **8** (1959), 161–184.

[38] ———, "Free boundary problems for parabolic equations I. Melting of solids," *J. Math. and Mech.*, **8** (1959), 499–518.

[39] ———, "Free boundary problems for parabolic equations II. Evaporation or condensation of a liquid drop," *J. Math. and Mech.*, **9** (1960), 19–66.

[40] ———, "Free boundary problems for parabolic equations III. Dissolution of a gas bubble in liquid," *J. Math. and Mech.*, **9** (1960), 327–345.

[41] ———, "Remarks on Stefan-type free boundary problems for parabolic equations," *J. Math. and Mech.*, **9** (1960), 885–904.

[42] ———, "On quasi-linear parabolic equations of the second order II," *J. Math. and Mech.*, **9** (1960), 539–556.

[43] ———, "Asymptotic behavior of solutions of parabolic equations of any order," *Acta Math.*, **106** (1961), 1–43.

[44] ———, *Generalized Functions and Partial Differential Equations* (Prentice-Hall, 1963).

[45] K. O. Friedrichs, "The identity of weak and strong extensions of differential operators," *Trans. Amer. Math. Soc.*, **55** (1944), 132–151.

[46] ———, "On the differentiability of solutions of linear elliptic differential equations," *Comm. Pure Appl. Math.*, **6** (1953), 299–326.

[47] L. Gårding, "Dirichlet's problem for linear elliptic partial differential equations," *Math. Scand.*, **1** (1953), 55–72.

[48] M. Gevrey, "Sur les équations aux dérivées partielles du type parabolique," *J. de Math.*, (6) **10** (1913), 105–148.

[49] G. Giraud, "Équation à integrales principales d'ordre quelconque," *Ann. Sci. l'École Norn. Super.*, **53** (1936), 1–40.

[50] L. M. Graves, *The Theory of Functions of Real Variables* (2nd ed.; McGraw-Hill, 1956).

[51] E. Hopf, "Elementare Bemerkungen Über die Lösungen partieller Differentialgleichungen zweiter Ordnung vom elliptischen Typus," *Sitber. Preuss. Akad. Wiss. Berlin*, **19** (1927), 147–152.

[52] ———, "Über den funktionalen, insbesondere den analytischen Charakter der Lösungen elliptischer Differentialgleichungen zweiter Ordnung, *Math. Zeit.*, **34** (1931), 194–233.

[53] ———, "A remark on linear elliptic differential equations of the second order," *Proc. Amer. Math. Soc.*, **3** (1952), 791–793.

[54] A. M. Ilin, A. S. Kalashnikov, and O. A. Oleinik, "Linear second order parabolic equations," *Uspehi Math. Nauk SSSR,* **17,** no. 3 (1962), 3–146.

[55] S. Itô, "A boundary value problem of partial differential equations of parabolic type," *Duke Math. J.*, **24** (1957), 299–312.

[56] ———, "Fundamental solutions of parabolic differential equations and boundary value problems," *Japan J. Math.*, **27** (1957), 55–102.

[57] ———, "A remark on my paper 'A boundary value problem of partial differential equations of parabolic type' in Duke Mathematical Journal," *Proc. Japan Acad.*, **34** (1958), 463–465.

[58] F. John, "General properties of solutions of linear elliptic partial differential equations," *Proceedings of the Symposium on Spectral Theory and Differential Problems,* Stillwater, Oklahoma, 1951, 113–175.

[59] ———, "Derivatives of continuous weak solutions of linear elliptic equations," *Comm. Pure Appl. Math.*, **6** (1953), 327–335.

[60] ———, *Plane Waves and Spherical Means* (Interscience Publishers, 1955).

[61] S. Kaplan, "On the growth of solutions of quasi-linear parabolic equations," *Comm. Pure Appl. Math.*, **16** (1963), 305–330.

[62] O. D. Kellogg, "On the derivatives of harmonic functions on the boundary," *Trans. Amer. Math. Soc.*, **33** (1931), 486–510.

[63] ———, *Foundation of Potential Theory* (Dover, New York, 1953).

[64] I. I. Kolodner, "Free boundary problem for the heat equation with applications to problems of change of phase," *Comm. Pure Appl. Math.*, **9** (1956), 1–31.

[65] M. Krzyzanski, "Sur les solutions de l'équation linéare du type parabolique déterminées par les conditions initiales," *Ann. Soc. Polon. Math.*, **18** (1945), 145–156.

[66] ———, "Certaines inégalites relatives aux solutions de l'équation parabolique linéare normale," *Bull. Acad. Polon. Sci. math. astr. phys.* **7** (1959), 131–135.

[67] ———, "Une propriété de solution de l'équation linéare du type parabolique à coefficients non bornés," *Ann. Polon. Math.*, **12** (1962), 209–212.

[68] W. T. Kyner, "An existence and uniqueness theorem for a nonlinear Stefan problem," *J. Math. and Mech.*, **8** (1959), 483–498.

[69] ———, "On free boundary value problem for the heat equation," *Quart. Appl. Math.*, **17** (1959), 305–310.

[70] O. A. Ladyzhenskaja, "On the uniqueness of solutions of the Cauchy problem for linear parabolic equations," *Math. Sbornik,* **27** (1950), 175–184.

[71] P. D. Lax, "On Cauchy's problem for hyperbolic equations and the differentiability of solutions of elliptic equations," *Comm. Pure Appl. Math.*, **8** (1955), 615–633.

[72] P. D. Lax and A. Milgram, "Parabolic equations," Contributions to the Theory of Partial Differential Equations, *Ann. Math. Studies*, no. 33 (Princeton University Press, 1954), 167–190.

[73] M. Lees and M. H. Protter, "Unique continuation for parabolic differential equations and inequalities," *Duke Math. J.*, **28** (1961), 369–382.

[74] J. Leray and J. Schauder, "Topologie et équations fonctionelles," *Ann. Sci. l'École Norm. Sup.*, **51** (1934), 45–78.

[75] E. E. Levi, "Sulle equazioni lineari totalmente ellittiche alle derivate parziali," *Rend. del. Circ. Mat. Palermo*, **24** (1907), 275–317.

[76] J. L. Lions, "Sur les problème mixtes pour certains systèmes paraboliques dans des ouverts non cylindriques," *Ann. l'Institute Fourier*, **7** (1957), 143–182.

[77] ———, *Equations differentielles. Operationelles et Problèmes aux Limites* (Springer-Verlag, 1961).

[78] J. L. Lions and B. Malgrange, "Sur l'unicité rétrograde dans les problèmes mixtes parabolique," *Math. Scand.*, **8** (1960), 277–286.

[79] B. Ya. Lipko, "A mixed problem with oblique derivatives for a parabolic equation of the second order," *Doklady Akad. Nauk SSSR* (N.S.), **132** (1960), 279–282.

[80] S. Mandelbrojt, *Séries de Fourier et Classes Quasi-Analytique de Fonctions* (Gauthier-Villars, Paris, 1935).

[81] E. J. McShane, "Extension of range of functions," *Bull. Amer. Math. Soc.*, **40** (1934), 837–842.

[82] C. Miranda, *Equazioni Alle Derivate Parziali di Tipo Ellittico* (Springer-Verlag, 1955).

[83] S. Mizohata, "Le problème de Cauchy pour les équations paraboliques," *J. Math. Soc. Japan*, **8** (1956), 269–299.

[84] ———, "Hypoellipticité des équations parabolique," *Bull. Soc. Math. France*, **85** (1957), 15–49.

[85] ———, "Le Problème de Cauchy le Passé pour Quelques Équations Paraboliques," *Proc. Japan Acad.*, **34** (1958), 693–696.

[86] C. B. Morrey, "Second order elliptic systems of differential equations," Contributions to the Theory of Partial Differential Equations, *Ann. Math. Studies*, no. 33 (Princeton University Press, 1954), 101–159.

[87] ———, "On the analyticity of the solutions of analytic nonlinear elliptic systems of partial differential equations I, II," *Amer. J. Math.*, **80** (1958), 197–218, 219–238.

[88] C. B. Morrey and L. Nirenberg, "On the analyticity of the solutions of linear elliptic systems of partial differential equations," *Comm. Pure Appl. Math.*, **10** (1957), 271–290.

[89] L. Nirenberg, "A strong maximum principle for parabolic equations," *Comm. Pure Appl. Math.*, **6** (1953), 167–177.

[90] ———, "Remarks on strongly elliptic partial differential equations," *Comm. Pure Appl. Math.*, **8** (1955), 648–674.

[91] ———, *Existence Theorems in Partial Differential Equations*, New York University Notes.

[92] O. A. Oleinik, "On properties of some boundary problems for equations of elliptic type," *Math. Sbornik*, **30** (72), (1952), 695–702.

[93] ———, "On a method of solving general Stefan problems," *Doklady Akad. Nauk SSSR* (N.S.), **135** (1960), 1054–1057.

[94] M. Pagni, "Su un problema contorno tipico per l'equazione de calore," *Ann. Scuola Norm. Sup. Pisa*, **11** (1957), 73–115.

[95] I. G. Petrowski, "Über das Cauchysche Problem für ein System linearen partialler Differentialgleichungen in Gebiet der nichtanalytischen Funktionen," *Bull. Univ. d'Etat Moscow*, **1A**, no. 7 (1938), 1–74.

[96] W. Pogorzelski, "Étude de la solution fondamental de l'équation parabolique," *Ricerche di Mat.*, **5** (1956), 25–57.

[97] ———, "Étude de la solution fondamental de l'équation elliptique et des problèmes aux limites," *Ann. Polon. Math.*, **3** (1957), 247–284.

[98] ———, "Propriétés des intégrales de l'équation parabolique normale," *Ann. Polon. Math.*, **4** (1957), 61–92.

[99] ———, "Problèmes aux limites pour l'équation parabolique normale," *Ann. Polon. Math.*, **4** (1957), 110–126.

[100] ———, "Propérités de dérivées tangentielles d'une intégrales de l'équation parabolique," *Ricerche di Mat.*, **6** (1957), 162–194.

[101] ———, "Étude d'un fonction de Green et du problème aux limites pour l'équation parabolique normale," *Ann. Polor. Math.*, **4** (1958), 288–307.

[102] ———, "Étude de la matrice des solution fondamentales du système parabolique d'équations aux dérivées partielles," *Ricerche di Mat.*, **7** (1958), 153–185.

[103] ———, "Propriétés de solution du système parabolique d'équation aux dérivées partielles," *Math. Scand.*, **6** (1958), 237–262.

[104] ———, "Sur quelques propriétés de potentiels généralisés et un problème aux limites pour l'équation elliptique," *Ann. Polon. Math.*, **11** (1961), 177–197.

[105] M. H. Protter, "Properties of solutions of parabolic equations and inequalities," *Canad. J. Math.*, **13** (1961), 331–345.

[106] E. Rothe, "Über die Grundlösung bei parabolischen Gleichungen," *Math. Zeit.*, **33** (1931), 488–504.

[107] J. Schauder, "Der Fixpunktsatz in Funktionalraümen," *Studia Math.*, **2** (1930), 171–180.

[108] ———, "Über lineare elliptische Differentialgleichungen zweiter Ordnung," *Math. Zeit.*, **38** (1934), 257–282.

[109] ———, "Numerische Abschätzungen in elliptischen linearen Differentialgleichungen," *Studia Math.*, **5** (1934), 34–42.

[110] G. Sestini, "Sul problema non lineare di Stefan in uno strato piano indefinito," *Annali Mat. Pura Appl.*, **51** (1960), 203–224.

[111] ———, "Sul problema non lineare di Stefan in strati cilindrici o sferici," *Annali Mat. Pura Appl.*, **56** (1961), 193–207.

[112] L. N. Slobodetski, "On the fundamental solution and the Cauchy problem for parabolic systems," *Math. Sbornik*, **46** (1958), 229–258.

[113] S. Sobolev, "On a theorem of functional analysis," *Math. Sbornik,* **4** (1938), 471-497.

[114] P. E. Sobolevski, "On equations of parabolic type in Banach space," *Trudy Moscow Math. Obsch.*, **10** (1961), 298–350.

[115] F. Trèves, "Relations de domination entre opérateurs différentiels," *Acta Math.*, **101** (1959), 1–139.

[116] A. N. Tychonov, "Uniqueness theorems for the heat equation," *Math. Sbornik,* **42** (1935), 199–216.

[117] R. Viborni, "On properties of solutions of some boundary value problems for equations of parabolic type," *Doklady Akad. Nauk SSSR* (N.S.), **117** (1957), 563–565.

[118] M. I. Vishik, "The method of orthogonal and direct decomposition in the theory of elliptic partial differential equations," *Math. Sbornik,* **25** (1949), 189–234.

[119] ———, "On strongly elliptic systems of differential equations," *Math. Sbornik,* **29** (1951), 617–676.

[120] H. Westphal, "Zur Abschätzung der Lösungen nichtlinearer parabolischer Differentialgleichungen," *Math. Zeit.*, **51** (1949), 690–695.

[121] H. Weyl, "The method of orthogonal projection in potential theory," *Duke Math. J.* **7** (1940), 411–444.

INDEX